CHEMISTRY AND CHEMICAL ENGINEERING OF CATALYTIC PROCESSES

NATO ADVANCED STUDY INSTITUTES SERIES

Proceedings of the Advanced Study Institute Programme, which aims at the dissemination of advanced knowledge and the formation of contacts among scientists from different countries.

The series is published by an international board of publishers in conjunction with NATO Scientific Affairs Division

A	Life Sciences	Plenum Publishing Corporation
B	Physics	London and New York
C	Mathematical and Physical Sciences	D. Reidel Publishing Company Dordrecht and Boston
D	Behavioural and Social Sciences	Sijthoff & Noordhoff International Publishers B.V.
E	Applied Sciences	Alphen aan den Rijn, The Netherlands and Germantown, Maryland, USA

Series E: Applied Sciences - No. 39

CHEMISTRY AND CHEMICAL ENGINEERING OF CATALYTIC PROCESSES

edited by

ROEL PRINS
Professor of Inorganic Chemistry and Catalysis
Eindhoven University of Technology
Eindhoven, The Netherlands

and

GEORGE C. A. SCHUIT
Professor of Inorganic Chemistry and Catalysis
Eindhoven University of Technology
Eindhoven, The Netherlands

Department of Chemical Engineering
University of Delaware, Newark, USA

SIJTHOFF & NOORDHOFF   1980
Alphen aan den Rijn, The Netherlands
Germantown, Maryland, USA

Proceedings of the NATO Advanced Study Institute on
Chemistry and Chemical Engineering of Catalytic Processes
Noordwijkerhout, The Netherlands
August 19-31, 1979

ISBN 90 286 0730 7

Copyright © 1980 Sijthoff & Noordhoff International Publishers B.V., Alphen aan den Rijn, The Netherlands.

All rights reserved. No part of this book may be reproduced, stored in a retrieval system, or transmitted, in any form or by any means, electronic, mechanical, photocopying, recording, or otherwise, without the prior permission of the copyright owner.

Printed in The Netherlands.

INTRODUCTION

Catalysis is a most complex subject, encompassing solid and surface structure, thermodynamics, reaction mechanisms, heat and mass transfer, chemical analysis, and the design of chemical reactors. A good understanding of the chemistry and chemical engineering of catalytic processes is therefore a prerequisite for further development of the field of catalysis. It is for this reason that a NATO Advanced Study Institute on 'Chemistry and Chemical Engineering of Catalytic Processes' was organised at the Leeuwenhorst Congress Centre in Noordwijkerhout, The Netherlands, from August 19-31, 1979. The lectures presented during this Advanced Study Institute are collected in this volume.

A good understanding of the chemistry and chemical engineering of catalytic processes is therefore a prerequisite for further development of the field of catalysis. It is for this reason that a NATO Advanced Study Institute on 'Chemistry and Chemical Engineering of Catalytic Processes' was organised at the Leeuwenhorst Congres Centre in Noordwijkerhout, The Netherlands, from August 19-31, 1979. The lectures presented during this Advanced Study Institute are collected in this volume.

The rapid progress of the many branches of chemistry and chemical engineering which contribute to catalysis tends to widen the communication gap between chemists, physicists, and engineers involved in catalysis. It is important for the advancement of catalysis that this gap does not widen and that chemists and physicists involved understand and appreciate the technology associated with catalytic processes. Applied catalysis is after all, a chemical engineering undertaking. In a similar manner engineers taking part in the development of catalysts or catalytic processes must be aware of the chemistry involved and be able to assimilate, understand and ultimately implement the sophisticated new results which come out of the laboratory.

For these reasons, an important objective of the Advanced Study Institute was to provide an integrated presentation of chemistry and chemical engineering. In our opinion such an integrated presentation is the only way to teach applied catalysis.

In the past decade, two other Advanced Study Institutes on Catalysis have been held. In 1971, an Institute on 'Fundamental Principles in Heterogeneous Catalysis' was held in Venice, Italy, which covered subjects vital to a good understanding of applied catalysis and which was interesting because it was taught by outstanding chemists. Also the proceedings of the NATO Study Institute on 'Catalysis' in Santa Margherita di Pula, Sardinia, 1972 offer many extremely interesting lectures. Both courses, however, had difficulties with organising the subject matter in a coherent way.

At universities in the U.S.A., and especially in their chemical engineering departments, there have been some attempts at constructing courses in applied catalysis during the past decade. A tendency to emphasize the chemical engineering aspects, such as reactor engineering, usually resulted in a logical series of presentations, but frequently at the cost of ignoring essential and extensive chemical information. An exception must be made for the course created at the chemical engineering department of the University of Delaware. Here an entirely novel method was generated that enabled presentation of a satisfactory survey of catalytic chemistry and chemical engineering. The secret was the concentration on some five fields of catalytic chemistry, each field representing a class of reactions which were treated in relation to their mechanisms and catalysts. Each class was selected in such a way that it contained at least one of the major commercial catalytic processes. The strategy was to first discuss the organic reaction mechanisms, followed by a description of the structure of the catalyst employed. One special process was then selected for a more penetrating analysis. At this point the chemical engineering was discussed, not only in general terms, but, more importantly, in its relation to the chemistry involved.

The Delaware course is taught in a time span of one week (five days). To allow sufficient coverage of the subjects the attendants were required to have a considerable advance knowledge of both chemistry and chemical engineering. Moreover, in order to maintain coherence and uniformity, each set of reactions was discussed in one day by, at most, two teachers. The entire course was taught by only four teachers.
Five years of testing this course led to its final form laid down

in the book 'Chemistry of Catalytic Processes' by B.C. Gates, J.R. Katzer and G.C.A. Schuit.

The NATO Advanced Study Institute, the proceedings of which are given in this book, follows the model of the Delaware course, in particular in the selection of the reaction classes. It deviates from the model in three respects:
In the first place the Institute was planned at a more advanced level and a scientific background comparable to at least a M.Sc. degree in chemistry or chemical engineering was required. To further ensure an equal footing of chemists and engineers the book 'Chemistry of Catalytic Processes' was sent to the participants of the Study Institute as advance reading material. (Readers of the present book should bear this point in mind!) To present all the subjects at an advanced level, it was deemed necessary to increase the number of lecturers and have each lecturer discuss only his own particular fields of research. Although this policy increased the risk of losing some of the coherence obtained in the Delaware course, it added to the liveliness of the Institute. The resulting lectures proved to be most stimulating. The inevitable tendency to veer off from the main track was, we believe, sufficiently counterbalanced by general introductions to the main fields. These introductions, which are not included in the present book, placed the subsequent lectures in perspective within the particular field of catalysis under discussion. Prof. J.R. Katzer and Prof. B.C. Gates deserve special mention for their meritorious service in this area.
Secondly, the subject of hydrodesulfurisation was enlarged to encompass hydrodenitrogenation and coal liquefaction, contributions on coal gasification, the ZSM-Mobil process, and on Fischer-Tropsch synthesis. All this added up to the subject of coal conversion in the Institute, whereas only hydrodesulfurisation is taught in the Delaware course.
Thirdly, since the Study Institute was set up as a two-week course (ten working days), some introductory lectures covering advanced general topics in chemistry and chemical engineering could be included. This was considered essential for a good understanding of what was to follow.

The ordering of the lectures in this book follows that of the programme of the Institute to a large extent. After the introductory chapter there are five chapters, each concerned with an industrial process or class of processes, namely, catalytic cracking, transition-metal-complex catalysis, catalytic reforming, selective oxidation of hydrocarbons and coal conversion. The processes illustrate the major classes of catalysts: acids, transition-metal complexes, transition metals, metal oxides, and metal sulfides; they also illustrate the main types of reactors: fluidized bed, fixed bed and moving bed. The sequence proceeds roughly from the best-understood to the least-well-understood chemistry.

All lecturers are to be complimented and thanked for the clarity of their oral and written contributions.

We express our deepest gratitude to the Scientific Affairs Division of NATO, the main sponsor of this Study Institute. Also their special subsidy that enabled a considerable number of Turkish and Portugese scientists to participate, is gratefully acknowledged.

During a two-week meeting many things have to be organised and problems have to be solved to guarantee a smoothly running programme, scientific as well as nonscientific. In so far as our objectives have been met, this certainly has been due to the help of Dr. V.H.J. de Beer, Drs. J.P. van den Berg, Ir. H.F.J. van 't Blik, Ir. T. Huizinga, and especially the secretary, Miss C.J. van 't Blik. Before, during, and after the Institute, she not only managed to solve the problems of the participants, but also those of the lecturers and ourselves. We all thank her.

We acknowledge the help of the people of the Leeuwenhorst Congress Centre, who made our stay such a pleasant one.

Last, but not least, we want to thank the participants for their contributions to the Study Institute. It was their active participation in the discussions that formed the basis of success of the Institute.

Roel Prins
George C.A. Schuit

TABLE OF CONTENTS

Part I. General introduction to chemical engineering   1

J.R. Katzer
    Chemical kinetics. The first step to reaction modeling
    and reaction engineering   3

J.R. Katzer
    Mass transfer in reacting systems   49

G.F. Froment
    Fixed bed catalytic reactors   115

Part II. Catalytic cracking   135

D.M. Brouwer
    Reactions of alkylcarbenium ions in relation to
    isomerization and cracking of hydrocarbons   137

J.H.C. van Hooff
    Cracking catalysts   161

P.N. Rowe
    Basic fluidisation   181

P.N. Rowe
    Fluidisation of fine powders such as fcc   203

H.S. van der Baan
    Catcracker operations. Reaction network and kinetics   217

Part III. Reforming of hydrocarbons on metals and alloys   235

R. Prins
    Chemical bonding   237

V. Ponec
    Bonding in and on metals   257

G. Ertl
    Surface science and catalysis on metals   271

R.F. Willis
    Surface electron spectroscopy   281

W.M.H. Sachtler
    Surface composition of binary alloys   317

H.C. de Jongste and V. Ponec
    Catalysis by metals and alloys
    Reforming of hydrocarbons and some other reactions   337

F. Garin and F.G. Gault
    Skeletal isomerization of hydrocarbons on metals   351

H.S. van der Baan
    Catalytic reforming, the reaction network   381

R. Prins
    Modern processes for the catalytic reforming of
    hydrocarbons   389

Part IV. Homogeneous catalysis   405

R. Prins
    Reaction mechanisms in homogeneous catalysis   407

B.C. Gates
    Catalysis by metal clusters   427

B.C. Gates
    Polymer supported catalysts   437

Part V. Partial oxidation of hydrocarbons and the
    acrylonitrile process   459

G.C.A. Schuit and B.C. Gates
    Catalytic oxidation, an introduction   461

F.S. Stone
    Oxide crystal chemistry and catalysis   477

J.H.C. van Hooff
    Industrial catalytic partial oxidation processes   507

H.S. van der Baan
    The acrylonitrile process   523

G.F. Froment
    Hot spots and runaway in fixed bed tubular   535

Part VI. Coal conversion   561

J.R. Katzer
    Catalysis in coal gasification   563

W.M.H. Sachtler
    Mechanism of hydrocarbon synthesis over Fischer-
    Tropsch catalysts   583

J.H.C. van Hooff
    The conversion of methanol to hydrocarbons using a
    new type of zeolite as catalyst (Mobil process)   599

B.C. Gates
    Liquefied coal by hydrogenation   621

J.R. Katzer and R. Sivasubramanian
    Process and catalyst needs for hydrodenitrogenation   635

Upon arrival in Noordwijkerhout to attend the NATO Advanced Study Institute on 'Chemistry and Chemical Engineering of Catalytic Processes' we were shocked to hear of the death of Professor Dr. F.G. Gault on August 4, 1979.
Professor Gault had accepted our invitation, in July 1978, to participate as a lecturer in the Study Institute and to present a lecture on 'Mechanisms of hydrocarbon reactions on metals'. Just before the summer of 1979 he had sent us his preliminary manuscript, which we had distributed among the participants as prereading material. Then, at the beginning of the Study Institute, the sad news of his death reached us.

Being faced with the problem of finding someone to replace him at short notice, we were fortunate that Dr. F. Garin, a former student of his, was also attending the Study Institute and was prepared to present the lecture and to complete the manuscript. We would like to thank Dr. Garin for the excellent way in which he executed this difficult task.

R. Prins
G.C.A. Schuit

ORGANIZING COMMITTEE

V.H.J. de Beer
C.J. van 't Blik-Quax
J.P. van den Berg
T. Huizinga
H.F.J. van 't Blik

Laboratory for Inorganic Chemistry
Eindhoven University of Technology
P.O. Box 513
5600 MB Eindhoven
The Netherlands

LECTURERS

H.S. van der Baan	Laboratory for Chemical Technology, Eindhoven University of Technology, Eindhoven, The Netherlands.
D.M. Brouwer	Koninklijke/Shell Laboratory, Amsterdam, The Netherlands
G. Ertl	Institut für Physikalische Chemie, Universität München, D 8000 München 2, Federal Republic Germany.
G.F. Froment	Lab. voor Petrochemische Techniek, Universiteit van Gent, 9000 Gent, Belgium.
B.C. Gates	Department of Chemical Engineering, University of Delaware, Newark (Del.) 19711, U.S.A.
J.H.C. van Hooff	Laboratory for Inorganic Chemistry, Eindhoven University of Technology, Eindhoven, The Netherlands.
J.R. Katzer	Department of Chemical Engineering, University of Delaware, Newark (Del.) 19711, U.S.A.
V. Ponec	Gorleaus Laboratory, University of Leiden, Leiden, The Netherlands.
R. Prins	Laboratory for Inorganic Chemistry, Eindhoven University of Technology, Eindhoven, The Netherlands.
P.N. Rowe	Department of Chemical and Bio-chemical Engineering, University College, London, United Kingdom.
W.M.H. Sachtler	Koninklijke/Shell Laboratory, Amsterdam, The Netherlands.
G.C.A. Schuit	Laboratory for Inorganic Chemistry, Eindhoven University of Technology, Eindhoven, The Netherlands.
F.S. Stone	School of Chemistry, University of Bath, Bath BA2 7AY, United Kingdom.
R.F. Willis	University of Cambridge, Department of Physics, Cambridge CB3 0HE, United Kingdom.

LIST OF PARTICIPANTS

A. Akgerman, Kimya Fakültesi, Ege University, Bornova, Izmir,
    Turkey.
J. Akyurtlu, Chemical Engineering Department, Middle East Technical
    University, Ankara, Turkey.
E. Alper, Chemical Engineering Department, Hacettepe University,
    Ankara, Turkey.
M.C. Alvim Ferraz, Faculdade de Engenharia do Porto, Centro de
    Engenharia Quimica, 4099 Porto Codex, Portugal.
V. Amir-Ebrahimi, Laboratoire de Catalyse, Université de Strasbourg,
    67000 Strasbourg, France.
P. Antonucci, University of Messina 98100 Messina, Italy.
A. Anundskas, Research Centre, Norsk Hydro, 3900 Porsgrunn, Norway.
M. Baerns, Technische Chemie, Ruhr-Universität Bochum, D-4630 Bochum,
    Federal Republic Germany.
N.N. Bakhshi, Department of Chemistry and Chemical Engineering,
    University of Saskatchewan, Saskatoon, Saskatchewan,
    Canada S7N 0W0.
E. Banks, Department of Chemistry, Polytechnic Institute of N.Y.,
    Brooklyn, N.Y. 11201, U.S.A.
W.A.A. van Barneveld, Gorlaeus Laboratory, Catalysis Department,
    University of Leiden, Leiden, The Netherlands.
H.J. Brockmeyer, W.C. Heraeus GmbH, 6450 Hanau, Federal Republic
    Germany.
G.D.L. Carter, B.P. Research Centre, The British Petroleum Co. Ltd.,
    Sunbury-on-Thames, Middlesex TW16 7LN, United Kingdom.
B.H. Cooper, Research and Development Division, Haldor Topsøe A/S,
    2800 Lyngby, Denmark.
D. Cornet, Institut des Sciences de la Matière et du Rayonnement,
    Université de Caen, F 14000-Caen, France.
R. Covini, Research Center, Montedison SpA, 20021 Bollate (Milano),
    Italy.
R.S. McDaniel, Alberta Research Council, Edmonton, Alberta,
    Canada T6G 2C2.
P.J. Denny, Research Department, ICI Agricultural Division,
    Billingham, Cleveland TS23 1LD, United Kingdom.
A. Dosumu, Department of Chemical Engineering, University,
    Newcastle upon Tyne NE1 7RU United Kingdom.
E. Drent, Koninklijke/Shell Laboratory, Amsterdam, The Netherlands.
J.C. Duchet, Laboratoire de Catalyse, I.S.M.R.A., Université de
    Caen, 14032 Caen Cedex, France.
P. Dufresne, Lab. de Catalyse et de Physico Chimie des Surfaces,
    Université des Sciences et Techniques de Lille,
    59655 Villeneuve d'Ascq Cedex, France
L. Fiermans, Lab. voor Kristallografie en Studie van de Vaste Stof
    Rijksuniversiteit Gent, B-9000 Gent, Belgium.
J.L.C.C. Figueiredo, Centro de Engenharia Quimica, Faculdade de
    Engenharia do Porto, 4099 Porto Codex, Portugal.

G. Franz, Chemische Werke Hüls A.G., 4370 Marl, Federal Republic
   Germany.
P. Frenken, Central Laboratory DSM, Geleen, The Netherlands.
F. Garin, Laboratoire de Catalyse, Université de Strasbourg,
   67000 Strasbourg, France.
D.G. Gavin, Coal Research Establishment (N.C.B.) Cheltenham,
   United Kingdom.
P.J. Gellings, Department of Chemical Technology, Twente University
   of Technology, Enschede, The Netherlands.
E. Giamello, Istituto di Chimica Fisica, Universitá di Torino,
   10100 Torino, Italy.
A. Greco, Assoreni Scientific Research, S. Donato Milanese (Milano)
   Italy.
M. Gruia, Laboratoire de Chimie des Surfaces, Université Pierre
   et Marie Curie, 75230 Paris cedex 05, France.
G. Gubitosa, Donegani Research Institute, Montedison SpA, 28100
   Novara, Italy.
C. Gueguen, Centre de Recherches de Solaize, Société Nationale
   Elf Aquitaine, 69360 Saint Symphorien d'Ozon, France.
J.J.L. Heinerman, Koninklijke/Shell Laboratory, Amsterdam,
   The Netherlands.
L. Imre, Bayer A.G., D-5090 Leverkusen 1, Federal Republic Germany.
C.S. John, Koninklijke/Shell Laboratory, Amsterdam, The Netherlands.
K.P. de Jong, Laboratory for Analytical Chemistry, University of
   Utrecht, Utrecht, The Netherlands.
F. Kapteijn, Laboratory for Chemical Technology, University of
   Amsterdam, Amsterdam, The Netherlands.
F. King, Johnson Matthey Research Centre, Reading RG4 9NH,
   United Kingdom.
J. Koch, Hoechst A.G., 5040 Brühl, Federal Republic Germany.
S. Kolboe, Department of Chemistry, University of Oslo, Blindern,
   Norway.
M. Kotter, Institut für Chemische Verfahrenstechnik, Universität
   (TH) Karlsruhe, 7500 Karlsruhe, Federal Republic Germany.
H.J. Krebs, Institut für Grenzflächenforschung und Vakuumphysik,
   Kernforschungsanlage Jülich GmbH, 5170 Jülich 1, Federal
   Republic Germany.
L.A. Kristiansen, Norsk Hydro A/S, 3900 Porsgrunn, Norway.
J.G. Lammers, Unilever Research Laboratory, Vlaardingen, The
   Netherlands.
S.P. Lankhuijzen, Laboratory for Chemical Technology, Eindhoven
   University of Technology, Eindhoven, The Netherlands.
M. Legendre, Centre de Recherches Total, Compagnie Française de
   Raffinage, 76700 Harfleur, France.
W. Liebelt, Süd Chemie A.G., D-8000 München 2, Federal Republic
   Germany.
R.J. Litchfield, Department of Chemical Engineering, University
   of Surrey, Guildford, Surrey GU2 5XH, United Kingdom.

M. LoJacono, Consiglio Nazionale delle Ricerche, Centro di Studio
   S.A.C.S.O., Istituto di Chimica Generale ed Inorganica,
   Università di Roma, Roma, Italy.
J. Lucien, Centre de Recherche, Shell Française, 76530 Grand
   Couronne, France.
P. Lusman, Research and Development, B.P. Chemicals Ltd.,
   Grangemouth, Stirlingshire, United Kingdom.
K.E. Mack, Hoechst A.G., 6230 Frankfurt 80, Federal Republic
   Germany.
J. Medema, Prins Maurits Laboratory, National Defence Research
   Organization TNO Rijswijk, The Netherlands.
R.Z.C. van Meerten, Faculty of Sciences, University of Nijmegen,
   Nijmegen, The Netherlands.
L. Mendolovici, Department of Inorganic Chemistry, Hebrew
   University of Jerusalem, Jerusalem, Israel.
L.M. Moroney, School of Chemistry, University of Bradford,
   Bradford, W. Yorkshire BD7 1DP, United Kingdom.
H. Mousty, Facultés Universitaires Notre-Dame de la Paix,
   Département de Chimie, Laboratoire de Catalyse, B-5000 Namur,
   Belgium.
R.W. Naylor, Department of Chemistry, Potomac State College of
   West Virginia University, Keyser, W.V. 26726, U.S.A.
B.P. Nilsen, Central Institute for Industrial Research, Oslo 3,
   Norway.
O.T. Onsager, Laboratory of Industrial Chemistry, University of
   Trondheim NTH, N-7034 Trondheim NTH, Norway.
J.P.E. Pirard, Institut de Chimie, 4000 Liège, Belgium.
M.J. Pires, Laboratory for Chemical Technology, Instituto Superior
   Tecnico, Lisboa, Portugal.
P. Pomonis, Department of Chemistry, University of Ioannina,
   Ioannina, Greece.
M.F. Portela, Instituto Superior Tecnico, 1100 Lisboa, Portugal.
A.V. Ramaswamy, Indian Institute of Petroleum, Dehradun, India.
M. Ramirez de Agudelo, School of Chemistry, University of Bath,
   Bath BA2 7AY, United Kingdom.
C.A.M. van Reisen, Akzo Zout Chemie (Netherlands) B.V., Hengelo
   The Netherlands.
L. Rizzuti, Istituto di Ingegneria Chimica, Università di Palermo
   90100 Palermo, Italy.
L. Roet, Departement Scheikunde, Universitaire Instellingen
   Antwerpen, B-2610 Wilrijk, Belgium.
O.A. Rokstad, SINTEF, The Norwegian Institute of Technology,
   N-7034 Trondheim - NTH, Norway.
S. de Rossi, Consiglio Nazionale delle Ricerche, Centro di Studio
   S.A.C.S.O., Istituto di Chimica Generale ed Inorganica,
   Università di Roma, Roma, Italy.
D.M. Ruthven, Department of Chemical Engineering, University of
   New Brunswick, Fredericton, N.B., Canada.

XVIII

E. Santacesaria, Consiglio Nazionale delle Ricerche, 20139 Milano,
    Italy.
R.A.R. Sassoulas, Pechiney Ugine Kuhlmann, Produits Chimiques
    Ugine Kuhlmann, 59110 La Madeleine, France.
J.W.M. Schoof, Dow Chemical (Netherlands) B.V., Terneuzen, The
    Netherlands.
J.H. Schutten, Laboratory for Inorganic Chemistry, Eindhoven
    University of Technology, Eindhoven, The Netherlands.
R. Sigg, Kalie Chemie Aktiengesellschaft, 3000 Hannover 1,
    Federal Republic Germany.
T.H.M. van Sint Fiet, Dow Chemical (Netherlands) B.V., Terneuzen,
    The Netherlands.
A. Skov, Research and Development Division, Haldor Topsøe A/S,
    2800 Lyngby, Denmark.
A. Vaccari, Faculty of Industrial Chemistry, University of Bologna,
    I 40136 Bologna, Italy.
J.A. Valero, Department of Chemistry, University of Reading,
    Reading RG6 2AD, United Kingdom.
P. Villa, Istituto di Chimica Industriale, Politecnico di Milano,
    20133 Milano, Italy.
W.J.J. van der Wal, Laboratory for Inorganic Chemistry, University
    of Utrecht, Utrecht, The Netherlands.
J.R. Walls, Chemical Engineering Dept., Teeside Polytechnic,
    Middlesborough, Cleveland TS1 3PA, United Kingdom.
T.M. Wortel, Essochem Benelux B.V. Rotterdam, The Netherlands.
Y. Yürüm, Department of Chemistry, Hacettepe University, Ankara,
    Turkey.
L. Zanderighi, University of Milano, Milano, Italy.
H. Zimmerman, Zintl Institut, Technische Hochschule Darmstadt,
    61 Darmstadt, Federal Republic Germany.

Part I

GENERAL INTRODUCTION TO CHEMICAL ENGINEERING

Part 1

CHEMICAL KINETICS - THE FIRST STEP TO REACTION MODELING AND REACTION ENGINEERING

James R. Katzer

Center for Catalytic Science and Technology
Department of Chemical Engineering
University of Delaware
Newark, Delaware 19711 U.S.A.

INTRODUCTION

The first step in our ability to model chemical reactors is to be able to describe the rate and behavior of chemical reactions quantitatively. This is one of the most important steps in chemical reaction engineering for reactor design and operation. In many cases the expressions for reaction kinetics can be derived from a knowledge of the elementary steps involved in a chemical reaction and the rate-determining step. Whenever it is possible to derive a kinetic expression in this manner, it should be done because the expression should give the best fit to a full range of kinetic data, give constants with physical meaning and allow confident extrapolation beyond the regions in which data were available for parameter estimation. Similarly since different reaction mechanisms with different rate-determining steps give different kinetic expressions, it is possible to use chemical kinetics to eliminate potential reaction mechanisms and rate-determining steps. However if a given reaction mechanism and rate-determining step produces a kinetic expression that fits the observed data well, it does not prove that the mechanism is correct, only that it is one of possibly several mechanisms that could be operative.

When there is insufficient information on a reaction available, kinetic expressions not based on fundamental chemical information may be applied. Such expressions include power law kinetics, and Langmuir-Hinshelwood or Hougan-Watson kinetics. These kinetic forms frequently represent well the kinetic behavior and might well have been derived if the appropriate chemistry were known. However because they are not necessarily based on fundamental chemistry, they must be used with caution when they are extrapolated beyond the range over which the kinetic parameters were fitted.

In the following we will assume that one is familiar with the basic definitions and relations in chemical kinetics and with the analysis of simple reaction networks, based on elementary reactions. If needed the reader may consult the references quoted at the end of the present article. Here we will discuss complex reaction networks and in particular the models of the rate determining step and of the steady state.

DEVELOPMENT OF RATE EQUATIONS FOR COMPLEX NON-ELEMENTARY REACTION NETWORKS

When dealing with elementary reactions one can write rate equations for the reactions directly from the reaction stoichiometry. For non-elementary reactions the appropriate rate equation does not necessarily follow from the stoichiometry. Now let us consider some of the details of non-elementary reactions, of how their rate equations come about, and of how the kinetics may be used to say something about the reaction mechanism.

Every reaction proceeds via elementary molecular reactions from reactants to products. For an elementary reaction there is only one elementary reaction involved. For non-elementary reactions there is a series of reactions involved in going from reactants to products; each of the reactions is an elementary reaction. When the sequence is summed it leads to the overall reaction. For example consider:

$$A \rightleftarrows R$$

which could be elementary, and then

$$-\frac{dC_A}{dt} = kC_A - k'C_R$$

If it were not elementary it could be composed of a series of elementary reactions such as:

$$A \underset{k_1'}{\overset{k_1}{\rightleftarrows}} B \qquad -\frac{dC_A}{dt} = k_1 C_A - k_1' C_B$$

$$B \underset{k_2'}{\overset{k_2}{\rightleftarrows}} C \qquad \frac{dC_B}{dt} = k_1 C_A - k_1' C_B - k_2' C_B$$

$$C \underset{k_3'}{\overset{k_3}{\rightleftarrows}} D, \quad D \underset{k_4'}{\overset{k_4}{\rightleftarrows}} R \text{ etc. for each reaction.}$$

Let us consider a real reaction system, for example, consider the decomposition of nitrous acid in aqueous solution.

$$3HNO_2 \underset{k'}{\overset{k}{\rightleftarrows}} H^+ + NO_3^- + 2NO(g) + H_2O$$

The <u>experimentally observed</u> rate equation is

$$\frac{dC_{NO_3^-}}{dt} = \frac{k \, C_{HNO_2}^4}{(P_{NO})^2} - k'(C_{H^+})(C_{NO_3^-})(C_{HNO_2})$$

where P_{NO} is the partial pressure of NO: P_{NO} appears because this was undoubtedly the way that the concentration of NO was varied and measured. There should be an essentially constant proportionality between P_{NO} and C_{NO} (Henry's Law).

The overall equation for the reaction suggests that there are three HNO_2 molecules interacting simultaneously in the decomposition reaction; this is unlikely although not entirely impossible. However if we look at the reverse reaction the equation suggests a simultaneous interaction of five species at once; this is an impossible occurrence. Therefore we know that this reaction is not an elementary reaction. If we have

A→R+S+T+U+V

it cannot be an elementary reaction.

We must therefore attempt to pose a series of plausible elementary reactions for nitrous acid decomposition, the sum of which forms the overall reaction; this is the reaction mechanism. Below is one possible set of elementary reactions which give the products observed for HNO_2 decomposition.

1). $2HNO_2 \underset{k_{-1}}{\overset{k_1}{\rightleftarrows}} NO_2 + NO + H_2O \qquad$ rapid

2). $2NO_2 \underset{k_{-2}}{\overset{k_2}{\rightleftarrows}} N_2O_4$ \hspace{2cm} Rapid

3). $N_2O_4 + H_2O \underset{k_{-3}}{\overset{k_3}{\rightleftarrows}} H^+ + NO_3^- + HNO_2$ \hspace{1cm} Slow

We could start by writing rate equations for each of the elementary reactions and attempt to eliminate the intermediate species that are not observed, e.g., N_2O_4 and NO_2. This becomes very complex. However if we know something or at least can postulate something more about the reaction it will frequently simplify the mathematical manipulations that we need to make.

Before going any further, let us consider the concept of a rate-determining step.

<u>Rate-determining step</u>: The reaction in a series of consecutive reactions is rate determining if the rate of the entire series of reactions is determined by the rate of the slow reaction. Or, the first reaction in a sequence of reactions that has a gross forward rate equal to the net rate of the overall reaction is the rate-determining step. <u>Concept</u>: Consider people moving down a crowded hall or cars down a crowded highway, and then consider a construction activity in the hall or an accident on the highway. The total rate of movement of people or of cars is determined by the rate that they can get by the construction activity or the accident. The situation is exactly analogous in chemical kinetics—one step in a reaction sequence can control the overall rate if it is slower than all of the other steps in the mechanism. The overall rate is thus determined by the rate of this one step.

Considering for the moment that the third reaction in the HNO_2 decomposition mechanism is slow relative to the rates of Reactions 1 and 2. Taking the reverse reaction only first

$$-\frac{dC_{NO_3^-}}{dt} = k_{-3}(C_{H^+})(C_{NO_3^-})(C_{HNO_2})$$

This is an elementary reaction, and therefore we can treat it as such and write the equation directly from the stoichiometry. Furthermore, this form fits exactly the form of the experimental rate equation for the reverse reaction giving us considerable confidence that the mechanism written is good. Because of the similarity of the two forms $k_{-3}=k'$. Therefore we will now write the rate equation for the forward step,

$$\frac{dC_{NO_3^-}}{dt} = k_3 C_{N_2O_4}$$

We did not include H_2O or C_{H_2O} here because its value is essentially constant, and it thus ends up in the rate constant k_3. The reaction is in aqueous solution. We could have written

$$\frac{dC_{NO_3^-}}{dt} = k_3' C_{H_2O} C_{N_2O_4} = k_3 C_{N_2O_4}$$

We do this because we know from practical experience that we cannot vary C_{H_2O} over more than a few percent and thus could not determine its functionality even if we wanted. The overall rate of charge in nitrate is given by the rate of the rate-determining step

$$\frac{dC_{NO_3^-}}{dt} = k_3 C_{N_2O_4} - k_{-3}(C_{H^+})(C_{NO_3^-})(C_{HNO_2})$$

We most probably do not know the concentration of N_2O_4 and cannot determine it easily, therefore, we apparently have not yet solved the entire problem. But note that we assumed that the first two reactions were fast; if this were really true than these reactions can be assumed to be at equilibrium, and we can write for Equation 2;

$$K_2 = \frac{C_{N_2O_4}}{C_{NO_2}^2}$$

and Equation 1;

$$K_1' = \frac{C_{NO_2} C_{NO} C_{H_2O}}{C_{HNO_2}^2}$$

Again however since the concentration of H_2O is not variable, we can, if we wish, include it in the equilibrium constant. At the same time we want to express C_{NO} in terms of the partial pressure of NO over the solution since we know that the final rate expression has been determined with this functionality in it. Therefore:

$$P_{NO} = K^* C_{NO}$$

$$\frac{K_1'}{C_{H_2O}} = \frac{C_{NO_2} K^* P_{NO}}{C_{HNO_2}^2}$$

$$\frac{K_1'}{C_{H_2O} K^*} = K_1 = \frac{C_{N_2O} P_{NO}}{C_{HNO_2}^2}$$

$$K_1 = \frac{C_{NO_2} P_{NO}}{C_{HNO_2}^2}$$

$$K_2 = \frac{C_{N_2O_4}}{C_{NO_2}^2} \quad \text{and} \quad C_{N_2O_4} = K_2 C_{NO_2}^2$$

We do this because N_2O_4 is an intermediate in the reaction which we cannot easily measure.

But NO_2 is another species which we may not be able to measure—it does not occur in the overall reaction; therefore, we should probably also eliminate it. NO_2 is found in the equation for K_1

$$C_{NO_2} = \frac{K_1 C_{HNO_2}^2}{P_{NO}}$$

substituting for C_{NO_2} in the above equation we obtain:

$$C_{N_2O_4} = k_2 C_{NO_2}^2 = K_2 K_1^2 \frac{C_{HNO_2}^4}{P_{NO}^2}$$

Then substituting for $C_{N_2O_4}$ in the rate equation above we obtain:

$$\frac{dC_{NO_2^-}}{dt} = k_3 K_2 K_1^2 \frac{C_{HNO_2}^4}{(P_{NO})^2} - k_{-3}\, C_{H^+}(C_{H^+})(C_{NO_3^-})(C_{HNO_2})$$

This has exactly the same form as the experimentally determined rate equation where we note that

$$k = k_3 K_1^2 K_2$$

and

$$k' = k_{-3}$$

and thus

$$\frac{dC_{NO_3^-}}{dt} = k \frac{C_{HNO_2}^4}{(P_{NO})^2} - k'(C_{H^+})(C_{NO_3^-})(C_{HNO_2})$$

Note that we will not determine k_{+3} in the experimentally determined rate law but will determine k which is

$$k_3 K_1^2 K_2$$

Consider equilibrium in the system, in which case

$$\frac{dC_{NO_3^-}}{dt} = 0$$

$$k \frac{C_{HNO_2}^4}{(P_{NO})^2} = k'(C_{H^+})(C_{NO_3^-})(C_{HNO_2})$$

$$\frac{k}{k'} = \frac{(C_{H^+})(C_{NO_3^-})(C_{HNO_2})(P_{NO})^2}{C_{HNO_2}^4} = \frac{(C_{H^+})(C_{NO_3^-})(P_{NO})^2}{C_{HNO_2}^3} = K_{eq}$$

Therefore the rate law that we derived and which was consistent with the experimentally-determined rate law, gives the correct expression for the equilibrium condition, i.e., for the equilibrium constant.

Because of the agreement observed between the experimentally-determined kinetic expression and that derived from the proposed reaction mechanism, it is reasonable to assume that the reaction mechanism proposed is a very probable one. However this agreement does not prove that the reaction mechanism is the correct one. All it says is that the agreement is good, that it is a probable mechanism, and that it is consistent with the kinetic data. There may be other reaction mechanisms which give the same kinetic or rate expressions and thus additional evidence is needed to confirm a reaction mechanism. This additional evidence can take the form of additional chemical information such as other chemistry that is known on the system, the results of tracer studies, or the identification of reaction intermediates by spectroscopic or other means.

If the proposed mechanism had produced a kinetic rate expression that did not agree with the experimentally-observed kinetic rate expression, then the proposed mechanism would have clearly been wrong. Thus kinetics can eliminate potential reaction mechanisms, but it cannot provide proof of a mechanism. <u>Kinetic agreement is necessary but not sufficient</u>.

The kinetic expression can also provide some useful evidence about the nature of the activated complex formed in the rate-determining step. For HNO_2 decomposition the forward kinetic expression was

$$r_f = k \frac{C_{HNO_2}^4}{(P_{NO})^2}$$

This suggests

$$4(HNO_2) - 2(NO) \pm n(H_2O)$$

or

$\{H_4N_2O_6 \pm nH_2O\}^{\ddagger}$ as being the activated complex

Next consider the reverse rate

$$r_r = k'[H^+][NO_3^-][HNO_2]$$

Chemical kinetics

$$(H^+) + (NO_3^-) + (HNO_2)$$

giving

$$\{H_2N_2O_5 \pm mH_2O\}^{\ddagger}$$

If m=0 and n=1, the activated complex becomes

$$\{H_2N_2O_5\}^{\ddagger}$$

and also if m−n=1 we continue to have the same activated complex; we cannot define it any more completely than this. The principle of microscopic reversibility requires that the activated complex be the same for the forward reaction and the reverse reaction.

Now let us attempt to investigate another mechanism for HNO_2 decomposition. We wish to demonstrate how a proposed reaction mechanism can be ruled out, based on kinetic data. We will take the same reactions and the same sequence of overall steps, but now let us assume a different rate-determining step.

1). $2HNO_2 \underset{k_{-1}}{\overset{k_1}{\rightleftarrows}} NO_2 + NO(g) + H_2O$ slow or rate determining

2). $2NO_2 \underset{k_{-2}}{\overset{k_2}{\rightleftarrows}} N_2O_4$ rapid (K_2)

3). $N_2O_4 + H_2O \underset{k_{-3}}{\overset{k_3}{\rightleftarrows}} H^+ + NO_3^- + HNO_2$ rapid (K_3)

First write the rate equation for the forward reaction

$$-\frac{dC_{HNO_2}}{dt} = +k_1 C_{HNO_2}^2$$

and then for the reverse reaction

$$-\frac{dC_{HNO_2}}{dt} = -k_{-1} C_{NO_2} C_{NO_g} C_{H_2O}$$

We will leave water in here, and see what happens to it. If we carried the reaction out in aqueous solution, we would know that the concentration of H_2O remains constant and would drop it out. The overall rate equation becomes:

$$-\frac{dC_{HNO_2}}{dt} = +k_1 C_{HNO_2}^2 - k_{-1} C_{NO_2} C_{NO_g} C_{H_2O}$$

If all the other steps are fast, then they are at equilibrium, and we can proceed as we did earlier and write equilibrium expressions for them.

For 2).

$$K_2 = \frac{C_{N_2O_4}}{C_{NO_2}^2}$$

and for 3).

$$K_3 = \frac{(C_{H^+})(C_{NO_3^-})(C_{HNO_2})}{C_{H_2O} C_{N_2O_4}}$$

In the rate equation above we have NO_2 which we do not have in the overall reaction, and we have assumed its value to be small; therefore we want to eliminate it from the kinetic equation. Therefore,

$$C_{NO_2}^2 = \frac{C_{N_2O_4}}{K_2}, \quad \text{giving} \quad C_{NO_2} = \frac{C_{N_2O_4}^{\frac{1}{2}}}{K_2^{\frac{1}{2}}}$$

We need to eliminate N_2O_4 also and:

$$C_{N_2O_4} = \frac{(C_{H^+})(C_{NO_3^-})(C_{HNO_2})}{C_{H_2O} K_3}$$

Substituting for $C_{N_2O_4}$ we obtain

$$C_{NO_2} = \frac{(C_{H^+}^{\frac{1}{2}})(C_{NO_3^-}^{\frac{1}{2}})(C_{HNO_2}^{\frac{1}{2}})}{C_{H_2O}^{\frac{1}{2}} K_3^{\frac{1}{2}} K_2^{\frac{1}{2}}}$$

We can now substitute for C_{NO_2} into the kinetic equation

$$-\frac{dC_{HNO_2}}{dt} = k_1 C_{HNO_2}^2 - \frac{k_{-1}}{K_2^{1/2} K_3^{1/2}} \frac{(C_{H^+}^{1/2})(C_{NO_3^-}^{1/2})(C_{HNO_2}^{1/2})(C_{NO_g})(C_{H_2O})}{C_{H_2O}^{1/2}}$$

$$-\frac{dC_{HNO_2}}{dt} = k_1 C_{HNO_2}^2 - \frac{k_{-1}}{K_2^{1/2} K_3^{1/2}} (C_{H_2O}^{1/2})(C_{H^+}^{1/2})(C_{NO_3^-}^{1/2})(C_{NO_g})(C_{HNO_2}^{1/2})$$

$$-\frac{dC_{HNO_2}}{dt} = k \, C_{HNO_2}^2 - k_{-1} (C_{H^+}^{1/2})(C_{NO_3^-}^{1/2})(P_{NO})(C_{HNO_2}^{1/2})$$

This kinetic expression bears no functional similarity to the form of the rate expression determined empirically indicating clearly that this mechanism is not the correct mechanism. This says that the mechanism with the rate-determining step where it was assumed to be is not correct, and thus the proposed overall scheme is not correct. However it does not tell us whether the proposed mechanism is incorrect or whether it is just the location of the rate-determining step. If no rate determining steps fit, it is the mechanism which is not correct.

<u>From kinetics it is possible to rule out a given reaction mechanism, it is possible to show that a reaction mechanism is probable, but it is not possible to prove a reaction mechanism.</u>

If we assume that the system is at equilibrium and set
$$\left(-\frac{dC_{HNO_2}}{dt}\right) = 0$$

we obtain

$$\frac{k_1}{k_{-1} K_2^{-1/2} K_3^{-1/2}} = \frac{(C_{H^+}^{1/2})(C_{NO_g})(C_{NO_3^-}^{1/2})(C_{HNO_2}^{1/2})(C_{H_2O}^{1/2})}{C_{HNO_2}^2}$$

This is not the correct equilibrium constant, but if it is squared, we obtain:

$$\left(\frac{k_1}{k_{-1}K_2^{-\frac{1}{2}}K_3^{-\frac{1}{2}}}\right)^2 = \frac{(C_{H^+})(C_{NO_g}^2)(C_{NO_3^-})(C_{H_2O})}{C_{HNO_2}^3}$$

Let us next consider a second example, the reaction of hypochlorite ion with iodide ion:

$$I^- + OCL^- \rightarrow OI^- + CL^-$$

At a constant OH^- concentration the empirical rate expression is

$$\frac{dC_{OI^-}}{dt} = k'(C_{I^-})(C_{OCL^-})$$

Therefore it looks like a simple reaction, possibly even an elementary reaction. But if the pH or OH^- concentration is changed, we find that the rate varies inversely with it and therefore that the empirical rate expression is actually

$$\frac{dC_{OI^-}}{dt} = k\frac{(C_{I^-})(C_{OCL^-})}{C_{OH^-}}$$

Therefore the reaction is not an elementary reaction. Let us propose the following mechanism:

1). $OCL^- + H_2O \underset{}{\overset{K_1}{\rightleftarrows}} HOCL + OH^-$ rapid

2). $I^- + NOCL \underset{}{\overset{k_2}{\rightleftarrows}} HOI + CL^-$ slow

3). $OH^- + HOI \underset{}{\overset{K_3}{\rightleftarrows}} H_2O + OI^-$ rapid

For the indicated rate-determining step the kinetic expression becomes:

$$\frac{dC_{OI^-}}{dt} = k_2(C_{I^-})(C_{HOCL})$$

We can write this in terms of OI^- because the last reaction is fast and is at equilibrium; therefore HOI formed immediately decomposes into OI. It is assumed that the concentration of HOI is small relative to that of OI^- ultimately and thus that a build up in its concentration does not significantly affect the rate of appearance of OI$^-$ in solution. Using Equation 1 which is assumed to be at equilibrium we can eliminate the HOCL concentration which it is assumed that we cannot measure

$$K_1' = \frac{(C_{HOCL})(C_{OH^-})}{(C_{OCL^-})(C_{H_2O})}$$

$$C_{HOCL} = K_1 \frac{C_{OCL^-}}{C_{OH^-}}$$

where C_{H_2O} is a constant and has been incorporated with K_1'. Substituting into the rate equation gives:

$$\frac{dC_{OI^-}}{dt} = k_2(C_{I^-})(C_{HOCL}) = k_2 k_1 \frac{C_I \, C_{OCL^-}}{C_{OH^-}}$$

This agrees very well with the experimentally-observed kinetic behavior. We had to use no kinetic or equilibrium information from the third reaction in any formal manner. Thus we have gotten no information on what happens after the second step.

If we had A+B $\rightleftarrows$ M+N which involved the following mechanism

1). A+B $\rightleftarrows$ C+D   Fast

2). C+D $\rightleftarrows$ E+F   Fast

3). E+F $\rightleftarrows$ G+H   Slow

4). G+H $\rightleftarrows$ I+J   Fast

5). I+J $\rightleftarrows$ K+L   Fast

6). K+L $\rightleftarrows$ M+N   Fast

we could from kinetic data say that the proposed mechanism up to step #3 was reasonable if we obtained agreement between observed kinetics and derived kinetics, but we could not have drawn any conclusions about the mechanism of the reaction following the third step.

STEADY STATE APPROXIMATION

Above we discussed reactions where the sequence of reactions had one step which was much slower than the rest so that the remaining steps were at equilibrium, and the overall rate was governed by the rate of the slow step or the rate-determining step. In this case all intermediates may have large or small values, and we can treat the sequence as we did. If there is no one rate-determining step and if there is a condition in which none of the reactions in the sequence can be considered to be at equilibrium, the reaction must be treated differently.

Let us take as an example the overall reaction between nitrogen dioxide and fluorine

$$2NO_2 + F_2 \xrightarrow{k} 2NO_2F$$

The reaction has been shown to be first order in each of its reactants at low pressures. From this we can begin to postulate the following mechanism

1). $NO_2 + F_2 \underset{k_2'}{\overset{k_2}{\rightleftarrows}} NO_2F + F$

The fluorine atom so generated then can be removed by several possible mechanism.

2). the reverse of reaction 1

3). $F + NO_2 + M \xrightarrow{k_3} NO_2F + M^*$

4). $F + F + M \xrightarrow{k_4} F_2 + M^*$

Rewriting this in the more standard manner

$$NO_2 + F_2 \xrightarrow{k_2} NO_2F + F \tag{1}$$

$$F + NO_2F \xrightarrow{k_2'} NO_2 + F_2 \tag{2}$$

$$F + NO_2 + M \xrightarrow{k_3} NO_2F + M^* \tag{3}$$

$$F + F + M \xrightarrow{k_4} F_2 + M^* \tag{4}$$

Chemical kinetics

Reactions (3) and (4) are bimolecular recombinations in the presence of a third body; the third body removes the energy of reaction so that the other two species can remain together.

At this point we can look more closely at the reactions and make the following statement. Under normal conditions of reaction we know that a fluorine atom, being as highly reactive as it is, will react with anything in sight and will also be present in a very low concentration. The fact that it will be present in very low concentration relative to other species means that the probability of an F+F collision will be very small and as such reaction (4) can be assumed to be inconsequential; therefore, we will eliminate it.

The rate of formation of NO_2F then becomes

$$\frac{dC_{NO_2F}}{dt} = k_2 C_{NO_2} C_{F_2} - k_2' C_F C_{NO_2F} + k_3 C_F C_{NO_2} C_M$$

This rate equation adequately describes the rate of NO_2F formation except that we have no way of measuring the F concentration. Therefore let's write a rate equation for F

$$\frac{dC_F}{dt} = k_2 C_{NO_2} C_{F_2} - k_2' C_F C_{NO_2F} - k_3 C_F C_{NO_2} C_M$$

But C_F always remains much, much less than C_{NO_2}, C_{F_2}, C_{NO_2F}, or C_M. We can thus apply the steady-state approximation to the rate of change of C_F with respect to time. With this we write

$$\frac{dC_F}{dt} = 0$$

which is essentially true when compared with

$$\frac{dC_{NO_2}}{dt}, \frac{dC_{F_2}}{dt}, \frac{dC_{NO_2F}}{dt} \gg \frac{dC_F}{dt}$$

Therefore

$$0 = k_2 C_{NO_2} C_{F_2} - k_2' C_F C_{NO_2F} - k_3 C_F C_{NO_2} C_M$$

We can now solve this for C_F, and we see that all other terms

(C's) involve measurable species.

$$C_F(k_2' C_{NO_2F} + k_3 C_{NO_2} C_M) = k_2 C_{NO_2} C_{F_2}$$

$$C_F = \frac{k_2 C_{NO_2} C_{F_2}}{k_2' C_{NO_2F} + k_3 C_{NO_2} C_M}$$

Rewriting the equation for NO_2F formation we obtain

$$\frac{dC_{NO_2F}}{dt} = k_2 C_{NO_2} C_{F_2} + \{-k_2' C_{NO_2F} + k_3 C_{NO_2} C_M\} C_F$$

and then substituting in for C_F gives

$$\frac{dC_{NO_2F}}{dt} = k_2 C_{NO_2} C_{F_2} + \{k_2' C_{NO_2F} + k_3 C_{NO_2} C_M\} \frac{k_2 C_{NO_2} C_{F_2}}{k_2' C_{NO_2F} + k_3 C_{NO_2} C_M}$$

$$\frac{dC_{NO_2F}}{dt} = k_2 C_{NO_2} C_{F_2} \left\{ \frac{k_2' C_{NO_2F} + k_3 C_{NO_2} C_M - k_2' C_{NO_2F} + k_3 C_{NO_2} C_M}{k_2' C_{NO_2F} + k_3 C_{NO_2} C_M} \right\}$$

This is the derived rate equation. This rate expression could be first order in NO_2 and F_2, but it can also be second order in NO_2 or can show intermediate behavior. However, if

$$k_3 C_{NO_2} C_M \gg k_2' C_{NO_2F} ,$$

we have

$$\frac{dC_{NO_2F}}{dt} = 2k_2 C_{NO_2} C_{F_2} \left\{ \frac{k_3 C_{NO_2} C_M}{k_3 C_{NO_2} C_M} \right\}$$

and thus

$$\frac{dC_{NO_2F}}{dt} = 2k_2 C_{NO_2} C_{F_2}$$

Chemical kinetics

This is the behavior observed. Furthermore looking back at the original equation if this assumption about relative magnitude were not true, then we would be able to check it by operating with a large amount of NO_2F present and observing an inhibiting effect by NO_2F. However even under conditions of very high concentrations of NO_2F the rate is uneffected indicating that k_2' is really very small; that is the reverse reaction is not important. If the reverse reaction were more important, the rate equation would have been more complex. Consider the case where

$$F + F + M \rightarrow F_2 + M$$

is important. Then

$$\frac{dC_F}{dt} = k_2 C_{NO_2} C_{F_2} - k_2' C_F C_{NO_2F} - k_3 C_{NO_2} C_F C_M - k_4 C_F^2 C_M$$

and when this is set equal to zero we have a quadratic to solve. This would result in a rate equation which would be even more complicated and which would not fit the experimentally determined results; therefore this can be ruled out.

Thus the final equation which best represents the mechanism and the data is:

$$\frac{dC_{NO_2F}}{dt} = 2k_2 C_{NO_2} C_{F_2}$$

This tells us that the rate of formation of NO_2F is twice the rate of the first reaction because from the constraints which we have arrived at for this reaction sequence each time one NO_2F is formed by the first reaction, the F formed reacts to give NO_2F. If other steps had been important, the rate equation would have been more complex, and the above conclusion would not have held.

Because we do not see any effect of the third body in the kinetics or rate expression for the system we can furthermore conclude that the system really is of the much simpler form

$$NO_2 + F_2 \rightarrow NO_2F + F \qquad \text{slow}$$
$$F + NO_2 + M \rightarrow NO_2F + M^* \qquad \text{fast}$$

HETEROGENEOUS CATALYSIS

Heterogeneous catalysts generally operate by chemisorbing one or several reactants and usually produce some rearrangement or pertubation in the electronic or physical structure of the species; then chemical reaction occurs either as a result of interaction with another species adsorbed or by rearrangement or decomposition. Physical adsorption is a weak adsorption much like a two-dimensional fluid, and because it does not perturb the adsorbed species significantly, it does not result in catalyzing reactions.

Chemisorbed species are actually chemically bonded at specific surface sites to the surface of the solid. If the adsorbed species is too weakly bonded there will not be sufficient rearrangement to promote appreciable chemical reaction. If the species is too strongly bonded then it will be almost inert to reaction with other molecules or to decomposition or rearrangement, since it is just like a very stable compound. There is only a range of sorption strengths which are useful in terms of catalysis.

Since chemisorption of at least one of the reactants is required for a solid to catalyze a reaction, we must consider chemisorption and adsorption isotherms. Adsorption isotherms are most important in terms of relating partial pressures to surface concentrations and through this relating heterogeneous kinetics to gas-phase concentrations of reactants and products.

We will consider first the Langmuir isotherm since it is one of the simplest, the most frequently used, and illustrates well the principles involved. The assumptions involved include:

- molecules from the vapor are adsorbed on the surface in a monolayer only at the final level of adsorption,
- or they are adsorbed only on specific sites on the surface-- one molecule per site. (Thus the Langmuir isotherm can apply to either physical or chemical adsorption.)
- The surface is homogeneous, i.e., all sites have the same strengths and the differential heat of adsorption is independent of surface coverage.
- There is no interaction between adsorbed molecules or at least no interaction which affects either the rate of adsorption or the adsorption equilibrium.
- All adsorption occurs by the same mechanism, i.e., all sites are the same and all adsorbed species of the same type are the same.

The assumption of uniformity of active sites is open to the most serious question. Surfaces for chemisorption are most often observed to be heterogeneous; ΔH_{ad} almost always varies with coverage. This can frequently be handled and thus at times may not be a serious problem. For example, adsorption sites having a narrow range of adsorption energies are probably responsible for catalyzing a given reaction, although the true range of adsorption energies may be much broader.

The following is one development for chemisorption. It assumes that the chemisorption velocity depends upon:

i) Pressure since it determines the number of collisions.
ii) Activation energy for chemisorption E_A--since this gives the fraction of colloiding molecules which possess sufficient energy to be chemisorbed. The activation energy is frequently zero.
iii) Fractional coverage of surface, $f(\theta)$, since we assume that the rate is dependent on fraction bare. $f(\theta) = 1-\theta$ for one molecule adsorbing per site.
iv) Steric factor or condensation coefficient, σ, which defines the fraction of colliding molecules having sufficient energy, E_A, which do adsorb as a result of collision. This is not to be confused with the accommodation coefficient, α, which defines the fraction of all collisions resulting in adsorption.

The rate of adsorption can then be expressed as

$$r_{ads} = \sigma \frac{P}{(2\pi mkT)^{\frac{1}{2}}} \exp(-E_a/RT) \, f(\theta)$$

| condensation coefficient | No. of collisions with surface per unit time, from kinetic theory of gases | fraction of collisions having sufficient energy for chemisorption | fraction of surface uncovered |

The rate of desorption can be written as shown below

$$r_{des} = \nu_{-1} \, \phi(\theta) \, \exp(-E_d/RT)$$

ν_{-1} = frequency of vibration normal to the surface
$\phi(\theta)$ = fractional no. of sites available for desorption,
$\phi(\theta) = \theta$ for desorption from single site.
E_d = activation energy of desorption.
θ_i = fraction of surface covered with species i
$1-\theta_i$ = fraction of surface bare, when species i only is present.

At equilibrium

$$r_{ads} = r_{des}$$

Equating and then rearranging gives:

$$\frac{\sigma P}{(2\pi mkT)^{\frac{1}{2}}} f(\theta) \exp(\frac{-E_a}{RT}) = \nu_{-1} \phi(\theta) \exp(\frac{-E_d}{RT})$$

$$P = (2\pi mkT)^{\frac{1}{2}} \frac{\nu_{-1}}{\sigma} \exp(\frac{-E_d + E_a}{RT}) \frac{\phi(\theta)}{f(\theta)}$$

where $-E_d + E_a = +\Delta H_{ads}$

For one molecule on a single site $\phi(\theta) = \theta$
$f(\theta) = 1-\theta$

$$P = \frac{1}{K} \frac{\theta}{(1-\theta)}$$

where $\frac{1}{K} = (2\pi mkT)^{\frac{1}{2}} \frac{\nu_{-1}}{\sigma} \exp(\frac{+\Delta H_{ads}}{RT})$

This can then be rearranged to give

$$KP(1-\theta) = \theta$$

and

$$\theta = \frac{KP}{1+KP}$$

where

$$K_A = \frac{\sigma}{\nu_{-1}(2\pi mkT)^{\frac{1}{2}}} \exp(\frac{-\Delta H_{ads}}{RT})$$

This is the Langmuir adsorption isotherm for one component. Statistical mechanics allows us to calculate σ/ν_{-1} and then all that is needed is ΔH_{ads}.

For a molecule adsorbing and dissociating on the surface such as

$$A_2 + 2S \rightleftarrows 2(A-S)$$

The rate of adsorption becomes

$$r_{ads} = \frac{\sigma P_A}{(2\pi mkT)^{\frac{1}{2}}} \exp(-E_a/RT)(f(\theta))$$

If it is assumed that the A_2 must impinge on a site which has a vacant adjacent site, then

$f(\theta) = (1-\theta)^2$ to fairly high coverages.

Chemical kinetics

For desorption we can write

$$r_{des} = \nu_{-1} \exp(-E_d/RT) \phi(\theta) = \nu_{-1} \exp(-E_d/RT) \theta^2$$

If two sites (A-S) containing A atoms must be close together to have an A_2 desorb, $\phi(\theta) = \theta^2$. Equating the rate of adsorption to the rate of desorption gives

$$\frac{\sigma P}{(2\pi mkT)^{\frac{1}{2}}} \exp(-E_a/RT) (1-\theta)^2 = \nu_{-1} \exp(-E_d/RT) \theta^2$$

Upon rearrangement we obtain

$$P = 1/K_A \, (\theta/1-\theta)^2$$

$$(K_A P_A)^{\frac{1}{2}} = \theta/1-\theta$$

$$\theta = (K_A P_A)^{\frac{1}{2}}/1+(K_A P_A)^{\frac{1}{2}}$$

where $K_A = \dfrac{\sigma}{\nu_{-1}(2\pi mkT)^{\frac{1}{2}}} \exp(-\Delta H_{ad}/RT)$

This expression holds only to fairly high coverages if atoms are not mobile. It holds for all coverages if the atoms are mobile on the surface. Surface mobility is not uncommon at reaction conditions.

The Langmuir isotherm

$$\theta = K_A P_A/1+K_A P_A$$

where θ = fraction coverage = v/v_m where v = volume (STP) adsorbed at pressure P and v_m = volume adsorbed (STP) at monolayer coverage or at saturation

shows the following characteristics:

- At low values of $K_A P_A$ it is a straight line of slope K on a θ vs P_A plot (Curve 1).
- At high values $K_A P_A$ it is a curve when plotted as θ vs P_A, and the curve becomes flat and asymptotic to $\theta = 1.0$ with increasing P_A (Curve 3).

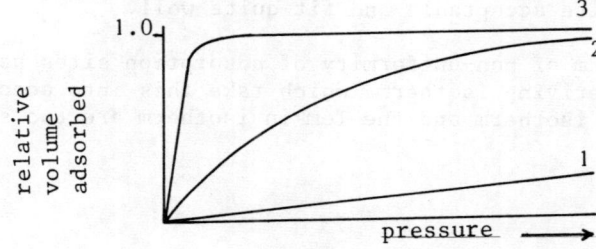

For a given pressure and temperature curve shape is determined by the value of K_A. Thus strong adsorption and high ΔH_{ads} result in a rectangular-shaped curve because K_A is large. If adsorption is weak, ΔH_{ads} is small, and the isotherm is like Curve 1.

There are a number of problems associated with the Langmuir isotherm. The major ones are:

- The shape of a measured adsorption isotherm never follows the theoretical shape over the full range of pressure.

- A true saturation value is never clearly observed.

- If a reasonable fraction of the isotherm is used to predict the saturation value, v_m usually represents only a small fraction of the total surface, indicating non-uniform sites.

- v_m, the monolayer or saturation adsorption is often observed to decrease with increasing temperature.

These effects are due mainly to heterogeneity or non-uniformity of the surface sites.

This problem can be handled by splitting the sites into groups, each group having a small range of ΔH_{ads}.

$$\theta_i = \frac{K_{Ai} P_A}{1 + K_{Ai} P_A}$$

$$\theta = \frac{\sum_i \theta_i N_i}{\sum_i N_i} = \sum_{i=1}^{n} \frac{K_{Ai} P_A}{1 + K_{Ai} P_A} N_i / \sum_i N_i$$

Frequently chemical reaction is catalyzed only by sites having a narrow range of heats of adsorption. Thus a Langmuir isotherm fitted to reaction data will have adsorption constants appropriate to range of sites catalyzing the reaction, and although these constants will be different from those obtained from a purely adsorption experiment and a Langmuir isotherm will not truly fit broadly-ranging adsorption data, its application to catalytic data may be quite acceptable and fit quite well.

The problem of non-uniformity of adsorption sites has been addressed by deriving isotherms which take this into account. The Freundlich isotherm and the Temkin isotherm are two such examples.

Chemical kinetics

The <u>Freundlich Isotherm</u> assumes that the heat of adsorption decreases logarithmically with coverage. To obtain the Freundlich isotherm the Langmuir isotherm with a logarithmic decrease in the heat of adsorption in the K term is integrated from $-\Delta H_{ads} = 0$ to $-\Delta H_{ads} = \infty$.
The result is

$$\ln \theta = RT/Q_o \ln P + \text{constant } k$$

or

$$\theta = k \cdot P^{RT/Q_o} = kP^{1/n} \quad \text{where } n > 1$$

The equation has two adjustable parameters rather than one, and therefore it should fit data better. The significance of n is $n = Q_o/RT$ where Q_o is the heat of adsorption at zero coverage. The Freundlich isotherm is good only at low coverages because this is where the integration performed is valid.

The <u>Temkin Isotherm</u> assumes that the heat of adsorption decreases linearly with coverage which is the behavior typically observed. Therefore it assumes that

$$Q = Q_o(1-\alpha\theta)$$

where Q_o = heat of adsorption at zero coverage
α = proportionality constant.

Applying this to the Langmuir isotherm we obtain

$$\theta = \frac{RT}{Q_o \alpha} \ln A_o P$$

where

$A_o = a_o e^{-q_o/RT}$ is independent of coverage.

This isotherm like the Freundlich isotherm has two adjustable parameters for correlating data. Thus the fit will undoubtedly be better regardless of the validity of the model.

The <u>Brunauer-Emmett-Teller</u> (BET) isotherm is valid mainly for physical adsorption and is presented only briefly here. Langmuir considered only monolayer adsorption, which must be the maximum for chemisorption. However, physical adsorption is never single layer adsorption except at very low coverages. At these low coverages thus the Langmuir equation might hold true but for higher coverage there is multilayer adsorption. The BET isotherm treats this by doing essentially a series of Langmuir isotherms.

The result is:

$$\frac{P}{V(P_o - P)} = \frac{1}{v_m C} + \frac{C-1}{v_m C}\left(\frac{P}{P_o}\right)$$

where

P = partial pressure of component of interest
P_o = saturation pressure of the sorbate at the temperature of the experiment
v = volume adsorbed at T and P at STP
v_m = volume of gas adsorbed in one monolayer at STP
c = a constant which should be quite large.

This equation fits experimental-physical adsorption data well over a range of $\theta = 0.05$ to 0.40 and is used to determine the surface area of porous catalysts and other materials.

SURFACE REACTION RATES

For homogeneous elementary reactions one can write the rate expression directly from the stoichiometric equation. For example, for $A + B \rightarrow D$

$$\text{rate} = kC_A C_B = A\, e^{-E/RT} C_A C_B$$

In the foregoing we showed how complex reactions composed of a series of elementary reactions could be treated. In this case it was possible to derive the kinetic expressions by first writing the overall reaction as a set of elementary reactions. By assuming that one elementary reaction in the sequence was rate determining or by assuming that certain of the intermediates reached a low steady-state concentration and by applying the pseudo-steady state approximation it was easy to derive kinetic expressions for these more complex reaction networks. The same concepts can be applied when considering reaction on surfaces. This will be discussed below; connection between the surface and the gas phase is usually made through application of an isotherm.

By analogy with the gas-phase case we might write for a reaction catalyzed by a surface:

$$\text{rate}\left[\frac{\text{molecule reacted}}{(\text{sec})(\text{cm}^2)}\right] = \frac{k_o}{S} e^{-E/RT} f(C_A C_B)$$

$$r = \frac{k_o}{S} e^{-E/RT} C_A C_B = k\, C_A C_B$$

This is a simple power rate law. This works well for some reactions and sometimes over a wide enough range to make it useful. However, in many more cases this model does not fit experimental results well. Typically activation energies and reaction orders are observed to change. Thus we have to look for a better kinetic model.

Since reaction is occurring on the surface of the catalyst we postulate that the rate must be dependent upon the concentration of the species on the surface of the catalyst which is proportional to θ, the fractional surface coverage.

For a unimolecular reaction, for example

$$A \rightarrow B$$

via

$$A + S \rightarrow A\text{-}S \rightarrow B\text{-}S \rightarrow B + S$$

$$\text{rate} = k\theta_A$$

In general we cannot measure concentration as θ_A on the surface. We must therefore relate gas-phase pressure to surface coverage; we do this by applying an adsorption isotherm, for example the Langmuir isotherm to the surface rate expression.

$$\theta_A = \frac{K_A P_A}{1 + K_A P_A}$$

Substitution for θ_A gives

$$\text{rate} = \frac{k\, K_A P_A}{1 + K_A P_A}$$

When this rate expression is plotted vs. gas-phase partial pressure of reactant the following behavior ensues.

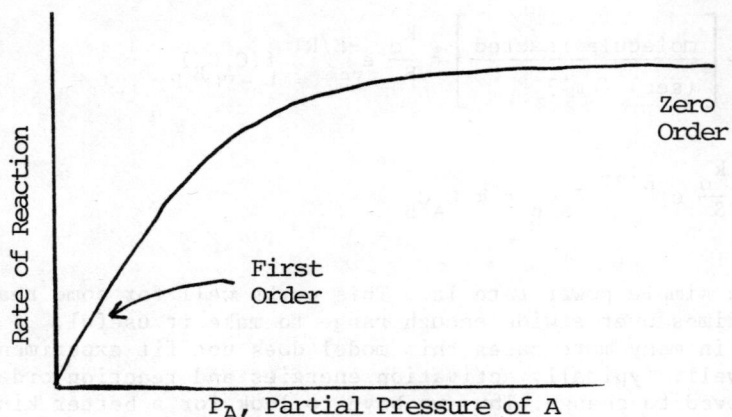

P_A, Partial Pressure of A

This type of kinetic expression is called Langmuir-Hinshelwood kinetics because Hinshelwood applied the Langmuir isotherm to a surface rate model. Langmuir-Hinshelwood kinetics hold for a large number of reaction systems and over a wide enough range of operating variables to allow the expression to be of value.

In addition to the assumptions present in the Langmuir isotherm, the following assumptions are present in a Langmuir-Hinshelwood kinetic model.

- Adsorption equilibrium is assumed to be established; that is adsorption-desorption rates are <u>much</u> faster than reaction rates. This allows surface concentration to be expressed by the isotherm.

- Reaction involves a single adsorbed species or occurs between adsorbed species on the surface of the catalyst. If a single component is decomposed on the surface, the process can be assumed to be either unimolecular or bimolecular: If there is not strong evidence to the contrary, a decomposition reaction is usually unimolecular. If reaction occurs between A and B molecules, these must be adsorbed on neighboring sites, and the probability of reaction is taken as being proportional to the product $\theta_A \theta_B$.

Let us next derive the Langmuir-Hinshelwood kinetic expression for a general case.

$A \rightarrow B + C + I.$

We will assume where I is a chemically inert species for the reaction

- A, B, I and C are all appreciably adsorbed.
- the reaction rate is proportional to the quantity of adsorbed A or concentration of A on the surface.
- adsorption occurs on only one type of site.
- reverse reaction is negligible.

We have just stated that the rate can be written as

$$\text{rate} = k\theta_A$$

Therefore we need θ_A, but since we have more than one species adsorbed, we cannot use the expression that we derived earlier. Therefore we will derive an expression for θ_A which is more general. We will first equate the rate of adsorption to that of desorption for A.

rate of adsorption of A = rate of desorption of A

$$k_A(1 - \sum_i \theta_i) P_A = k'_A \theta_A$$

θ_i = fraction of surface covered by species i
$\sum_i \theta_i = \theta_A + \theta_B + \theta_C + \theta_I$ = total fraction of surface sites covered.
$1 - \sum_i \theta_i$ = fraction of sites on the surface which are bare.

Therefore we obtain from the above

$$\theta_A = k_A/k'_A \, P_A(1-\sum_i \theta_i)$$

$$\theta_A = K_A P_A(1-\sum_i \theta_i)$$

where $k_A/k'_A = K_A$ where K_A is an adsorption equilibrium constant for species A.
By the same token

$$\theta_B = K_B P_B(1-\sum_i \theta_i)$$

$$\theta_C = K_C P_C(1-\sum_i \theta_i)$$

$$\theta_I = K_I P_I(1-\sum_i \theta_i)$$

Summing the above three equations gives:

$$\theta_A + \theta_B + \theta_C + \theta_I = (K_A P_A + K_B P_B + K_C P_C + K_I P_I)(1-\sum_i \theta_i)$$

$$\sum_i \theta_i = (K_A P_A + K_B P_B + K_C P_C + K_I P_I)(1-\sum_i \theta_i)$$

If we then subtract both sides of the above equation from 1, we obtain:

$$1 - \sum_i \theta_i = 1 - (K_A P_A + K_B P_B + K_C P_C + K_I P_I)(1 - \sum_i \theta_i)$$

$$1 - \sum_i \theta_i = \frac{1}{1 + K_A P_A + K_B P_B + K_C P_C + K_I P_I}$$

This expression can then be substituted into the expressions above for θ_A, θ_B, θ_C and θ_I giving

$$\theta_A = \frac{K_A P_A}{1 + K_A P_A + K_B P_B + K_C P_C + K_I P_I}$$

$$\theta_B = \frac{K_B P_B}{1 + K_A P_A + K_B P_B + K_C P_C + K_I P_I}$$

$$\theta_C = \frac{K_C P_C}{1 + K_A P_A + K_B P_B + K_C P_C + K_I P_I}$$

If other components were present in the mixture they would also appear in the denominator as $K_X P_X$, along with the other terms.

We can now apply these expressions for surface coverage to the rate expression giving

$$r = k\theta_A = \frac{k K_A P_A}{1 + K_A P_A + K_B P_B + K_C P_C + K_I P_I}$$

Let us examine the rate behavior that this expression predicts. For example if B and C are weakly adsorbed, we have

$$\text{rate} = \frac{k K_A P_A}{1 + K_A P_A}$$

This is the expression that we derived above.

If A and C weakly adsorbed, then we have

$$\text{rate} = \frac{k K_A P_A}{1 + K_B P_B}$$

Further if $K_B P_B \gg 1$, we have

$$\text{rate} = \frac{k K_A P_A}{K_B P_B}$$

Chemical kinetics

or

$$\text{rate} = k' \left(\frac{P_A}{P_B}\right)$$

where $k' = kK_A/K_B$

These latter cases are typical examples of product inhibition.

An example of this is the decomposition of NH_3 on platinum or copper catalysts

$$2NH_3 \rightarrow N_2 + 3H_2$$

where hydrogen strongly inhibits the rate

$$\text{rate} = \frac{k' P_{NH_3}}{P_{H_2}}$$

If the above reaction is such that there is relatively weak adsorption of A and weaker adsorption of products, the rate expression becomes

$$\text{rate} = \frac{k K_A P_A}{1 + K_A P_A}$$

If we are in either a low pressure region or a high enough temperature region the denominator $\cong 1.0$, and we have pseudo-first order behavior. The rate expression:

$$\text{rate} = kK_A P_A$$

Let us now consider the effect of temperature on the kinetic behavior. K_A is an equilibrium constant and therefore it can be written as

$$K_A = e^{-\Delta G_a/RT} = e^{+\Delta S_a/RT} e^{-\Delta H_A/RT}$$

k is a kinetic rate constant, and it can be written as shown

$$k = Ae^{-E/RT}$$

Finally we consider the rate of reaction as observed or measured experimentally.

$$\text{rate} = k' P_A$$

If we plot k' as observed vs $1/T$, an Arrhenius plot, we obtain an activation energy which we will refer to as an apparent activation energy E_{app}.

We have

$$k' = A_{app} e^{-E_{app}/RT}$$

and also

$$kK_A = A\, e^{-E/RT} e^{+\Delta S_a/R} e^{-\Delta H_a/RT}$$

Since

$$k' = kK_A$$

we see that

$$E_{app} = E + \Delta H_a$$

The apparent activation energy is lower than the true value by the heat of adsorption.

<u>True activation energy</u> is that activation energy associated with the chemical transformation of the reactant on the surface.

<u>Apparent activation energy</u> is the activation energy observed. It may contain other factors.

For a second-order reaction which is assumed to be bimolecular and irreversible for example

$$A + B \rightarrow C + D$$

we can write

$$r_A = k\theta_A \theta_B$$

which gives

$$r_A = \frac{k(K_A P_A)(K_B P_B)}{(1 + K_A P_A + K_B P_B + K_C P_C + K_D P_D)^2}$$

Chemical kinetics

It is possible that reactions involving two reactants can occur without having both reactants adsorbed on the surface. Such reaction mechanisms are referred to as Rideal or Rideal-Eley mechanisms.

An example is A + B → C where A may be adsorbed on the catalyst surface, and B may react with it from the gas phase. The rate expression would therefore become

$$\text{rate} = k\,\theta_A P_B$$

$$\text{rate} = \frac{k P_B K_A P_A}{1 + K_A P_A + K_B P_B}$$

giving

$$\text{rate} = \frac{k K_A P_A P_B}{(1 + K_A P_A)}$$

Kinetic parameters are easily obtained for such rate expressions assuming that sufficiently accurate data are available. Sophisticated non-linear regression techniques can be used or the equation and data can be rearranged and plotted. Given a simple A + B reaction on a surface which follows a rate expression of the form

$$\text{rate} = r = \frac{k K_A P_A}{1 + K_A P_A + K_B P_B}$$

we would first rearrange the rate expression to give:

$$1 + K_A P_A + K_B P_B = \frac{k K_A P_A}{r}$$

$$\frac{1}{k K_A} + \frac{P_A}{k} + \frac{K_B}{k K_A} P_B = \frac{P_A}{r}$$

The data would then be plotted as $\frac{P_A}{r}$ vs. P_A for data in which P_B was held constant. In this case the slope $\triangleq \frac{1}{k}$ and the

$$\text{intercept} = \frac{1}{kK_A} + \frac{K_B}{kK_A} P_B = I$$

This then gives

$$kK_A I = 1 + K_B P_B$$

If we now study the rate at constant P_A; vary P_B and plot $\frac{P_A}{r}$ vs P_B we obtain intercept $= \frac{1}{kK_A} + \frac{P_A}{k}$ and slope $= \frac{K_B}{kK_A}$.

From these equations K_A and K_B can be estimated.

We have already mentioned that surfaces are not homogeneous, but that they are heterogeneous with respect to heats of adsorption and specifically with respect to chemisorption. Yet Langmuir-Hinshelwood kinetics appear to hold for many cases. Intuitively it seems logical that if you have a set of sites with varying heats of adsorption then some sites would not bond strongly enough with the reactant to perturb its electron configuration enough to promote reaction. Other sites will bond so strongly that they will form a stable surface complex which will not decompose or react, and these sites will be effectively poisoned. Thus sites in a relatively narrow range of heats of chemisorption will be responsible for reaction. Tamaru, in studying the decomposition of formic acid over copper, found results consistent with this picture. He studied the rate of reaction and simultaneously determined the amount adsorbed. He found that the amount of formic acid adsorbed increased with formic acid pressure, but the reaction rate did not change, i.e., the rate was zero order, above a very low pressure. Clearly only a very small fraction of the sites--those which chemisorbed formic acid at very low pressures--take part in the reaction. Those sites which adsorb formic acid at higher pressures contribute nothing to the rate. Thus we are really considering a group of sites with a very narrow range of heats of adsorption, and the Langmuir isotherm should work well for the sites in this energy range.

This suggests that the values of K_A, K_B, and K_I determined from adsorption measurements may not be valid for use in the reaction rate expression. This has been found to be the situation in that it has not been possible to obtain a reasonable prediction of kinetic rates using parameters obtained from adsorption experiments.

Chemical kinetics

METHANATION

We will illustrate the application of these concepts for representing heterogeneous reactions by the following example concerning CO hydrogenation to form methane. In any experimental program it is important to determine the forms of kinetic expressions one might expect on the basis of a proposed reaction mechanism. Such an exercise is essential in designing experiments to help discriminate between different reaction mechanisms and between the rate-controlling steps in a proposed reaction mechanism. In this section, analyses of kinetic models based on the 'Carbide' theory and the 'Enol complex' theory are presented and their implications in light of the available experimental data are briefly discussed.

Carbide Theory: The carbide theory in its simplest form postulates that CO chemisorbs on the metal surface, and then undergoes dissociation. Surface carbon is then hydrogenated to form methane; adsorbed oxygen is removed as CO_2 and/or as H_2O. The reaction sequence can be represented as follows:

Adsorption and Dissociation

Step 1: $\quad CO(g) + * \underset{k'_{CO}}{\overset{k_{CO}}{\rightleftarrows}} CO_{ad}$ \hfill (A-1)

Step 2: $\quad H_2(g) + 2* \underset{k'_{H_2}}{\overset{k_{H_2}}{\rightleftarrows}} 2H_{ad}$ \hfill (A-2)

Surface Reactions

Step 3: $\quad CO_{ad} + * \underset{k'_1}{\overset{k_1}{\rightleftarrows}} C_{ad} + O_{ad}$ \hfill (A-3)

Step 4: $\quad C_{ad} + H_{ad} \underset{k'_2}{\overset{k_2}{\rightleftarrows}} CH_{ad} + *$ \hfill (A-4)

Step 5: $\quad CH_{ad} + H_{ad} \underset{k'_3}{\overset{k_3}{\rightleftarrows}} CH_{2ad} + *$ \hfill (A-5)

$\quad CH_{2ad} + H_{ad} \underset{k'_4}{\overset{k_4}{\rightleftarrows}} CH_{3ad} + *$ \hfill (A-6)

$$CH_{3_{ad}} + H_{ad} \underset{k_5'}{\overset{k_5}{\rightleftarrows}} CH_{4_{ad}} + * \qquad (A-7)$$

Step 6: $\quad O_{ad} + CO_{ad} \underset{k_6'}{\overset{k_6}{\rightleftarrows}} CO_{2_{ad}} + * \qquad (A-8)$

Step 7: $\quad O_{ad} + 2H_{ad} \underset{k_7'}{\overset{k_7}{\rightleftarrows}} H_2O_{ad} + 2* \qquad (A-9)$

Desorption

Step 8: $\quad CH_{4_{ad}} \underset{k_8}{\overset{k_8'}{\rightleftarrows}} CH_4(g) + * \qquad (A-10)$

$$H_2O_{ad} \underset{k_9}{\overset{k_9'}{\rightleftarrows}} H_2O(g) + * \qquad (A-11)$$

$$CO_{2_{ad}} \underset{k_{10}}{\overset{k_{10}'}{\rightleftarrows}} CO_2(g) + * \qquad (A-12)$$

where asterik represents a vacant surface site.

The analyses that follow assume that only one step is rate determining. Steps occurring after the rate-determining step are not kinetically important; and steps preceding the rate-determining step are in a state of dynamic equilibrium.

If CO adsorption (Step #1) is rate-determining:

$$rate_{CH_4} = k_{CO} \cdot P_{CO} \cdot \theta_v \qquad (A-13)$$

$$k_{H_2} \cdot P_{H_2} \cdot \theta_v^2 = k_{H_2}' \theta_H^2$$

or $\qquad \theta_H = K_{H_2}^{0.5} \cdot P_{H_2}^{0.5} \cdot \theta_v \qquad (A-14)$

where $\quad K_{H_2} = k_{H_2}/k_{H_2}'$, and

represents the H_2 adsorption equilibrium constant.
Other steps are assumed fast, and the surface is assumed to be predominantly covered with adsorbed hydrogen.

Chemical kinetics

then, $\theta_v + \theta_H = 1.0$

or $\quad \theta_v = \dfrac{1}{1 + K_{H_2}^{0.5} P_{H_2}^{0.5}}$ \hfill (A-15)

Equations (A-13) and (A-15) yield:

$$\text{rate}_{CH_4} = \dfrac{k_{CO} P_{CO}}{1 + k_{H_2}^{0.5} P_{H_2}^{0.5}} \tag{A-16}$$

If dissociative chemisorption of H_2 (Step #2) is rate determining:

$$\text{rate}_{CH_4} = k_{H_2} \cdot P_{H_2} \cdot \theta_v^2 \tag{A-17}$$

Equilibrium of CO adsorption (from Equation A-1) gives

$$\theta_{CO} = K_{CO} P_{CO} \theta_v \tag{A-18}$$

Equilibrium CO dissociation (Equation A-3) gives

$$k_1 \theta_{CO} \theta_v = k_1' \theta_C \theta_O \tag{A-19}$$

or $\quad \theta_C = \dfrac{K_1 \theta_{CO} \theta_v}{\theta_O}$

$$= \dfrac{K_1 K_{CO} P_{CO} \theta_v^2}{\theta_O} \tag{A-20}$$

At steady state

$$\dfrac{d\theta_O}{dt} = 0 = (k_1 \theta_{CO} \theta_v - k_1' \theta_C \theta_O - k_6 \theta_{CO} \theta_O$$
$$+ k_6' \theta_{CO_2} \theta_v - k_7 \theta_H^2 \theta_O + k_7' \theta_{H_2O} \theta_v^2)$$

It has been found that reaction products do not inhibit methanation over Ni. This suggests that the products of methanation, once formed on the Ni surface, readily desorb. Therefore,

$$k_1' \cong k_6' \cong 0$$

The value of θ_H would be small so that Equation (A-9) is not very important.

$$k_7 \cong k_7' \cong 0$$

This yields:

$$\frac{d\theta_O}{dt} = 0 = k_1 \theta_{CO} \theta_v - k_6 \theta_{CO} \theta_O$$

or $\quad \theta_O = \dfrac{k_1}{k_6} \cdot \theta_v$ \hfill (A-21)

Substituting Equation (A-21) into Equation (A-20) gives:

$$\theta_C = \frac{k_6}{k_1} K_1 K_{CO} P_{CO} \theta_v \quad \text{(A-22)}$$

If it is assumed that the catalyst surface is covered predominantly by CO, C and O:

$$\theta_v + \theta_{CO} + \theta_C + \theta_O = 1.0$$

or, $\quad \theta_v (1 + K_{CO} P_{CO} + \dfrac{k_6 K_1 K_{CO} P_{CO}}{k_1} + \dfrac{k_1}{k_6}) = 1.0$

or, $\quad \theta_v = \dfrac{1}{(1 + \dfrac{k_1}{k_6}) + (1 + \dfrac{k_6}{k_1'}) K_{CO} P_{CO}}$ \hfill (A-23)

Substituting Equation (A-23) into Equation (A-17) gives:

$$\text{rate}_{CH_4} = \frac{k_{H_2} P_{H_2}}{\left[(1 + \dfrac{k_1}{k_6}) + (1 + \dfrac{k_6}{k_1'}) K_{CO} P_{CO} \right]^2} \quad \text{(A-24)}$$

If CO dissociation (Step #3) is rate-determining

$$\text{rate}_{CH_4} = k_1 \theta_{CO} \theta_v \quad \text{(A-25)}$$

$$\theta_{CO} = K_{CO} P_{CO} \theta_v \quad \text{(A-26)}$$

$$\theta_H = K_{H_2}^{0.5} P_{H_2}^{0.5} \theta_v$$

If it is assumed that the catalyst surface is covered predominantly by CO and H:

$$\theta_v + \theta_{CO} + \theta_H = 1.0$$

$$\theta_v (1 + K_{CO} P_{CO} + K_{H_2}^{0.5} P_{H_2}^{0.5}) = 1.0$$

or,
$$\theta_v = \frac{1}{(1 + K_{CO} P_{CO} + K_{H_2}^{0.5} P_{H_2}^{0.5})} \tag{A-27}$$

On substituting Equations (A-27) and (A-26) into Equation (A-25), one gets:

$$rate_{CH_4} = \frac{k_1 K_{CO} P_{CO}}{(1 + K_{CO} P_{CO} + K_{H_2}^{0.5} P_{H_2}^{0.5})^2} \tag{A-28}$$

If surface reaction between C_{ad} and H_{ad} (Step # 4) is rate determining:

$$rate_{CH_4} = k_2 \theta_C \theta_H \tag{A-29}$$

If equilibrium exists for:

$$\theta_{CO} = K_{CO} P_{CO} \theta_v$$

$$\theta_H = K_{H_2}^{0.5} P_{H_2}^{0.5} \theta_v$$

$$\theta_C = \frac{K_1 K_{CO} P_{CO} \theta_v^2}{\theta_O}$$

at steady state, $\frac{d\theta_O}{dt} = 0$; assuming $k_1' = k_6' = 0$, one gets

$$\theta_O = \frac{(k_1 \theta_{CO} + k_7' \theta_{H_2O} \theta_v)}{(k_6 \theta_{CO} + k_7 \theta_H^2)} \theta_v$$

then,
$$\theta_C = \frac{K_1 K_{CO} (k_6 \theta_{CO} + k_7 \theta_H^2) P_{CO} \theta_v}{(k_1 \theta_{CO} + k_7' \theta_{H_2O} \theta_v)}$$

Assuming that

$$\theta_v + \theta_H + \theta_{CO} + \theta_C + \theta_O = 1.0$$

$$\theta_v = \frac{1}{1 + K_{H_2}^{0.5} P_{H_2}^{0.5} + K_{CO} P_{CO} \left[1 + \frac{K_1(k_6 \theta_{CO} + k_7 \theta_H^2)}{k_1 \theta_{CO} + k_7' \theta_{H_2O} \theta_v}\right] + \frac{k_1 \theta_{CO} + k_7' \theta_{H_2O} \theta_v}{k_6 \theta_{CO} + k_7 \theta_H^2}} \tag{A-30}$$

Assuming $\theta_{H_2O} \approx 0$, and substituting all the terms into Equation (A-29), one gets:

$$\text{rate}_{CH_4} = \frac{k_2 K_{H_2}^{0.5} K_1 \left[\dfrac{k_6 K_{CO}}{k_1} + \dfrac{k_7 P_{H_2} \theta_v}{k_1 P_{CO}}\right] P_{CO} P_{H_2}^{0.5}}{\left\{1 + K_{H_2}^{0.5} P_{H_2}^{0.5} + K_{CO} P_{CO}\left[1 + \dfrac{K_1(k_6 \theta_{CO} + k_7 \theta_H^2)}{k_1 \theta_{CO}}\right] + \dfrac{k_1 \theta_{CO}}{k_6 \theta_{CO} + k_7 \theta_H^2}\right\}^2}$$

or,

$$\text{rate}_{CH_4} = \frac{k''' f(P_{CO}, P_{H_2}) P_{CO} P_{H_2}^{0.5}}{\left[1 + K_{H_2}^{0.5} P_{H_2}^{0.5} + K_{CO} P_{CO}(1 + \Psi(P_{CO}, P_{H_2}))\right]^2} \quad (A-30)$$

If surface reaction between CH_{ad} and H_{ad} is rate determining, a rate expression of the following nature is obtained:

$$\text{rate}_{CH_4} = \frac{k'''' P_{CO} P_{H_2}}{(1 + K_{H_2}^{0.5} P_{H_2}^{0.5} + K_{CO} P_{CO} + \ldots)^2} \quad (A-31)$$

<u>Enol Complex Theory</u>: Several researchers have suggested the following reaction sequence for methanation involving an Enol Complex:

<u>Adsorption</u>

Step 1: $\quad H_2(g) + * \underset{k'_{H_2}}{\overset{k_{H_2}}{\rightleftarrows}} H_{2\,ad} \quad\quad\quad (A-32)$

Step 2: $\quad CO(g) + * \underset{k'_{CO}}{\overset{k_{CO}}{\rightleftarrows}} CO_{ad} \quad\quad\quad (A-33)$

<u>Surface Reaction</u>:

Step 3: $\quad CO_{ad} + H_{2\,ad} \underset{k'_1}{\overset{k_1}{\rightleftarrows}} HCHO_{ad} + * \quad\quad (A-34)$

Chemical kinetics 41

Step 4: $HCHO_{ad} + H_{2_{ad}} \underset{k_2'}{\overset{k_2}{\rightleftarrows}} CH_{2_{ad}} + H_2O_{ad}$ (A-35)

Step 5: $CH_{2_{ad}} + H_{2_{ad}} \underset{k_3'}{\overset{k_3}{\rightleftarrows}} CH_{4_{ad}} + *$ (A-36)

Desorption

Step 6: $CH_{4_{ad}} \underset{k_4}{\overset{k_4'}{\rightleftarrows}} CH_4(g) + *$ (A-37)

Step 7: $H_2O_{ad} \underset{k_5}{\overset{k_5'}{\rightleftarrows}} H_2O(g) + *$ (A-38)

If associative chemisorption of H_2 (Step #1) is the rate determining step:

$$rate_{CH_4} = k_{H_2} P_{H_2} \theta_v$$

$$\theta_{CO} = K_{CO} P_{CO} \theta_v$$

If it is assumed that most of the surface is covered with CO only then,

$$\theta_v = \frac{1}{1 + K_{CO} P_{CO}}$$

and

$$rate_{CH_4} = \frac{k_{H_2} P_{H_2}}{(1 + K_{CO} P_{CO})} \quad (A-39)$$

If CO adsorption (Step #2) is rate determining:

$$rate_{CH_4} = k_{CO} P_{CO} \theta_v$$

$$\theta_{H_2} + \theta_v = 1.0$$

$$\theta_{H_2} = K_{H_2} P_{H_2} \theta_v$$

so,

$$\theta_v = \frac{1}{1 + K_{H_2} P_{H_2}}$$

and

$$\text{rate}_{CH_4} = \frac{k_{CO} P_{CO}}{(1 + K_{H_2} P_{H_2})} \qquad (A-40)$$

If surface reaction between CO_{ad} and $H_{2_{ad}}$ (Step #3) is rate determining:

$$\text{rate}_{CH_4} = k_1 \theta_{CO} \theta_{H_2}$$

$$\theta_{CO} = K_{CO} P_{CO} \theta_v$$

$$H_2 = K_{H_2} P_{H_2} \theta_v$$

assume

$$\theta_v + \theta_{CO} + \theta_{H_2} = 1.0$$

$$\theta_v = \frac{1}{(1 + K_{CO} P_{CO} + K_{H_2} P_{H_2})}$$

then

$$\text{rate}_{CH_4} = \frac{k_1 K_{CO} K_{H_2} P_{CO} P_{H_2}}{(1 + K_{CO} P_{CO} + K_{H_2} P_{H_2})^2} \qquad (A-41)$$

If formation of $CH_{2_{ad}}$ complex (Step #4) is rate determining:

$$\text{rate} = k_2 \theta_{CH_2O} \theta_{H_2}$$

$$\theta_{CH_2O} = \frac{K_1 \theta_{CO} \theta_{H_2}}{\theta_v} = K_1 K_{CO} K_{H_2} P_{CO} P_{H_2} \theta_v$$

$$\theta_v + \theta_{CO} + \theta_{H_2} + \theta_{CH_2O} = 1.0$$

$$\theta_v = \frac{1}{(1 + K_{CO} P_{CO} + K_{H_2} P_{H_2} + K_1 K_{CO} K_{H_2} P_{CO} P_{H_2})}$$

Chemical kinetics

or

$$\text{rate}_{CH_4} = \frac{k_2 K_1 K_{CO} K_{H_2}^2 P_{CO} P_{H_2}^2}{(1 + K_{CO} P_{CO}(1 + K_1 K_{H_2} P_{H_2}) + K_{H_2} P_{H_2})^2} \quad (A-42)$$

If dissociative chemisorption of H_2 on the catalyst surface is assumed in the enol complex theory, then Step #2 (Equation A-32) should be replaced by Equation (A-2). Basically, the results remain unchanged on doing so. There is a possibility, however, that surface reaction between CO_{ad} and H_{ad} to form CHO_{ad} may be the rate determining step. The following rate expression is obtained:

$$\text{rate}_{CH_4} = \frac{K P_{CO} P_{H_2}^{0.5}}{(1 + K_{CO} P_{CO} + K_{H_2}^{0.5} P_{H_2}^{0.5})^2} \quad (A-43)$$

Many important conclusions can be derived merely by an inspection of the rate-expressions obtained above. Large differences exist among various rate-controlling steps for the dependence of methanation rate on partial pressures of H_2 and CO. Intrinsic kinetic data over a wide range of reactant partial pressures can provide useful information regarding the rate-controlling step.

The general forms of the rate expressions are qualitatively the same in the 'Carbide' and the 'Enol' complex theories for similar rate-controlling steps. For example, if surface reaction is the rate-controlling step, both mechanisms predict a positive dependence of methanation rate on the H_2 partial pressure and a negative dependence of methanation rate on CO partial pressure at large CO partial pressures. The dependence of methanation rate on P_{H_2} is different in the two theories primarily because the carbide theory assumes dissociative chemisorption of H_2 on the metal surface. Using Equation (A-2) instead of Equation (A-32) in the enol complex theory would yield an identical H_2 partial pressure dependence of the methanation rate in both theories. Recent surface studies (not kinetic) of methanation provide strong evidence for the carbide theory.

This comparison, nevertheless, underlines the difficulties involved in proposing a reaction mechanism on the basis of kinetic data alone. While a disagreement between the kinetic model and the observed rate data can be used in rejecting the model and thus the reaction mechanism upon which it is based, the contrary is not true. Acceptance of a rate model must be supported by experiments which deal with the micromolecular processes. Unfortunately, such definitive information is usually inadequate in the literature, and research in this area is far from the goal of outlining a reaction mechanism based on direct fundamental information.

Until recently, little data have been obtained on the intrinsic kinetics of methanation because of the presence of heat- and mass-transfer limitations in the experiments. The most widely studied transition metal for methanation is nickel supported on alumina. Experimental conditions have varied in different studies over a wide range for the operating parameters, including pressure (1-70 atm), temperature (170-480^{0}C), H_2/CO ratio (3-100), reactor type (fixed bed vs. fluidized bed), and reactor operation (integral vs. differential reactor operation). Likewise, the rate equations obtained from data-fitting widely differ in their basic nature as well as their implications. In too many cases the range of experimental variables is too small to lead to a statistically sound rate-equation. However, reviewing the available literature for Ni shows that the rate of methanation shows a positive order in H_2 pressure, probably $P_{H_2}^{0.5}$. The CO partial pressure dependence is positive at low pressures and becomes negative at higher CO partial pressures. These results allow mechanisms involving CO adsorption and H_2 adsorption as rate determining steps to be eliminated and suggest that surface reaction is rate controlling. The P_{H_2} dependence suggests that the rate determining step is reaction between adsorbed H and adsorbed C. However more detailed kinetic information and more surface studies are required to define more carefully this information.

NOMENCLATURE

A_j	chemical species j in chemical reaction
a_j, b, c	numerical constants
C_j	concentration of species j
k_i	kinetic rate constant for reaction i
K_{eq}	equilibrium constant for reaction
N_j	number of moles of species j, integer
r_i	intrinsic reaction rate for species i
t	time
V	volume of batch reactor

Greek

α_j stoichiometric coefficients for species j in chemical reaction

τ   time, space time

Subscripts

o   value at initial or zero time
eq   equilibrium

Superscripts

α,β,γ   reaction orders

SELECTED REFERENCES IN CHEMICAL KINETICS, CATALYSIS AND REACTOR DESIGN

● Basic Material for Review:

E. L. King "How Chemical Reactions Occur", Benjamin, 1964 (paperback).
F. Daniels and R. A. Alberty, "Physical Chemistry" Wiley, Ch. 10.

● Introductory Texts Emphasizing Chemical Aspects of Subject:

A. A. Frost and R. G. Pearson, "Kinetics and Mechanism", 2nd Edition, Wiley, 1961.
W. C. Gardiner, Jr., "Rates and Mechanisms of Chemical Reactions", Benjamin, 1969.
K. J. Laidler, "Chemical Kinetics", 2nd edition, McGraw-Hill, 1965.
K. J. Laidler, "Reaction Kinetics", Volumes I and II, Pergamon Press, 1963 (paperback).
C. N. Hinshelwood, "Kinetics of Chemical Change", Claredon Press, 1940 (somewhat out of date, but contains a good qualitative discussion of basics).
M. Boudart, "Kinetics of Chemical Processes", Prentice Hall 1968.

● More Advanced Treatments:

K. J. Laidler, "Theories of Chemical Reaction Rates", McGraw-Hill, 1969.
S. W. Benson, "Foundations of Chemical Kinetics", McGraw-Hill, 1960.
V. N. Kondratiev, "Kinetics of Chemical Gas Reactions", 1958, translated from the Russian, Pergamon Press, 1964.
S. Glasstone, K. J. Laidler and H. Eyring, "The Theory of Rate Processes", McGraw-Hill, 1941.

● Texts Emphasizing Engineering Aspects of Kinetics and Reactor Design:

O. Levenspiel, "Chemical Reaction Engineering" Wiley, 1962.
J. M. Smith, "Chemical Engineering Kinetics", Second Edition, McGraw-Hill, 1970.
S. M. Walas, "Reaction Kinetics for Chemical Engineers", McGraw-Hill, 1959.
O. A. Hougen and K. M. Watson, "Chemical Process Principles", Volume III, Wiley, 1947.
R. Aris, "Introduction to the Analysis of Chemical Reactors", Prentice Hall, 1965.
K. G. Denbigh and J. C. R. Turner, "Chemical Reactor Theory", Second Edition, Cambridge University Press, 1971.
M. Kramer and K. R. Westerterp, "Elements of Chemical Reactor Design and Operation", Academic Press, 1963.

Chemical kinetics

A. R. Cooper and G. V. Jeffreys "Chemical Kinetics and Reactor Design", Prentice Hall, 1971.
H. Scott Fogler, "The Elements of Chemical Kinetics and Reactor Calculat-ons", Prentice-Hall, 1974.
J. J. Carberry, "Chemical and Catalytic Reaction Engineering", McGraw-Hill, 1976.
C. D. Holland and R. G. Anthony, "Fundamentals of Chemical Reaction Engineering", Prentice Hall, 1979.
G. F. Froment and K. B. Bischoff, "Chemical Reactor Analysis and Design" John Wiley, 1979.
Peterson, E. E., "Chemical Reactor Analysis", Prentice-Hall, 1965.

● Experimental Techniques:

H. W. Melville and B. G. Gowenlock, "Experimental Methods in Gas Reactions", Macmillan, 1964.
S. L. Friess and A. Weissberger (editors) "Investigations of Rates and Mechanism of Reactions", Volumes I and II (Volume 6, parts 1 and 2, of "Technique of Organic Chemistry"), Interscience, 1953 and 1963.
R. B. Anderson, "Experimental Methods in Catalytic Research", Academic Press, 1968.

● Heterogeneous Catalysis

J. M. Thomas and W. J. Thomas, "Introduction to the Principles of Heterogeneous Catalysis", Academic Press, 1967.
E. E. Peterson, "Chemical Reaction Analysis", Prentice Hall, 1965.
P. G. Ashmore, "Catalysis and Inhibition of Chemical Reactions", Butterworths, 1963.
G. C. Bond, "Catalysis by Metals", Academic Press, 1962.
O. V. Krylov, "Catalysis by Nonmetals", Academic Press, 1970.
S. J. Thomson and G. Webb, "Heterogeneous Catalysis", Oliver and Boyd, 1968.
C. N. Satterfield, "Mass Transfer in Heterogeneous Catalysis", MIT Press, 1970.
C. L. Thomas, "Catalytic Processes and Proven Catalysts", Academic Press, 1970.
A. Clark, "The Theory of Adsorption and Catalysis", Academic Press, 1970.
B. C. Gates, J. R. Katzer and G.C.A. Schuit, "Chemistry of Catalytic Processes", McGraw-Hill, 1979.
J. R. Anderson, "Structure of Metallic Catalysts," Academic Press, 1975.
B. Delmon, P. A. Jacobs, and G. Poncelet, "Preparation of Catalysts," Elsevier, 1976.
J. A. Cusumano, R. A. Dalla Betta, and R. B. Levy, "Catalysis in Coal Conversion", Academic Press, 1978.

MASS TRANSFER IN REACTING SYSTEMS

James R. Katzer

Center for Catalytic Science and Technology
Department of Chemical Engineering
University of Delaware
Newark, Delaware 19711 U.S.A.

I. INTRODUCTION

The transport of reactants and products to and from interfaces can be very important in chemical reactions and can markedly affect both the rate and the selectivity of the desired reaction. In many processes, diffusion is rate controlling.

Consider the burning of carbon (coal), for example. Figure 1.1 shows the rate behavior observed for a carbon particle held in a stream of hot air as a function of air temperature, air flow and carbon particle diameter. Only at lower temperatures is the rate strongly dependent on temperature, i.e., kinetically controlled. In the regions in which the rate becomes essentially temperature independent, the rate is no longer under kinetic control but is, in fact, controlled by the rate of mass transfer to the surface of the carbon particle. The temperature at which this transition from kinetic control to mass transfer control occurs is dependent on air flow rate and on particle size. Note that the rate is normalized to "per cm^2 surface area".

Figure 1.2 is an Arrhenius-type plot of the rate of carbon combustion versus temperature. At low temperatures, where the rate of reaction is controlling, the activation energy observed is that for the chemical reaction, a value of order 80 kcal per mole. When the rate of mass transfer is controlling, the activation energy becomes characteristic of the value for mass transfer or for diffusion E_D which is typically 2 to 4 kcal per mole (Figure 1.2). There is, of course, a transition region between these two limiting cases where both diffusion and reaction influence the rate.

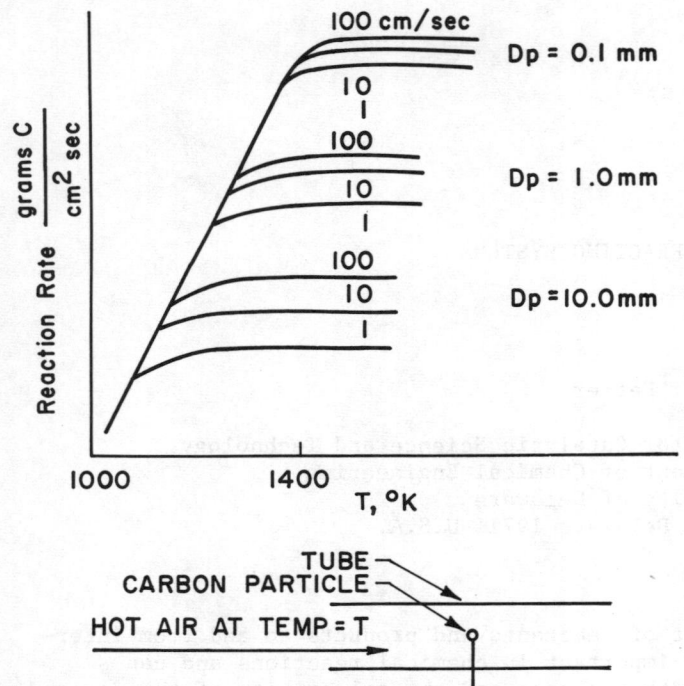

Figure 1.1 Rate of combustion of carbon as a function of air temperature, air velocity and particle diameter.

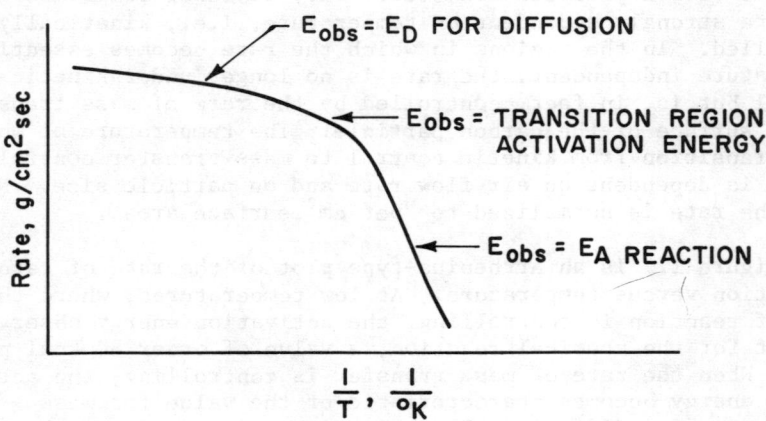

Figure 1.2 Arrhenius plot of rate of carbon combustion for one set of conditions.

Mass transfer 51

In industrial furnaces burning coal, it has been estimated that the rate of combustion at 1500 K (a reasonable temperature) should be 0.020 g carbon per cm^2 per sec (150 lb/ft^2-hr) if reaction kinetics were controlling, but the actual combustion rate is only a few percent of this rate because oxygen mass transfer rate is controlling.

The analysis of the carbon burning situation in terms of concentration gradients near the surface of the carbon is similar to that for the analysis of reaction at a solid catalytic surface and is shown in Figure 1.3. Conceptually, it is assumed that the bulk gas phase is well mixed and that there exists a gas film (δ thick) near the surface across which all diffusion must occur and that all concentration gradients occur across this film. If the rate of reaction is rapid enough as in the case of carbon burning at high temperatures, steep concentration gradients are required to provide the driving force for the transfer of reactants and products to and from the solid surface. The presence of such gradients shows immediately what the problem is. The concentration of reactant immediately next to the surface $C_{A,i}$ is much less than the bulk concentration for the reactant, and thus the surface concentration is not that which would be expected considering the bulk concentration and an adsorption isotherm of some sort. The lower surface concentration results in a lower observed rate of reaction. If the rate of reaction is rapid enough, the surface reaction may be essentially at equilibrium (or the surface concentration may be close to zero) and thus the rate is controlled solely by the rate of transport of reactants to the surface. This is exactly the condition that is present at higher temperatures in the carbon burning case.

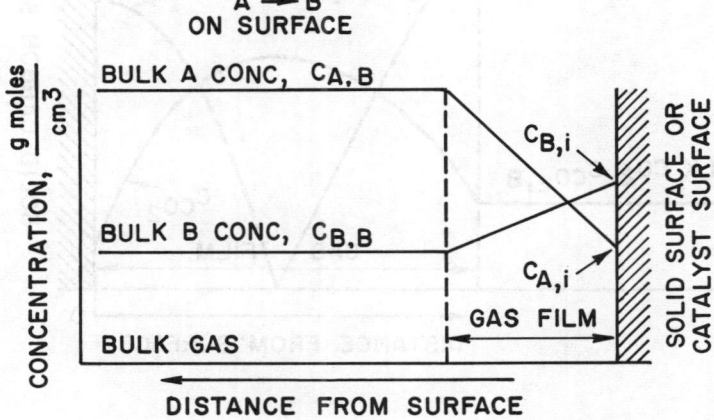

Figure 1.3 Simplified picture of concentration gradients across an imaginary boundary layer gas film near a reacting solid surface or catalytically active surface.

A second good example of this situation is the industrial production of nitric acid (or really NO) by the oxidation of ammonia using a Pt gauze catalyst. The temperature is about 90°C, and the rate of reaction is very rapid. The rate of NH_3 oxidation can be calculated with no knowledge of the chemical kinetics; the only information needed is the values of the mass transfer coefficients operative in the system and the diffusion coefficients in the system. This information allows the rate of oxidation to be accurately predicted, and an ammonia oxidation reactor can be adequately designed without additional information.

The oxidation of carbon is, however, more complex than shown in Figure 1.3. Under strongly diffusion-controlled conditions, the concentration profiles shown in Figure 1.4 actually apply as imposed by the kinetics and thermodynamics of the system in conjunction with the mass transfer limitation. The actual transporter of oxygen atoms to the surface is CO_2, where $CO_2 + C \longrightarrow 2CO$ occurs. The CO formed diffuses away from the surface under a concentration gradient, but its concentration is zero at the interface between the CO-containing and the O_2-containing regions. CO_2 concentration goes through a maximum because of the driving forces required to transfer it to the carbon surface for carbon oxidation and to transfer it to the bulk gas region. Molecular oxygen under these reaction conditions never actually reaches the carbon surface.

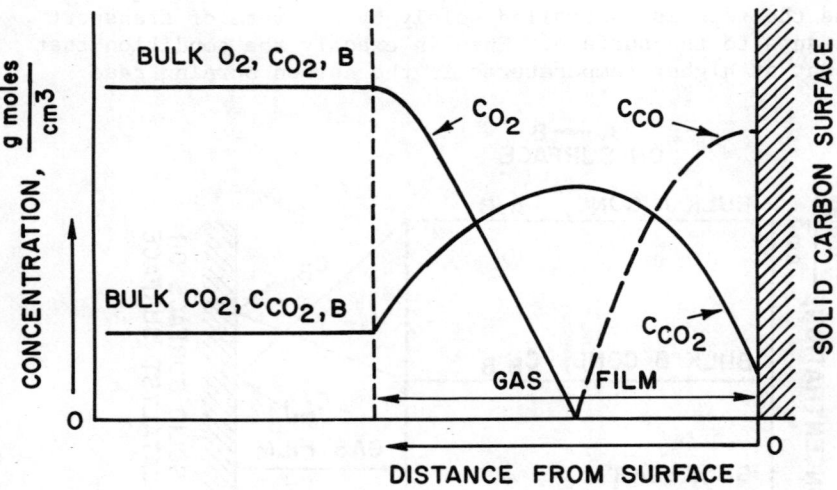

Figure 1.4 Schematic of concentration profiles which exist near carbon surface under high-temperature mass-transfer controlled oxidation conditions.

Mass transfer

Absorption and reaction across liquid interface is analyzed in a similar manner with the solid surface being replaced by a liquid phase. Such systems have large industrial inportance and include gas absorption such as nitric acid formation by NO absorption into water and gas-liquid reaction systems both heterogeneously catalyzed and homogeneously catalyzed. Figure 1.5 shows schematically the concentration gradients experienced in a gas-liquid absorption system. If reaction occurs between A and the liquid or another component in the liquid phase, the concentration gradient of A in the liquid film is increased further.

In the case of a porous catalyst particle, we need to add one more diffusional resistance and one more concentration gradient. This resistance is the one in the porous structure of the solid itself. This situation is shown schematically in Figure 1.6. We now have a situation in which we can have fluid

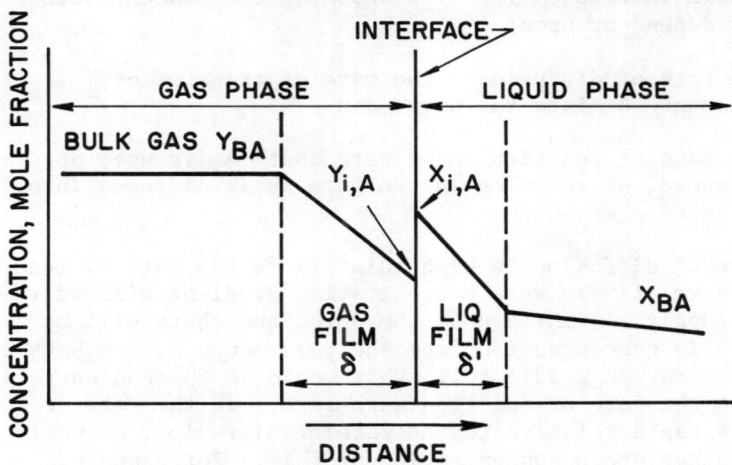

Figure 1.5   Schematic of the concentration gradients for gas-liquid absorption system with A being absorbed.

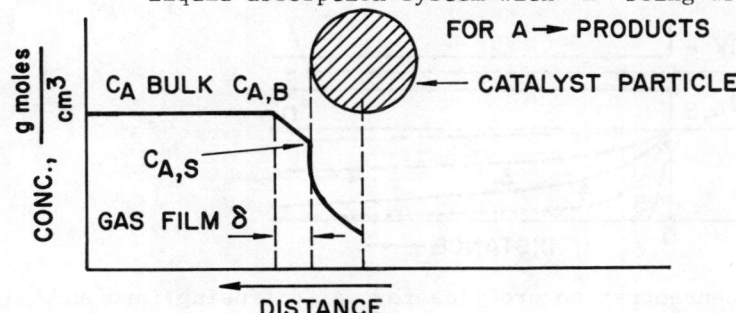

Figure 1.6   Concentration gradients involved for reaction in a porous catalyst particle.

film diffusional resistance exterior to the catalyst particle, and we also have a diffusional resistance within the pores of the solid catalyst particle. This is a common case. To illustrate, consider the following idealized pore model in which A is

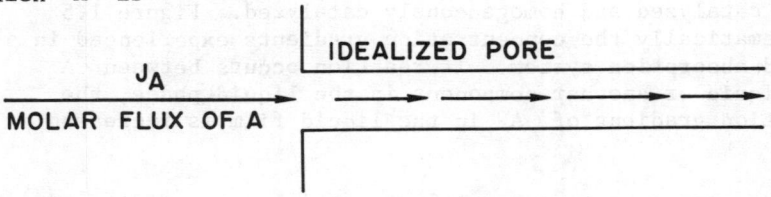

diffusing back to the catalyst pore and reacting at the same time removing A and effectively reducing its concentration. This is shown in Figure 1.7. The shape of the concentration profile is dependent upon:

- the rate of diffusion - the rate of transfer of reactant A into the pore, and

- the rate of reaction - the rate that A is used up, consumed, by reaction in the pore as it diffuses into the pore.

If the rate of diffusion is high relative to the rate of reaction, then only a relatively weak concentration gradient will be required to supply A throughout the pore, and there will be little drop in concentration from the pore mouth to its back end, i.e., concentration gradient 1. What would be the concentration gradient if the rate of reaction were zero? If the rate of reaction is rapid relative to the rate of diffusion, we would expect a rather steep concentration profile. For example, concentration profiles 4 or 5. Note that the word relative is used in referring to the rates of reaction and diffusion; this is the key word. Each rate is coupled with the other.

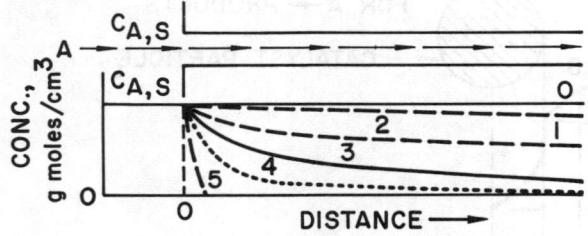

Figure 1.7 Concentration profiles for A diffusing into an idealized catalyst pore with reaction occurring on the walls of the pore.

Further, if we assume, for example, that this pore is all metal in nature, and that the surface is uniformly active, then we could form a flat sheet of metal of the same surface area as the interior surface area (circumference × length) of the pore. We could then study the reaction of A over this flat metal sheet. The total rate of conversion of A, referred to as the "global" rate of reaction A, can then be determined for a given uniform gas-phase concentration of A, $C_{A,S}$, contacting the entire surface uniformly. The global rate for the pore will be the same as that for the flat sheet if the concentration is the same throughout the pore, i.e., is equal to $C_{A,S}$ along the entire pore length (Profile 0 in Figure 1.7). However, if the reaction rate shows a positive dependence on concentration, and if there is any decrease in concentration along the pore (Profiles 1 through 5 in Figure 1.7), the global rate of reaction observed for the pore will be less than that for the flat sheet which sees the same concentration at all points.

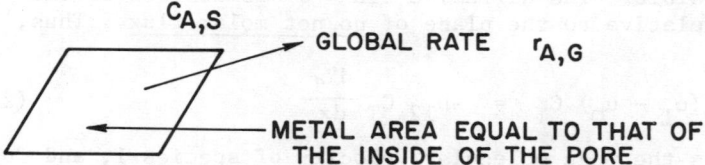

In this case the pore is not being used as effectively as it could, and we say that it has an effectiveness factor which is less than 1. The effectiveness factor is defined as the actual observed global rate divided by what the global rate would have been if the entire pore had seen concentration $C_{A,S}$ i.e., the global rate for our flat sheet.

$$\eta = \frac{\text{Global Rate Observed}}{\text{Global Rate for } C_A = C_{A,S} \text{ everywhere}} \quad (1.1)$$

We will now develop the principles needed to analyze the situations above more adequately. This will involve first, a review of diffusion fundamentals and then an integration with reaction systems.

II. DIFFUSION FUNDAMENTALS

A. Diffusion in gases

<u>Diffusion and Transport Mathematics</u>. Diffusion in gases occurs because of the presence of concentration differences between regions in the gas and occurs via the random walk motion of the gas molecules. In a stagnant binary gas mixture, the molar

flux J (g mole/sec cm^2) is proportional to the concentration gradient in the direction of diffusion.

$$J_1 = -D_{12} \frac{dC_1}{dX} = -D_{12} C_T \frac{dY_1}{dX} \tag{2.1}$$

D_{12} is the binary diffusion coefficient for species 1 diffusing through 2. Equation (2.1) is Fick's first law.

The diffusion coefficient D_{12} is a function of molecular properties of the two gases; it increases with temperature; it decreases with increasing pressure (C_T), and it is a weak function of mole fractions.

Motion must be related to a coordinate system to be uniquely described. A uniform composition of bulk gas can flow with respect to fixed coordinates in bulk motion without diffusion being responsible. The diffusive flux J is defined as the molar flux relative to the plane of <u>no net molar flux</u>. Thus,

$$J_1 = (u_1 - u_m) C_1 = -D_{12} C_T \frac{dY_1}{dx} \tag{2.1}$$

where u_1 is the mean molecular velocity of species 1, and u_m is the velocity of a plane of no net molar flux. The total molar flux of species 1 relative to a fixed coordinate system is $u_1 C_1$ and can be obtained from (2.1).

$$N_1 = u_1 C_1 = u_m C_1 + J_1 \tag{2.2}$$

$$N_1 = u_m C_T \left(\frac{C_1}{C_T}\right) + J_1 \tag{2.2'}$$

$U_m C_1$ is the total molar flux relative to the fixed coordinate system and is thus equal to $(N_1 + N_2)$. Equation (2.2') becomes

$$N_1 = (N_1 + N_2) Y_1 - D_{12} C_T \frac{dY_1}{dx} \tag{2.3}$$

In a non-steady-state situation,

$$\frac{\partial Y_1}{\partial t} + u_m \frac{\partial Y_1}{\partial x} = D_{12} \frac{\partial^2 Y_1}{\partial x^2} \tag{2.4}$$

or

$$\frac{\partial C_1}{\partial t} + u_m \frac{\partial C_1}{\partial x} = D_{12} \frac{\partial^2 C_1}{\partial x^2} \tag{2.4'}$$

which in vector form is

$$\frac{\partial Y_1}{\partial t} + (\vec{u}_m \cdot \nabla)Y_1 = D_{12}\nabla^2 Y \tag{2.5}$$

or

$$\frac{\partial Y_1}{\partial t} + (\vec{u}_m \cdot \nabla)C_1 = D_{12}\nabla^2 C_1 \tag{2.5'}$$

which in vector form is

$$\frac{\partial Y_1}{\partial t} + (\vec{u}_m \cdot \nabla)Y_1 = D_{12}\nabla^2 Y \tag{2.5}$$

or

$$\frac{\partial Y_1}{\partial t} + (\vec{u}_m \cdot \nabla)C_1 = D_{12}\nabla^2 C_1 \tag{2.5'}$$

This is Fick's second law.

In the case of chemical reaction, a source or a sink term must be added to Equation (2.5). For generality, the equation should be written in terms of concentration allowing application to liquid systems as well. Equation (2.5) then becomes

$$\underbrace{\frac{\partial C_1}{\partial t}}_{\substack{\text{Accumulation}\\\text{Term}}} + \underbrace{(u_m \cdot \nabla)C_1}_{\substack{\text{Convection}\\\text{Term}}} + \underbrace{r}_{\substack{\text{Reaction}\\\text{Rate}\\\text{Term}}} = \underbrace{D_{12}\nabla^2 C_1}_{\substack{\text{Molecular}\\\text{Diffusion}\\\text{Term}}} \tag{2.5'}$$

Equation (2.3) can be rearranged and applied to many non-reacting binary gas systems at constant C_T. For species 1 diffusing in stagnant species 2, $N_2 = 0$ and,

$$N_1 = \frac{C_T D_{12}}{(1-Y_1)} \frac{dY_1}{dx} \tag{2.6}$$

which upon integration between x_1 and x_2, and Y_1 and Y_2 gives

$$N_1 = \frac{C_T D_{12}}{(x_2-x_1)} \ln\left[\frac{(1-(Y_1)_2)}{(1-(Y_1)_1)}\right] \tag{2.7}$$

This form is frequently used to evaluate D_{12}, for example, in the Stefan experiment.

For equal-molar diffusion fluxes, $N_1 = -N_2$

$$N_1 = J_1 = -D_{12} C_T \frac{dY_1}{dx} \tag{2.8}$$

and upon integration

$$N_1 = J_1 = \frac{C_T D_{12}}{(x_2-x_1)} \left[(Y_1)_1 - (Y_2)_2 \right] \tag{2.9}$$

If the two species diffuse at constant but unequal rates, integration of (2.3) gives

$$N_1 + N_2 = \frac{C_T D_{12}}{(x_2-x_1)} \ln \left\{ \frac{1-(Y_1)_2 N_1 - (Y_1)_2 N_2}{1-(Y_1)_1 N_1 - (Y_1)_1 N_2} \right\} \tag{2.10}$$

For diffusion of species in mixtures, the theory becomes more complex. However, to a good approximation, it is possible to estimate the diffusion coefficient of species 1 diffusing in a mixture, D_{1m}, from the binary diffusion coefficients for species 1 in the components present according to:

$$D_{1m} = (1-Y_1) \left[\sum_{j=2}^{n} \frac{Y_j}{D_{1j}} \right]^{-1} \tag{2.11}$$

<u>Gas-Phase diffusion coefficients</u>. The best method of estimating diffusion coefficients in binary gas mixtures or of extrapolating available experimental values to conditions other than those at which they were determined is a theoretical equation based on the kinetic theory of gases and Lennard-Jones force expressions.

$$D_{12} = \frac{0.001858 \, T^{3/2} \left[(M_1 + M_2)/M_1 M_2 \right]^{1/2}}{P \sigma_{12}^2 \Omega_D} \tag{2.12}$$

T = absolute temp (°K)
M_1, M_2 = molecular weights of the two species
P = total pressure (atm)
Ω_D = "collision integral" (See Table 2.2)
ε, σ = force constants in the Lennard-Jones potential function. (See Tables 2.3 and 2.4)
k = Boltzmann constant

Table 2.1 gives representative values of D_{12}, and Tables 2.2, 2.3 and 2.4 give values required to calculate values of D_{12}. When the proper data are not available, the parameters may be estimated by the empirical equations

$$\frac{kT}{\varepsilon} = 1.30 \frac{T}{T_C} \qquad (2.13)$$

$$\sigma = 1.18 \left(V_b\right)^{1/3} \qquad (2.14)$$

where T_C = critical temp. in K and V_b is the molar volume (cm^3/gm-mole) at the normal boiling point. Table 2.4 gives the additive volume increments for various constituents to be used in Kopp's law for additive volumes.

For binary systems ε_{12} and σ_{12} are given by

$$\sigma_{12} = \tfrac{1}{2}(\sigma_1 + \sigma_2) \qquad (2.15)$$

$$\varepsilon_{12} = \sqrt{\varepsilon_1 \varepsilon_2} \; ; \quad \frac{kT}{\varepsilon_{12}} = \frac{kT}{\sqrt{\varepsilon_1 \varepsilon_2}} \qquad (2.16)$$

Equation (2.12) gives values of D_{12} that are usually within 10 percent of measured values and even considerably better with ε and σ obtained from viscosity data. It is accurate to roughly 20 atmospheres and holds at higher pressures for higher temperatures. Note that D_{12} is inversely proportional to pressure. The average values of D_{12} are 0.2 to 0.07 cm^2/sec for gases at one atmosphere. Note also, that the temperature dependence is $T^{3/2}$; it is not exponential. The apparent activation energy for diffusion is thus much less than that normally observed for chemical reactions. Activation energies for diffusion are typically 1 to 3 kcal/mole as compared with 15 to 50 kcal/mole for activation energies for chemical reaction.

A more complete list of published D_{12} values is given by Eerkens, J. W., and Grossman, L. M., (Tech. Rept. HE-150-150, University of California, Institute of Engineering Research, December 5, 1957). Values of ε and σ are reported by Svehla, R. A. (Tech. Report R-132, Lewis Research Center, NASA, Cleveland, 1962) for over 200 compounds and by Hirschfelder and co-workers (Chem. Rev. 44, 205 (1949); "Molecular Theory of Gases and Liquids", Wiley, New York, 1954).

Table 2.1. Diffusion Coefficients for Binary Gas Systems
Experimental Values of $D_{12}P$, where D_{12} is in cm²/sec and
P is in atm.

Gas Pair	T(°K)	$D_{12}P$	Gas Pair	T(°K)	$D_{12}P$
Air-ammonia	273	0.198	Ethane-methane	293	0.163
-benzene	298	0.0962	-propane	293	0.0850
-carbon dioxide	273	0.136			
	1000	1.32	Helium-argon	273	0.641
-chlorine	273	0.124	-benzene	298	0.384
-diphenyl	491	0.160	-ethanol	298	0.494
-ethanol	298	0.132	-hydrogen	293	1.64
-iodine	298	0.0834			
-methanol	298	0.162	Hydrogen		
-mercury	614	0.473	-ammonia	298	0.783
-naphthalene	298	0.0611	-benzene	273	0.317
-oxygen	273	0.175	-ethanol	340	0.578
-sulfur dioxide	273	0.122	-ethylene	298	0.602
-toluene	298	0.0844	-methane	288	0.694
-water	298	0.260	-nitrogen	293	0.760
	1273	3.253	-oxygen	273	0.697
Argon-neon	293	0.329	-propane	300	0.450
			Nitrogen		
Carbon dioxide-benzene	318	0.0715	-ammonia	298	0.230
-ethanol	273	0.0693	-ethylene	298	0.163
-hydrogen	273	0.550	-iodine	273	0.070
-methane	273	0.153	-oxygen	273	0.181
-methanol	299	0.105	Oxygen-ammonia	293	0.253
-nitrogen	298	0.167	-benzene	296	0.0939
-propane	298	0.0863	-carbon tetrachloride	298	0.071
Carbon monoxide			-ethylene	293	0.182
-ethylene	273	0.151			
-hydrogen	273	0.651	Water-hydrogen	307.2	1.020
-nitrogen	288	0.192	-helium	307	0.902
-oxygen	273	0.185	-methane	307.6	0.292
			-ethylene	307.7	0.204
Dichlorodifluoromethane			-nitrogen	307.5	0.256
-ethanol	298	0.0475	-oxygen	352	0.352
-water	298	0.105	-carbon dioxide	307.4	0.198

Table 2.2. Values of the Collision Integral Ω_D Based on the Lennard-Jones Potential.

$\dfrac{kT}{\varepsilon_{12}}$	Ω_D	$\dfrac{kT}{\varepsilon_{12}}$	Ω_D	$\dfrac{kT}{\varepsilon_{12}}$	Ω_D
0.30	2.662	1.65	1.153	4.0	0.8836
0.35	2.476	1.70	1.140	4.1	0.8788
0.40	2.318	1.75	1.128	4.2	0.8740
0.45	2.184	1.80	1.116	4.3	0.8694
0.50	2.066	1.85	1.105	4.4	0.8652
0.55	1.966	1.90	1.094	4.5	0.8610
0.60	1.877	1.95	1.084	4.6	0.8568
0.65	1.798	2.00	1.075	4.7	0.8530
0.70	1.729	2.1	1.057	4.8	0.8492
0.75	1.667	2.2	1.041	4.9	0.8456
0.80	1.612	2.3	1.026	5.0	0.8422
0.85	1.562	2.4	1.012	6	0.8124
0.90	1.517	2.5	0.9996	7	0.7896
0.95	1.476	2.6	0.9878	8	0.7712
1.00	1.439	2.7	0.9770	9	0.7556
1.05	1.406	2.8	0.9672	10	0.7424
1.10	1.375	2.9	0.9576	20	0.6640
1.15	1.346	3.0	0.9490	30	0.6232
1.20	1.320	3.1	0.9406	40	0.5960
1.25	1.296	3.2	0.9328	50	0.5756
1.30	1.273	3.3	0.9256	60	0.5596
1.35	1.253	3.4	0.9186	70	0.5464
1.40	1.233	3.5	0.9120	80	0.5352
1.45	1.215	3.6	0.9058	90	0.5256
1.50	1.198	3.7	0.8998	100	0.5130
1.55	1.182	3.8	0.8942	200	0.4644
1.60	1.167	3.9	0.8888	400	0.4170

Table 2.3. Lennard–Jones Force Constants Calculated from Viscosity Data

Compound	ε/k (°K)	σ (Å)
Acetone	560.2	4.600
Acetylene	231.8	4.033
Air	78.6	3.711
Ammonia	558.3	2.900
Argon	93.3	3.542
Benzene	412.3	5.349
Bromine	507.9	4.296
i-Butane	330.1	5.278
Carbon dioxide	195.2	3.941
Carbon disulfide	467	4.483
Carbon monoxide	91.7	3.690
Carbon tetrachloride	322.7	5.947
Carbonyl sulfide	336	4.130
Chlorine	316	4.217
Chloroform	340.2	5.389
Cyanogen	348.6	4.361
Cyclohexane	297.1	6.182
Cyclopropane	248.9	4.807
Ethane	215.7	4.443
Ethanol	362.6	4.530
Ethylene	224.7	4.163
Fluorine	112.6	3.357
Helium	10.22	2.551
n-Hexane	339.3	5.949
Hydrogen	59.7	2.827
Hydrogen cyanide	569.1	3.630
Hydrogen chloride	344.7	3.339
Hydrogen iodide	288.7	4.211
Hydrogen sulfide	301.1	3.623
Iodine	474.2	5.160
Krypton	178.9	3.655
Methane	148.6	3.758
Methanol	481.8	3.626
Methylene chloride	356.3	4.898
Methyl chloride	350	4.182
Mercury	750	2.969
Neon	32.8	2.820
Nitric oxide	116.7	3.492
Nitrogen	71.4	3.798
Nitrous oxide	232.4	3.828
Oxygen	106.7	3.467
n-Pentane	341.1	5.784
Propane	237.1	5.118
n-Propyl alcohol	576.7	4.549
Propylene	298.9	4.678
Sulfur dioxide	335.4	4.112
Water	809.1	2.641

Table 2.4. Additive (Atomic) Volume Increments[a] for the Estimation of the Molal Volume (V_a) at the Normal Boiling Temperature.

Carbon	14.8
Hydrogen	3.7
Oxygen, generally	7.4
in methyl esters and ethers	9.1
in ethyl esters and ethers	9.9
in higher esters and ethers	11.0
in acids	12.0
joined to S, P, N	8.3
Nitrogen, doubly bonded	15.6
in primary amines	10.5
in secondary amines	12.0
Bromine	27.0
Chlorine	24.6
Fluorine	8.7
Iodine	37.0
Sulfur	25.6
Ring, three-membered	-6.0
four-membered	-8.5
five-membered	-11.5
six-membered	-15.0
naphthalene	-30.0
anthracene	-47.5

[a] The additive-volume method of obtaining V_b should not be used for simple molecules. The following approximate values may be employed in the estimation of σ by Equation 1.18; H_2, 14.3; O_2, 25.6; N_2, 31.2; air, 29.9; CO, 30.7; CO_2, 34.0; SO_2, 44.8, NO, 23.6; N_2O, 36.4; NH_3, 25.8; H_2O, 18.9; H_2S, 32.9; COS, 51.5; Cl_2, 48.4; Br_2, 53.2. However, the values of σ given in Table 2.3 are preferable if listed.

B. Diffusion in Liquids

The molecular theory of liquids is not as well developed as is that for gases, and it is not as easy to treat diffusion in liquid systems. The problems encountered in predicting diffusion in liquids include:

- The theory is not as well based as with gases.

- D_{12} changes considerably with concentration.

- Mass density is more nearly constant than the molar density for liquids.

Technical calculations usually employ Equation (2.1)

$$J_1 = -D_{12} \frac{dC_1}{dx} \qquad (2.1)$$

with a constant diffusion coefficient calculated for the average concentration or with a variable $D_{12} = f(C_1)$.

The best known correlation for liquid-phase diffusion is the Wilke and Chang equation for <u>dilute</u> solutions, [Wilke and Chang, A.I.Ch.E. JL4 1, 264 (1955)].

$$D_{12} = 7.4 \times 10^{-10} \frac{T(XM_2)^{\frac{1}{2}}}{\mu (V_{b_1})^{0.6}}$$

D_{12} = Diff. coeff., cm^2/sec
T = Temp. °K
M_2 = Molecular weight of <u>solvent</u>
μ = Viscosity of solution in <u>poises</u>
V_b = Molar volume of diffusing solute, cm^3/g mole
X = "Association Parameter", - 2.6 for H_2O; 1.9 for methanol; 1.5 for ethanol, and 1.0 for benzene, ether heptane and other undissociated solvents.

The units given must be used; Equation (2.17) is not dimensionally consistent. Table 2.5 gives values of D_{12} for liquid systems. D_{12} for liquids is typically 1×10^{-5} to 0.3×10^{-5} cm^2/sec and is thus about 10^{-4} times D_{12} for gases. Since D_{12} varies as T/μ, and since μ decreases with increasing temperature, D_{12} increases roughly as the square of the absolute temperature. Thus, low activation energies are also observed for liquid-phase diffusion. For example, the apparent activation energy for typical solutes in hexane is 2 to 3 kcal/g mole.

Table 2.5. Experimental Diffusion Coefficients for Binary Liquid Systems at Low Solute Concentrations

	Solute	$T(°K)$	$D^o_{12} \times 10^5$, cm^2/sec
In Water	Helium	298	6.3
	Hydrogen	298	4.8
	Oxygen	298	2.41
	Carbon dioxide	298	2.00
	Ammonia	285	1.64
	Chlorine	298	1.25ª
	Methane	293	1.49
	Propane	277	0.55
		333	2.71
	Propylene	298	1.44
	Benzene	298	1.09
	Methanol	288	1.26
	Ethanol	283	0.84
	n-Propanol	288	0.87
	n-Butanol	288	0.77
	i-Butanol	288	0.77
	i-Pentanol	288	0.69
	Ethylene glycol	293	1.04
	Glycerol	293	0.82
	Acetic acid	293	1.19
	Benzoic acid	298	1.21
	Glycine	298	1.05
	Ethyl acetate	293	1.00
	Acetone	288	1.22
	Furfural	293	1.04
	Urea	293	1.20
	Diethylamine	293	0.97
	Aniline	293	0.92
	Acetonitrile	288	1.26
	Pyridine	288	0.58
In Benzene	Acetic acid	298	2.09
	Carbon tetrachloride	298	1.92
	Ethylene chloride	281	1.77
	Ethanol	288	2.25
	Methanol	298	3.82
	Naphthalene	281	1.19
In Acetone	Acetic acid	298	3.31
	Benzoic acid	298	2.62
In Ethanol	Carbon dioxide	290	3.20
	Pyridine	293	1.10
	Urea	285	0.54
	Water	298	1.13
In Toluene	Acetic acid	293	2.00
	Acetone	293	2.93
	Benzoic acid	293	1.74
	Ethanol	288	3.00

Equation (2.17) does not apply for the diffusion of large molecules such as polymer molecules; in this situation the Stokes-Einstein equation

$$D_{12} = \frac{1.05 \times 10^{-9}\ T}{\mu (V_b)^{1/3}} \tag{2.18}$$

must be used. Equation (2.17) should not be used if D_{12} predicted by it is greater than the value predicted by Equation (2.18).

C. Transport to interfaces

The transport of reactants to interfaces either solid or liquid is the most fundamental of processes in heterogeneous systems. We must first develop methods of predicting the transport of materials to interfaces before we can continue.

In considering this problem, we must consider two flow regimes. In one flow regime, the laminar flow regime, the hydrodynamics are well behaved, and it is possible to do complete mass transfer calculations from first principles. In the second flow regime, the turbulent flow regime, the hydrodynamics are not accurately describable, and it is not possible to calculate mass transfer rates from first principles. In this case, various models and empirical correlations must be utilized to enable the calculation to be done. There are few industrial situations in which laminar flow prevails, since in most situations the transport rates needed require turbulent flow, and/or equipment configurations lead to turbulent flow. Thus we will ultimately be more interested in the turbulent-flow regime.

The simplest model for transport to interfaces is the film-theory model presented schematically in Figures 1.3 through 1.6. It is assumed that close to any fluid interface there is a stagnant film of thickness δ through which mass transport occurs by molecular diffusion. The driving force (concentration gradient) is all expended over the film.

If it is assumed that A is diffusing across the film of thickness δ and B (from $A \rightarrow B$) is diffusing back, then $N_1 = -N_2$, and Equation (2.8) applies, and upon integration from $C_{A,1}$ to $C_{A,2}$ with $(x_2-x_1) = \delta$ Equation (2.8) becomes (2.9)

$$N_A = \frac{D_{AB}}{\delta} (C_{A,2} - C_{A,1}) \tag{2.9'}$$

Mass transfer

Example 2.1[a] Estimation of Diffusion Coefficient in a Binary Gas Mixture

Using Equation 2.12, estimate D_{12} for thiophene in hydrogen at 660 K (730°F) and 30 atm (425 psig). (Thiophene is taken as an example since it is representative of the organic sulfur compounds that are hydrogenated in the commercial hydrodesulfurization of petroleum naptha.)

From Table 2.4, V_b for thiophene is $4 \times 14.8 + 4 \times 3.7 + 25.6 - 11.5 - 88.1$. From Equations (2.14) and (2.15) and the value 2.827 for σ for hydrogen from Table 2.3, σ_{12} is 4.04. T_c for thiophene is estimated to be 579°K. Using $\varepsilon/k = 57.9$ for hydrogen (Table 2.3), kT/ε_{12} is found from Equations (2.13) and (2.16) to be 4.04. Table 2.2 then gives Ω_D as 0.8817. Substituting in Equation 1.12, we have

$$D_{12} = \frac{(0.001858)(660)^{3/2}(2+84)/(2 \times 84)^{1/2}}{(30)(4.04)^2(0.8817)} = 0.052 \; \frac{cm^2}{sec}$$

Example 2.2[a] Estimation of Diffusion Coefficient in a Liquid System

Estimate D_{12} for thiophene in dilute solution in hexane at 40°C.

Hexane is presumed not to associate, X is taken as 1.0, M_2 is 86 (hexane), and V_b was found in the previous numerical example to be 88.1 (thiophene). The viscosity of hexane at 40°C is 0.262 cP. Substituting in Equation (2.17), we obtain

$$D_{12} = 7.4 \times 10^{-10} \frac{313(86)^{1/2}}{0.00262(88.1)^{0.6}} = 5.6 \times 10^{-5} cm^2/sec$$

Since V_b is evidently much less than $0.27 \, (XM_2)^{1.87}$, Equation (2.17) should apply.

Perhaps the most significant feature of D_{12} for liquid systems is that at ordinary temperatures it is of the order of 10^{-4} times D_{12} for typical gas systems at atmospheric pressure, or one percent of D_{12} for gas systems at 100 atm.

---

[a] Examples taken from C.N. Satterfield.

The mass transfer coefficient for this process is defined by

$$k_m = \frac{\text{Flux}}{\text{Driving Force}} = \frac{N_A}{(C_{A,2} - C_{A,1})} \qquad (2.24)$$

Substituting Equation (2.9') into (2.24) we obtain

$$k_m = \frac{D_{AB}}{\delta} \qquad (2.25)$$

If we are considering an absorption process for species 1 through a stagnant film, $N_2 = 0$ and Equation (2.7) becomes

$$N_1 = \frac{D_{12} C_T}{\delta} \ln \left[\frac{(1 - (\frac{C_1}{C_T})_2)}{1 - (\frac{C_1}{C_T})_1} \right] \qquad (2.7)$$

and the mass transfer coefficient becomes

$$k_m = -\frac{D_{12}}{\delta} \frac{\ln \frac{(Y_2)_2}{(Y_2)_1}}{(Y_2)_2 - (Y_2)_1} \qquad (2.26)$$

$$k_m = \frac{D_{12}}{\delta (\Delta Y)_{1m}} \qquad (2.26')$$

Neither Equation (2.25) nor (2.26) provide us with an ability to calculate a mass transfer coefficient k_m because the film thickness is not known. Although δ might be estimated, the results are not very satisfactory. What Equation (2.25) provides is an indication of the dependence that k_m should show with respect to the molecular diffusivity D_{12} and the hydrodynamics as they may appear in δ. Shortly we will return and discuss this result.

In 1935 Higbie proposed that mass transfer near an interface occurs by a mechanism in which the interface is made up of a variety of small liquid elements which are continuously brought to the surface from the bulk of the fluid. Thus the interface continually receives fresh fluid elements which have the bulk fluid concentration; mass transfer is then considered as a transient diffusion process from a stagnant fluid element initially of uniform concentration. Mass transfer from that particular fluid element occurs only for the time that it resides in the vicinity of the interface. This mechanism is referred to as the <u>penetration theory</u> for mass transport.

The predicted mass transfer coefficient utilizing penetration theory becomes

$$k_m = 2(d/\pi t^*)^{\frac{1}{2}} \qquad (2.27)$$

Equation (2.27) predicts that the mass transfer coefficient varies as the square root of D_{12} and not as the first power of D_{12}.

Let us now see how the predicted mass transfer behavior compares with the observed mass transfer dependence on system parameters. The engineering approach to obtaining correlations that would allow mass transfer (heat also) calculations to be done before the first-principles analysis had been completed was to correlate mass (and heat) transfer data with respect to dimensionless groups which were considered to be important to the mass transfer process and which contained all of the relevant fluid properties and physical dimensions. Since the rate of heat and mass transfer depends on the hydrodynamics, the Reynolds Number, $Re = d\, u_{Ave}\, \rho/\mu$ became an important independent variable in the correlations. Similarly mass transfer must depend on the molecular transport properties of the fluid, and thus the Schmidt No., $Sc = \mu/\rho D_{12} = \nu/D_{12}$, became another dependent variable. The dimensionless mass transfer group, $k_m d/D_{12}$ containing the mass transfer coefficient k_m became the dependent group. Thus a relationship such as

$$\frac{k_m d}{D_{12}} = a_1 \cdot (Re)^{a_2} \cdot (Sc)^{a_3} \qquad (2.28)$$

was correlated with a large amount of data over a broad range of fluid physical properties, fluid transport properties and system dimensions. The data were generally correlated well by an equation of this form allowing a_1, a_2, and a_3 to be determined and allowing mass transfer coefficients for completely new systems and situations to be calculated with accuracy in the absence of experimental data. The first-principles analysis of mass transfer in laminar flow systems, which was done considerably after the correlations were well developed, gave the same functional form as Equation (2.28) and often gave a good approximation of the constants in the equation. Such first-principles analysis has not been achieved for turbulent flow systems, and thus the correlations are our only predicting tool, e.g., Equation (2.29).

$$\frac{k_m d}{D_{12}} \left(\frac{C_{1m}}{C_t} \right) = 0.023 \, (Re)^{0.83} (Sc)^{0.44} \qquad (2.29)$$

The mass transfer coefficient in this case is dependent on $(D_{12})^{0.56}$ which is not in good agreement with the prediction of film theory but which is in good agreement with the predicted square root dependence of Higbie's <u>penetration theory</u>. More recent studies in this system support a 0.50 power dependence.

Mass transfer data is frequently correlated in terms of j_D factors, the j-factor for mass transfer. This expression ($j_D = f/2$) is a statement of the Colburn-Chilton analogy which says that there is an analogy between mass transfer and momentum transfer and that $j_D = f/2$ where f is the Fanning friction factor determined from frictional loss-pressure drop measurements. If this relation holds, it becomes possible to estimate the mass transfer coefficient for broadly ranging cases where correlations of mass transfer are not available. This is important in considering mass transfer in packed-bed catalytic reactors where there is a large amount of pressure drop (f data) data available but where there is less direct mass-transfer data. Figure 2.1 is a friction factor plot for packed beds.

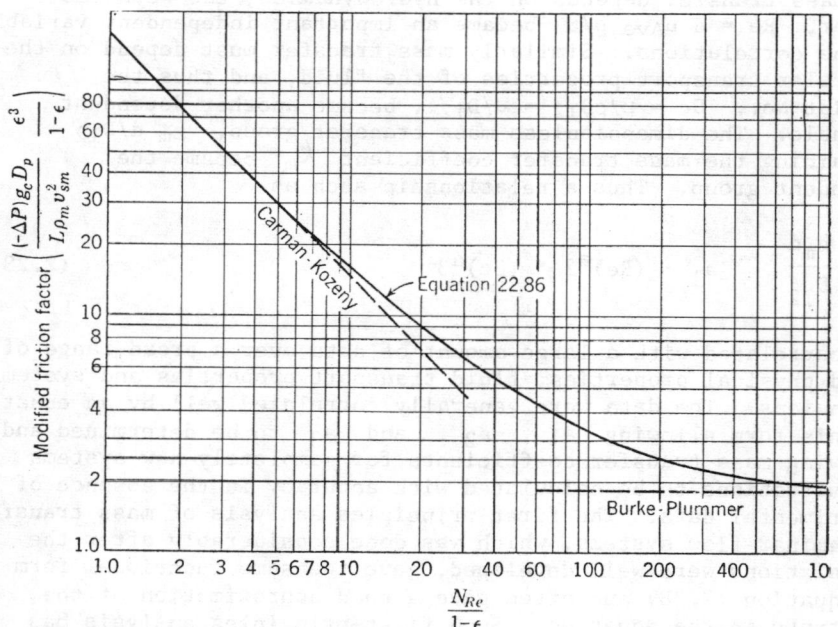

Figure 2.1 Friction factor relationship for flow through packed beds. (Figure from Foust, <u>et al</u>., "Principles of Unit Operation," p. 475.)

Mass transfer

Next let us review the data that is available on mass transfer in packed beds. In this case the Reynolds number $Re = d_p u \rho / \mu$ is best written in terms of G, the mass flow rate of the fluid in g/sec cm^2 of total or superficial bed cross section normal to the mean flow. d_ρ is the diameter of the catalyst particles in the case of spheres; for cylinders d_ρ represents the diameter of a sphere having the same surface area. For cylinders of length X_c and diameter d_c, d_ρ is

$$d_\rho = [d_c X_c + \tfrac{1}{2} d_c]^{\tfrac{1}{2}} \tag{2.30}$$

and

$$j_D = \frac{k_m \rho}{G} Sc^{2/3} \tag{2.31}$$

Recommended correlations are

$$\varepsilon j_D = (0.357)/Re^{0.359} \quad \text{(gases)} \tag{2.32}$$

for gases in the Reynolds number range of

$$3 < Re < 2000$$

where ε is the void fraction of the bed including only the interparticle void spaces. For ε from 0.416 to 0.778 and Re from 100 to 2000 j_D was inversely proportional to ε for gases as indicated by Equation (2.32).

Data for mass transfer to liquids in packed beds is correlated by Equation (2.33) which is recommended for other calculations:

$$\varepsilon j_D = (0.250)/Re^{0.31} \quad \text{(liquids)} \tag{2.33}$$

for

$$55 < Re < 1500$$

and

$$0.35 < \varepsilon < 0.75$$

For

$$0.0016 < Re < 55$$

Equation (2.34) is recommended:

$$\epsilon j_D = (1.09)/Re^{2/3} \qquad \text{(liquids)} \qquad (2.34)$$

For higher Reynolds numbers ϵj_D becomes a function of Re only and is the same for both liquids and gases. Equations (2.32) and (2.33) differ very little so either can be used. For very high Reynolds numbers the friction factor correlation given above should be used.

The correlations show that the mass transfer coefficient varies as $(D_{12})^{2/3}$ which is closer to the penetration theory prediction of $(D_{12})^{1/2}$ than to the film model prediction of D_{12}. To actually calculate mass transfer rates in a packed bed it is necessary to calculate the total exterior surface area of the particles in the bed in addition to knowing the value of k_m.

Heat and mass transfer occur by quite similar mechanisms, and thus data on heat transfer in packed beds are correlated in the same way as data on mass transfer. Thus

$$j_H \equiv \frac{h}{C_p G} Pr^{2/3} \qquad (2.35)$$

$$h \equiv \frac{q}{T_S - T_B} \qquad (2.36)$$

$$P_R \equiv \frac{C_p \mu}{k} \qquad (2.37)$$

where

h = heat transfer coefficient
q = the heat flux per unit pellet surface area
C_p = heat capacity per unit mass of the fluid
T_S = pellet surface temperature
T_B = bulk fluid temperature
k = thermal conductivity of the fluid.

j_D is equal to j_H within the experimental error of most data although j_H has frequently been reported to be 20 - 30% higher at times.

Let us now consider mass transfer around separated spherical objects, bubbles, etc. For a sphere at rest in a stagnant fluid the rate of mass transfer is easily derived. For steady state the flux at any distance r from the center of the sphere times the area of a sphere with radius r must be a constant. Thus

Mass transfer

$$N_1 (\pi d_p^2) = -D_{12} (4\pi r^2) \left.\frac{dC_1}{dr}\right|_r \qquad (2.38)$$

diffusion rate from the
entire sphere at the = diffusion rate through any sphere of radius r at r
surface of the sphere

Integrating Equation (2.38) gives

$$N_1 d_p^2 \left(\frac{2}{d_p} - \frac{1}{r}\right) = D_{12}(C_{1,s} - C_1) \qquad (2.39)$$

For $C_1 = C_{1,\infty}$ at $r = \infty$ we have

$$N_1 = \frac{2D_{12}(C_{1,s} - C_{1,\infty})}{d_p} \qquad (2.40)$$

and upon substitution into Equation (2.24) we obtain

$$k_m = \frac{N_1}{(C_{1,s} - C_{1,\infty})} = \frac{2D_{12}}{d_p} \qquad (2.41)$$

$$\frac{k_m d_p}{D_{12}} = 2.0 \qquad (2.42)$$

for a stagnant fluid. This limit is indicated in Figure 2.2. For flow the Sherwood number will be greater than 2.0 and will increase with flow rate. Exact theoretical analysis gives for Pe = Re . Sc less than 10,000

$$\frac{k_m d_p}{D_{12}} = (4.0 + 1.21 \, Pe^{2/3})^{1/2} \qquad (2.43)$$

This equation fits well the solid line in Figure 2.2. For Pe greater than 10,000

$$\frac{k_m d_p}{D_{12}} = 1.0 \, Pe^{1/3} \qquad (2.44)$$

best represents the theoretical results and the data that are available. This equation is represented by the dotted line in Figure 2.2. Thus again theory and experiment are in good agreement.

Example 2.3[a] Estimation of Coefficient of Mass Transfer from
Gas Stream to Pellets in a Fixed Bed

A hydrodesulfurization reactor employs catalyst pellets in the form of cylinders 3.2 mm (1/8 in.) long and 3.2 mm (1/8 in.) in diameter. Operation is at 30 atm (425 psig) and at 600°K (728°F) with feed containing 82.8 mole percent hydrogen and 17.2 mole percent of 49.4° A.P.I. naphtha. The mixed vapor has a density of 0.0168 g/cm^3 (1.043 lb/ft^3), an average molecular weight of 30.3, and is fed to the reactor at a total rate corresponding to a mass velocity of 0.188 g/sec·cm^2 (1390 lb/hr·ft^2).

It is assumed that thiophene is representative of the small quantities of sulfur compounds to be removed. Using the data of Maxwell and available procedures for estimating physical properties of mixtures we estimate the viscosity of the gas mixture to be 0.00038 P, and N_{Sc} for thiophene in this gas mixture to be 2.96. From Equation (2.30), the equivalent sphere diameter d_p is 0.392 cm.

$$Re = \frac{d\rho G}{\mu} = \frac{0.392 \times 0.188}{0.00038} = 194$$

Estimating $\varepsilon \approx 0.04$, j_D is given by Equation (2.32)

$$j_D = \frac{0.357}{\varepsilon Re^{0.359}} = \frac{0.357}{(0.4)(6.6)} = 0.135$$

$$j_D = 0.135 = \frac{k_m \rho}{G} (Sc)^{2/3}$$

$$k_m = \frac{(0.135)(G)}{\rho (Sc)^{2/3}}$$

$$k_m = \frac{(0.135)(0.188)}{(0.0168)(2.96)^{2/3}} = 0.733 \text{ cm/sec}$$

$$k_m = 0.733 \text{ cm/sec}$$

[a] Example from C. N. Satterfield

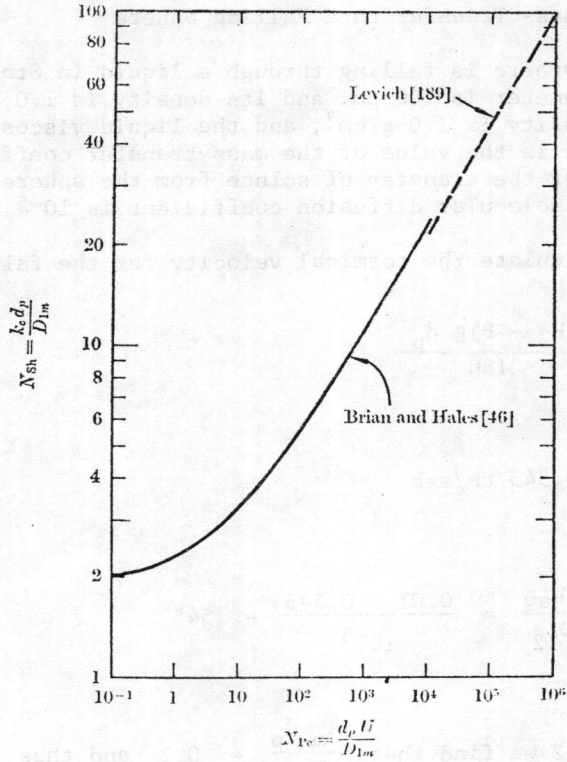

Figure 2.2 Mass transfer to a single sphere moving in a liquid at low velocity. (Figure from C. N. Satterfield.)

Example 2.4 Mass Transfer to a Falling Sphere

A solid sphere is falling through a liquid in Stokes' flow. The sphere diameter is 100 μm, and its density is 1.0 g/cm³. The liquid density is 1.0 g/cm³, and the liquid viscosity is 1.0 cP. What is the value of the mass transfer coefficient k_m (cm/sec) for the transfer of solute from the sphere to the liquid if the molecular diffusion coefficient is 10^{-5} cm²/sec?

First calculate the terminal velocity for the falling sphere.

$$u_{ts} = \frac{(P_2 - P) g\, d_p}{18\mu}$$

$$u_{ts} = 0.545 \text{ cm/sec}$$

$$Pe = \frac{d_p u_{av}}{D_{12}} = \frac{0.01 \times 0.545}{10^{-5}} = 545$$

From Figure 2.2 we find that $\frac{k_m d_p}{D_{12}} = 0.2$ and thus $k_m = \frac{9.2 \times 10^{-5}}{0.01} = 0.0092$ cm/sec.

Mass transfer

III. DIFFUSION IN POROUS MEDIA

Catalysts of industrial interest are normally hard particles for which their exterior surface area comprises only a minute fraction of their total surface area. The majority of the surface area present is distributed throughout the entire particle and is reached via a very small, complicated pore structure through which diffusion occurs.

A. Bulk diffusion

When the pores of a consolidated solid are quite large and the gas is relatively dense, diffusion within the pores occurs via the bulk or ordinary diffusion mechanism. The requirement for bulk diffusion is that the pore diameter be larger than the mean free path of the molecule. Thus the diffusing molecule collides with other molecules more often than it collides with the pore wall. This is the situation for most liquid filled pores.

If θ = free cross section of a porous solid and if θ can be expressed as the volume fraction voids, the flux per unit total cross-section of the solid is θ times that without the solid under similar conditions. If a factor, τ, called a "tortuosity" factor is defined to account for the tortuous path a diffusing molecule must travel and if other complicating factors affecting the diffusion are lumped into τ, we can write

$$D_{12,eff} = \frac{D_{12}\theta}{\tau} = \frac{D_{12}\theta}{L'S'} \qquad (3.1)$$

where D_{12} = bulk diffusion coefficient, cm^2/sec
θ = void fraction of the porous mass
τ = tortuosity factor
$D_{12,eff}$ = effective diffusion coefficient for diffusion in the consolidated solid, cm^2/sec
L' = length or angle factor
S' = shape factor

Both L' and S' are greater than one; neither has been adequately developed theoretically for actual consolidated powders.

Consider Figure 3.1 to help clarify Equation (3.1). For Figure 3.1A diffusion occurs through the pores only; the solid material here is considered non-porous. Thus $D_{12,eff} = D_{12}\theta$, obviously. In Figure 3.1B the pores are still uniform but are at an angle such that the diffusion distance is now greater by a factor q than the thickness of the slab.

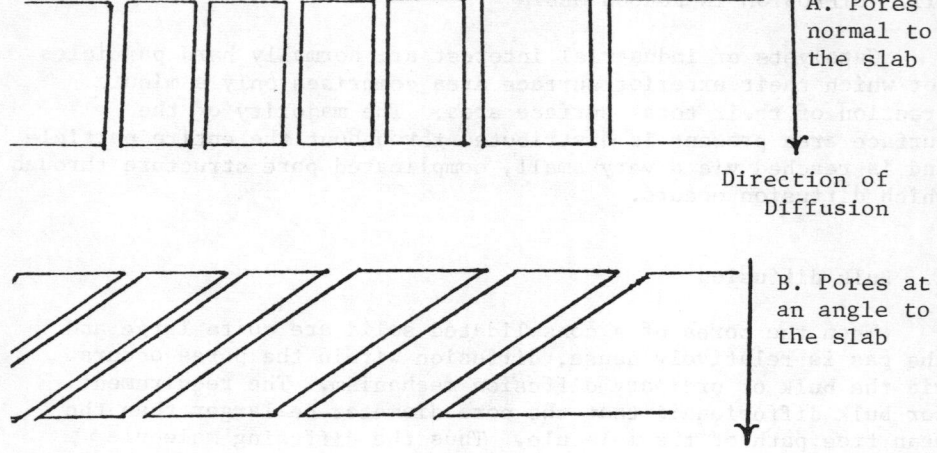

Figure 3.1 Schematic pore structure in a slab illustrating effect of diffusion path length; θ is fraction of area or volume occupied by pores.

$$q = \frac{\text{pore or diffusion length}}{\text{slab thickness}}$$

Since the flux is determined by the concentration gradient, the flux per unit pore cross section is reduced by $1/q$. Further the flux that is of interest is that normal to the slab surface, whereas the calculated flux is at an angle reducing the normal flux by $1/q$ giving

$$D_{12,\text{eff}} = \frac{D_{12}\theta}{q^2} \tag{3.2}$$

Using this approach and assuming that the pores are of uniform diameter and randomly oriented they would have an average angle of 45°. With this Wheeler showed that $q = \sqrt{2}$ and thus $q^2 = 2$. This gives an interpretation of L'. Real pores also vary in shape having large diameter regions and constrictions. The large diameter regions do not make up for the constrictions. This has been treated by several investigators leading to values of $1/S'$ varying from 1.0 to 5. Thus $L S = \tau = S'q^2$ could vary from 2 to 10. Typical τ values vary from 3 to 7. Because of the complexity of the pore structure of consolidated powders it is not possible to calculate τ <u>a priori</u> although theory adequately predicts the range of values. τ must be determined empirically for each catalyst. If a value must be assumed in order to calculate an effective diffusivity, either 3 or 4 is appropriate. τ increases as the density of the particle increases and as the void fraction decreases; a more tightly packed particle has a more tortuous diffusion path; thus larger values of τ should be chosen for catalyst particles with a lower void fraction.

As noted above diffusion in liquid-filled pores is always bulk diffusion and as such calculations are liquid-phase diffusivity calculations. The effective diffusivity is calculated using Equation (3.1). Two situations exist where the calculations are more complex or cannot be accurately done. One situation is diffusion in zeolites which have very fine pores. It is not possible to estimate the diffusivity of materials in zeolites, and there is very little data available. This topic will not be further pursued.

The other situation involves diffusion of large molecules in small pores, however not as small as in zeolites. Here it is found that there is a restricted diffusion that occurs as the diameter of the molecule approaches that of the pore. Thus the rate of diffusion, D_{12}, becomes less than that predicted for liquid diffusion, and this must be accounted for to predict the correct rate of mass transfer in the porous solid. Precisely stated the effective diffusion coefficient should be

$$D_{12,eff} = \frac{D_{12} \, \theta \, K_r}{\tau} \quad (3.3)$$

where K_r is the fractional reduction in the effective diffusion coefficient due to restricted diffusion in the small pores.

Figure 3.2 summarizes the data available and shows that the reduction can be in excess of 10-fold if the critical molecular diameter is one half the pore diameter. It has also been shown however that for non-rigid molecules such as polymers the statistical molecular diameter does not predict the reduction present. A better way of calculating the critical molecular diameter for non-rigid molecules is not currently available. For rigid molecules the critical diameter is the smallest circle through which the molecule can pass, and it is easily calculated using van der Waals radii and bond lengths.

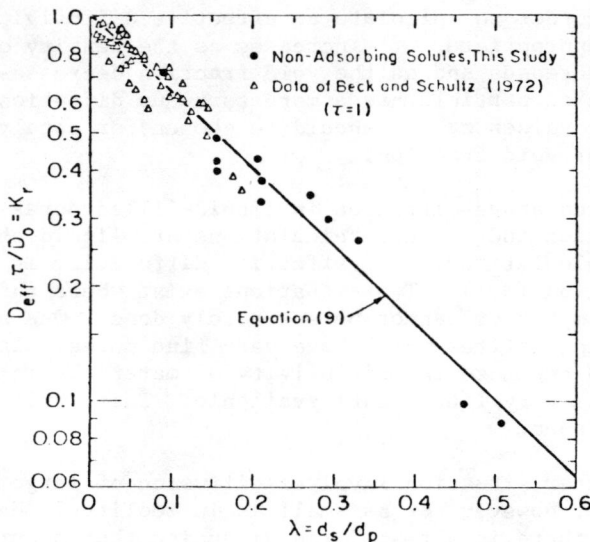

Figure 3.2 Effect of solute critical diameter to pore diameter ratio on effective diffusivity. (Figure from Satterfield, Pitcher and Colton, A.I.Ch.E.J., 19, 632 (1973).)

B. Knudsen diffusion

If the gas density is quite low or if the pores are quite small in diameter or both, the diffusing molecules strike the pore wall much more frequently than they collide with each other. This occurs if the mean free path is much larger than the pore diameter. The molecule apparently momentarily adsorbs on the pore wall, and then diffusely desorbs from the wall in a random direction. Such diffusion is called "Knudsen diffusion" or Knudsen flow.

From kinetic theory for diffusion in a straight round pore the relationship of variables describing Knudsen flow can be derived. By definition for straight round pores

$$N_1 = D_1 \frac{(c_{1,1} - c_{1,2})}{X_o} = \frac{D_{K_{11}}}{RT} \frac{((p_1)_1 - (p_1)_2)}{X_o} \quad (3.4)$$

$\qquad\qquad\quad$ Driving force term $\qquad\qquad$ Driving force term

Mass transfer

From kinetic theory

$$N_1 = \frac{2r_e u_{ave}}{3RT} \frac{((p_1)_1 - (p_1)_2)}{X_o} \qquad (3.5)$$

Equating (3.3) and (3.4) with $u_{ave} = (8RT/M)^{\frac{1}{2}}$

$$\frac{D_{K,1}}{RT} \frac{((p_1)_1 - (p_1)_2)}{X_o} = \frac{2\,r_e}{3} \left(\frac{8RT}{\pi M}\right)^{\frac{1}{2}} \frac{((p_1)_1 - (p_1)_2)}{RT\,X_o} \qquad (3.6)$$

and upon rearranging and substituting for known terms

$$D_{k,1} = 9700\, r_e \sqrt{T/M} \qquad (3.7)$$

where

D_K = Diffusion coefficient for Knudsen diffusion, cm^2/sec
r_e = Equivalent radius of pore, cm
T = temperature, $°C$
M = molecular weight

    Note that the units given must be used and that the equation applies to a single component only. This is because collisions between molecules are infrequent, and thus each species diffuses independently of any other component present in the system. The temperature dependence of D_K is $(T)^{\frac{1}{2}}$, and thus the observed activation energy will be no more than 1 kcal per mole.

For Knudsen diffusion the above equation works very well for predicting diffusion rates in long, smooth capillaries. Porous catalysts do not have pore systems which are well represented by long, smooth tubes. For solids (porous) we define an equivalent pore radius, r_e, which has the same volume to surface ratio as the solid.

$d_e \updownarrow \quad r_e = d_e/2 \qquad$ Pore

$$V = \pi r_e^2 L \qquad (3.8)$$

$$S = \pi\, 2 r_e L \qquad (3.9)$$

$$r_e = \frac{2V}{S} \qquad (3.10)$$

For real porous materials

$$r_e = \frac{2 V_g}{S_g} = \frac{2\theta}{S_g \rho_p} \tag{3.11}$$

S_g = total surface area of porous solid per gram, cm^2/g
V_g = total pore volume per gram, cm^3/g
ρ_p = bulk pellet density, gm/cm^3
θ = porosity, cm^3/cm^3
V_g = θ/ρ

Substituting Equation (3.13) into Equation (3.7) and eliminating D_K by substituting for it from Equation (3.12)

$$D_{K,1,eff} = \frac{D_{K,1} \theta_K}{\tau_K} \tag{3.12}$$

gives

$$D_{K,1,eff} = \frac{D_{K,1} \theta_K}{\tau_K} = \frac{8 \theta_K^2}{3 \tau_K S_g \rho_p} \left(\frac{2RT}{\pi M}\right)^{1/2} \tag{3.13}$$

which upon eliminating all known quantities gives

$$D_{k,1,eff} = 19,400 \frac{\theta_K^2}{\tau_K S_g \rho_p} \sqrt{T/M} \tag{3.14}$$

$D_{K,1,eff}$ is the effective diffusion coefficient based on the entire cross section of the solid as for bulk diffusion. Tortuosity factors, τ_K, have values that are similar to those for bulk diffusion. However, they should not be the same as τ for bulk diffusion, even for the same porous medium, except for the case of uniform pore diameters and straight tubular pores. This can be shown theoretically. The best data obtained on Knudsen diffusion is in Vycor glass with an effective pore radius of 25 to 60 Angstroms. Here τ_K was 5.9 from several studies. τ_K might be expected to be larger than τ because of the line-of-site nature of Knudsen diffusion.

C. Transition region

As in other situations in which there are two limiting regions, there is a transition region between bulk and Knudsen diffusion regimes. For binary gas diffusion in a porous solid at constant total pressure we have

$$N_1 = \frac{-(P/RT)(dY_1/dx)}{\dfrac{1-(1+N_2/N_1\,Y_1)}{D_{12,eff}} + \dfrac{1}{D_{K,1,eff}}} \tag{3.15}$$

$$N_1 = -D_{eff}\frac{P}{RT}\frac{dY_1}{dx} = -D_{eff}\frac{dC_1}{dx} \tag{3.16}$$

where

D_{eff} = the effective diffusion coefficient = $\dfrac{\text{flux}}{\text{concentration gradient}}$

$D_{12,eff}$ and $D_{K,1,eff}$ = the effective diffusion coefficients for species 1 for bulk and Knudsen diffusion respectively

Y_1 = mole fraction species 1

$\dfrac{N_2}{N_1}$ = ratio of fluxes, it is positive for concurrent flow and negative for countercurrent diffusion.

The coordinate system used in the definition is fixed in relation to the solid through which flux is occurring. For non-equimolar fluxes $N_1 \neq J_1$ as defined relative to equimolar flow coordinates, and D_{eff} is a function of (N_2/N_1) which in turn is a function of the physical situation.

The diffusion flux is dependent upon the ratio $D_{12}/D_{K,1}$ and not solely dependent upon the pore size or pressure. D_{12} is proportional to pressure; $D_{K,1}$ is proportional to pore diameter. Whether bulk diffusion or Knudsen diffusion predominates depends on the ratio of $D_{12}/D_{K,1}$. If $D_{12}/D_{K,1}$ is large

$$N_1 = -D_{K,1,eff}\frac{dC_1}{dx} \tag{3.17}$$

If $D_{12}/D_{K,1}$ is small and $N_2 = 0$

$$N_1 = \frac{D_{12,eff} \, P}{p_2} \frac{dC_1}{dx} \tag{3.18}$$

where p_2 = partial pressure of component 2. For self-diffusion or for equimolar counterdiffusion of gases, $N_1 = -N_2$; at constant P Equation (3.15) reduces to

$$N_1 = -\frac{1}{\left(\dfrac{1}{D_{12,eff}}\right)\left(\dfrac{1}{D_{K,1,eff}}\right)} \frac{dC_1}{dx} \tag{3.19}$$

where the reciprocal diffusion coefficients are treated like resistances

$$\frac{1}{D_{eff}} = \frac{1}{D_{12,eff}} + \frac{1}{D_{K,1,eff}} \tag{3.20}$$

This relation has not been verified experimentally, but it agrees closely with a much more rigorous derivation based on kinetic theory; a theory for which there appears to be experimental evidence for support.

D. Surface diffusion

Molecules adsorbed on surfaces, particularly if they are not strongly chemisorbed, exhibit considerable surface mobility. Movement is in the direction of the surface concentration gradient which is usually in the same direction as the gas-phase concentration gradient. Thus surface diffusion is a parallel-transport process with the vapor-space diffusion. For surface diffusion

$$J_s = \frac{-D_s}{\tau_s} \rho_p S_g \frac{dC_s}{dx} \tag{3.21}$$

where

- J_s = surface flux per unit cross section of porous material, g moles/cm^2-sec
- D_s = surface diffusion coefficient, (on a cross section basis), cm^2/sec
- C_s = surface concentration, moles/cm^2
- $\rho_p S_g$ = S_v = surface area per unit volume, cm^2/cm^3
- τ_s = tortuosity factor—not necessarily equal to τ or τ_K

Example 3.1[a] Estimation of D_{eff} for Gas Diffusion in a Porous Catalyst

Estimate D_{eff} for the diffusion of thiophene in hydrogen at 600°K and 30 atm in a catalyst having a B.E.T. surface of 180 m^2/g, a void volume of 40 percent, a pellet density of 1.40 g/cm^3, and exhibiting a narrow pore-size distribution.

For this system, D_{12} was found to be 0.052 cm^2/sec by the procedure illustrated by Example 2.1. Substituting in Equation (3.1), we obtain

$$D_{12,eff} = \frac{D_{12}\theta}{\tau} = \frac{0.052 \times 0.40}{\tau} = \frac{0.0208}{\tau} \ cm^2/sec$$

Substituting in Equation (3.14), we have

$$D_{K,eff} = \frac{19400 \times 0.4^2}{\tau_k \times 1800000 \quad 1.40} \sqrt{660/84} = \frac{0.00344}{\tau_k} \ cm^2/sec$$

Taking $\tau_k = 2$, $D_{12,eff} = 0.0104$ and $D_{K,1,eff} = 0.00172 \ cm^2/sec$ Knudsen diffusion may be expected to predominate, since $D_{12,eff}$ is so much larger than $D_{K,1,eff}$. As a good approximation, Equation (3.20) may be used to give

$$\frac{1}{D_{eff}} = \frac{1}{D_{K,eff}} + \frac{1}{D_{12,eff}} = 673 \quad D_{eff} = 0.0015 \ cm^2/sec$$

---

[a] Example from C. N. Satterfield

$$D_s = D_{s,o}\, e^{-E_s/RT} \tag{3.22}$$

Values of D_s range from 10^{-3} to 10^{-5} cm²/sec for small, physically adsorbed molecules at room temperature. E_s is usually about ½ the heat of adsorption of the diffusing molecule, but it sometimes approaches it. This would indicate that the diffusing molecule truly does remain associated with the surface as it diffuses. J_s should thus usually decrease with temperature because of a decreasing surface concentration.

Figure 3.3 shows a correlation of surface diffusion data from the literature developed by Sladek. D_s is correlated against the parameter q/mRT where q is the heat of adsorption, and m is an estimate of the bond order and of the energy barrier which must be surmounted in diffusing; it is however really an empirical fitting parameter. The correlation is remarkable in that it covers data spanning eleven orders of magnitude in D_s, a range of q/mRT of 60 covering both van der Waals physical adsorption and chemisorption, and temperatures from -230 to +600°C. Figure 3.3 can be used to estimate surface diffusion coefficients in a range of systems or for conditions for which data are not available.

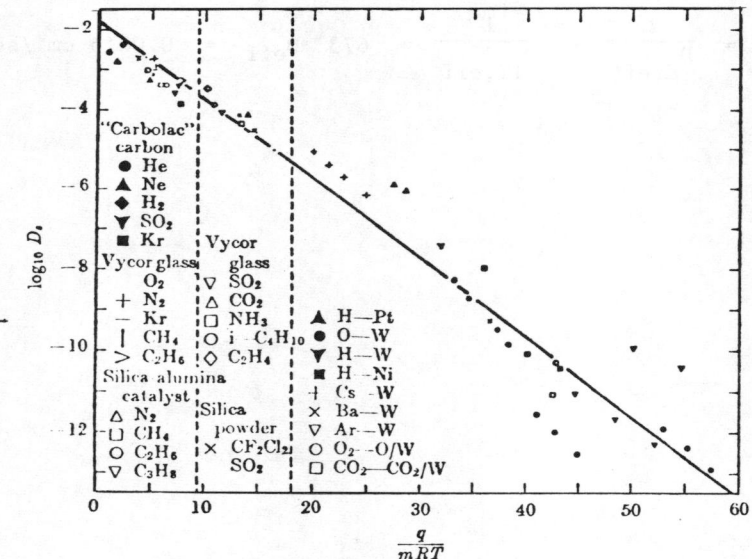

Figure 3.3 Correlation of surface diffusion coefficients by Sladek. Figure from C. N. Satterfield.

Mass transfer

At the temperature and pressure of important industrial processes, it is doubtful that surface diffusion is important to the overall diffusion process although there is little evidence either way at the moment. It would appear to have the greatest chance of being important for very high surface area and thus very small pore pellets. Surface diffusion may also appear more important where a component in a highly dilute state is diffusing. In this case there is very little driving force for diffusion (dC_1/dx is small because C_1 at its largest value is small) in the bulk phase; yet reasonably strong adsorption can produce a surface concentration gradient $dC_{1,s}/dx$ which is large enough to result in more transport than occurs through the bulk phase. For this same system at higher concentrations $dC_{1,s}/dt$ would be small because of surface saturation and most transport would occur through the bulk phase (pore) transport mode.

E. Transport models and pore models

<u>Applicability to reacting systems</u>: If the value for the effective diffusion coefficient which has been derived truly represents the proper value of D_{eff} for that porous medium then it will be applicable to predict transport in a reacting system in accordance with the following comments:

i) The value for the non-reacting system assumes that the derivative of C_1 with respect to x or y (distance) is zero at all solid interface; this comes in a more sophisticated derivation. Thus we have a given concentration profile in the solid which is not distorted or perturbed by solid interfaces.

ii) If reaction is present obviously the concentration profile at the solid interfaces is not zero because reaction is occurring at these interfaces and reactants must be transported to these surfaces and products away from them.

iii) However as a practical consideration there are few cases in which reaction occurs so rapidly that there is appreciably large conversion in passing the first several particles making up a pellet, i.e., the first several microns. Thus the concentration derivative normal to the surfaces, while not equal to zero, is typically so small that it does not distort the concentration profile in the pores appreciably.

iv) Even if reaction is quite rapid the mixing time in the distance of the pore diameter is so rapid in comparison with the diffusion time for the pellet radius that it is very difficult for an appreciable concentration gradient to exist across a pore or near an interface. This can be approximated by:

$$t_{D_o} = a^2/2D_{12}$$

For a 1000 Å pore diameter and $D_{12} = 10^{-2}$ cm²/sec

$$t_D = \frac{a^2}{2 D_{12}}\overset{a}{=} \frac{(10^{-5} \text{cm})^2}{2 \times 10^{-2} \text{cm}^2/\text{sec}} \tag{3.23}$$

$$t_D \cong 5 \times 10^{-9} \text{ sec} \tag{3.23}$$

For a pore length of 0.2 cm (radius of a small catalyst particle)

$$t_D = \frac{L^2}{2D_{12}} = \frac{(0.2 \text{ cm})^2}{2 \times 10^{-2} \text{cm}^2/\text{sec}} \tag{3.24}$$

$$t_D \cong 2 \text{ sec} \tag{3.24}$$

Thus relaxation in the pore is so rapid that appreciable gradients could not exist in most cases.

v) For Knudsen diffusion which predominates in most catalysts with gas-phase reactions, gradients in the pores near interfaces have no physical meaning and thus such questions have no relevance in this case.

Figure 3.4 below shows the effect of pressure upon the diffusion flux for the steady-state counterdiffusion of ethylene and hydrogen through a porous plug 1.0 cm thick with pure ethylene on one side at P and pure hydrogen on the other at P. Low pressures represent the region in which Knudsen diffusion predominates and at the higher pressures diffusion begins to exhibit the behavior of ordinary or bulk diffusion. The flux in the Knudsen regime increases with pressure because D_K is independent of pressure, but the concentration gradient dC/dx increases with pressure; thus the flux increases. The bulk diffusion coefficient exhibits an inverse first-order dependence on pressure, and thus the increased driving force with increasing pressure is exactly countered by the decreasing D_{12}. The flux therefore is pressure independent. Bulk diffusion becomes predominant at higher pressures because the gas density in the pores increases to such a level that the mean free path of the molecules becomes less than the pore diameter, i.e., molecule-molecule collisions begin to predominate over molecule-pore-wall collisions.

---

[a] For unsteady-state diffusion processes it is possible to estimate the time required to approach equilibrium by $Dt/a^2 = 0.5$ for a sphere with 99% approach to equilibrium being desired at the center.

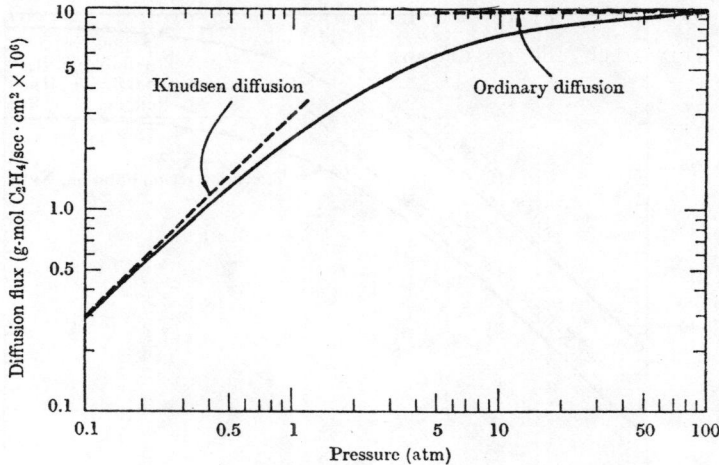

Figure 3.4 Steady-state counterdiffusion of ethylene and hydrogen in the transition region calculated by Equation (3.19) for a porous plug 1.0 cm thick with pure ethylene at P on one side and pure H_2 at P also on the other; for the plug $S_g = 10$ m^2/g, $T = 298°K$, $\tau = 1.0$, $\theta = 0.04$, $\rho_p = 1.4$ 2/cm, $r_e = 570$ Å and $D_{12}P = 0.602$ cm^2 sec. (From C. N. Satterfield)

Figure 3.5 shows the effect of pore radius upon the mode of transport within the system. At very small pore radii the flux is almost exclusively by the Knudsen mode, whereas the pore radius increases in size, the diffusion flux passes through the transition region and then becomes predominated by ordinary or bulk diffusion, at which point it is independent of pore radius. The transition from Knudsen to bulk diffusion occurs because with increasing pore diameter the diameter becomes larger than the molecular mean free path and molecule-molecule collisions begin to predominate. The ratio of the two fluxes is inversely proportional to the ratio of the square root of their molecular weights in all three diffusion regimes.

Note that Knudsen diffusion predominates to pore diameters of 500 Å under these conditions. Thus unless industrial operations are at elevated pressures Knudsen diffusion will predominate in the fine pore structure of most industrial catalysts.

Estimation of D_{eff}: The diffusion data available on consolidated porous solids show the following points:

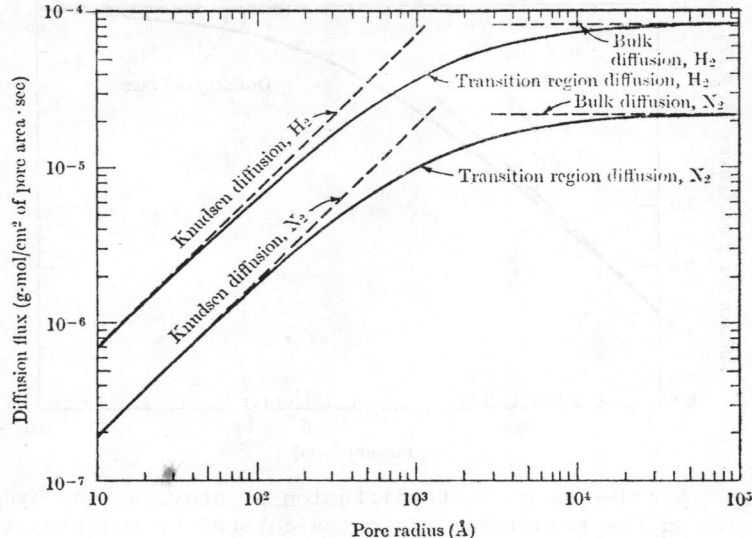

Figure 3.5 Calculated counterdiffusion flux, N_1 of hydrogen and nitrogen as a function of pore size at a total pressure of 1 atm and 290°K. (Figure from C. N. Satterfield.)

- Porous Vycor glass studied by many different workers gives uniformly similar results. For most studies, tortuosity factors were between 4.7 and 6.6 with an average value of 5.9.

- τ for silica-aluminas range around 2 to 2.5 from steady-state flow studies; they range around 2 to 6 from unsteady-state flow measurements, and they range about 3.5 to 6 from reaction studies. Silica-aluminas have narrow pore size distributions and fine pores leading to expected values from steady-state flow studies. For transient studies the dead-end pores are sampled also, and a higher τ is found; for reaction the dead-end pores are also sampled, and interestingly enough, similar values are found for τ.

- For catalysts which have a bimodal pore-size distribution τ values are not reasonable indicating the need to consider pore size distribution to make such data useful and to allow prediction. The diffusion in the larger pores outweighs that in the small pores, leading to a value of $\tau < 1.0$.

Mass transfer

Parallel-path pore model. To more adequately represent porous solids many pore models have been proposed. A simple model making much sense is the parallel-path pore model which assumes that the pore structure is composed of parallel pores having the pore size distribution measured with a constant tortuosity factor across the size range.

In this case integration of Equation (3.15) for constant pressure P over the pore length X_o and between $(Y_1)_1$ and $(Y_1)_2$ leads to

$$N_1 = \frac{D_{12,\text{eff}} \, P}{RTX_o(1+N_2/N_1)} \ln \left[\frac{1-(1+N_2/N_1)(Y_1)_2 + (D_{12,\text{eff}}/D_{K,1,\text{eff}})}{1-(1+N_2/N_1)(Y_1)_1 + (D_{12,\text{eff}}/D_{K,2,\text{eff}})} \right] \quad (3.24)$$

$$N_1 = \frac{-D_{\text{eff}} \, P}{RTX_o} \left((Y_1)_2 - (Y_1)_1 \right) \quad (3.25)$$

This equation is then integrated over the range of pore sizes with the proper weighting function to give:

$$N_1 = \frac{D_{12}\theta P}{_p RTX_o(1+N_2/N_1)} \int_o^\infty \ln \left[\frac{1-(1+N_2/N_1)(Y_1)_2+(D_{12}/D_{K,1})}{1-(1+N_2/N_1)(Y_1)_1+(D_{12}/D_{K,1})} \right] f(r) dr \quad (3.26)$$

where

$D_{12,\text{eff}}$ has been replaced by $D_{12}\theta/\tau_p$
$D_{K,1,\text{eff}}$ has been replaced by $D_{K,1}\theta/\tau_p$

and $f(r)$ is the fraction of void fraction θ in pores of radii between r and $r+dr$. The integration is readily accomplished by dividing the pore size distribution into increments and calculating $D_{12}/D_{K,1}$ for the average r of each increment and then integrating numerically.

The value of τ_p obtained using this model is given in Table 3.4. It is 4 for about half of the catalysts and falls between 2.8 and 7.3 for all others, except two that had been calculated at very high temperature.

In summary a value of D_{eff} can be estimated which in most instances is accurate enough to be of considerable value in predicting diffusion rates. Furthermore if a value of τ can be obtained for a given porous media, this value should allow fairly accurate estimates of the value D_{eff} to be determined for different conditions. Careful attention must however be paid to the possible transition from one diffusion regime to another in any extrapolation. Further the theory we have just considered clearly outlines the diffusion behavior which can be expected in porous systems and how changes in the system will produce changes in the behavior which will be observed. In cases where a pore distribution exists it is necessary to invoke more detailed theories, but these theories again allow good estimates of the behavior expected for the system.

DIFFUSION AND REACTION, EFFECTIVENESS FACTORS

A. Derivation of the basic relationships

Now let us return to the problem of diffusion and reaction in porous solids. The effect of mass transfer upon observed reaction characteristics in porous media was first analyzed by Thiele in the U.S., Damköhler in Germany and Zerdovitch in Russia between 1937 and 1939, each working independently. The basic theory as proposed by Thiele has not been altered since but has been extensively refined and extended. The result of all this is a quantitative description of the factors which determine the effectiveness of a porous catalyst.

Let us now return to our old model of a pore and a flat sheet with the same active surface area as the pore walls. With this concept we define the concept of an <u>effectiveness factor</u>, η

$$\eta = \frac{\text{rate of reaction for the pore, the actual reaction rate (Global Reaction Rate)}}{\text{rate of reaction for flat surface, reaction rate if entire surface of catalyst were exposed to the reactant concentration and the temperature at the surface of the particle, } C_{SA} \text{ and } T_S}$$

If $\eta = 1.0$, there are no diffusional limitations; if $\eta < 1.0$ there are diffusional limitations.

Let us now take a model and derive the desired relationships. The model and the assumptions we will use are:

i) A spherical porous catalyst particle of radius R.

ii) The particle is assumed to be isothermal throughout.

iii) The complicated diffusion process is assumed to be represented by a single overall effective diffusion coefficient, D_{eff}.

Mass transfer

iv) Furthermore it is assumed that Knudsen type diffusion predominates and that Fick's law describes the process. The Knudsen diffusion assumption is not required if there is no volume change upon reaction.

v) The catalytic activity of the porous material is uniform throughout.

vi) For simplicity of analysis consider the reaction to be A⟶B with no volume changes upon reaction. This relaxes iv) above.

vii) The rate of reaction (g-moles/sec · cm surface) can be expressed by a simple power function of the concentration A, i.e.,

$$\text{rate} = k_s c_a^m \quad (3.26)$$

k_s = surface reaction rate constant, based on unit surface area

Consider the spherical pellet shown in Figure 3.6 and focus attention on the differential shell of thickness dr. At steady state a mass balance about the shell gives

$$\begin{pmatrix} \text{rate of diffusion} \\ \text{inward at } r = r+dr \end{pmatrix} - \begin{pmatrix} \text{rate of diffusion} \\ \text{inward at } r = r \end{pmatrix} = \begin{pmatrix} \text{rate of reaction} \\ \text{in the shell} \end{pmatrix}$$

$$(3.27)$$

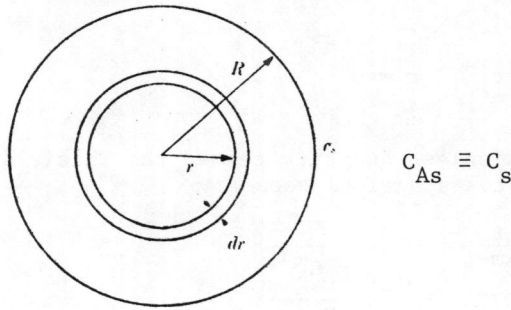

$C_{As} \equiv C_s$

Figure 3.6 Spherical model for simultaneous diffusion and reaction.

Substituting into Equation (3.27)

$$4\pi(r+dr)^2 D_{eff}\left[\frac{dC}{dr}+\frac{d^2C}{dr^2}dr\right] - 4\pi r^2 D_{eff}\frac{dC}{dr} =$$

$$(4\pi r^2 dr)(S_v)k_s C^m \qquad (3.28)$$

where

- $4\pi r^2$ = inner superficial surface area of spherical shell at r
- $\frac{dC}{dr}$ = concentration gradient on spherical shell at r
- S_v = pore surface area in cm^2 per unit volume of particle in cm^3
- C = C_A ; the subscript A has been dropped for ease of writing

Upon rearranging Equation (3.28) we obtain

$$D_{eff}\frac{d^2C}{dr^2}(4\pi r^2)dr + 8\pi r D_{eff}\frac{dC}{dr}dr = 4\pi r^2 S_v k_s C^m dr \qquad (3.29)$$

or

$$\frac{d^2C}{dr^2} + \frac{2}{r}\frac{dC}{dr} = \frac{S_v k_s C^m}{D_{eff}} \qquad (3.30)$$

The boundary conditions are:

$$C = C_S \quad \text{at} \quad r = R \qquad (3.31)$$

$$-\frac{dC}{dr} = 0 \quad \text{at} \quad r = 0 \qquad (3.32)$$

Next define a dimensionless quantity termed the Thiele diffusion modulus ϕ_S ; S denotes spherical geometry

$$\phi_S = R\sqrt{\frac{S_v k_s C^{m-1}}{D_{eff}}} = R\sqrt{\frac{k_v C^{m-1}}{D_{eff}}} \qquad (3.33)$$

where

- $k_v = S_v k_s$ = the intrinsic reaction rate constant per unit gross volume of catalyst
- m = reaction order

Mass transfer

If we now make the further assumption that the reaction order, m, is one, the Thiele Modulus, ϕ_S, becomes independent of concentration, and we can proceed easily with the solution.

$$\frac{d^2C}{dr^2} + \frac{2}{r}\frac{dC}{dr} = \frac{\phi_S^2}{R^2} C \tag{3.34}$$

Equation (3.34) is of the form

$$\frac{d^2y}{dx^2} + \frac{2}{x}\frac{dy}{dx} = b^2 y \ ; \ \text{where} \ b = \frac{\phi_S}{R} \tag{3.35}$$

To solve, set $y = v/x$; then Equation (3.35) becomes

$$\frac{d^2v}{dx^2} = b^2 v \tag{3.36}$$

The solution of Equation (3.36) is

$$v = C_1 e^{bx} + C_2 e^{-bx} \tag{3.37}$$

where C_1 and C_2 are integration constants. Eliminating v and expressing in terms of C_A

$$C_A = \frac{1}{r}\left[C_1 e^{br} + C_2 e^{-br} \right] \tag{3.38}$$

Applying the B.C. that $dC_A/dr = 0$ at $r = 0$ gives $C_1 = -C_2$. Therefore Equation (3.38) becomes

$$C_A = \frac{C_1}{r}\left[e^{br} - e^{-br} \right] = \frac{2C_1}{r} \text{Sinh}(br) \tag{3.39}$$

Applying the B.C. that $C = C_S$ at $r = R$ gives

$$C_{A,S} = \frac{2C_1}{R} \text{Sinh}(br) \quad , \quad C_1 = \frac{C_{AS} R}{2 \text{Sinh}(br)} \tag{3.40}$$

Therefore

$$C_A = \frac{C_{AS} R}{r} \frac{\text{Sinh}(br)}{\text{Sinh}(bR)} \tag{3.41}$$

and since $b_s = \phi/R$

$$C_A = \frac{C_{A,S} R \, \text{Sinh}(\phi_s(r/R))}{\text{Sinh}(\phi_s)} \tag{3.42}$$

Equation (3.42) describes the concentration profile in a catalyst particle, and from this it is seen that the profile is a function of the dimensionless parameter ϕ_s and the dimensionless distance (r/R). The important fact is that the concentration at any point r in the particle is dependent only upon the dimensionless parameter ϕ_s which contains the pertinent parameters for the particle: the diffusion rate parameter D_{eff}, the reaction rate parameter k_v, and the particle radius.

For our particle we can obtain the overall reaction rate by integrating the rate $r_A = k_v C_A$ over the radius of the particle using the calculated concentration profile or by calculating the rate of flux of A into the particle using the concentration gradient at the surface of the particle. Doing the latter

$$\text{Rate} = 4\pi R^2 D_{eff} \left(-\frac{dC_A}{dr}\right)_{r=R} \tag{3.43}$$

$$= \frac{C_{A,S}}{R}\left[\frac{\phi_s}{\text{Tanh}(\phi_s)} - 1\right]$$

$$\left(-\frac{dC_A}{dr}\right)_{r=R} = \frac{\phi_s C_{A,S}}{R}\left[\frac{1}{\text{Tanh}(\phi_s)} - \frac{1}{\phi_s}\right] \tag{3.44}$$

Substituting Equation (3.44) into Equation (3.43) gives

$$\text{Rate} = 4 \phi_s \pi R D_{eff} C_{A,S} \left[\frac{1}{\text{Tanh}(\phi_s)} - \frac{1}{\phi_s}\right], \text{ g-moles/sec} \tag{3.45}$$

which is the rate of reaction for our particle or the particle global rate.

If the concentration throughout the entire pellet was the same as the concentration at the surface, $C_{A,S}$, the rate of reaction would have been

$$\text{Rate} = \frac{4}{3}\pi R^3 k_v C_{A,S}, \text{ g-mole/sec.} \tag{3.46}$$

The effectiveness factor, η, is the ratio of the two rates

$$\eta = \frac{\text{Equation (3.45)}}{\text{Equation (3.46)}} \tag{3.47}$$

$$\eta = \frac{3}{\phi_s}\frac{1}{\text{Tanh}(\phi_s)} - \frac{1}{\phi_s}, \quad \begin{pmatrix}\text{spherical}\\ \text{geometry and}\\ \text{1st order reaction}\end{pmatrix} \tag{3.48}$$

where

$$\phi_s = R\sqrt{\frac{k_v}{D_{eff}}}$$

This form is for first-order reaction in a sphere. The hyperbolic tangent approaches unity as ϕ_S increases and is 0.99 at $\phi_S = 2.65$. Thus η approaches $3/\phi_S$ at large values of ϕ_S.

Note that the effectiveness factor is a function only of ϕ_S and that

$$\phi_s^2 = \frac{R^2}{D_{eff}} = \frac{4/3\,\pi R^3 k_v C_{A,S}}{1/3(4\pi R^2)D_{eff}(C_{A,S}/R)} \tag{3.50}$$

which is the ratio of the maximum reaction rate to the maximum diffusion rate for the particle

where

$$\phi_s = R\sqrt{\frac{k_v}{D_{eff}}} \tag{3.51}$$

Thus since $\eta = f(\phi_s)$, η, the effectiveness factor, is a function of the three parameters comprising ϕ_S and is not a function of any one parameter. In fact the value of ϕ_S is not determined by fast reaction or by slow diffusion but only by the ratio of the reaction rate to the diffusion rate (Equation (3.50));

fast reaction is not a problem if we have fast diffusion. Thus the important quantity is the value of ϕ_s. The effectiveness factor approaches unity as the particle radius is reduced, as the volumetric reaction rate constant decreases or as the effective diffusivity increases. Conversely the effectiveness factor is small for large particles, large reaction rate constants, and small effective diffusivities.

A small effectiveness factor means that the catalyst is not working up to its potential and that the concentration drops rather sharply toward the interior of the particle. Note therefore that <u>very active</u> catalysts tend to have <u>low</u> effectiveness factors, while <u>inactive</u> catalysts tend to have <u>high</u> effectiveness factors. The most effective use of the catalyst can be made by using the smallest acceptable particle size. Further, since effectiveness factors of very active catalysts are low, it is desirable to place expensive components such as platinum or palladium in a thin layer around the exterior surface of the pellet. If the expensive metal is placed near the pellet center, it may see very little reactant.

The solution to the diffusion-reaction equation has been carried out for reactions of other orders, e.g., 0 and 2nd and for other simple geometries. The results are summarized by Satterfield and others.

For a flat plate with the edges and one side sealed and reactant contacting the other side, the resulting equation is

$$D_{eff} \frac{d^2 C}{dx^2} = k_v C_A^m \qquad (3.52)$$

$$\phi_L = L \sqrt{\frac{k_v C_{S,A}^{m-1}}{D_{eff}}} \qquad (3.53)$$

where L is the plate thickness. For first order reaction in a flat plate

$$\eta = \frac{\tan \phi_L}{\phi_L} \qquad \text{at high values of } \phi_L \qquad (3.54)$$

$$\eta = \frac{1}{L} \qquad (3.55)$$

For a Thiele modulus ϕ_L which is greater about 2, that is $\eta < 0.45$, <u>the functions all converge to one line if the Thiele modulus is defined in terms of the ratio of the volume of the</u>

Mass transfer

particle to the surface area through which the reactants diffuse into the particle. The fit is not bad for the entire range of ϕ_L though. This results in the relation $\phi_L = 1/3 \, \phi_S$ or $\phi_S = 3\phi_L$ which allows use of the same plot for several geometries. Figure 3.7 is a plot of the effectiveness factor vs. $3\phi_L$ or ϕ_S plotted logarithmically. Note in Figure 3.7 that the flat plate and sphere do fall quite close to one another. If ϕ_S is substantially less than about 1.0 ($\phi_L < 1/3$); η is essentially equal to 1.0, and the concentration does not decrease significantly toward the interior of the particle; that is diffusional effects are not important. In this case the kinetics of the reaction are not affected, and the rate is not reduced by the diffusional process.

However if the Thiele modulus is such that the effectiveness factor, η, is less than 0.1, then $\eta \cong 1/\phi_L$. This is the region in which diffusional effects are important.

Without diffusional effects the reaction rate per unit catalyst volume is

$$-\frac{1}{V_c} \frac{dn_A}{dt} = \text{rate} = k_v \, C_{A,S}^m \tag{3.56}$$

where V_c = volume of catalyst and

$-\frac{1}{V_c} \frac{dn_A}{dt}$ = observed rate of reaction per unit catalyst volume $\left[\dfrac{\text{moles A reacted}}{\text{sec} \cdot \text{cm}^3 \text{ pellet volume}}\right]$

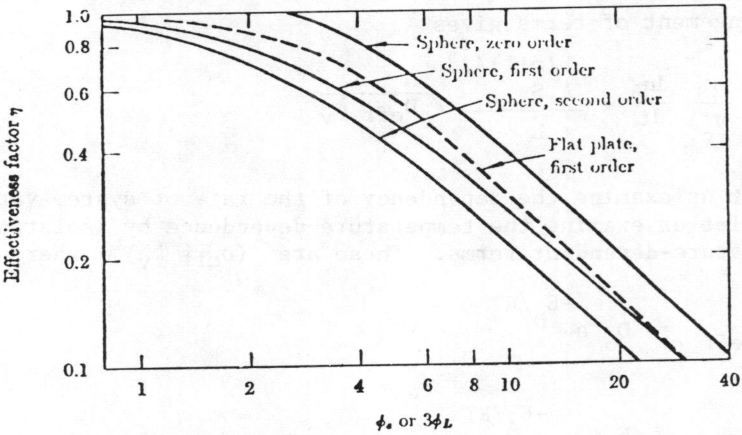

Figure 3.7 Effectiveness factors for reaction in spheres and in flat plates with power-law kinetics. For spheres use ϕ_S and for flat plates use $3\phi_L$ where ϕ_S is defined by Equation (3.33) and ϕ_L is defined by Equation (3.53). (Figure from C. N. Satterfield.)

Since the effectiveness factor η is defined as

$$\eta = \frac{\text{observed rate}}{\text{rate without any transport limitations}} \quad (3.57)$$

(the observed rate) = (rate without any transport limitations)(η)

$$(3.57')$$

Equation (3.56) is rewritten for the more general case where diffusional limitations <u>may</u> be important to give

$$-\frac{1}{V_c}\frac{dn_A}{dt} = k_v c_{A,S}^m \quad (3.58)$$

Equation (3.58) is the general equation used in all design calculations.

In the situation for which substitution into Equation (3.58) gives $\eta < 0.1$, $\eta = 1/\phi_L$

$$-\frac{1}{V_c}\frac{dn_A}{dt} = k_v c_{A,S}^m \frac{1}{\phi_L} = k_v \frac{c^m \sqrt{D_{eff}}}{L\sqrt{k_v c_{A,S}^{m-1}}} \quad (3.59)$$

Rearrangement of terms gives

$$-\frac{1}{V_c}\frac{dn_A}{dt} = \frac{c_{A,S}^{[(m+1)/2]}}{L}\sqrt{D_{eff}\, k_v} \quad (3.60)$$

Now let us examine the dependency of the rate of system variables. First let us examine the temperature dependence by isolating the temperature-dependent terms. These are $(D_{eff}\, k_v)^{1/2}$ where

$$D_{eff} = D_o\, e^{-E_D/RT} \quad (3.61)$$

and

$$k_v = k_{vo}\, e^{-E_A/RT} \quad (3.62)$$

Thus the temperature dependence of the rate now becomes

$$\text{rate} \propto D_o\, k_{vo}\, e^{-(E_A+E_D)/2RT} \quad (3.63)$$

The activation energy as determined from an Arrhenius plot of rate vs. 1/T is the arithmetic mean of that for diffusion and for chemical reaction on the surface.

$$E_{obs} = \frac{(E_A + E_D)}{2} \qquad (3.64)$$

Typically E_D = 1 - 3 kcal/mole and E = 50 kcal/mole for example. Thus the observed activation energy equals about one-half of the true or "intrinsic" activation energy. Thus the presence of diffusional limitations can easily lead to determination of an incorrect activation energy for chemical reaction. However by studying the reaction over a broad temperature range the appearance of a decrease in the activation energy to ½ of its value at a lower temperature is a good indication of the presence of diffusional limitations. An activation energy which is lower than experience would lead you to believe that it should is an indication of diffusional limitations and should always be suspect.

However let us examine the reaction order or the effect of concentration on the observed rate.

$$\text{rate} \propto C_{A,S}^{[(m+1)/2]} \qquad (3.65)$$

Note that the observed reaction order is no longer the same as the true or intrinsic reaction order for surface reaction. Thus a first-order reaction is 1st order, but a zero order reaction becomes ½ order and a 2nd order reaction becomes 3/2 order.

True Order	Observed Order (in the presence of significant diffusional limitations $\eta < 0.1$)
0	1/2
1	1
2	3/2

In the majority of catalysts most of the active surface is located in micropore regions where Knudsen diffusion prevails, and the basic assumptions apply. For bulk diffusion there may be added complications, but for A → B where the total concentration in the pores is constant, the above behavior is valid.

Note that diffusional limitations alter the observed kinetics of a reaction. This was not generally realized, although known by a few people much earlier, and thus much of the catalytic literature prior to the mid-50's is rather useless because the kinetics observed may have been altered by diffusion. Note also that even at very low η values diffusion does not "control" since the diffusion-reaction process is not a series process as is the case for mass transfer to the surface but is a coupled process. Again there is a transition region in which the activation energy and reaction order change.

B. Determination of effectiveness factors

ϕ_S and ϕ_L contain the quantities k_v and D_{eff} both of which are difficult to obtain quantitatively. Furthermore we usually have a measured value of the reaction rate and want to know how important diffusion is and then what the intrinsic reaction rate constant k_v is. This can be handled as follows. Note that $\phi_S = R\sqrt{k_v/D_{eff}}$ and that three quantities R, D_{eff}, and k_v are important in determining η; no one single parameter alone determines η. R is the one parameter which for a given catalyst is the most easily changed and is accurately known. k_v and D_{eff} are much more difficult. R can be changed simply by crushing and screening the catalyst to different size fractions and then using these in kinetic studies. The kinetic studies should determine the rate of reaction per gram of catalyst or per unit volume of catalyst.

The best method of determining the importance of diffusional limitations is to crush and screen the catalyst particles to smaller and smaller sizes until further reduction in size produces no further increase in rate of reaction per unit of catalyst. When this condition is found the ratio of the rate per unit volume of catalyst for the larger size to that for the smaller size catalyst is equal to the effectiveness factor, η, of the larger size particles. Show this from the above equations.

If you have data on two particle sizes only, then for size 1

$$\left(-\frac{1}{V_c} \frac{dn_A}{dt}\right)_1 = k_{v_1} C_{A,S_1}^m \eta_1 \qquad (3.66)$$

for size 2

$$\left(-\frac{1}{V_c} \frac{dn_A}{dt}\right)_2 = k_{v_2} C_{A,S_2}^m \eta_2 \qquad (3.67)$$

Mass transfer

Take the ratio of the two, noting that

$$k_{v_1} \equiv k_{v_2} \quad \text{and} \quad C^m_{A,S_1} \equiv C^m_{A,S_2} \tag{3.68}$$

$$\frac{\left(-\frac{1}{V_c}\frac{dn_A}{dt}\right)_1}{\left(-\frac{1}{V_c}\frac{dn_A}{dt}\right)_2} = \frac{\eta_1}{\eta_2} \tag{3.69}$$

Thus we have information on the ratio of effectiveness factors directly from rate information and in fact only a single rate measurement is needed for each particle size. If rate 2 is for very small particles such that there are no diffusional limitations present, $\eta_2 = 1.0$, and the effectiveness factor for the larger particle size is simply

$$\eta_1 = \frac{\left(-\frac{1}{V_c}\frac{dn_A}{dt}\right)_1}{\left(-\frac{1}{V_c}\frac{dn_A}{dt}\right)_2} \tag{3.70}$$

For the same catalyst D_{eff} and k_v are the same independent of size fraction and thus

$$\frac{\phi_1}{\phi_2} = \frac{R_1}{R_2} \tag{3.71}$$

For cases in which the smallest particle size catalyst may still have diffusional limitations present the effectiveness factor can be determined by combining Equation (3.69) and (3.71) with Figure 3.10 and solving for η_1 and η_2 by trial and error. This approach does not apply for a region in which the effectiveness factor is very small for both sizes (that is in the linear region of η vs. ϕ).

If you have data on only one catalyst particle size and wish to estimate the possible role of diffusion, you have the following problem: k_v, the intrinsic reaction rate constant, probably is not known, is not estimatable and is extremely hard to determine accurately if there is any mass transfer limitations involved. We therefore set out to eliminate k_v. We start with

$$-\frac{1}{V_c}\frac{dn_A}{dt} = k_v^1 C_{A,S}^m \eta \tag{3.72}$$

and

$$\phi_S = R\sqrt{k_v C_{A,S}^{m-1}/D_{eff}} \tag{3.73}$$

These are combined to give

$$-\frac{1}{V_c}\frac{dn}{dt} = \frac{\phi_s^2 D_{eff} C_{A,S}^m}{R^2 C_{A,S}^{m-1}} \tag{3.74}$$

Rearranging Equation (3.74)

$$\left[\frac{R^2}{D_{eff}}\overbrace{\left(-\frac{1}{V_c}\frac{dn}{dt}\right)}^{\text{(observed rate)}}\frac{1}{C_{A,S}}\right] = \theta_s^2 \eta = \Phi_s \tag{3.75}$$

This is defined as Φ_S and is called a <u>modified Thiele Modulus</u> for a spherical particle. For spherical geometry

$$\Phi = \frac{R^2}{D_{eff}}\left(-\frac{1}{V}\frac{dn}{dt}\right)\frac{1}{C_{A,S}} \tag{3.76}$$

for flat-plate geometry

$$\Phi_L = \frac{L^2}{D_{eff}}\left(-\frac{1}{V}\frac{dn}{dt}\right)\frac{1}{C_{A,S}} \tag{3.77}$$

These expressions are now in terms of values which are either directly measurable or are easily calculable. For low values of η, $\eta < 0.10$ and flat-plate geometry

$$\Phi_L = \phi_L^2 \eta \quad \text{and} \quad \eta = \frac{1}{\phi_L} = \frac{1}{\Phi_L} \tag{3.78}$$

for spherical geometry

$$\Phi_S = \phi_S^2 \eta \quad \text{and} \quad \eta = \frac{3}{\phi_S} = \frac{9}{\Phi_S} \tag{3.79}$$

Figure 3.8 is a plot of effectiveness factor vs. modulus for first-order reaction in a sphere. For now, only the $\beta = 0$ line is of interest. Figures 3.12 and 3.13 are the same for $\beta = 0$ but are for reactions having different activation energies and thermal effects ΔH_R and are required in considering non-isothermal reactions.

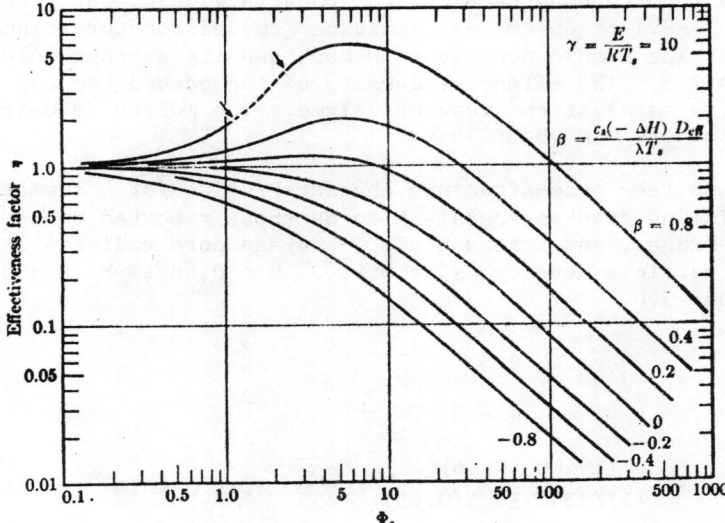

Figure 3.8 Effectiveness factor η as a function of the modified Thiele modulus Φ_s for first-order reaction in a sphere and $\gamma = 10$. (Figure from C. N. Satterfield.)

Example 3.2ᵃ Estimation of Effectiveness Factor η :
Isothermal, First-Order Kinetics

Archibald, May, and Greensfelder studied the rate of catalytic cracking of a West Texas gas oil at 500° and 630° C and 1 atm pressure by passing the vaporized feed through a packed bed containing a silica-alumina cracking catalyst of each of several sizes ranging from 8 to 14 mesh to 35 to 48 mesh. They report that at 630° C the apparent catalyst activity was directly proportional to the external surface for the three larger catalyst sizes studies. This implies that the catalyst is operating at a relatively low effectiveness factor. We can check this by suitable calculations based on their run on 8 to 14 mesh catalyst at 630° C. The average particle radius may be taken as 0.088 cm.

They report 50 percent conversion at a liquid hourly space velocity (LHSV) of 60 cm³ of liquid per cm³ of reactor volume per hour. The liquid density is 0.869, and its average molecular weight is 255. The effective density of the packed bed was about 0.7 g catalyst/cm³ reactor volume. The molecular weight of the products was about 70.

Let us take pore-structure characteristics for a commercial silica-alumina cracking catalyst to be those reported by Johnson, Kreger, and Erickson viz., average pore radius = 28 Å , catalyst particle density $\rho_p = 0.95$, $\theta = 0.46$, $S_g = 338$ m²/g , τ is about 3.0.

$$D_{eff} = 19400 \; \frac{\theta^2}{\tau S_g \rho_p} \; \sqrt{T/M}$$

$$= \frac{(19400)(0.46)^2}{(3)(338 \times 10^4)(0.95)} \sqrt{\frac{903}{255}} = 8.0 \times 10^{-4} \; cm^2/sec.$$

The rate of reaction is 50 percent of the rate of feed

$$\left(-\frac{1}{V_c} \frac{dn}{dt}\right) = \frac{(60)(0.869)}{255} \left(\frac{1}{3600}\right) \left(\frac{1}{0.7}\right) (0.95)(0.5)$$

$$= 3.86 \times 10^{-5} \; g\text{-mol/sec} \cdot cm^3 \; \text{pellet volume}$$

For the average concentration of reactant outside the pellets, an arithmetic average of inlet and exit concentrations is sufficiently precise for this example. This corresponds to conditions at 25 percent conversion.

Example 3.2 (Continued)

Each mole of gas oil produces 255/70 = 3.64 moles of products. Per 100 moles entering, at 25 percent conversion there remain 75 moles of gas oil and (25)(3.64) = 91 moles of products, for a total of 166 moles.

The average reactant concentration is therefore

$$\frac{1}{22400} \left(\frac{273}{903}\right) \left(\frac{75}{166}\right) = 0.61 \times 10^{-5} \text{ g-mole/cm}^3$$

The average particle radius is 0.088 cm. Using Equation (3.76)

$$\Phi_s = \frac{(0.088)^2 \, (3.86 \times 10^{-5})}{8 \; 10^{-4} \; (0.61 \times 10^{-5})} = 61$$

For moderate degrees of conversion the rate is approximately first-order. From the $\beta = 0$ curve of Figure 3.8 the predicted value of η is 0.12. The presence of relatively severe diffusional limitations is confirmed.

C. Diffusion and reaction, the non-isothermal case

To determine the situation for non-isothermal conditions we consider a spherical catalyst particle as before and assume

i) That diffusion can be represented by a single effective diffusion coefficient, D_{eff}, which is constant.

ii) That the thermal conductivity is constant throughout and is represented by λ.

iii) Reaction is $A \to B$ with a heat reaction ΔH_R.

Then the rate at which reactant A is transported across the spherical boundary at r determines the rate of heat generation within the sphere bounded by r.

$$\text{reactant flux in} = -D_{eff} \, (4\pi r^2) \, \left.\frac{dC_A}{dr}\right|_r \qquad (3.80)$$

---

[a] Example from C. N. Satterfield

$$\text{heat generation rate} = -D_{eff}(4\pi r^2)(\Delta H_R)\left.\frac{dC_A}{dr}\right|_r \tag{3.81}$$

where ΔH_R = enthalpy change upon reaction, cal/mole.

The rate of heat generaged within the sphere of radius r equals the rate at which heat must be transported out of the sphere of radius r through the boundary at r ; at steady state

$$\text{heat flux out} = -\lambda(4\pi r^2)\left.\frac{dT}{dr}\right|_r \tag{3.82}$$

λ = thermal conductivity (cal/sec cm · K)

Equating the rate of heat generation with the rate of heat transport out gives

$$-D_{eff}(4\pi r^2)(\Delta H_R)\frac{dC_A}{dr} = -\lambda(4\pi r^2)\frac{dT}{dr} \tag{3.83}$$

which gives

$$D_{eff}(\Delta H_R)\frac{dC_A}{dr} = \lambda\frac{dT}{dr} \tag{3.84}$$

at $r = R$ $C = C_S$ and $T = T_S$ T_S & C_S = values at surface

Integrating Equation (3.84) gives

$$D_{eff}(\Delta H_R)(C_A - C_{AS}) = \lambda(T - T_S) \tag{3.85}$$

This equation only gives a relation between temperature and concentration; it does not say anything about the temperature as a function of position in the particle. This equation does give, however, the maximum temperature which can be reached in the pellet; this occurs when the concentration drops to zero at some point in the pellet. In that case Equation (3.85), which gives the temperature difference

$$\frac{D_{eff}(-\Delta H_R)(C_{A,S} - C_A)}{\lambda} = (T - T_S) = \Delta T \tag{3.85'}$$

becomes

$$\Delta T_{MAX} = \frac{D_{eff}(-\Delta H_R) C_{A,S}}{\lambda} \tag{3.86}$$

A fractional temperature increase above the surface temperature can be defined as:

$$\frac{(D_{eff})(-\Delta H_R)C_{AS}}{\lambda T_S} \equiv \beta \equiv \frac{T - T_S}{T_S} \qquad (3.87)$$

which represents the maximum temperature which exists in the pellet and is a function of the rate of heat generation. This is really the potential which the system exhibits for temperature increase which of course affects the rate.

It is then necessary to define a second quantity $\gamma \equiv E_A/RT_S$ which is the exponent in the Arrhenius term and which determines the rate of change of reaction rate with temperature change. This term quantifies the effect of the potential as expressed in β (Equation (3.87)) on the rate of reaction. These two added terms, β and γ, now allow Figure 3.8 or others to be used for calculation of the effectiveness factor for non-isothermal reactions.

Figure 3.8 shows that the effectiveness factor is no longer restricted to values below 1.0 but may have values which can be 5 or more. Factors of 100 or more are possible. This is because rate is a function of temperature and the temperature within the catalyst particle has increased due to the heat of reaction. Note the definition of η

$$\eta = \frac{\text{observed rate of reaction}}{\text{rate of reaction if entire interior surface area of catalyst were at the temperature of the surface } T_S \text{ and had a uniform concentration equal to } C_S, \text{ the concentration at the surface.}}$$

Thus if temperature increases cause the rate of reaction to increase faster than concentration decreases (decrease the rate), the effectiveness factor will be greater than 1.

In the isothermal case the observed activation energy is about ½ the true intrinsic value when diffusional limitations are severe, i.e., $\eta < 0.1$. For the non-isothermal case with $\eta > 1$ the observed activation energy may be higher than the true intrinsic value. Heat effects become less important as the reaction order is increased.

TEXTS EMPHASIZING MASS TRANSFER

Satterfield, C. N., "Mass Transfer in Heterogeneous Catalysis," M.I.T. Press, 1970. Now available only through Professor Satterfield, Dept. Chem. Eng., M.I.T., Cambridge, Mass.

Sherwood, T. K., Pigford, R. L., and Wilke, C. R., "Mass Transfer," McGraw-Hill Book Co., 1975.

Astarita, G., "Mass Transfer with Chemical Reaction," Elsevier Pub. Co., 1967. Out of print—available through Dept. Chem. Eng., University of Delaware, Newark, Delaware.

Peterson, E. E., "Chemical Reaction Analysis," Prentice-Hall, 1965.

Carberry, J. J., "Chemical Reaction Engineering," McGraw-Hill, 1976.

Faust, A. S., Wenzel, L. A., Clump, C. W., Maus, L., Anderson, L. B., "Principles of Unit Operations," John Wiley, 1960.

Bird, B. R., Stewart, W. E., and Lightfoot, E. N., "Transport Phenomena," John Wiley, 1960.

Mass transfer 111

NOMENCLATURE

- a = area per unit volume, cm^2/cm^3
- C_i = concentration of i, g-moles/cm^3
- C_p = heat capacity, cal/g-mole
- d = diameter, cm
- D_{12} = binary diffusion coefficient of 1 in 2
- f = Fanning friction factor
- G = mass velocity of fluid; g/sec·cm^2 of superficial bed cross-section
- g = gravitational acceleration rate, cm/sec^2
- H = gas hold-up, cm^3/cm^3 of expanded slurry bed
- h = heat transfer coefficient, g-cal/(sec·cm^2·C)
- J_i = molar flux of i, relative to a no-net flux coordinate system, g-moles/sec – cm^2
- k = Boltzmann constant
- $\mathscr{k}_m \equiv k_m$ = mass transfer coefficient, cm/sec
- k_r = fractional reduction in the effective diffusivity
- k_s = reaction rate constant per unit surface area, (g-moles)$^{m-1}$(cm)$^{3m-2}$(sec)$^{-1}$
- k_t = thermal conductivity, g-cal/(sec·cm^2)(C/cm)
- $\mathscr{k}_v \equiv k_v$ = reaction rate constant per unit volume of catalyst, (g-moles)$^{1-m}$(cm)$^{3m-2}$(sec)$^{-1}$
- M_i = molecular weight of species i
- m_i = molar flow rate of , moles i/sec
- N_i = molar flux of i relative to a fixed coordinate system, g-moles/sec·cm^2
- P = total pressure, atm
- p = partial pressure, atm

Q = liquid flow rate, cm^3/sec

r = radius, cm

r_e = equivalent radius of the pore, cm

r_i = rate of reaction of species i, g-moles/g cat-sec

R = sphere radius, cm

S_g = total surface area of porous solid per gram, cm^2/g

S_v = pore surface area per unit volume of porous solid, cm^2/cm^3

T = temperature, °K

T_c = critical temperature, °K

U = velocity, cm/sec

u = mean molecular velocity, cm/sec

u_m = velocity of a plane of no net molar flux, cm/sec

V_b = molar volume at the normal boiling point, cm^3/g-moles

V_g = total pore volume of a porous solid per gram, cm^3/g

V_s = reactor volume, cm^3

W_c = weight of the catalyst, g

X = distance, cm

X_i = mole fraction species i in the liquid phase

$\overline{X}_i$ = fractional conversion of i

Y_i = mole fraction species in the gas phase

$\left(-\dfrac{1}{Vc}\dfrac{dn}{dt}\right)$ = rate of reaction per unit volume of catalyst, g-moles/cm^3 cat-sec

GREEK

δ = film thickness, cm

ε = Lennard-Jones Force Constant

ε = void fraction (between particles) of the packed bed

Θ = porosity of porous solid

η = effectiveness factor

ρ = density g/cm^3

ϕ = Thiele diffusion modulus

Φ = modified Thiele modulus

ϕ = Lennard-Jones force constant

Ω = Collision Integral for evaluating diffusion coefficients

μ = viscosity, poises

ν = kinematic viscosity, μ/ρ, cm/sec^2

SUBSCRIPTS

B = bulk

b = bubble

g = gas

i = species i

i = interface, as a second subscript

K = Knudsen Diffusion

M = molar

o = initial time or position

p = pellet or particle

S = surface

T = total

DIMENSIONLESS GROUPS

Nusselt No. $= hd/k_t$

Peclet No. $= Pe = d_p U_{ave}/D_{12} = Re \cdot Sc$

Reynolds No. $= Re = \dfrac{d\, U_{ave}\, \rho}{\mu} = \dfrac{d\, U_{ave}}{\nu} = \dfrac{d\, G}{\mu}$

Schmidt No. $= Sc = \mu/\rho \cdot D_{12} = \nu/D$

Sherwood No. $= Sh = k_m d_p / D_{12}$

Stanton No. $= St = k_m / U_{ave}$

$\qquad\quad = (k_m/U_{ave})\, Sc^{2/3}$

$\qquad\quad = (h/Cp \cdot P \cdot \mu_{ave})\, Pr^{2/3}$

SUPERSCRIPTS

m = reaction order

FIXED BED CATALYTIC REACTORS

Prof.dr.ir. G.F. Froment

Laboratorium voor Petrochemische Techniek
Rijksuniversiteit, Gent, Belgium.

The major part of the industrial catalytic processes is carried out in fixed bed reactors. This type of reactor is generally relatively simple in construction and operation as compared with fluidized or moving bed reactors. Also, the scaling up rules are rather well understood, so that the risk of extrapolating from laboratory or pilot reactors to the industrial size is smaller than with the other types. Laboratory kinetic studies of catalytic processes, on the other hand, are also generally carried out in fixed bed reactors, since the flow pattern is close to ideality and the data treatment relatively easy.
This brief review considers both industrial and laboratory fixed bed reactors for a single fluid phase. The discussion of industrial reactors is not model oriented [2,3,4] but instead centered around the factors determining the selection of one type of fixed bed reactor rather than another. The discussion of the laboratory reactors mainly deals with the precautions which have to be taken in order to obtain accurate kinetic data.

I. INDUSTRIAL FIXED BED CATALYTIC REACTORS

The simplest type of reactor to be considered is the adiabatic reactor. The construction and operation are straightforward : the catalyst is contained in a large diameter vessel and the heat exchanged with the surroundings is only a very small fraction of the heat

carried by the gases flowing through the bed.
For a single reaction and steady state conditions such a reactor is generally modeled by the following equations :

$$-u_s \frac{dC}{dz} = \rho_B r_A \qquad (1)$$

$$u_s \rho_g C_p \frac{dT}{dz} = (-\Delta H) \rho_B r_A \qquad (2)$$

$$-\frac{dp_t}{dz} = \frac{2 f \rho_g u_s^2}{g d_p} \qquad (3)$$

with initial conditions : at $z=0$: $C=C_o ; T=T_o ; p_t = p_{t_o}$ (4).

This model corresponds to model A.I in Froment's classification [1,2,3,4]. The first equation is the continuity equation for the feed component A, the second results from a heat balance over a longitudinal increment of the reactor, the third expresses the pressure drop according to Fanning. Pressure drop equations for packed beds have been proposed by several authors[5,6]. A recent review led Hicks to the following equation for the friction factor for flow through packed beds of spheres [7]

$$f = 3.4 \frac{(1-\varepsilon)^{1.2}}{\varepsilon^3} \cdot Re^{-0.2} \qquad (5)$$

Leva and also Brownell's equations [5,6] are valid for other shapes, too.
In equation (1) r_A is the true chemical rate of the reaction. The model therefore ignores temperature and concentration differences between the gas and the solid. If such differences would occur over the film surrounding the solid, separate continuity and energy equations for the gas phase and the solid would have to be written :
Fluid :

$$-u_s \frac{dC}{dz} = k_g a_v (C-C_s^s) \qquad (6)$$

$$u_s \rho_g C_p \frac{dT}{dz} = h_f a_v (T_s^s - T) \qquad (7)$$

Solid :

$$\rho_B r_A = k_g a_v (C-C_s^s) \qquad (8)$$

$$(-\Delta H) \rho_B r_A = h_f a_v (T_s^s - T) \qquad (9)$$

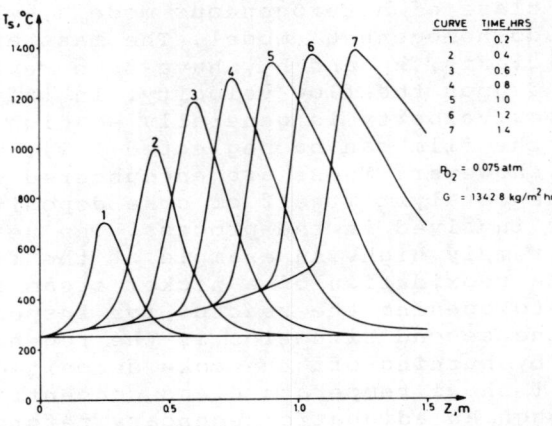
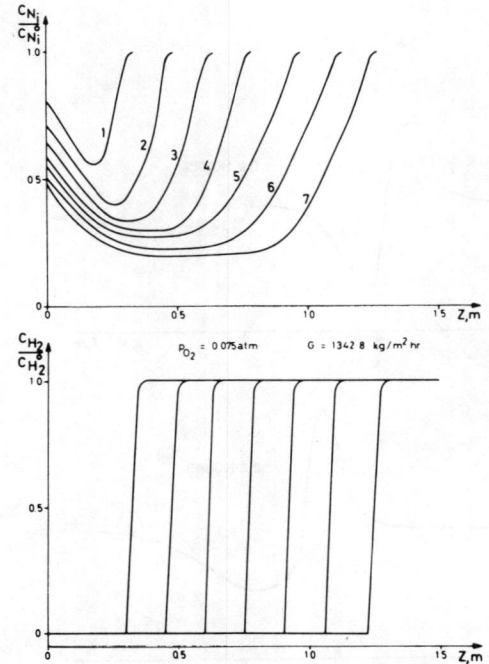

Fig.1. Reoxidation of a steam reforming catalyst in an adiabatic secondary reformer [10]
a-Solid temperature profiles
b-Nickel- and hydrogen concentration profiles.

with the initial conditions (4).
This is model (B.I) of Froment's classification. It belongs to the class of heterogeneous models, whereas (A.I) is a pseudo-homogeneous model. The mass and heat transfer coefficients, k_g and h_f, have been reviewed[8]. They both depend upon the flow velocity. In industrial reactors the flow velocity is generally so high that the ΔT and ΔC over the film can be neglected[1,9]. There are exceptions, however. These are encountered when a component of the catalyst itself or coke deposited on the catalyst is involved in the process. The heat effect may then be extremely high. An example of the first situation is the reoxidation of a nickel steam reforming catalyst prior to opening the reactor for inspection[10]. An example of the second situation is the regeneration of a fixed bed by burning off the coke deposited on the catalyst. Fig. 1 shows temperature and concentration transients through an adiabatic secondary reformer in which the catalyst is reoxidized [10].

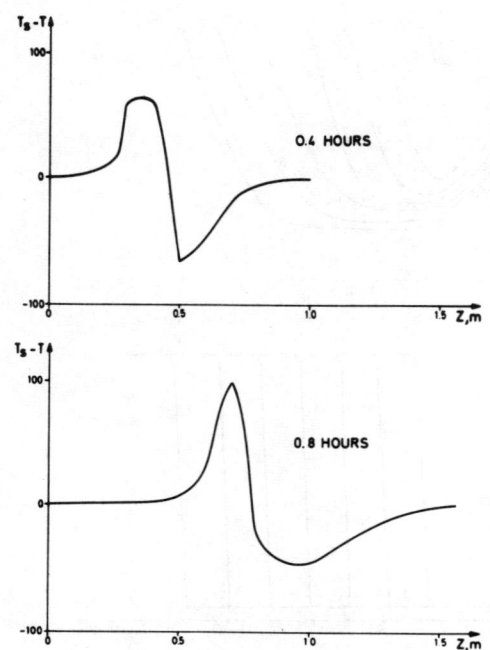

Fig. 2. Temperature differences between solid and gas in the reactor after 0.4 and 0.8 hours [10].

Fixed bed catalytic reactors

Fig. 2 shows the temperature difference between the gas and the solid. Gradients, mainly of concentration, are much more common inside the catalyst particle, however. To account for this again a distinction has to be made between the gas and the solid. The gas phase equations are unchanged with respect to (6) and (7).

The solid phase equations have to account for the finite rate of mass and heat transfer inside the particle. Bulk flow, molecules and Knudsen diffusion contribute to the mass transfer. The three mechanisms are lumped into a single mechanism called effective diffusion and considered to obey Fick's law. The heat transfer is expressed in terms of an effective conduction obeying Fourrier's law.

At steady state the solid state equations may then be written :

$$\frac{D_e}{\xi^2} \frac{d}{d\xi} \left(\xi^2 \frac{dC_s}{d\xi} \right) - \rho_s r_A (C_s, T_s) = 0 \qquad (10)$$

$$\frac{\lambda_e}{\xi^2} \frac{d}{d\xi} \left(\xi^2 \frac{dT_s}{d\xi} \right) + \rho_s (-\Delta H) r_A (C_s, T_s) = 0 \qquad (11)$$

with boundary conditions :

$$C = C_o \; ; \; T = T_o \quad \text{at } z = 0 \qquad (12)$$

$$\frac{dC_s}{d\xi} = \frac{dT_s}{d\xi} = 0 \quad \text{at } \xi = 0 \qquad (13)$$

$$k_g (C_s^s - C) = -D_e \frac{dC_s}{d\xi} \quad \text{at } \xi = \frac{d_p}{2} \qquad (14)$$

$$h_f (T_s^s - T) = -\lambda_e \frac{dT_s}{d\xi} \quad \text{at } \xi = \frac{d_p}{2} \qquad (15)$$

In practical cases temperature gradients inside the particle can be neglected [1], so that (11) may be dropped from the system of model equations. When interfacial gradients are negligible the equations linking the gas and solid phase equations (14) and (15) are replaced by $C_s^s = C$ and $T_s^s = T$.

For a simple first order equation and an isothermal particle equation (10) can be integrated analytically to yield the well known relation first derived by Thiele for $C_s^s = C$ and $T_s^s = T$:

$$\eta = \frac{3}{\phi^2} (\phi \coth \phi - 1) \qquad (16)$$

in which η is the effectiveness factor and $\emptyset = \dfrac{d_p}{2}\sqrt{\dfrac{k(T_s^s)}{D_e}}$ the modulus. $\eta = 1$ when there is no concentration gradient inside the particle and for $\emptyset = 0$ and gradually decreases as $\emptyset$ increases. As defined here η is a factor multiplying the reaction rate at the particle surface conditions to yield the rate which is actually experienced when the concentrations inside the particle differ from those at the surface. The modulus is related to the ratio of the reaction rate at surface conditions to the mass transfer rate towards the inside.
The use of η reduces the system (6),(7),(10),(11) to the system comprising (6),(7) and

$$k_g a_v (C - C_s^s) = \eta \rho_B r_A (C_s^s, T_s^s) \qquad (17)$$

$$h_f a_v (T_s^s - T) = \eta (-\Delta H) \rho_B r_A (C_s^s, T_s^s) \qquad (18)$$

for the solid field equations.
Through the modulus the effectiveness factor depends upon the local conditions. It therefore varies through the reactor and has to be computed in each node of the grid used in the numerical integration of the fluid field equations (6),(7). When there is no analytical solution for η there is no gain in its use from the reactor design and computational point of view. The only advantage of η is then the possibility of characterizing the situation inside the particle at a given point in the reactor by means of a single number. Notice that in the absence of interfacial gradients the system (6),(7),(17) and (18) reduces to :

$$- u_s \dfrac{dC}{dz} = \eta \rho_B r_A (C, T) \qquad (19)$$

$$u_s \rho_g C_p \dfrac{dT}{dz} = \eta_G \rho_B r_A (C, T) \qquad (20)$$

Many industrial reactions are carried out with such catalysts and in such conditions that $\eta < 1$:methanol synthesis, ammonia synthesis in many cases, ethylbenzene dehydrogenation into styrene, SO_3 synthesis, 1-butene dehydrogenation into butadiene. From the definition it is clear that one way of decreasing the modulus and thereby increasing η and making a better use of the catalyst is to reduce the particle size. There are limits to this, however : the pressure drop through the bed rapidly increases when the catalyst size is reduced and the compression costs may become excessive.
In fixed bed reactors the flow is generally axial and downwards rather than upwards, to avoid fluidization.

Fixed bed catalytic reactors

Some ammonia synthesis reactors and catalytic reformers have radial flow, however. This is done to limit the pressure drop over the bed, a factor which is particularly important for processes operating with recycle. The larger the pressure drop the larger the centrifugal recycle compressor. Radial flow converters enable the catalyst to be spread over a wide area, thereby reducing the bed depth. For isothermal conditions at least this does not affect the conversion: from (1) it can easily be shown that this is determined by the ratio W/F_{A_o}, not by the bed thickness. Reasoning in terms of residence or contact time, as is done too often, can be quite misleading when dealing with catalytic processes. Evidently, the bed depth has to remain sufficient to avoid chanelling otherwise the conversion will be affected.

The model equations for radial flow reactors are the same as those written above with axial flow in mind. Of course, the gas velocity depends upon the radial position, but this does not affect the conversion when there are no interfacial gradients. These are normally not encountered in industrial reactors, as mentioned above already.

Many processes are plagued with catalyst deactivation caused by coke deposition, poisoning or sintering. Guard reactors can help in limiting poisoning, but the coke deposition results from undesired side reactions which cannot be avoided and which can lead to very fast deactivation. In that case two or more fixed bed reactors are linked in parallel so that one is being regenerated while the others are in production. One example of such an operation is 1-butene dehydrogenation into butadiene. Catalytic cracking of gasoil used to be another example many years ago, before the cyclic operation was replaced by fluidized bed operation, allowing continuous removal of the coke in a regenerator. Accounting for coke deposition in the design of a new reactor or in the analysis of an existing one necessitates kinetic equations for the coke formation and these have to be coupled with the model equations given above. The non steady state behavior introduced by the coking in fixed bed reactors has been analyzed by Froment and Bischoff [11,12] and by Butt [13]. Weekman and coworkers have applied the theory to industrial situations [14]. Returning now to (1) and (2) it is easily seen that their combination leads to a linear relation:

$$T = A \pm B \, \Delta C \tag{21}$$

provided that ΔH may be considered a constant and taking an average value for the pressure.

Equation (21) clearly reveals the limitations of an

adiabatic bed when the reaction has a pronounced heat effect. When the reaction is endothermic the temperature drop in the bed may extinguish the reaction. When the reaction is very exothermic temperatures may be reached which may be detrimental either for the catalyst or for the selectivity or which may lead into explosive ranges. In such cases heat has to be exchanged and this can be done discretely or continuously.
The first solution leads to a multibed adiabatic reactor (MBAR), consisting of several adiabatic beds with intermediate cooling. The number of beds depends upon the allowable ΔT. Normally the number of beds is limited to four. Examples of MBAR are encountered in catalytic reforming for increasing the octane number of gasoline, in ethylbenzene dehydrogenation into styrene, in sulfuric acid, ammonia and methanol synthesis, in CO-conversion. The intermediate cooling can be indirect or direct. In the first case use is made of heat exchangers, which are internal as in Fauser-type ammonia or methanol synthesis reactors or external as in sulfuric acid synthesis reactors.
In both cases there is a problem of redistribution of the gas over the downstream bed. If this is not adequately achieved certain parts of the bed may get insufficient flow, causing runaway e.g. when the reaction is exothermic.
Intermediate heat exchange can also be achieved by injection of feed or diluent. In the exothermic sulfuric acid synthesis air is injected, but this is limited to the two first stages, to avoid too strong a dilution which would prevent reaching the desired conversion at the exit. The problem of mixing the effluent of a bed with the external shot to come to a uniform feed to the next bed requires careful engineering. Modern ammonia synthesis MBAR consist of three to four beds with intermediate cooling by cold feed injection. After condensation of the ammonia produced in the pass the effluent is recycled.
The intermediate heat exchangers provide a certain flexibility to a MBAR as a whole, but there is no grasp on the individual beds as such. If the amount of catalyst in a given bed has not been accurately calculated the temperature effect in the bed itself may be detrimental and there is no way the downstream heat exchanger can interfere with this. This is particularly true for the first bed, where the reaction is generally fast and the reactor sensitive to small perturbations in the feed conditions. The optimal sizing of the different beds and of the intermediate heat exchangers can be calculated with considerable riguor and sophisti-

cation by modern methods of optimization [15].
The continuous exchange of heat requires dividing the catalyst over a number of tubes with a relatively small diameter. The heat exchange can be achieved in several ways, depending upon the temperature level. In steam reforming for hydrogen production the tubes have an internal diameter of 10 cm. They are located in a gas-fired furnace ensuring heat fluxes of the order of 85 $kJ/m^2 S$ at temperatures inside the tubes ranging from 550° to 800°C. In phthalic anhydride synthesis by o.xylene oxidation, a strongly exothermic reaction, a molten salt bath is used to transfer the heat of reaction to a waste heat boiler. In Fisher-Tropsch synthesis steam is generated directly around the tubes. In the last two examples great care has to be taken to have radial uniformity in the heat exchanging fluid at any vertical position in the vessel to avoid runaway in some of the tubes. This is not so easily achieved in a vessel containing e.g. 10.000 tubes of 2.5 cm internal diameter for phthalic anhydride synthesis. Another problem encountered with multitubular reactors is the maldistribution of the fluid over the tubes. Extreme care has to be taken to uniformly fill the tubes so that the pressure drop is the same over each of the tubes.
In modeling multitubular reactors it is generally assumed that all tubes behave in exactly the same way, which implies that at any height in the vessel the heat exchanging fluid is at a uniform temperature. The system of modeling equations differs from (1)-(2)-(3) only in the energy equation, which now takes the form :

$$u_s \rho_g C_p \frac{dT}{dz} + 4 \frac{U}{d_t}(T-T_r) = (-\Delta H) \rho_B r_A \qquad (22)$$

Heat transfer between the catalyst bed and the tube-wall has been studied by a number of investigators[1,8]. When the heat exchanging fluid exhibits a temperature gradient along the tube T_r becomes a function of z and the following equation has to be added to (1)-(22)-(3):

$$m_r C_{p_r} \frac{dT_r}{dz} = U \pi d_t (T-T_r) \qquad (23)$$

McGreavy & Dunbobbin recently investigated the merits of co-current or counter-current heat exchange in multi-tubular reactors [16].
With some exothermic reactions the heat effect is sufficient to preheat the feed to the reaction temperature. The reactor is then said to operate in an autothermic

way. Ammonia synthesis and o.xylene oxidation are just a couple of examples. The feed can be preheated in two ways : by means of an external exchanger or inside the reactor itself. In modern MBAR ammonia synthesis units the feed is preheated by the effluent in an external heat exchanger. This is also done in multitubular o.xylene oxidation reactors. The reason for this is that the cooling of the reactor tubes by means of the gaseous feed would be insufficient : the high heat transfer rates required for safe operation can only be achieved with molten salts. In ammonia synthesis it is feasible to control the temperature profile in the catalyst bed by a continuous exchange of heat between the feed and the reacting mixture. The energy equation for the reacting mixture, (22), is unchanged, while (23) is slightly modified in the symbols only. The boundary conditions become :

$$C(0) = C_i \quad \text{at } z = 0$$
$$T_f = T \quad \text{at } z = 0 \quad (24)$$
$$T_f = T_i \quad \text{at } z = Z$$

Mathematically this is now a two-point boundary value problem and its solution is not straightforward. The feed-back of heat introduces the possibility of multiplicity of steady state profiles in the reactor. If the reactor were not appropriately started up e.g. drastic temperature excursions could be experienced because of this.

With exothermic equilibrium reactions the higher the temperature the lower the equilibrium conversion will be. A compromise between kinetics and equilibrium location has to be achieved here. The many types of multitubular ammonia synthesis reactors with internal feed preheat which have been proposed reflect the concern of achieving an "ideal" temperature profile that minimizes the amount of catalyst required to come to a given conversion. The relation T-x that would maximize the rate in each point of the reactor is represented in Fig. 3 by the curve Γ_m. This curve is best approximated when the feed and the reacting mixture flow co-currently. With countercurrent flow there is a considerable temperature overshoot in the catalytic bed near the inlet, while the cooling is too strong near het exit and the reaction slows down too much. These considerations have led to a rather complicated internal construction. In addition the effluent is sometimes cooled by an internal waste heat boiler, to recuperate the heat at a high temperature level. Fig. 3 also shows how the line Γ_m

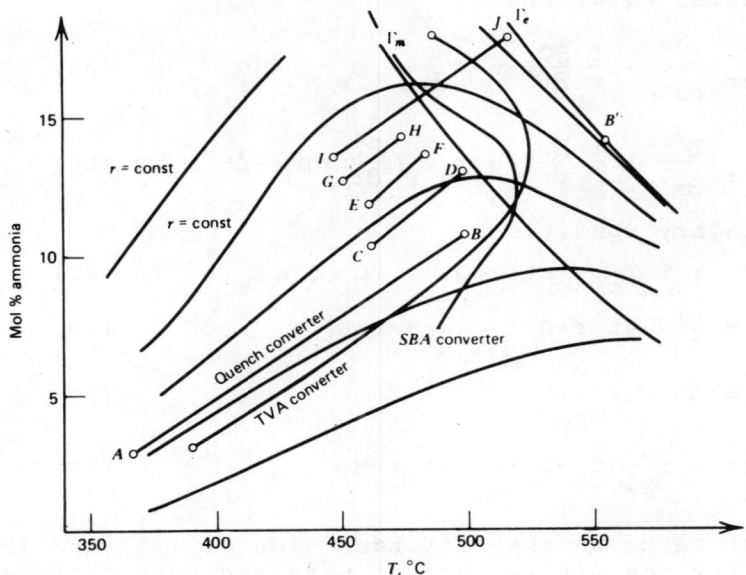

Fig. 3. Ammonia synthesis.
Optimal ways of operation [1,17].

is approximated in a MBAR.
The selection of the tube diameter is quite often a difficult problem. Steam reformer tubes have an internal diameter of 10 cm ; in o.xylene and naphtalene oxidation tubes of one inch are used. The heat effect of the reaction is the main factor determining the diameter. With strong heat effects radial temperature gradients inside the tube are inevitable. With endothermic reactions a cold core leads to a poor utilization of the catalyst. With exothermic reactions excessive temperatures in the tube axis may lead to poor selectivity, catalyst deterioration or runaway. The optimal tube diameter cannot be selected on the basis of simulations using equations (1)-(2)-(3) i.e. on the basis of the ideal one dimensional model, which assumes uniform temperature in a cross section. A more detailed model like the pseudo-homogeneous two dimensional model (A.III in Froment's classification) is

required here. The continuity equation for a reacting component and the energy equation are now partial differential equations:

$$\varepsilon D_{er}\left(\frac{\partial^2 C}{\partial r^2} + \frac{1}{r}\frac{\partial C}{\partial r}\right) - u_s \frac{\partial C}{\partial z} - \rho_B r_A = 0 \qquad (25)$$

$$\lambda_{er}\left(\frac{\partial^2 T}{\partial r^2} + \frac{1}{r}\frac{\partial T}{\partial r}\right) - u_s \rho_g C_p \frac{\partial T}{\partial z} + \rho_B(-\Delta H) r_A = 0 \qquad (26)$$

with boundary conditions

$$C = C_o \; ; \; T = T_o \quad \text{at } z = 0 \qquad 0 \leq r \leq R_t \qquad (27)$$

$$\frac{\partial C}{\partial r} = 0 \quad \text{at } r = 0 \quad \text{and } r = R_t \qquad (28)$$

$$\frac{\partial T}{\partial r} = 0 \quad \text{at } r = 0 \qquad (29)$$

$$\frac{\partial T}{\partial r} = -\frac{\alpha_w}{\lambda_{er}}(T_R - T_w) \quad \text{at } r = R_t \qquad (30)$$

The first terms in the left hand side of (25) and (26) account for the finite rate of mass and heat transfer in radial direction in the tube. The fluxes of these phenomena are expressed in terms of diffusion and conduction-like mechanisms. The mass transfer is described by a Fick-type law, with a proportionality factor called effective diffusivity, D_{er}. Note D_{er} relates to the bed as a whole and has no connection with the effective diffusivity for transport inside a particle, used in (10). The bed is not isotropic for this effective diffusion, but only the component in radial direction is considered here. In industrial reactors the component in axial direction is negligible with respect to the flux resulting from the flow. The heat flux in radial direction results from several mechanisms operating both in the fluid and the solid phase and which are lumped into a conduction-like mechanism described by a Fourrier-type equation, with a proportionality factor called effective conductivity, λ_{er}, not to be confused with that defined for heat transfer inside a particle. The heat transfer in the immediate vicinity of the wall is expressed in terms of a wall heat transfer coefficient, α_w, which is essentially different from that contributing to U in (22). The boundary conditions (28) and (29) express the radial symmetry in the tube and also that no mass can diffuse through the wall. The condition (30) expresses that the heat transferred to the vicinity of the wall by effective conduction is

Fixed bed catalytic reactors

transmitted to the wall by a convection-like mechanism. In industrial reactors the component of the effective conduction in axial direction is negligible with respect to the heat flux by the overall convection. Experimental correlations for λ_{er} and D_{er} are available in the literature and have been reviewed [1]. Fundamental models have also been developed [19,20]. The pressure drop equation, finally, still has the form (3), but local values of the temperature are used.

Little has been done so far to account for radial variation of the velocity, caused by the non uniform void fraction. Fig. 4 shows a comparison of radial temperature profiles in an o.xylene oxidation reactor, computed on the basis of resp. uniform flow velocity and of a flow velocity inversely proportional to the local void fraction [18]. For very severe reaction conditions considerable sophistication is required in the modeling to come close to the complexity of the real reactor. Accurate kinetic equations are prerequisites for the success of such efforts, however.

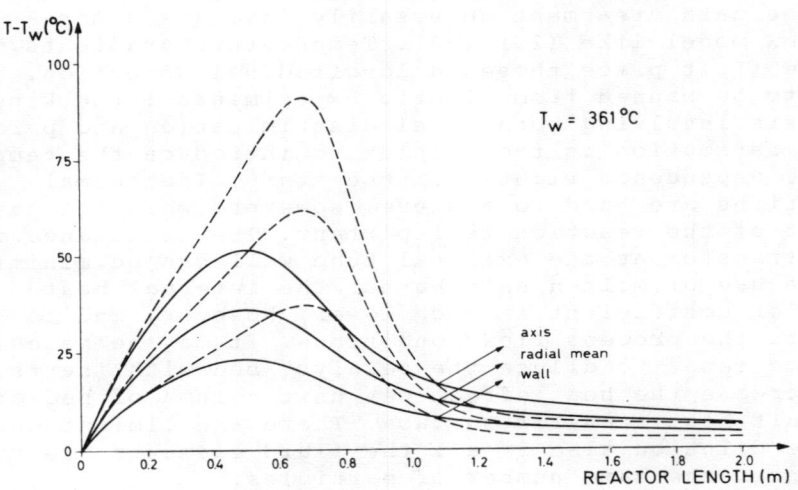

Fig. 4. O.xylene oxidation in a multitubular reactor [18]. Longitudinal temperature profiles (d_p=0.0045m, d_t=0.025m, T_o=361°C, ---, model with radial velocity profile; ___, model with uniform velocity).

II. BENCH SCALE FIXED BED REACTORS FOR KINETIC ANALYSIS

Bench scale reactors for kinetic studies of catalytic reactions are not necessarily scaled down replicas of the reactor type ultimately used in the industrial operation. Complete similarity with the large scale equipment is impossible, anyway. The laboratory reactor has to furnish the kinetic data with the greatest possible accuracy and it should be devised with that sole purpose in mind. The industrial reactor is then simulated through the mathematical model accounting for the specific aspects of flow, heat and mass transfer. Kinetic experiments on reactions catalyzed by solids are preferably carried out in flow reactors, generally of the tubular type. To limit the complexity of the data treatment the flow has to approximate as much as possible the ideal conditions of plug flow which lead to a simple continuity equation like (1). A rule of thumb requires ratios of the tube diameter to the particle diameter larger than ten to minimize the deviation from straight line flow paths and to avoid too much short-circuiting along the wall, where the void fraction is larger than in the core. If, however, the ratio is too high, important radial temperature gradients develop and the data treatment necessarily involves a more complex model like (25)-(26). Temperature gradients, in the first place those in longitudinal direction, have to be banned from kinetic experiments : the kinetic analysis involving both model discrimination and parameter estimation is too complex to introduce the temperature dependence right from the start. Isothermal conditions are hard to achieve, however, when the heat effect of the reaction is important. The resistance to heat transfer at the external tube wall can be minimized by the use of molten salt baths. The internal heat transfer coefficient is much lower, however, and is tied to the process flow conditions. It may be necessary in some cases to dilute the catalyst bed with inerts, to decrease the heat effect per unit volume of bed or per unit heat transfer surface. There are limitations on the dilution also if all the fluid elements are to encounter the same number of particles.

Transport phenomena rates can seriously interfere with the chemical reaction rate. Great care should be taken to eliminate such effects. Whereas gradients over the external film are negligible in industrial equipment they may be quite a problem in laboratory reactors in which the flow velocity is usually very low. Temperature gradients in particular are difficult to avoid. Internal concentration gradients can be eliminated by crushing

the catalyst. Then the effective diffusivity can be
determined from experiments with larger particle sizes
so that extrapolation to the size used in the industrial
reactor becomes possible. The crushing technique is only
valid when the active component(s) are uniformly distributed over the entire catalyst particle, which is
not guaranteed, or sometimes not even desired. It is
unlikely, though, that the industrial size particle
could be investigated in the laboratory reactor to
directly determine ηr_A, since the ratio d_t/d_p would be
far too low and the plug flow hypothesis would not be
valid any more because of excessive by passing along
the wall.

Catalyst deactivation introduces more complications.
Tubular reactors of the integral type are probably not
the most adequate type of reactors to study the kinetics of deactivation, as mentioned already. The preferred equipment for the kinetics of coking e.g. is
probably the electrobalance, but the discussion of
this technique falls outside the scope of this review.
De Pauw & Froment reported on a kinetic study of
pentane isomerization in an integral fixed bed reactor
of the tubular type [25]. The catalyst activity decayed
by coke deposition. The gas phase was sampled at
several heights and at various times to determine the
effect of the deactivation on the reactions. The catalyst was unloaded in well defined sections to determine
the coke profile throughout the reactor.

In differential reactors the amount of catalyst is very
small with respect to the feed rate so that the conversion is kept at a low level. The rate is directly obtained as such so that the data treatment is strongly
simplified. Isothermality is easily arrived at, but the
analysis has to be extremely accurate. Also, care has
to be taken to avoid channelling. It is probably better
to dilute the bed to a certain extent than to have a
bed consisting of a lew layers only. Since the conversion is very low mixed feeds are required for an
accurate kinetic analysis [27] but this may be a problem
with complex feedstocks. An integral reactor may be
used as a feed supplier [21].

Other fixed bed reactors can be used. Their merits and
shortcomings have been evaluated by Weekman[22]. Berty
has developed a fixed bed reactor with internal
recycle [23]. The recycle ratio has to be high so as
to achieve gradientless operation in the bed and the
gas velocity has to be sufficient to eliminate interfacial gradients. Here too the rate is obtained as such
from the data. The Carberry reactor has several fixed
beds contained in baskets rotating on a common axis in

the reacting gas [24]. Insufficient turbulence in the
voids between the particles have been reported.
The methodology of kinetic analysis and the data
treatment involved has been discussed by Froment [26,27].

LIST OF SYMBOLS

a_v	external particle surface area per unit reactor volume	m_p^2/m_r^3
C	molar concentration of species in the bulk gas phase	$Kmol/m^3$
C_p	specific heat of fluid	$kJ/kg\ K$
d_p	particle diameter	m
d_t	tube diameter	m
D_e	effective diffusivity in particle	$m_f^3/m_p\ s$
D_{er}	radial effective diffusivity in packed bed	$m_f^3/m_r\ s$
f	friction factor in Fanning equation	
g	acceleration of gravity	m/s^2
h_f	heat transfer coefficient fluid/particle	kJ/m^2sK
$(-H)$	heat of reaction	$kJ/kmol$
k_g	mass transfer coefficient fluid/particle	$m_f^3/m_p^2 s$
P_t	total pressure	N/m^2
r_A	rate of reaction of component A per unit catalyst mass	$Kmol/kgcat\ s$
R_t	tube radius	m
T	temperature	K
u_s	superficial velocity	$m_f^3/m_r^2 s$
U	overall heat transfer coefficient	kJ/m^2sK
z	axial coördinate in reactor	m_r
Z	total length of reactor	m_r
α	convective heat transfer coefficient	kJ/m^2sK
ε	void fraction of bed	m_f^3/m_r^3
η	effectiveness factor for solid catalyst	
λ_e	effective thermal conductivity of particle	$kJ/m\ s\ K$
λ_{er}	radial effective thermal conductivity of bed	$kJ/m\ s\ K$
ξ	reduced radial position in a particle radial position in a particle	

LIST OF SYMBOLS (continued)

ρ_g fluid density kg_f/m_f^3

ρ_s particle density kg_p/m_p^3

$\emptyset$ Thiele modulus

SUBSCRIPT

s   solid
o   initial

SUPERSCRIPT

s   surface

REFERENCES

1. G.F. Froment & K.B. Bischoff, Chemical Reactor Analysis and Design, J. Wiley N.Y., 1979.
2. G.F. Froment, Adv.Chem.Ser. $\underline{109}$, pp. 1-34 ; Amer.Chem.Soc., Washington, 1972.
3. G.F. Froment, Proc.5th Eur.Symp.Chem.React.Engng., Amsterdam, 1972, Elsevier 1972.
4. G.F. Froment, Chem.Ing.Techn., $\underline{46}$, 374, 1974.
5. M. Leva, Chem.Engng., $\underline{56}$, 115, May 1959.
6. L.E. Brownell, H.S. Dombrowsky, C.A. Dickey, Chem. Eng.Progr., $\underline{46}$, 415, 1950.
7. R.E. Hicks, Ind.Eng.Chem.Fund., $\underline{9}$, 500, 1970.
8. E.U. Schlünder, Chemical Reaction Engineering Reviews, Houston, A.C.S. Symp.Ser. 72, A.C.S. Washington, 1972.
9. R.F. Baddour, P.L.T. Brian, B.A. Logeais, J.P. Eymery, Chem.Engng.Sci., $\underline{20}$, 281, 1965.
10. W. Hatcher, L. Viville & G.F. Froment, Ind.Eng. Chem. Proc.Des.&Devpt., $\underline{17}$, 491, 1978.
11. G.F. Froment & K.B. Bischoff, Chem.Eng.Sci., $\underline{16}$, 189, 1961.
12. G.F. Froment & K.B. Bischoff, Chem.Eng.Sci., $\underline{17}$, 105, 1962.
13. J.B. Butt & R.M. Billimoria, Chemical Reaction Engineering Reviews, Houston, A.C.S. Symp.Ser. 72, A.C.S. Washington, 1972.
14. V.W. Weekman & D.M. Nace, A.I.Ch.E.J., $\underline{16}$, 397, 1970.
15. R. Aris, The optimal design of chemical reactors, Acad. Press, 1960.
16. C. McGreavy & B.R. Dunbobbin, A.C.S. Symp.Ser. 65, Chemical Reaction Engineering, Houston, p. 550, A.C.S. Washington, 1978.
17. L.M. Shipman & J.B. Hickman, Chem.Engng.Progr., $\underline{64}$, 59, May 1968.
18. J.J. Lerou & G.F. Froment, Chem.Eng.Sci., $\underline{32}$, 853, 1977.
19. D. Kunii & J.M. Smith, A.I.Ch.E.J., $\underline{6}$, 71, 1960.
20. P. Zehner & E.U. Schlünder, Chem.Ing.Techn., $\underline{44}$, 1303, 1972.
21. P.J. Lunde & F.L. Kester, Ind.Eng.Chem.Proc.Des.& Devpt., $\underline{13}$, 27, 1974.
22. V.W. Weekman, A.I.Ch.E.J., $\underline{20}$, 833, 1974.
23. J.M. Berty, A.I.Ch.E.J., 66th Ann.Meeting, Philadelphia 1973.
24. J.J. Carberry, Ind.Eng.Chem., $\underline{56}$, 39, 1964.
25. R. De Pauw & G.F. Froment, Chem.Engng.Sci., $\underline{30}$, 789, 1975.
26. G.F. Froment, A.I.Ch.E.J., $\underline{21}$, 1041, 1975.
27. J. Franckaerts & G.F. Froment, Chem.Eng.Sci., $\underline{19}$, 807, 1964.

Part II

CATALYTIC CRACKING

REACTIONS OF ALKYLCARBENIUM IONS IN RELATION TO ISOMERIZATION AND CRACKING OF HYDROCARBONS

D.M. Brouwer

Koninklijke/Shell Laboratory, (Shell Research B.V.)
Amsterdam, The Netherlands

1. INTRODUCTION

Alkylcarbenium ions* are ionic species that contain a positively charged (electron-deficient) carbon atom bonded to three alkyl groups and/or hydrogen atoms. The electron-deficient carbon has three sp_z-orbitals which are used for the C-R bonds - thus giving a planar structure -, and an empty p_z-orbital.

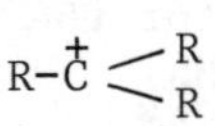

The alkylcarbenium ions play an important role as intermediates in many organic reactions, including the well-known acid-catalysed reactions in petroleum chemistry such as isomerization, alkylation and catalytic cracking [1]. The concept of the occurrence of alkylcarbenium ions as unstable, transient intermediates in organic reactions dates back to 1922 (Meerwein) and was later generalized by Whitmore in 1932. In the period following, it has been successfully used to describe, among a host of other organic reactions, the aluminium halide- and sulfuric acid-catalysed reactions of alkanes and olefins, such as isomerization and isobutane-isobutene alkylation.

---

*According to IUPAC nomenclature rules, the ions that were formerly called carb_o_nium ions have to be called carb_e_nium ions. The name carbocation, which has come into use recently, covers both carbenium ions and the positively charged ions that contain so-called pentacoordinated carbon.

Extension of the concepts of alkylcarbenium ion reactions to catalytic cracking, a gas-phase reaction over a solid acidic catalyst, came in 1947. Greensfelder, Voge and Good then showed that the typical product pattern obtained in catalytic cracking, which differs widely from that obtained in thermal cracking, can be rationalized by assuming that catalytic cracking involves carbenium ion reactions as opposed to the freeradical reactions taking place in thermal cracking.

2. FROM SHORT-LIVED INTERMEDIATES TO OBSERVABLE SPECIES

Until the early sixties, alkylcarbenium ions in solutions were hypothetical species, occurring as unstable, short-lived intermediates at such low concentrations that they were unobservable. This status changed completely with the introduction into this field of two, new developments. The first of these was the use of extremely strong acids (for which later the name 'superacids' was coined), which enables us to obtain stable solutions of alkylcarbenium ions, in which these species are long-lived and can be present at concentrations as high as one or more moles/litre. Thus it became possible to study alkyl-carbenium ions directly by spectroscopic and other techniques. The second development was that of nuclear magnetic resonance (NMR) spectroscopy as one of the most powerful tools for the organic chemist. The combined use of superacids and NMR spectroscopy was first successfully applied, in 1958, to alkylbenzenium ions and a few years later also to alkylcarbenium ions.

In order to obtain stable solutions of alkylcarbenium ions one has to make use of superacids because it is necessary to shift the equilibrium

$$R^+ \rightleftarrows H^+ + \text{olefin}$$

very far to the left-hand side. For unless the concentration of the olefin is reduced to an extremely low value, the rapid reactions between R^+ and olefins (*cf*. section 12) will prevent any alkylcarbenium ion from being long-lived.

The superacids systems that have been employed most frequently are solutions of SbF_5 in liquid HF, FSO_3H or SO_2ClF.*

---

*In solutions of SbF_5 in HF the 'acid proton' occurs as H_2F^+ formed according to the equilibrium $2HF + SbF_5 \rightleftarrows H_2F^+ + SbF_6^-$. The acid strength, which depends on the SbF_5:HF ratio, can easily be made more than 10^{10} times that of concentrated sulfuric acid ($-H_o > 20$)[2].

NMR spectroscopy (^{1}H and ^{13}C NMR) has made by far the largest contribution to our present knowledge of alkylcarbenium ions and their reactions. It provides direct information on the structures of the ions and it can be used to determine the equilibrium ratio's - and hence the relative stabilities - of isomeric ions and to measure the rates at which alkylcarbenium ions undergo transformations. These rates can be measured over a fantastically wide range of reaction rates. Firstly, by monitoring the disappearance (and appearance) of starting (and product) ions it is possible to follow reactions with half-lives that are longer than a few minutes (*i.e.*, (pseudo-)first-order rate constants anywhere down from ca $10^{-2}s^{-1}$). Secondly, with solutions that contain two (or more) rapidly equilibrating carbenium ions, NMR line broadening measurements can be used to determine rate constants from 1-2 s^{-1} up to as high as $10^5 s^{-1}$ (with ^{1}H-NMR) and even $10^8 s^{-1}$ (with ^{13}C-NMR). This technique can also be applied in those cases where the equilibrating ions are chemically identical, such as in the so-called degenerate rearrangements,

$$e.g. \quad \underset{+}{C-\overset{\overset{C}{|}}{C}-C-C} \rightleftarrows C-C-\underset{+}{\overset{\overset{C}{|}}{C}-C}$$

Other, non-spectroscopic measurements using solutions of alkylcarbenium ions in superacids have included studies of the isomerization of alkanes with $R^+SbF_6^-/HF$ (*cf.* section 6) and calorimetric measurements of the heats of (trans)formation of alkylcarbenium ions.

3. NOMENCLATURE OF INDIVIDUAL ALKYLCARBENIUM IONS

In view of the confusing nature of the names of the individual alkylcarbenium ions, the following may be useful to the reader.

According to one procedure, the alkylcarbenium ions are regarded as CH_3^+ (the 'parent' carbenium ion) in which one or more hydrogens have been replaced by alkyl groups (in the same way as is done with, *e.g.* alkylammonium ions). Next to this systematic nomenclature, alkylcarbenium ions are also often named after the complete alkyl residue (including C^+), followed by the word cation. For example:

CH_3^+	('carbenium ion')	methyl cation
$CH_3-CH_2^+$	methylcarbenium ion	ethyl cation
$CH_3-\overset{+}{C}H-CH_3$	dimethylcarbenium ion	iso-(or 2-)propyl cation
$CH_3-CH_2-CH_2^+$	ethylcarbenium ion	1-propyl cation
$CH_3-\underset{\substack{|\\CH_3}}{\overset{\substack{CH_3\\|}}{\overset{+}{C}}}-CH_3$	trimethylcarbenium ion	tert-butyl cation
$CH_3-\underset{\substack{|\\CH_3}}{\overset{\substack{CH_3\\|}}{C}}-CH_2^+$	tert-butylcarbenium ion	neopentyl cation

It should be emphasized that $\overset{+}{CH_3}$ (methyl cation) is <u>not</u> methylcarbenium ion and that $(CH_3)_3C^+$ (tert-butyl cation) is <u>not</u> tert-butylcarbenium ion.

4. STABILITIES OF ALKYLCARBENIUM IONS [3]

It is common knowledge that the stabilities of alkyl-Carbenium ions (relative to the corresponding alcohols or alkanes) decrease rapidly in the order $R_3C^+ \gg R_2CH^+ \gg RCH_2^+ \gg CH_3^+$, the stabilizing effect of the alkyl groups stemming from a combination of hyperconjugation and inductive effect. Earlier estimates (based on mass spectrometric data and estimated heats of solvation) gave values of the order of 10 kcal/mole for the difference in stabilization between tertiary and secondary ions and of 25 kcal/mole for that between secondary and primary ions.

The former value is in fair agreement with the result of a recent calorimetric measurement [5] of the heat that is evolved by the rearrangement

$$CH_3-CH_2-\overset{+}{C}H-CH_3 \longrightarrow (CH_3)_3C^+ \quad \Delta H_R = -14.5 \pm 0.5 \text{ kcal/mole}$$

Allowing for a difference in energy of ca 1 kcal/mole between the n-butyl and isobutyl carbon skeletons, this leaves ca 13 kcal/mole for the difference in stability between tertiary and secondary ions that have the same carbon skeleton. As the entropy effect is likely to be quite small, the difference in free enthalpy, which determines the equilibrium constant via $\Delta G_R = -RT \ln K$, will be nearly equal to ΔH_R. This gives an equilibrium constant of the order of 10^{10} for the equilibrium between tertiary and secondary ions with the same carbon skeleton.

There are no significant differences in stabilizing effect between methyl, ethyl, isopropyl, tert-butyl, etc. groups.

This is shown by the equilibrium ratio's of isomeric tertiary $C_6H_{13}^+$, $C_7H_{15}^+$ and some $C_8H_{17}^+$ cations. For example, at equilibrium the three tertiary $C_6H_{13}^+$ ions are present in the following proportions

```
      C                         C                    C C
      |                         |                    | |
C-C-C-C-C                 C-C-C-C-C              C-C-C-C
  +                           +                    +
   32%                         38%                   30%
```

Thus, the equilibrium constant for

$$R_1H + R_2^+ \overset{K}{\rightleftharpoons} R_1^+ + R_2H$$

is close to 1 if R_1^+ and R_2^+ are both tertiary (or both secondary) alkylcarbenium ions, and of the order of 10^{10} if R_1^+ is a tertiary and R_2^+ a secondary ion.

5. REACTIONS OF ALKYLCARBENIUM IONS

Alkylcarbenium ions are formed from olefins by reversible protonation

$$>C = C< + H^+ \rightleftharpoons R^+$$

and from alkanes by hydride transfer

$$R_1-H + R_2^+ \rightleftharpoons R_1^+ + R_2-H$$

The alkylcarbenium ions can undergo:

rearrangement: $R_1^+ \rightleftharpoons R_2^+$
β-cleavage: $R_1^+ \rightarrow R_2^+ + >C = C<$
addition to olefin: $R_1^+ + >C = C< \rightarrow R_2^+$

These reactions are the key steps in isomerization, cracking and alkylation (or polymerization), respectively.

6. ISOMERIZATION OF ALKANES

The isomerization of alkanes comprises two elementary steps, *viz.* (1) hydride transfer, involving simultaneous conversion of starting alkanes into alkylcarbenium ions and of product alkylcarbenium ions into product alkanes, and (2) rearrangement of the intermediate carbenium ions.

$$R_1\text{-H} + R_2^+ \underset{k'_h}{\overset{k_h}{\rightleftarrows}} R_1^+ + R_2\text{-H} \qquad (1)$$

$$R_1^+ \underset{k'_r}{\overset{k_r}{\rightleftarrows}} R_2^+ \qquad (2)$$

$$R_1\text{-H} \underset{k'_{ov}}{\overset{k_{ov}}{\rightleftarrows}} R_2\text{-H}$$

Once an alkylcarbenium ion is formed, it starts a chain of alternating rearrangement and hydride transfer steps, which continues until the carbenium ion is destroyed by some reaction or other.

(The same scheme can be written for the cracking of alkanes if we replace the second step by a β-cleavage step, and for the alkylation reaction if we replace the second step by addition of an alkylcarbenium ion to an olefin molecule).

There is, of course, a direct relation between k_{ov}, which can be measured in the usual manner, and the rate constants k_r and k_h. This allows the calculation of k_r or k_h (depending on which step is rate determining) once the concentrations of R_1^+ and/or R_2^+ are known. This is the case if the isomerization is carried out with a solution of alkylcarbenium ions (as the hexafluoroantimonate salts, $R^+SbF_6^-$) in liquid HF in which the concentrations of tert-R^+ are 0.1-1 mole/litre.

It has long been known that the isomerization of hexanes involves three groups of reactions that proceed at widely different rates (Chart 1).

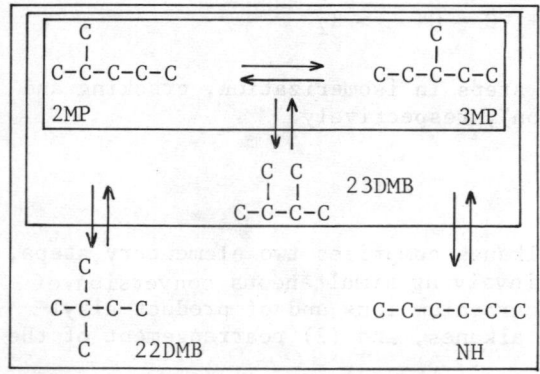

Chart 1

The interconversion of 2MP and 3MP is by far the fastest reaction. The next fastest one is that between 23DMB and the methylpentanes. The slowest reaction is the isomerizations of these three to 22DMB and NH, and vice versa.

When the hexanes are isomerized with $R^+SbF_6^-/HF$, all the reactions (hydride transfers as well as rearrangements) take place in the HF phase, which rapidly exchanges starting and product alkanes with the hydrocarbon phase. For this septem it was shown [6] that the rearrangement step is rate determining for the 23DMB-MP interconversion and that for the isomerization starting from, or leading to, NH and 22DMB, which have no tertiary carbons, the hydride transfer step is rate determining.

7. RATES AND ACTIVATION PARAMETERS

For the intramolecular reactions of alkylcarbenium ions (1,2-shifts, rearrangements) it is convenient to use, besides the rate constants k, the free enthalpies of activation, $\Delta G^{\neq}$, as a parameter which denotes how easily or difficultly the reactions take place. The relation between k and $\Delta G^{\neq}$ is given by Eyring's equation $k = (k^*T/h) \exp(-\Delta G^{\neq}/RT)$, where k^* and h are the Boltzmann and Planck constants, respectively and (k^*T/h) is ca 10^{13}.

It has consistently been found that for the above-mentioned intramolecular reactions of alkylcarbenium ions (and, similarly, also for those of alkylbenzenium ions) the entropy of activation, $\Delta S^{\neq}$, is practically zero*. Hence $\Delta G^{\neq}$ ($= \Delta H^{\neq} - T\Delta S^{\neq}$) is independent of the temperature, which facilitates comparison of reactions that have been studied at widely different temperatures. The use of $\Delta G^{\neq}$ also makes it easier to quantify the relations between reaction rates and the relative stabilities of starting, intermediate and product ions.

8. 1,2-HYDRIDE AND ALKYL SHIFTS [3]

The simplest reactions of alkylcarbenium ions are the 1,2-hydride and alkyl shifts

$$\underset{R}{>}\overset{+}{C}-C< \quad \rightleftarrows \quad \underset{R}{>}C-\overset{+}{C}< \qquad (R = H, \text{alkyl})$$

---

*$\Delta S^{\neq} = 0$ corresponds to log A = ca 13 where A is the pre-exponential factor in the Arrhenius equation $k = A \exp(-E_a/RT)$.

If the ions before and after the shift are of comparable stability, *i.e.* if they are both tertiary ions or both secondary ions, these shifts are extremely fast. The rates have been measured directly for the hydride and methyl shifts in tertiary alkylcarbenium ions [7], *e.g.*

$$\begin{matrix} \text{C} & \text{C} \\ | & | \\ \text{C-}\overset{+}{\text{C}}\text{-C-C} \\ & | \\ & \text{H} \end{matrix} \rightleftarrows \begin{matrix} \text{C} & \text{C} \\ | & | \\ \text{C-C-}\overset{+}{\text{C}}\text{-C} \\ | \\ \text{H} \end{matrix}$$

$$\begin{matrix} \text{C} & \text{C} \\ | & | \\ \text{C-}\overset{+}{\text{C}}\text{-C-C} \\ | \\ \text{C} \end{matrix} \rightleftarrows \begin{matrix} \text{C} & \text{C} \\ | & | \\ \text{C-C-}\overset{+}{\text{C}}\text{-C} \\ | \\ \text{C} \end{matrix}$$

$k = 10^7 - 10^8 \text{s}^{-1}/-120°\text{C}$

$\Delta G^{\neq} = 3 - 3.5 \text{ kcal/mole}$

and for the hydride shift in secondary alkylcarbenium ions

$$\begin{matrix} \text{H} & \text{H} \\ | & | \\ \text{C-}\overset{+}{\text{C}}\text{-C-C} \\ | \\ \text{H} \end{matrix} \rightleftarrows \begin{matrix} \text{H} & \text{H} \\ | & | \\ \text{C-C-}\overset{+}{\text{C}}\text{-C} \\ | \\ \text{H} \end{matrix} \quad \Delta G^{\neq} < 3 \text{ kcal/mole}$$

For the corresponding methyl shift, *e.g.*

$$\begin{matrix} \text{H} & \text{H} \\ | & | \\ \text{C-}\overset{+}{\text{C}}\text{-C-C} \\ | \\ \text{C} \end{matrix} \rightleftarrows \begin{matrix} \text{H} & \text{H} \\ | & | \\ \text{C-C-}\overset{+}{\text{C}}\text{-C} \\ | \\ \text{C} \end{matrix} \quad \Delta G^{\neq} = \text{ca 2 kcal/mole}$$

we refer to the next section.

In keeping with these very low free enthalpy barriers, shifts that start from a tertiary ion to yield a secondary ion have free enthalpy barriers that are only slightly higher than the difference in free enthalpy (ca 13 kcal/mole) between the tertiary and secondary ions. Thus, for the interconversion of the equally stable (K = 1.1) tertiary 2,4,4- and 2,3,4-trimethylpentyl ions, which proceeds by H and CH_3 shifts via an intermediate secondary ion, $\Delta G^{\neq}$ is only 13.5 kcal/mole.

$$\begin{matrix} \text{C} & & \text{C} \\ | & \text{H} & | \\ \text{C-}\overset{+}{\text{C}}\text{-C-}\overset{}{\text{C}}\text{-C} \\ | & \text{H} \\ \text{C} \end{matrix} \xrightarrow{\text{H}\sim} \begin{matrix} \text{C} & & \text{C} \\ | & \text{H} & | \\ \text{C-C-}\overset{+}{\text{C}}\text{-C-C} \\ | & \text{H} \\ \text{C} \end{matrix} \xrightarrow{\text{Me}\sim} \begin{matrix} \text{C} & \text{C} & \text{C} \\ | & | & | \\ \text{C-}\overset{+}{\text{C}}\text{-C-C-C} \\ \text{H} & \text{H} \end{matrix}$$

$k = 5 \times 10^{-4} \text{s}^{-1}/-88°\text{C} \qquad \Delta G^{\neq} = 13.5 \text{ kcal/mole}$

9. BRANCHING AND NON-BRANCHING REARRANGEMENTS [3]

It is necessary to make a distinction between rearrangements in which the degree of chain branching remains the same ('non-branching' rearrangements) and those in which it decreases or increases ('branching' rearrangements). Not only are the latter rearrangements much slower than the former, they also proceed by another mechanism.

Non-branching rearrangements

Examples of these are

$$\begin{array}{c} C \\ | \\ C-\overset{+}{C}-C-C \\ | \\ H \end{array} \overset{H}{\underset{}{}} \rightleftarrows \begin{array}{c} H \\ | \\ C-C-\overset{+}{C}-C \\ | \\ H \end{array} \overset{C}{\underset{}{}}$$

$k = 10\ s^{-1}/0°C \qquad \Delta G^{\neq} = 14.8\ \text{kcal/mole}$

$$\begin{array}{c} C \\ | \\ C-\overset{+}{C}-C-C-C \\ | \\ H \end{array} \overset{H}{\underset{}{}} \rightleftarrows \begin{array}{c} H \\ | \\ C-C-\overset{+}{C}-C-C-C \\ | \\ H \end{array} \overset{C}{\underset{}{}}$$

$k = 8 \times 10^{-4} s^{-1}/-78°C \qquad \Delta G^{\neq} = 14.3\ \text{kcal/mole}$

$$\begin{array}{c} C-C \\ | \\ C-C-\overset{+}{C}-C-C \\ | \\ H \end{array} \overset{H}{\underset{}{}} \rightleftarrows \begin{array}{c} H \\ | \\ C-C-C-\overset{+}{C}-C \\ | \\ H \end{array} \overset{C-C}{\underset{}{}}$$

$k = 10^{-3} s^{-1}/-74°C \qquad \Delta G^{\neq} = 14.2\ \text{kcal/mole}$

The classical mechanism for the non-branching rearrangements supposes them to proceed by 1,2-shifts of H and alkyl groups via secondary ions as intermediates:

$$\begin{array}{c} C \\ | \\ C-\overset{+}{C}-C-C \\ | \\ H \end{array} \overset{H}{\underset{H\sim}{\rightleftarrows}} \begin{array}{c} C \\ | \\ C-C-\overset{+}{C}-C \\ | \\ H\ H \end{array} \underset{Me\sim}{\rightleftarrows} \begin{array}{c} C \\ | \\ C-\overset{+}{C}-C-C \\ | \\ H\ H \end{array} \underset{H\sim}{\rightleftarrows} \begin{array}{c} H \\ | \\ C-C-\overset{+}{C}-C \\ | \\ H \end{array} \overset{C}{\underset{}{}}$$

This mechanism is reconcilable with the present experimental data. An overall $\Delta G^{\neq}$ of 14.5 - 15 kcal/mole and a difference of ca 13 kcal/mole between secondary and tertiary ions just leaves ca 2 kcal/mole for $\Delta G^{\neq}$ of the methyl(alkyl) shifts between the secondary carbons.

Branching rearrangements

Examples are the rearrangements of mono- to di-branched ions and of mono-branched to linear ions (and vice versa)

$$\text{C-}\underset{+}{\text{C}}\text{-C-C} \overset{\text{C C}}{\underset{\text{| |}}{}} \rightleftharpoons \text{C-}\underset{+}{\text{C}}\text{-C-C-C}$$

$k = 0.08\ s^{-1}/0°C$ $\qquad \Delta G^{\neq} = 17.3$ kcal/mole
$k = 4 \times 10^{-4} s^{-1}/-45°C$ $\qquad \Delta G^{\neq} = 16.8$ kcal/mole

$$\text{C-}\underset{+}{\overset{\text{C}}{\text{C}}}\text{-C-C} \overset{k}{\rightleftharpoons} \text{C-}\underset{+}{\text{C}}\text{-C-C-C}$$

$k = 600\ s^{-1}/120°C$ $\qquad \Delta G^{\neq} = 18.3$ kcal/mole

and also the interconversion of cyclohexyl and methylcyclopentyl ions

$k = 900\ s^{-1}/100°C$
$\Delta G^{\neq} = 17$ kcal/mole

In the past, the branching rearrangements, too, were supposed to proceed via 1,2-shifts of H and alkyl groups, examples being the rearrangements from linear to mono-branched ions

$$\text{C-}\underset{+}{\text{C}}\text{-C-C-R} \underset{\text{Me}\sim}{\rightleftharpoons} \underset{+}{\text{C}}\text{-}\overset{\text{C}}{\underset{\text{|}}{\text{C}}}\text{-C-R} \underset{\text{H}\sim}{\rightleftharpoons} \text{C-}\underset{+}{\overset{\text{C}}{\text{C}}}\text{-C-R}$$

and those between tertiary mono- and di-branched ions.

$$\text{C-}\underset{+}{\overset{\text{C}}{\text{C}}}\text{-C-C-C} \underset{\text{H}\sim}{\rightleftharpoons} \underset{+}{\overset{\text{C}}{\text{C-C-C-C-C}}} \underset{\text{Me}\sim}{\rightleftharpoons} \text{C-C-}\underset{+}{\text{C}}\text{-}\overset{\text{C C}}{\underset{\text{| |}}{\text{C}}} \underset{\text{H}\sim}{\rightleftharpoons} \text{C-C-}\underset{+}{\overset{\text{C C}}{\text{C}}}\text{-C}$$

The – unavoidable – intermediacy of primary carbenium ions in the branching rearrangements was thought to explain why branching rearrangements are much slower than non-branching rearrangements.

This classical mechanism, however, had to be rejected on two accounts. First, with the isomerization of alkanes with $R^+SbF_6^-/HF$ it was found that no isomerization of n-butane occurs under the conditions where n-pentane isomerizes readily. This

Reactions of alkylcarbenium ions

implied that the rearrangement of n-butyl to tert-butyl ion is very much slower than that of n-pentyl to tert-pentyl ion, which was later confirmed by direct NMR measurements (vide infra). According to the classical mechanism it should make no difference whether in the above-mentioned linear-to-mono-branched rearrangement R = H or R = CH_3. Secondly, the value of only 17-18 kcal/mole found for $\Delta G^{\neq}$ of branching rearrangements that start from a tertiary ion was irreconcilable with the intermediacy of a primary cation, since that would have required a $\Delta G^{\neq}$ of at least 30 kcal/mole.

The mechanism that is currently accepted for branching rearrangements involves the intermediacy of a protonated cyclopropane ring:

$$\underset{\overset{+}{C}}{\overset{\diagdown C \diagup}{}} \underset{\diagup \diagdown}{\overset{\diagdown \diagup}{C}} H \quad \longrightarrow \quad \underset{\diagdown \diagup}{\overset{\diagdown C \diagup}{C - C}} H^+ \quad \longrightarrow \quad \underset{\underset{+}{C}}{\overset{\diagdown C - H}{C - C}}$$

Attack by the C^+ carbon on the β-C atom leads to ring closure with displacement of a hydrogen at the latter carbon atom as a proton, which remains attached to the cyclopropane ring. The reverse process, but now on one of the other two sides of the ring, results in the formation of a carbenium ion with a different carbon skeleton*.

The interconversion between mono- and di-branched ions proceeds as follows

$$\underset{\underset{+}{C}}{\overset{C}{C}}-C-C-C \quad \rightleftarrows \quad \overset{C}{\underset{C}{\diagdown}}C\overset{H^+}{-}C-C \quad \rightleftarrows \quad \overset{C\ C}{\underset{+}{C-C-C-C}}$$

Thus, this branching rearrangement is accomplished without the intermediacy of any primary carbenium ion.

A similar mechanism is operative in the interconversion of linear and mono-branched ions:

$$\underset{\underset{+}{C}}{\overset{C}{C}}-C-C-R \quad \rightleftarrows \quad \overset{C}{C-C-C-R}\overset{H^+}{} \quad \rightleftarrows \quad \overset{C}{\underset{+}{C-C-C-R}} \quad \rightleftarrows \quad \overset{C}{\underset{+}{C-C-C-R}}$$

*More precisely, a system of rapidly interconverting so-called edge- and corner-protonated cyclopropane rings is involved. For details see refs. 4, 8.

Again, no primary ion is involved if R is an alkyl group, in other words, if the above ion is a pentyl or larger ion. However, if R = H, the rearrangement still has to proceed via a primary ion, which is the reason why the interconversion of n- and tert-butyl ions is very much slower than that of n- and tert-pentyl ions.

The rearrangement of the secondary n-butyl to the tertiary butyl ion has recently been found [5] to have a $\Delta G^{\neq}$ of 18 kcal/mole. Hence, in the reverse direction $\Delta G^{\neq}$ = 32 kcal/mole, which is ca 14 kcal/mole higher than $\Delta G^{\neq}$ for the rearrangement of tert-pentyl to the secondary n-pentyl cation. (This corresponds to a difference in rate by a factor of 10^{10} at room temperature).

Whereas rearrangement of n-butyl to tert-butyl ion is very difficult, the following n-butyl rearrangement proceeds without intermediacy of a primary ion and should be as fast as the rearrangement of n-pentyl to tert-pentyl ion.

$$C-\overset{C}{\underset{\ }{C}}\diagup\overset{\diagdown}{C^*} \rightleftarrows C-\overset{\overset{H^+}{\underset{\times}{C}}}{C}-C^* \rightleftarrows C-\underset{\ }{C}-\overset{C}{\diagup}C^*$$

In agreement with this it has been found that the 'isomerisation' of n-butane-1-^{13}C to n-butane-2-^{13}C, in contrast to the isomerization to isobutane, is as fast as the isomerization of n-pentane to isopentane.

10. β-CLEAVAGE [9]

The β-cleavage involves the migration of the two electrons of the β C-C bond towards the C^+ carbon to form an olefinic π-bond, the β carbon atom being left behind as the electron-deficient carbon of a smaller carbenium ion

$$\overset{+}{>}C-\overset{|}{C}-C\overset{\diagup}{\diagdown} \longrightarrow >C=C\overset{\diagup}{\diagdown} + {}^+C\overset{\diagup}{\diagdown}$$

When the cleavage takes place in a 'superacid', the product olefin is immediately protonated to give the corresponding carbenium ion.

The rate of β-cleavage depends, of course, on the relative stabilities of the starting and product ions.

If both starting and product ions are tertiary ions the β-cleavage takes place quite rapidly: the cleavage of the 2,4,4-trimethylpentyl cation has a half-life of only 20 minutes at -73^0C.

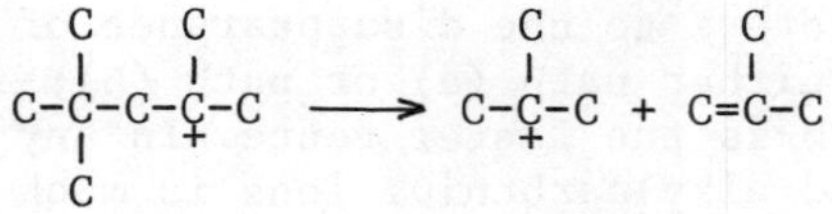

$$\begin{array}{c} \text{C} \quad \text{C} \\ | \quad | \\ \text{C-C-C-C-C} \\ {}^{+}| \\ \text{C} \end{array} \longrightarrow \begin{array}{c} \text{C} \\ | \\ \text{C-C-C} \\ {}^{+} \end{array} + \begin{array}{c} \text{C} \\ | \\ \text{C=C-C} \end{array}$$

$$k = 5 \times 10^{-4} \text{s}^{-1}/-73°\text{C}$$

Thus, this cleavage is faster than branching rearrangements and slower than non-branching rearrangements. This means that once a trimethylpentyl (or any larger tri-branched) cation is formed it will equilibrate rapidly with all the other trimethylpentyl (or larger tri-branched) cations, one of which undergoes cleavage to yield two branched fragmentation products (Chart 2).

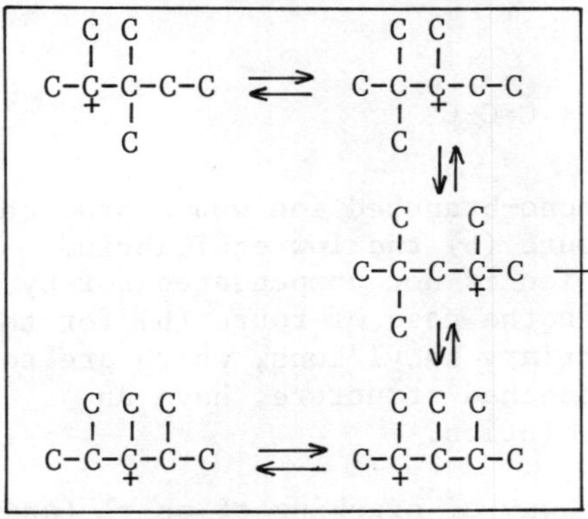

Chart 2

[Chart showing equilibration of trimethylpentyl cations leading to cleavage products C-C-C⁺ + C=C-C]

This whole process is faster than rearrangement to di- or tetra-branched structures.

Cleavage of di-branched ions is much slower. Path (a) is slow

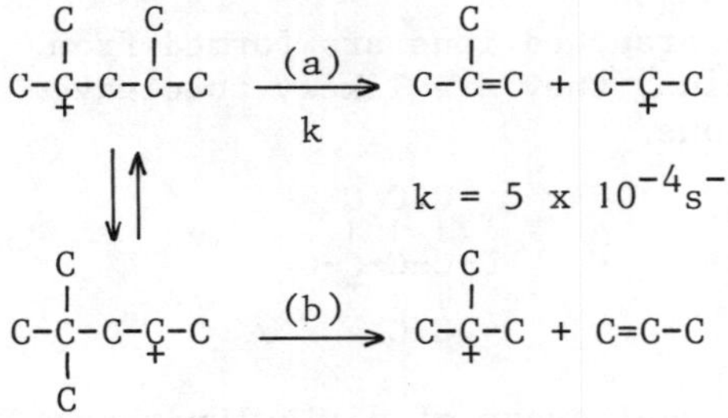

$$\begin{array}{c} \text{C} \quad \text{C} \\ | \quad | \\ \text{C-C-C-C-C} \\ {}^{+} \end{array} \xrightarrow{(a)} \text{C-C=C} + \begin{array}{c} \text{C} \\ | \\ \text{C-C-C} \\ {}^{+} \end{array}$$

$$k = 5 \times 10^{-4} \text{s}^{-1}/0°\text{C, or slower}$$

$$\begin{array}{c} \text{C} \\ | \\ \text{C-C-C-C-C} \\ {}^{+}| \\ \text{C} \end{array} \xrightarrow{(b)} \begin{array}{c} \text{C} \\ | \\ \text{C-C-C} \\ {}^{+} \end{array} + \text{C=C-C}$$

because the β-cleavage of a tertiary ion that produces a secondary ion is ca 13 kcal/mole more endothermic than the above-mentioned β-cleavage of the tert-2,4,4-trimethylpentyl ion; path (b) is slow because of the low equilibrium concentration of the secondary cation (ca 10^{-10} that of the tertiary ion), which is only partly compensated for by the much higher rate of β-cleavage of this secondary ion (ca 13 kcal/mole less endothermic than the β-cleavage of tert-2,4,4-trimethylpentyl cation).

The rate constant quoted above refers to the disappearance of the tert-2,4-dimethylpentyl ion via either path (a) or path (b); it is not known with certainty which is the faster route. In any case, the cleavage of di-branched alkylcarbenium ions is much slower than their rearrangement to mono- or tri-branched ions.

β-Cleavage of mono-branched alkylcarbenium ions is again much more difficult.

$$\begin{array}{c} \text{C} \\ | \\ \text{C-}\overset{+}{\text{C}}\text{-C-C-C} \end{array} \xrightarrow{(a)} \begin{array}{c} \text{C} \\ | \\ \text{C-C=C} \end{array} + \overset{+}{\text{C}}\text{-C}$$

$$\updownarrow$$

$$\begin{array}{c} \text{C} \\ | \\ \text{C-C-C-}\overset{+}{\text{C}}\text{-C} \end{array} \xrightarrow{(b)} \text{C-}\overset{+}{\text{C}}\text{-C} + \text{C=C-C}$$

Direct cleavage of a tertiary mono-branched ion would produce a primary cation, whereas in route (b) the low equilibrium concentration of the secondary ion is not compensated for by a very high rate of cleavage as in the case of route (b) for the di-branched ions. Thus, the tertiary hexyl ions, which are too small to give a cleavable di-branched structure, have an 'unlimited' life in superacid solution.

Chart 3 summarizes the pathway of cracking of octyl (and larger) cations, which shows how all linear, mono- and di-branched cations will disappear via rearrangement to tri-branched cations to give, eventually, two branched cleavage products.

When larger, more highly branched ions are formed from corresponding alkanes or olefins, they are likely to cleave faster than tri-branched cations.

$$\begin{array}{c} \text{C} \quad \text{C} \\ | \quad\quad | \\ \text{C-C-C-}\overset{+}{\text{C}}\text{-C} \\ | \\ \text{C} \end{array} \ll \begin{array}{c} \text{C} \quad \text{C} \quad \text{C} \\ | \quad | \quad | \\ \text{C-C-C-}\overset{+}{\text{C}}\text{-C} \\ | \\ \text{C} \end{array} \ll \begin{array}{c} \text{C} \quad \text{C} \quad \text{C} \\ | \quad | \quad | \\ \text{C-C-C-}\overset{+}{\text{C}}\text{-C} \\ | \quad | \\ \text{C} \quad \text{C} \end{array}$$

This is shown by data for the β-cleavage of dialkylhydroxy-carbenium ions that are formed by protonation of ketones.

$$\begin{array}{c} \text{C R C} \\ | \; | \; | \\ \text{C-C-C-}\overset{+}{\text{C}}\text{-OH} \\ | \; | \\ \text{C R} \end{array} \longrightarrow \begin{array}{c} \text{C} \\ | \\ \text{C-}\overset{+}{\text{C}}\text{-C} \end{array} + \text{R}_2\text{C=C-OH} \xrightarrow{\text{H}^+} \text{R}_2\text{HC-}\overset{+}{\text{C}}\text{-OH}$$

$$R = H \quad k = 8 \times 10^{-6} s^{-1}/70°C, \quad R = CH_3 \quad k = 8 \times 10^{-2} s^{-1}/70°C$$

Chart 3: Pathway of cracking of octanes

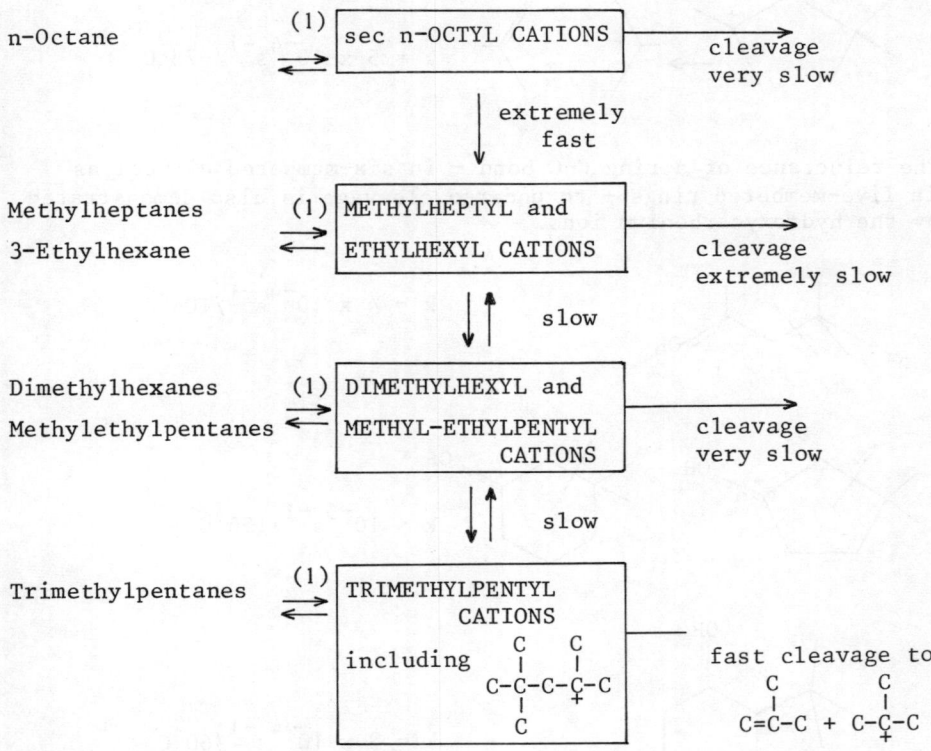

(1) Rapid hydride transfer

The large increase in rate of cleavage upon methyl substitution at the α-carbon atom is due to weakening of the β C-C bond as the result of mutual repulsion between the vicinal methyl groups.

β-Cleavage of C-C bonds that form part of the ring in cycloalkyl cations is a very difficult process. Whereas the tertiary 2,4,4-trimethylpentyl ion cleaves readily at -73^0C, the tertiary trimethylcyclopentyl cation is still stable at 0^0C (where it is in equilibrium with the other trimethylcyclopentyl and dimethylcyclohexyl cations).

$$k < 10^{-4} s^{-1} / 0^0 C$$

$k = 5 \times 10^{-4} s^{-1} / -73°C$

The reluctance of a ring C-C bond - in six-membered as well as in five-membered rings - to undergo cleavage is also demonstrated by the hydroxycarbenium ions.

$k = 4 \times 10^{-4} s^{-1} / 104°C$

$k < 10^{-5} s^{-1} / 156°C$

$k = 8 \times 10^{-4} s^{-1} / 60°C$

The easy cleavage of the exocyclic β C-C bond in the last cation contrasts sharply with the absence of observable cleavage of the endocyclic C-C bonds.

The fact that β-cleavage of ring C-C bonds is so slow can be easily understood if we take a closer look at what happens during the β-cleavage process (see Chart 4). When the C-R bond is broken, the alkyl group R leaves as R^+, leaving behind the two electrons of the C-R bond. At the same time, the sp_3-orbital of the α-C atom changes into a p_z-orbital, which is filled with the two electrons. Overlap of this filled p_z-orbital with the vacant p_z-orbital of the electron-deficient carbon atom results in the formation of the new olefinic π-bond. This energy-lowering overlap of the two p_z-orbitals already occurs in the transition state. In order to give the most efficient overlap, the two p_z-orbitals, and consequently the vacant p_z-orbital and the

Reactions of alkylcarbenium ions 153

chart 4

original C-R bond, must lie in the same plane. This situation is easily realized in acyclic alkylcarbenium ions in which there is free rotation about the C^+-C bond. However, in cyclopentyl ions the ring C-C bond is fixed in a plane perpendicular to the vacant p_z-orbital; in consequence, efficient overlap of the two p_z-orbitals in the transition state is impossible, resulting in a very high activation energy. To a smaller extent, this effect is also present in the β-cleavage of endocyclic bonds in cyclohexyl cations.

11. HYDRIDE TRANSFER [3]

The conversion of alkanes into alkylcarbenium ions and vice versa proceeds via the hydride transfer reaction

$$R_1^+ + H-R_2 \rightleftharpoons R_1-H + R_2^+$$

The hydride transfer from a tertiary carbon to a tertiary alkylcarbenium ion is an extremely fast reaction. The rate constant of the hydride transfer from isobutane to tert-butyl cation has been determined (by NMR line broadening in $SO_2ClF-CH_2Cl_2$ solution) to be $k = 10^4$ l.mole^{-1}.s^{-1} at $-40^\circ C$ with $E_a = 3.5$ kcal/mole.

$$\text{tert-}C_4H_9^+ + i\text{-}C_4H_{10} \rightleftharpoons i\text{-}C_4H_{10} + \text{tert-}C_4H_9^+$$

Rate constants of the hydride transfer from a secondary carbon to a tertiary carbenium ion, which reaction is endothermic by ca 13 kcal/mole, have been obtained from kinetic studies of the isomerization of pentanes and hexanes with $R^+SbF_6^-$ in liquid HF (c.f. section 6).

$$n-C_5H_{12} + \text{tert-}C_5H_{11}^+ \rightleftarrows \text{sec-}C_5H_{11}^+ + i-C_5H_{12}$$

$$n-C_6H_{14} + \text{tert-}C_6H_{13}^+ \rightleftarrows \text{sec-}C_6H_{13}^+ + (2MP, 3MP, 23DMB)$$

$$(CH_3)_3CCH_2CH_3 + t-C_6H_{13}^+ \rightleftarrows (CH_3)_3CCHCH_3^+ + (2MP, 3MP, 23DMB)$$

The rate constants of these reactions in the forward directions are 0.01-0.02 $l.mole^{-1}s^{-1}$ at $0°C$. The enthalpies of activation, calculated from the variation of k with the temperature, are $\Delta H^{\neq} = 13-14$ kcal/mole, which is hardly more than the increase in enthalpy that is associated with this reaction. From these data and the equilibrium constant ($\Delta G_R \sim \Delta H_R \sim 13$ kcal/mole) it follows that for the hydride transfer from a tertiary carbon to a secondary alkylcarbenium ion $k \sim 10^8$ $l.mole^{-1}s^{-1}$ and $\Delta H^{\neq} < 2$ kcal/mole.

12. REACTIONS OF ALKYLCARBENIUM IONS WITH OLEFINS

The reverse of the β-cleavage reaction is the addition of an alkylcarbenium ion to an olefin to produce a larger carbenium ion, e.g.

$$\underset{\substack{|\\C}}{C}-\underset{\substack{|\\+}}{C}-C + \underset{\substack{|\\C}}{C}=C-C \longrightarrow C-\underset{\substack{|\\C}}{C}-\underset{\substack{|\\+}}{C}-C-C \quad (\rightleftarrows \text{ other } C_8H_{17}^+ \text{ ions})$$

In combination with the hydride transfer between isobutane and iso-octyl cations this reaction leads to the isobutane-isobutene alkylation that produces iso-octanes

$$i-C_4H_{10} + i-C_4H_8 \longrightarrow i-C_8H_{18}$$

Repetition of the addition reaction leads to higher, and eventually polymeric, products.

Depending on the conditions, however, the reaction between alkylcarbenium ions and olefins can also take an entirely different course [10] and produce polyalkylcyclopentenyl cations

Reactions of alkylcarbenium ions

together with alkanes, e.g.

$$i\text{-}C_4H_8 + H^+ \rightleftharpoons tert\text{-}C_4H_9^+$$

$$tert\text{-}C_4H_9^+ + x\,C_4H_8 \longrightarrow [\text{polyalkylcyclopentenyl cation}] + \text{alkanes}$$

The formation of the stable, allylic polyalkylcyclopentenyl cations (containing from 8 to over 20 carbon atoms) consumes acid protons and, thus, leads to deactivation of acid catalysts. The formation of these polycyclopentenyl cations and alkanes can be quite fast; for example, the reaction is completed in a few minutes at room temperature when tert-butanol (a precursor of tert-butyl cations and isobutene) is dissolved in 96% sulfuric acid at 0.1M concentration [10].

The reaction involves the following steps:
1) Di-, tri- and oligomerization of the olefin, which produces larger olefins.
2) Hydride transfer from olefins to alkylcarbenium ions to produce allylic cations and alkanes

$$\mathrm{>C=C-C(H)<} + R^+ \longrightarrow \mathrm{>C \dot{=} C \dot{=} C<} + RH$$

3) Equilibration of allylic cations and conjugated dienes

$$\mathrm{>C \dot{=} C \dot{=} C(H)-C<} \rightleftharpoons \mathrm{>C=C-C=C<} + H^+$$

4) Hydride transfer from conjugated dienes to alkylcarbenium ions to produce dienyl cations and alkanes

$$\mathrm{>C=C-C=C-C(H)<} + R^+ \longrightarrow \mathrm{>C \dot{=} C \dot{=} \dot{C} \dot{=} C \dot{=} C<} + RH$$

5) Electrocyclization of dienyl ions to cyclopentenyl ions e.g.

$$\mathrm{C_2C\text{-}C\dot{=}C\dot{=}\overset{+}{C}\dot{=}C\dot{=}C\text{-}C_2} \longrightarrow [\text{pentamethylcyclopentenyl cation}] \quad (1)$$

6) Rearrangement of the initially formed cyclopentenyl ions to more stable polyalkylcyclopentenyl ions by 1,2-hydride and alkyl shifts and deprotonation/protonation,

$$(1) \xrightarrow{\text{Me}\sim} [\text{rearranged}] \xrightarrow{-H^+} [\text{diene}] \xrightarrow{+H^+} [\text{new cation}]$$

13. NOVEL REACTIONS OF ALKANES IN SUPERACIDS [3]

The study of the reactions of alkanes and alkylcarbenium ions in superacids have led to the discovery of a series of novel reactions. It turns out that the hydride transfer reaction is just one member of a family of related reactions that all involve attack of H^+ or R^+ on the C-H or C-C bonds of alkanes to form new C-H or C-C bonds with expulsion of H^+ or R^+.

$$R_1^+ + H-R_2 \rightleftarrows R_1-H + R_2^+ \qquad (1)$$

$$H^+ + H-R \underset{(b)}{\overset{(a)}{\rightleftarrows}} H-H + R^+ \qquad (2)$$

$$H^+ + R-H \rightleftarrows H-R + H^+ \qquad (3)$$

$$H^+ + R_1-R_2 \underset{(b)}{\overset{(a)}{\rightleftarrows}} H-R_1 + R_2 \qquad (4)$$

Reaction (1) is the hydride transfer reaction in which attack of an alkylcarbenium ion on the C-H bond of an alkane results in the formation of a new C-H bond with expulsion of another carbenium ion. Similarly, H^+ can attack a C-H bond with expulsion of an alkylcarbenium ion to give molecular hydrogen (2a). The reverse of this reaction is the reduction of carbenium ions with molecular hydrogen (2b). Attack of H^+ on a C-H bond can also lead to the formation of a new C-H bond with expulsion of a proton (hydrogen exchange, 3). C-C bonds, too, can be attacked by H^+ to form a smaller alkane and an alkyl carbenium ion (4a). The reverse of this reaction involves attack of R^+ on a C-H bond, just as in reaction (1), but this time with formation of a new C-C bond. The rates of all these reactions except (4b) have been determined in solutions of SbF_5 in liquid HF where the 'proton' is H_2F^+.

For reactions (2a) and (2b) with an iso-alkane/tert-alkyl cation the rate constants are approximately the same.

$$H^+ + \underset{H}{\underset{|}{C-\overset{\overset{C}{|}}{C}-C}} \underset{k_{2b}}{\overset{k_{2a}}{\rightleftarrows}} \underset{+}{\underset{|}{C-\overset{\overset{C}{|}}{C}-C}} + H_2$$

$k_{2a} = 2 \times 10^{-3}$ l.mole^{-1}.s^{-1}/20°C

$k_{2b} = 8 \times 10^{-3}$ l.mole^{-1}.s^{-1}/20°C

It is seen that these reactions are very much slower than the hydride transfer between isoalkanes and tert-alkyl cations ($k \sim 10^4$ l.mole^{-1}.s^{-1}).

Reaction (3) has been studied by means of H-D exchange:

$H^+ + CH_3D \longrightarrow CH_4 + D^+ \quad k = 2 \times 10^{-4} \text{ l.mole}^{-1}.s^{-1}/20°C$

$H^+ + C_2H_5D \longrightarrow C_2H_6 + D^+ \quad k = 2 \times 10^{-4} \quad /-10°C$

$H^+ + (CH_3)_3CD \longrightarrow (CH_3)_3CH + D^+ \quad k = 5 \times 10^{-3} \quad / 0°C$

The fact that this reaction is only 10^2 times faster for isobutane than for methane clearly shows that it does not involve any free carbenium ions. It is also seen that the H-D exchange of the tertiary hydrogen of isobutane is somewhat faster than the hydride abstraction that produces tert-butyl cation and molecular hydrogen.

The C-C bond splitting in neopentane by H_2F^+ is a much slower reaction than the C-H bond splitting in isobutane.

$(CH_3)_3C-CH_3 + H^+ \longrightarrow (CH_3)_3C^+ + CH_4$

$k = 5 \times 10^{-5} \text{ l.mole}^{-1}s^{-1}/20°C$

A similar C-C bond cleavage in methylcyclopentane produces the acyclic hexyl cations:

$k = 10^{-3} \text{ l.mole}^{-1}.s^{-1}/-15°C$

Surprisingly, this reaction, which first produces a secondary carbenium ion, is much faster than the C-C bond cleavage in neopentane which produces a tertiary carbenium ion. If this reaction is carried out in the presence of molecular hydrogen, the hexyl ions are reduced to give hexanes with regeneration of H^+. Thus, under the influence of $H_2F^+SbF_6^-/HF$ methylcyclopentane (MCP) undergoes a purely acid-catalysed 'hydrocracking' to hexanes:

$H^+ + MCP \longrightarrow C_6H_{13}^+$

$C_6H_{13}^+ + H_2 \longrightarrow C_6H_{14} + H^+$

———————————————— +

$MCP + H_2 \longrightarrow C_6H_{14}$

The occurrence of reaction (4) was first established in the 'oligocondensation' of methane to tert-butyl cations by FSO_3H-SbF_5 at elevated temperatures.

$$CH_4 \xrightarrow{FSO_3H-SbF_5} \text{tert-}C_4H_9^+$$

This reaction involves oxidation* of methane to CH_3^+, followed by:

$$CH_3^+ + CH_4 \longrightarrow H_3C-CH_3 + H^+$$

$$CH_3^+ + C_2H_6 \longrightarrow C_3H_8 + H^+$$

and/or:

$$C_2H_5^+ + CH_4 \longrightarrow C_3H_8 + H^+$$

and so on to C_4H_{10}, which is converted into tert-butyl ions that are too stable to react further with methane.

Very recently, similar reactions have been carried out starting with ethene in TaF_5-HF [11]. Addition of the ethyl cation to methane, ethane and n-butane was found to produce propane, n-butane and 3-methylpentane, respectively.

$$C=C + H^+ \longrightarrow C-\overset{+}{C}$$

$$C-\overset{+}{C} + CH_4 \longrightarrow C_3H_8$$

$$C-\overset{+}{C} + C_2H_6 \longrightarrow \text{n-}C_4H_{10}$$

$$C-\overset{+}{C} + \text{n-}C_4H_{10} \longrightarrow C-C-\overset{|}{\underset{C}{C}}-C-C$$

The secondary propyl cation, too, was demonstrated to be capable of reacting with methane:

$$C-C=C + H^+ \longrightarrow C-\overset{+}{C}-C$$

$$C-\overset{+}{C}-C + CH_4 \longrightarrow \text{i-}C_4H_{10}$$

---

*It should be noted that oxidation of alkanes to alkylcarbenium ions is effected not only by protons but also by many other oxidants including H_2SO_4, oleum, $ClSO_3H$, FSO_3H (S(VI) → S(IV)) and SbF_5 (Sb(V) → Sb(III)).

Reactions of alkylcarbenium ions

The reactions discussed in this section are no ordinary carbenium ion reactions. They are generally believed to proceed via structures (either as short-lived intermediates or as transition states) that contain so-called 'two-electron three-centre bonds', *i.e.* structures in which two electrons provide for a bond between two hydrogen atoms and one carbon atom or between two carbon atoms and one hydrogen atom. For example

$$\mathrm{\geq C-H + D^+ \rightleftharpoons \left[\geq C \!\!<\!\!\begin{array}{c}H\\D\end{array} \right]^+ \rightleftharpoons \geq C-D + H^+}$$

$$\mathrm{\geq C-C\leq + H^+ \rightleftharpoons \left[\geq C \cdots \underset{H}{\cdots} C\leq \right]^+ \rightleftharpoons \geq C-H + {}^+C\leq}$$

REFERENCES

1. For descriptions of these processes and the chemistry involved see, *e.g.*, P.H. Emmett (Ed.) 'Catalysis' (Reinhold Publ. Corp., New York, 1958), volume VI, chapters 1, 2 and 5 by R.M. Kennedy, F.E. Condon and H.H. Voge, respectively, and G.A. Olah (Ed.) 'Friedel-Crafts and related reactions' (Interscience, New York, 1964), volume II, chapters 25 and 28 by L. Schmerling, and by H. Pines and N.E. Hoffman, respectively.

2. J. Somner *et al.*, J. Amer. Chem. Soc. 100 (1978) 2576.

3. Most of the references pertaining to sections 4, 8, 9, 11 and 13 can be found in the review article of ref. 4. It is only for more recent work that separate references are given.

4. D.M. Brouwer and H. Hogeveen, Progress in Physical Organic Chemistry 9 (1972) 179.

5. E.W. Bittner, E.M. Arnett and M. Saunders, J. Amer. Chem. Soc. 98 (1976) 3734.

6. D.M. Brouwer and J.M. Oelderik, Preprints Div. of Petroleum Chem., ACS meeting San Francisco, April 1968; Rec. Trav. Chim. 87 (1968) 721.

7. M. Saunders and M.R. Kates, J. Amer. Chem. Soc. 100 (1978) 7082.

8. M. Saunders. P. Vogel, E. Hagen and J. Rosenfeld, Accounts Chemical Research 6 (1973) 53.

9. For details and references, see D.M. Brouwer and H. Hogeveen, Rec. Trav. Chim. 89 (1970) 211; D.M. Brouwer and J.A. van Doorn, Rec. Trav. Chim. 90 (1971) 535.

10. N.C. Deno *et al.*, J. Amer. Chem. Soc. 86 (1964) 1745.

11. M. Siskin, J. Amer. Chem. Soc. 98 (1976) 5413.

CRACKING CATALYSTS

J.H.C. van Hooff

Laboratory for Inorganic Chemistry, Eindhoven
University of Technology, Eindhoven, The Netherlands

INTRODUCTION

As mentioned before catalytic cracking reactions proceed via carbenium ion intermediates. A cracking catalyst has the function to assist in the formation of such carbenium ions from the feedstock molecules. Normally this can be done by acids, but in this special case the acid catalyst must have two special properties:
i. It must be possible to use the catalyst in a fluid bed process, and
ii. The catalyst must be stable at temperatures up to 700^0C.
Therefore the common acids like H_2SO_4, HNO_3, H_3PO_4 and HCl cannot be used.
Instead of these acids the so-called 'solid acids' have been developed. At first acid-treated clay minerals were used for this purpose but with little success. The development of the amorphous silica-alumina catalysts meant a big step forwards. However, the most important advance in catalytic cracking in the last three decades has been the development of zeolite catalysts. It is the purpose of this chapter to summarize the structures of cracking catalysts and their influence on cracking reactions.

AMORPHOUS SILICA ALUMINA CATALYSTS

Neither SiO_2 (silica) nor Al_2O_3 (alumina) nor a mechanical mixture of the two dry oxides has acid properties, but a cogelled mixture of silica and 10 to 25% alumina can be used as an active cracking catalyst. The incorporation of alumina into silica results in the formation of surface acid sites which catalyze

cracking reactions.
The preparation of these so-called silica-alumina catalysts [1] involves first the preparation of a porous silica hydrogel. This hydrogel consists of a coherent aggregate of primary spherical particles about 3 to 5 nm in diameter. The primary particles consist of a three-dimensional network of interconnected SiO_4 tetrahedra, each silicon atom being linked to four oxygen atoms and each oxygen atom linking two silicons. The surfaces of the particles are terminated with hydroxyl groups in the form of silanol (Si-OH) groups (see figure 1).

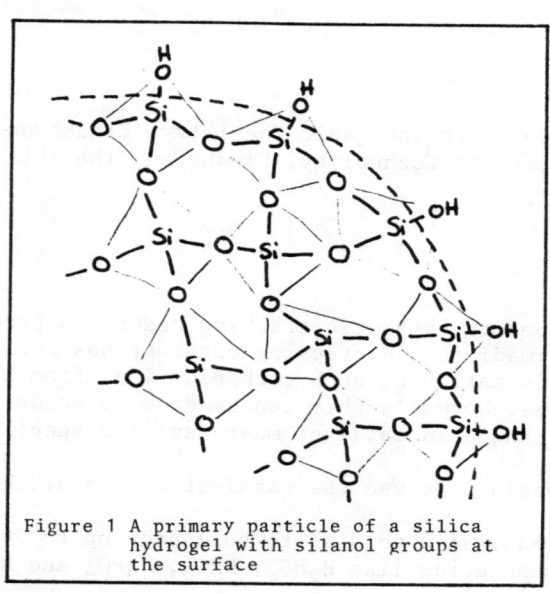

Figure 1 A primary particle of a silica hydrogel with silanol groups at the surface

These silanol groups vary from weakly acidic to alcoholic and show no cracking activity.

In the second step of the preparation an aluminum salt is added to the silicahydrogel so that the six coordinated aluminum ions react with the surface of the primary silica particles. Reaction involves a condensation of the aluminum hydroxide with the elimination of water. The result is that aluminum ions are incorporated into the surfaces of the silica particles in the form of AlO_4 tetrahedra (see figure 2).

Since the trivalent aluminum ion is bonded to four oxygen anions each AlO_4 tetrahedron has a residual charge of -1. Therefore each alumina tetrahedron requires a +1 charge from a cation in the structure to maintain electrical neutrality.
If the condensation of the aluminum hydroxide is carried out at

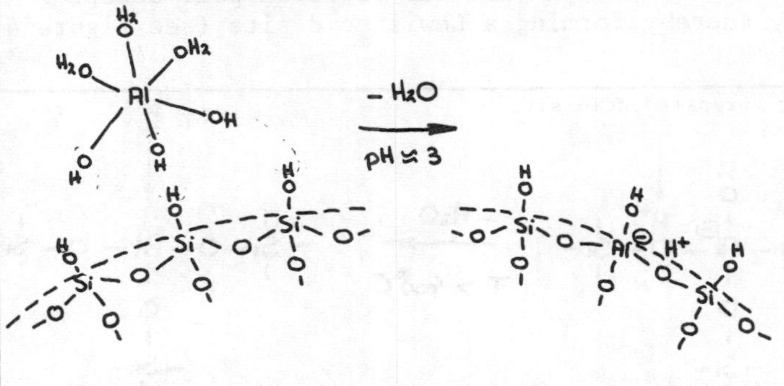

Figure 2 Condensation reaction of aluminumhydroxide with the surface of a silica hydrogel

low pH (pH ≈ 3) H^+ ions will take this place.
This results in the formation of surface OH groups having strong Brønsted acid character.
The protonic acidity evidently arises from the dissociative adsorption of H_2O on the aluminum ion. It is associated with the proton which moves to the oxygen bonded to a neighbouring silicon ion (see figure 3).

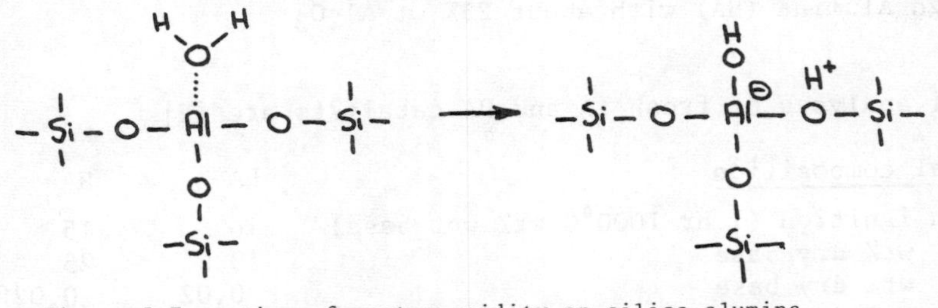

Figure 3 Formation of proton acidity on silica-alumina

The strongly electrophilic nature of the aluminum ion can be traced to its position in the surface, where it is surrounded by 3 fourvalent silicon cations. This accounts for the withdrawal of charge from the aluminum ion to make it more positive and develop a sufficiently strong field to allow it to acquire a hydroxyl group by splitting off a proton from water. The resultant
                 H
Si - O - Al group is strongly polarized, which induces a strong acidity in the group.

When the silica-alumina surface is heated to high temperature

(> about 400°C), water is removed from the Brønsted acid site, exposing the aluminum ion with its electron-pair-acceptor properties, thereby forming a Lewis acid site (see figure 4).

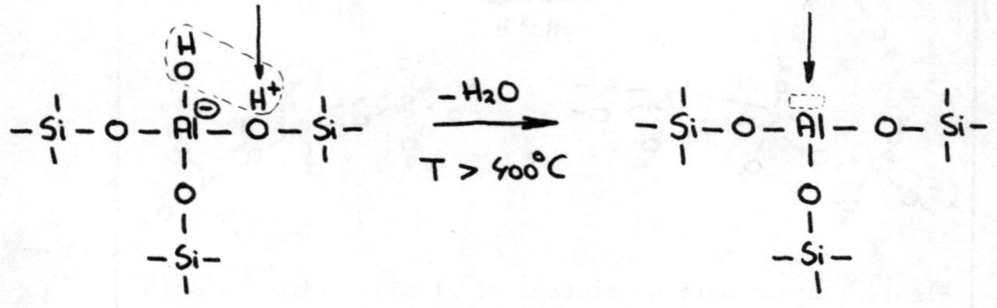

Figure 4 Conversion of a Brønsted acid site into a Lewis acid site by dehydration of silica alumina

As the actual reaction temperature in the catalytic cracking process is about 500°C especially the Lewis-acid sites of the silica-alumina catalyst will play an important role.

Commercially two types of amorphous silica-alumina catalysts are available:
  Low Alumina (LA) with about 13% wt Al_2O_3 and
  High Alumina (HA) with about 25% wt Al_2O_3

Typical analyses of fresh LA and HA catalysts are [2]

Chemical composition		LA	HA
Loss on ignition (1 hr 1000°C wt% wet base)		16	15
Al_2O_3	wt% dry base	13	25
Na_2O	wt% dry base	0,02	0,02
SO_4	wt% dry base	0,2	0,7
Fe	wt% dry base	0,02	0,03
SiO_2	wt% dry base	balance	balance

Physical properties			
Surface Area (1 hr 600°C)	m²/g	600	510
Pore Volume (H_2O)	ml/g	0,7-1,1	0,75
Apparent Bulk Density	g/ml	0,33-0,40	0,42
Attrition 1 hr	wt%	22	16
Average Particle Size	μm	65	60

Cracking catalysts

The thermal stability of these catalysts is limited especially in the presence of steam as is the case during stripping and regeneration. This causes sintering of the catalyst particles by which the pore volume and the surface area decrease.

An 'equilibrium' LA catalyst has the following physical properties

LA catalyst	fresh	equilibrium
Surface Area m^2/g	600	100
Pore Volume ml/g	0,9	0,45
Average Pore Diameter nm *	6	18
Apparent Bulk Density g/ml	0,35	0,70

$$* \; \overline{PD}(nm) = \frac{4 \times PV(ml/g)}{SA(m^2/g)} \times 10^3$$

Catalytic Cracking of Mid Continent Gasoil with a LA catalyst results in the following product distribution

Feed: Mid Continent Gasoil
Catalyst: Low-Alumina
WHSV: $2.0 \; hr^{-1}$
Reactor Temperature: $500°C$

Product distribution Total conversion 75%

Gas C_1-C_4:	15%	paraffins	10%
Gasoline C_5^+:	45%	olefins	45%
Light Cycle Oil:	10%	naphthenes	10%
Coke:	5%	aromatics	35%

As can be seen in this table besides gasoline relatively large amounts of 'gas' and 'coke' are formed, leaving much room for improvement. An important improvement could be realised by the introduction of cryatalline zeolite cracking catalysts.
A description of this type of catalysts will be given in the next paragraph.

CRYSTALLINE ZEOLITE CATALYSTS

About twenty years ago coworkers of Mobil and Exxon [3, 4] discovered that special molecular-sieve zeolites could be used as effective cracking-catalysts.
Further development of this finding at which besides Mobil and Exxon also Union Carbide has contributed, has resulted in improved cracking catalyst with the following advantages with respect to the amorphous silica-alumina catalysts

- higher activity
- higher gasoline selectivity
- higher stability.

To understand this behaviour we first have to explain the zeolite structure and the origin of its acidity.

Zeolites are crystalline alumino-silicates, with SiO_4 and AlO_4 tetrahedra as fundamental building blocks. These tetrahedra are arranged in such a way that each of the four oxygen anions is shared with another silica or alumina tetrahedron to form a threedimensional crystal lattice.

The zeolite that is most applicated in cracking catalysts (type X or Y) has a crystal-structure related to the mineral faujasite. In this structure 24 SiO_4 and/or AlO_4 tetrahedra form together a secondary building unit in the form of a cubo-octahedron the so-called sodalite-unit (see figure 5).

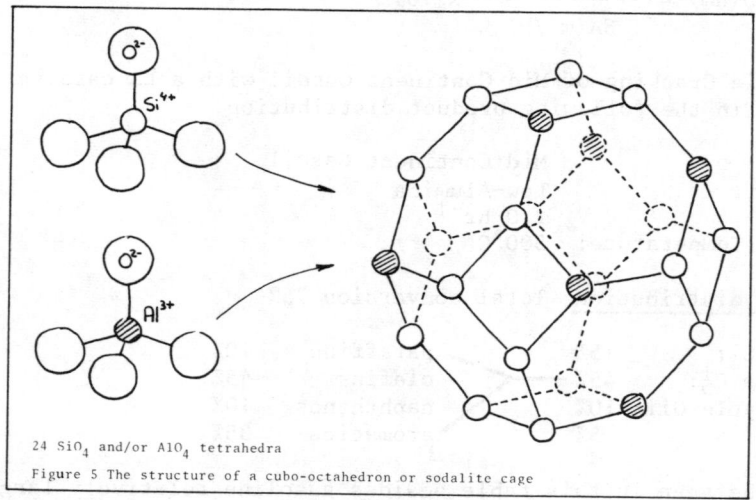

24 SiO_4 and/or AlO_4 tetrahedra

Figure 5 The structure of a cubo-octahedron or sodalite cage

In the sodalite-unit we can distinguish 6 four-membered rings (the faces of a cube) and 8 six-membered rings (the faces of an octahedron)

Coupling of the cubo-octahedra to form crystalline structures can occur in different ways

i. When the cubo-octahedra are stacked so that each four-membered ring is shared by two units, sodalite is formed (see figure 6).
ii. When the cubo-octahedra are connected by bridge oxygen atoms between the four-membered rings, zeolite A is formed (see figure 7).
iii. The third possibility is that the cubo-octahedra are connected by bridge oxygen atoms between the six-membered rings. In that case the faujasite-type zeolite is formed (see figure 8)

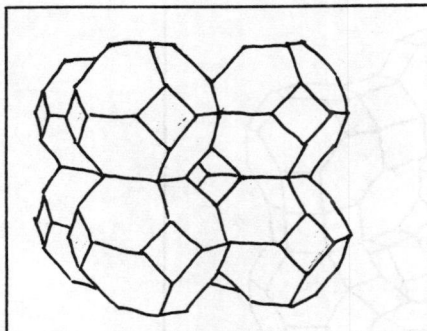

Figure 6 The arrangement of the cubo-octahedra in sodalite

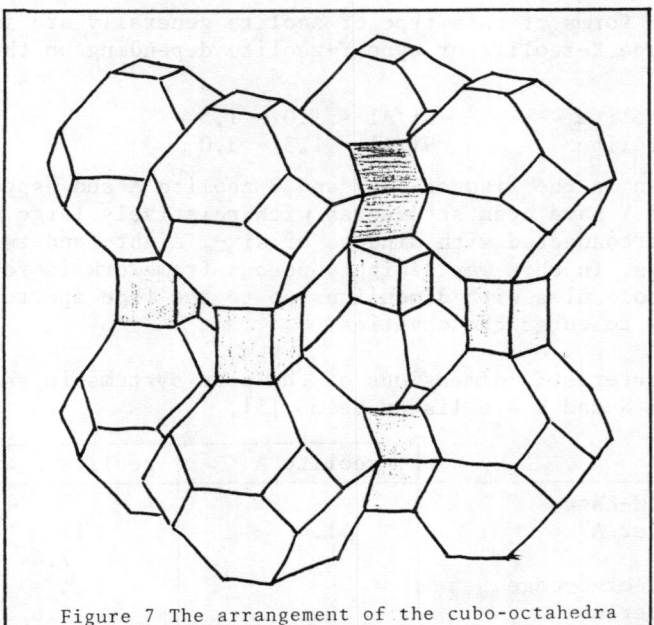

Figure 7 The arrangement of the cubo-octahedra in zeolite A

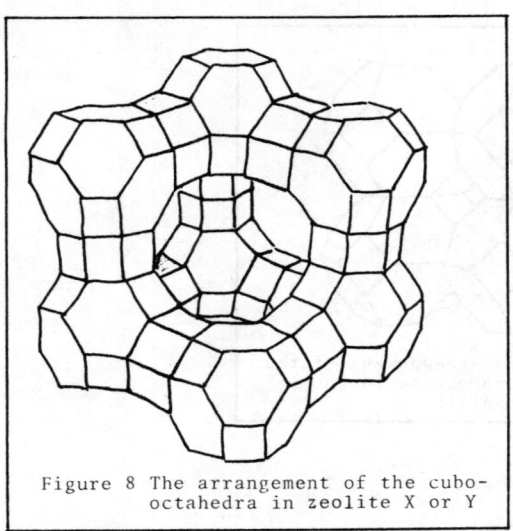

Figure 8 The arrangement of the cubo-octahedra in zeolite X or Y

The synthetic forms of this type of zeolite generally are indicated by type X-zeolite or type Y-zeolite depending on the Si/Al ratio.

    type X zeolite   Si/Al = 1,0 - 1,5
    type Y zeolite   Si/Al = 1,5 - 3.0

As can be seen in the figures 6, 7 and 8 zeolite A and especially zeolite X and Y have open structures with relatively large cavities that are interconnected with windows of six-, eight- and twelve-membered rings. In this way a highly porous framework is formed, that allows molecules with dimensions up to the free aperture of the window to enter the cavities.

The characteristic dimensions of the pore systems in zeolite A and zeolite X and Y are listed below [5].

	zeolite A	zeolite X and Y
Supercage or α-cage		
free diameter A^o	11.4	12
aperture A^o	4,2	7.4
Sodalite cage or β-cage		
free diameter A^o	6.6	6.6
aperture A^o	2.6	2.6
Hexagonal prism or γ-cage		
aperture A^o	-	2.6
density g/ml	1.33	1.30
void volume ml/g	0.30	0.35 (51%)

As was the case in the amorphous silica-alumina catalysts the presence of a Al^{3+} ion in a AlO_4 tetrahedron creates an effective negative charge. This charge is partially delocalized over the entire structure, but the extent of the delocalization is not known. The excess of negative charge is neutralized by positive ions that are located on special positions in the zeolite structure. In the X and Y zeolites, four distinct cation sites have been located as shown in figure 9 [6].

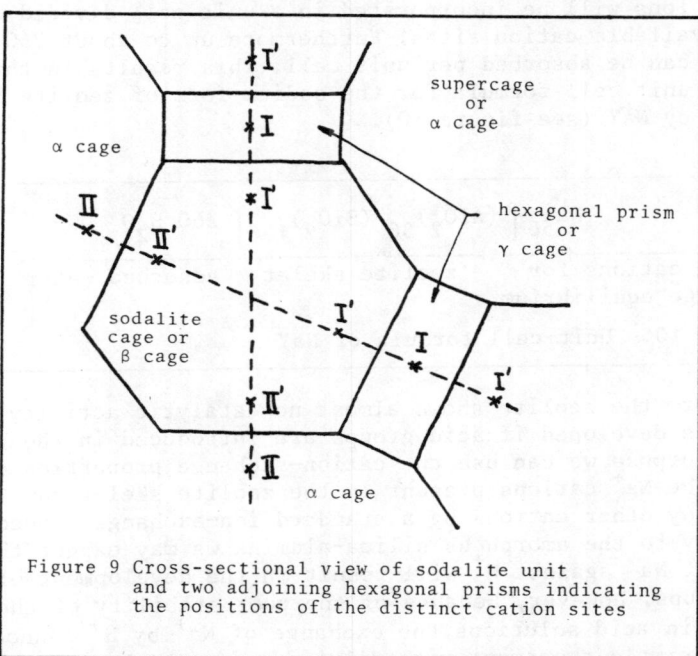

Figure 9 Cross-sectional view of sodalite unit and two adjoining hexagonal prisms indicating the positions of the distinct cation sites

The type I site is located at the center of the hexagonal prism formed by the six bridging oxygens; there are 16 type I sites per unit cell.
The type I' site is located in the sodalite-unit just on the other side of the shared hexagonal face from the type I site; there are 32 type I' sites per unit cell.
Type II and II' sites are located on the unjoined hexagonal faces of the sodalite units respectively outside and inside the sodalite cage; there are 32 of each of these sites per unit cell
Therefore the total number of cation site is 16 + 32 + 32 + 32 = 112.
This number exceeds the number of cations that are really present in zeolite X or Y so that the sites are only partly occupied.

This can be illustrated by the following example:
let us start from a Y zeolite with a Si/Al ratio of 2.5.
Per unit cell we have 8 cubo-octahedra and thus 8 x 24 = 192 SiO_4 and/or AlO_4 tetrahedra. Because Si/Al = 2.5 these 192

tetrahedra can be devided in $\frac{2.5}{3.5} \times 192 = 136$ SiO_4 tetrahedra and

$\frac{1.0}{3.5} \times 192 = 56$ AlO_4 tetrahedra.

For every AlO_4 unit we need a +1 charge in the form of a cation; thus totally a cation charge of +56 is needed.
As Y zeolites are synthesized in the presence of Na^+ ions, 56 of these Na^+ ions will be incorporated in a unit cell divided over the 112 available cation sites. Furthermore up to about 260 water molecules can be absorbed per unit cell. This results in the following unit cell formula for the sodium-form of zeolite Y also indicated by NaY (see figure 10).

$$Na_{56} \underbrace{\left[(AlO_2)_{56} (SiO_2)_{136} \right]}_{\text{zeolite skelet}} \underbrace{260\ H_2O}_{\text{adsorbed water}}$$
$\underbrace{\phantom{Na_{56}}}_{\substack{\text{cations for}\\ \text{charge equilibrium}}}$

Figure 10   Unit-cell formula of NaY

In this form the zeolite shows almost no catalytic activity. This activity is developed if acid groups are introduced in the zeolite. For this purpose we can use the cation-exchange properties of the zeolite. The Na^+ cations present in the zeolite skelet can be exchanged by other cations by a standard ion-exchange procedure. Analogously to the amorphous silica-alumina we may expect that exchange of Na^+ against H^+ will result in the development of acidic groups. however, because of the poor stability of the Y zeolite in acid solutions the exchange of Na^+ by H^+ cannot be done by a simple treatment with diluted acids. We have to use a circuitous route.
One method is to exchange Na^+ by NH_4^+ first, followed by a thermal decomposition of the NH_4^+ ions in H^+ and NH_3.
The proceeding reactions can be expressed by the following reaction equations

$Na_{56}[(AlO_2)_{56}(SiO_2)_{136}] + xNH_4^+$

ion exchange of
Na^+ by NH_4^+ $\updownarrow$

$Na_{56-y}(NH_4)_y[(AlO_2)_{56}(SiO_2)_{136}] + yNa^+ + (x-y)NH_4^+$

Thermal decomposition
of NH_4^+ in H^+ and NH_3 $\downarrow$

$Na_{(56-y)}H_y[(AlO_2)_{56}(SiO_2)_{136}] + yNH_3$

A second method to illustrate this modification is shown in figure 11

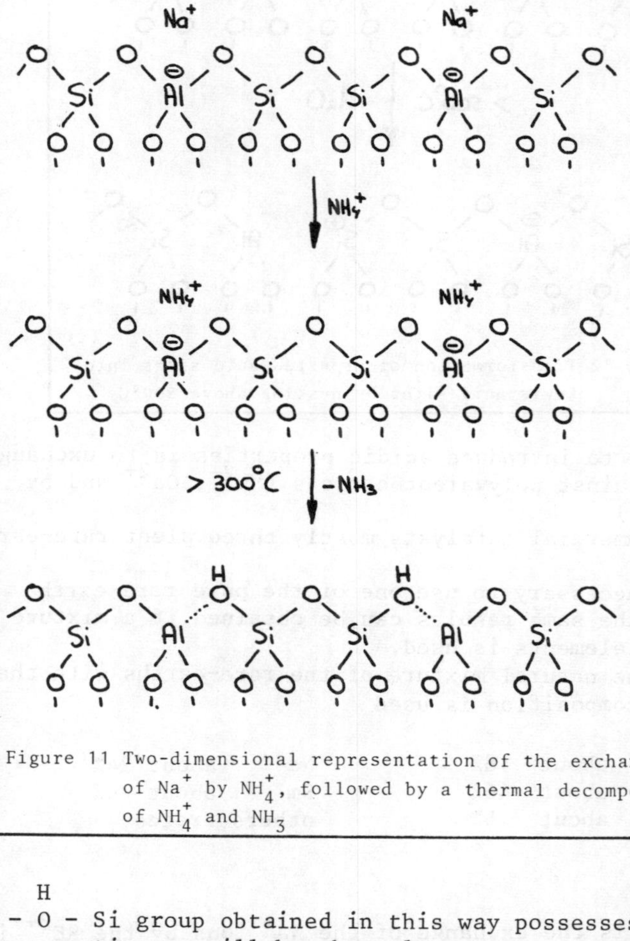

Figure 11 Two-dimensional representation of the exchange of Na^+ by NH_4^+, followed by a thermal decomposition of NH_4^+ and NH_3

The Al − O − Si group obtained in this way possesses Brønsted
        |
        H

acidic properties as will be shown later on.
At temperatures above about $600°C$ dehydroxylation occurs, resulting in the transformation of the Brønsted-acid sites into Lewis-acid sites (see figure 12).

At still increasing temperatures above about $750°C$ the crystal structure of the zeolite collapses resulting in an inactive amorphous material

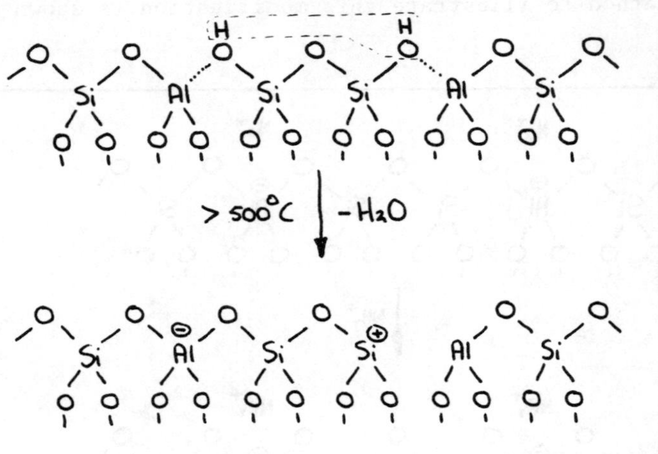

Figure 12 Transformation of Brønsted-acid sites into Lewis-acid sites by heating above 500°C

A second method to introduce acidic properties is to exchange the Na^+ ions against polyvalent cations (Mg^{2+}, Ca^{2+} and by preference La^{3+})
Nowadays in commercial catalysts mostly threevalent rare-earth ions are used.
    It is not necessary to use one of the pure rare-earth elements; the same results can be obtained if a mixture of rare-earth elements is used.
    Normally the natural mixture of the rare-earths with the following composition is used

La	about 18%	Nd	about 24%
Ce	about 46%	Sm	about 6%
Pr	about 5%	others	traces

The first step is the exchange of the Na^+ ions by the RE^{3+} ions by contacting the NaY zeolite with a aqueous solution of the rare earth chlorides. Three Na^+ ions are replaced by one RE^{3+} ion that under these circumstances will be hydrated.
The second step occurs by gently heating the exchanged zeolite. At about 250°C one of the H_2O molecules of the hydrate mantle will be hydrolyzed under the influence of the threevalent rare-earth ion. This is shown schematically in figure 13.

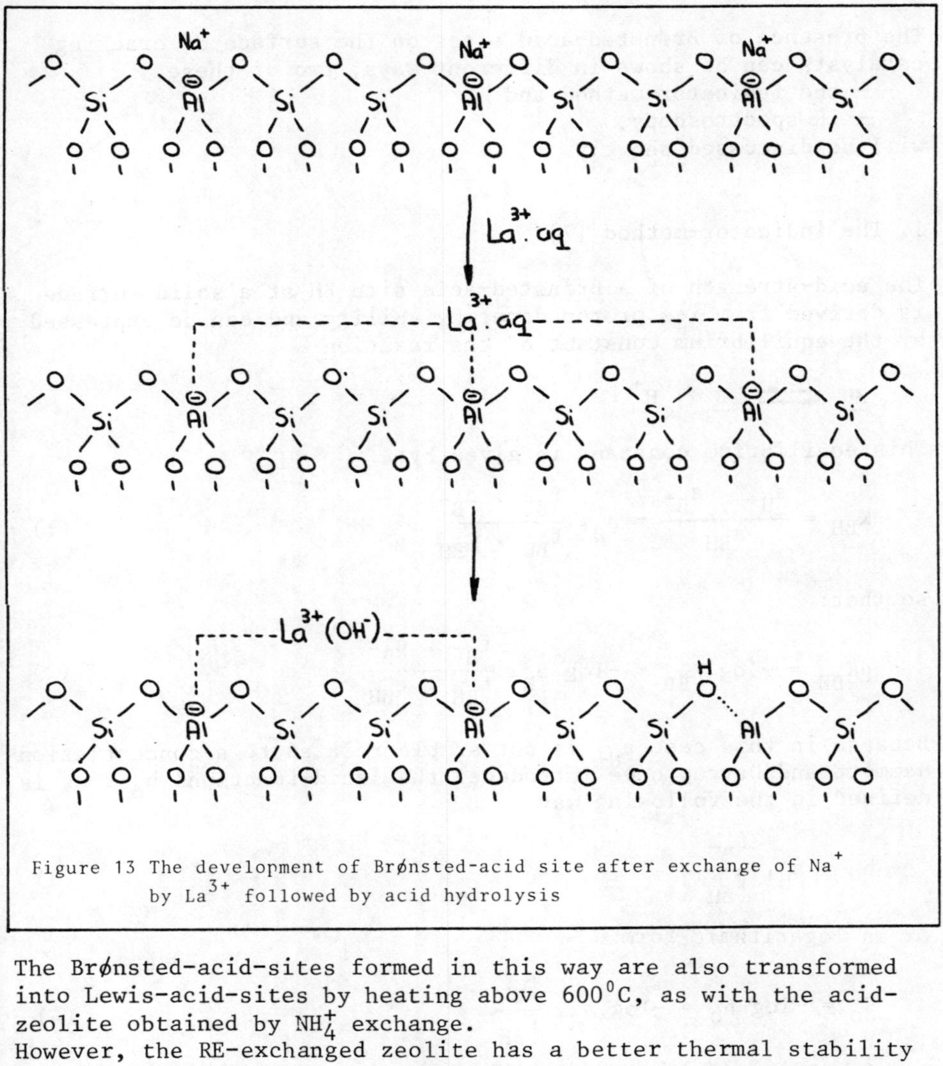

Cracking catalysts

Figure 13 The development of Brønsted-acid site after exchange of Na^+ by La^{3+} followed by acid hydrolysis

The Brønsted-acid-sites formed in this way are also transformed into Lewis-acid-sites by heating above $600°C$, as with the acid-zeolite obtained by NH_4^+ exchange.
However, the RE-exchanged zeolite has a better thermal stability than the NH_4-exchanged one. The collapse temperature can be increased up to $900°C$ depending on the rest Na content of the zeolite.
In literature often the following nomenclature is used to indicate the distinct steps

$$NaY \xrightarrow{NH_4^+} NH_4NaY \xrightarrow{300°C} HNaY \xrightarrow{600°C} \text{decationated Y}$$

$$NaY \xrightarrow{RE_{aq}^{3+}} RENaY \xrightarrow{300°C} \text{activated RENaY}$$

The presence of Brønsted-acid sites on the surface of cracking catalysts can be shown in different ways; two of these,
- the indicator-method and
- IR spectroscopy,

will be discussed shortly.

i. The indicator-method [7]

The acid-strength of a Brønsted-acid site BH at a solid surface is derived from its proton donating ability and can be expressed by the equilibrium constant of the reaction

$$\underline{BH} \rightleftarrows \underline{B^-} + H^+$$

This equilibrium constant is given by:

$$K_{BH} = \frac{a_{H^+} \cdot a_{B^-}}{a_{BH}} = a_{H^+} \frac{f_{B^-} \cdot c_{B^-}}{f_{BH} \cdot c_{BH}} \tag{1}$$

so that:

$$pK_{BH} = -\log K_{BH} = -\log a_{H^+} \frac{f_{B^-} \cdot c_{B^-}}{f_{BH} \cdot c_{BH}} \tag{2}$$

Because in this case a_{H^+} is not simple related to a concentration Hammett and Deyrup have introduced the 'acid-function' h_o that is defined in the following way

$$h_o = a_{H^+} \frac{f_{B^-}}{f_{BH}} \tag{3}$$

or in logarithmic form

$$H_o = -\log h_o = -\log a_{H^+} \frac{f_{B^-}}{f_{BH}} \tag{4}$$

Combination of equations (2) and (4) than result in:

$$H_o = pK_{BH} + \log \frac{c_{B^-}}{c_{BH}}$$

The determination of this so-called Hammett function is based upon the reaction of the acid sites with a set of indicators with decreasing basic strength. In the table a number of these indicators are given together with the Hammett function of the acids that is needed for a colour change.

Indicator	Basic colour	Acid colour	H_o	H_2SO_4 conc wt%
Neutral red	yellow	red	+6.8	8×10^{-8}
Phenylazonaphthylamine	yellow	red	+4.0	5×10^{-5}
Butter yellow	yellow	red	+3.3	$\times 10^{-4}$
Benzeneazophenylamine	yellow	purple	+1.5	0.02
Dicinnamalacetone	yellow	red	-3.0	48
Benzalacetophenone	colourless	yellow	-5.6	71
Anthraquinone	colourless	yellow	-8.2	90

The experimental method is as follows:
The zeolite from which the acid strength has to be determined is suspended in benzene and a few drops of a non-aqueous solution of one of the indicators is added. If a colour change is observed it can be concluded that the zeolite contains acid-sites with an acid-strength exceeding the Hammett function of the indicator used. It can be shown in this way that RENaY contains acid-sites with an acid-strength stronger than $H_o = -8.2$ (with anthraquinone a colour change can be observed).
The quantity of these strong acid-sites can be determined by a titration method as developed by Benesi. In this method increasing amounts of n-butylamine are added before the addition of the indicator.
From the amount of n-butylamine that is needed to obtain no colour change of a certain indicator, the number of acid groups exceeding the corresponding acid-strength can be determined.

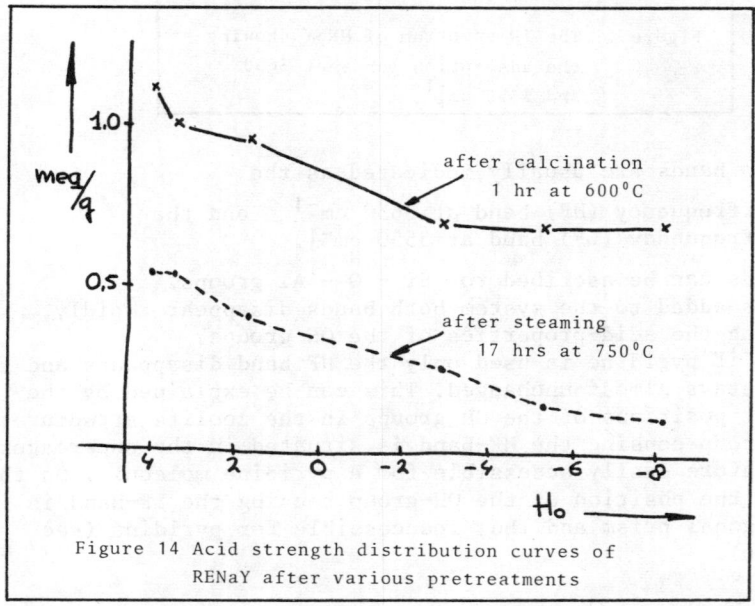

Figure 14 Acid strength distribution curves of RENaY after various pretreatments

By repeating this titration with the distinct indicators it is possible to obtain an acid-strength distribution curve. For RENaY the following distribution curve can be obtained [8] (see figure 14).

It is shown in this figure that the acid-sites are influenced by the pretreatment conditions. Hydrothermal treatment at 750^0C causes the disappearance of most of the very strong acid-sites while the sites with medium acid-strength are only slightly influenced.

ii. IR spectroscopy [9]

The IR spectrum of a HNaY zeolite obtained by heating NH_4NaY at $300^0C - 450^0C$ shows two characteristic absorption bands (see figure 15).

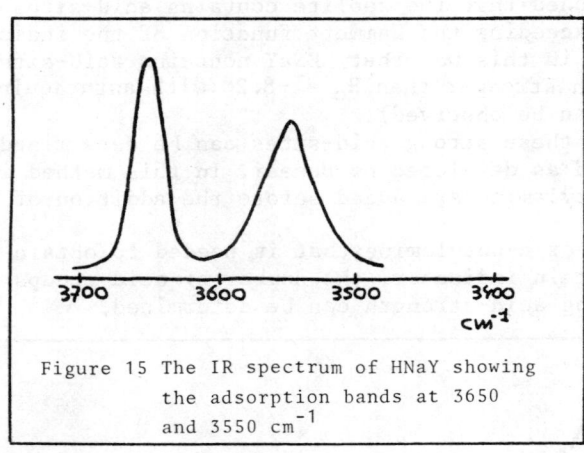

Figure 15 The IR spectrum of HNaY showing the adsorption bands at 3650 and 3550 cm^{-1}

These two bands are usually indicated as the

high frequency (HF) band at 3650 cm^{-1} and the
low frequency (LF) band at 3550 cm^{-1}.

Both bands can be ascribed to $Si - \overset{H}{O} - Al$ groups.

If NH_3 is added to the system both bands disappear rapidly, indicating the acid-properties of the OH groups.

However, if pyridine is used only the HF band disappears and the LF-band stays almost unchanged. This can be explained by the different positions of the OH groups in the zeolite structure; the OH group causing the HF-band is situated in the supercages and therefore easily accessible for a pyridine molecule. On the contrary the position of the OH-group causing the LF-band is in the hexagonal prism and thus inaccessible for pyridine (see figure 16).

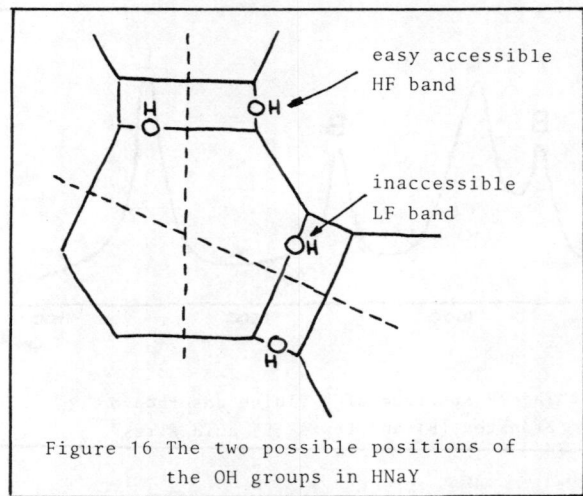

Figure 16 The two possible positions of the OH groups in HNaY

Lewis-acid groups cannot be detected directly by IR spectroscopy. However, after the adsorption of pyridine it is possible. On Brønsted-acid sites pyridine is adsorbed in the form of pyridinium ions while on Lewis-acid sites the adsorption takes place via a coordinative bond (see figure 17), leading to different IR spectra (see figure 18).

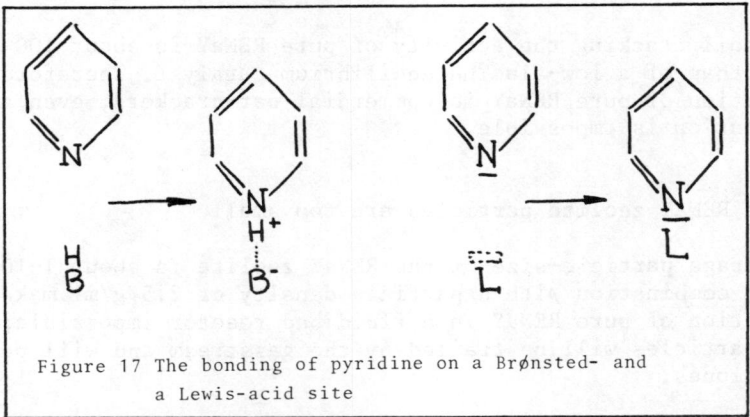

Figure 17 The bonding of pyridine on a Brønsted- and a Lewis-acid site

The pyridinium ion has absorption bands at 1540 and 1636 cm^{-1} and the coordinative bonded pyridine at 1450 and 1615 cm^{-1}

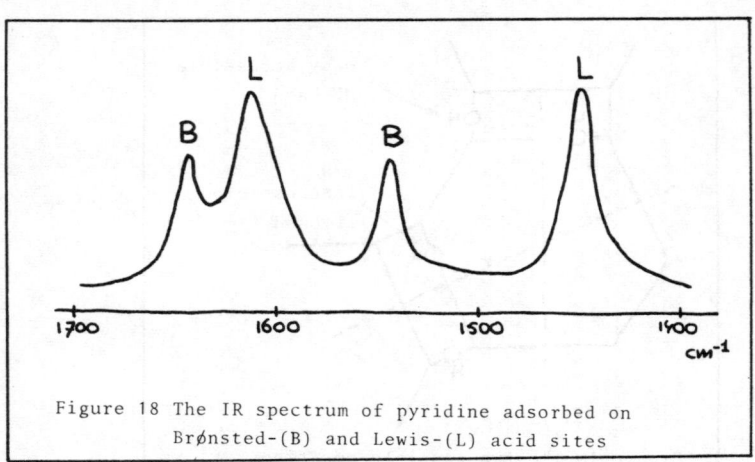

Figure 18 The IR spectrum of pyridine adsorbed on Brønsted-(B) and Lewis-(L) acid sites

APPLICATION OF RENaY AS CRACKING CATALYST

In the preceeding paragraphs it is shown that RENaY has acid properties and therefore it may be used as a cracking catalyst. However, two practical problems have to be solved first.

i. The activity of pure RENaY is too high

For gasoil cracking the activity of pure RENaY is about 100 times higher than of a low-alumina equilibrium catalyst. Therefore application of pure RENaY in commercial cat-crackers, even after modification is impossible.

ii. The RENaY zeolite particles are too small

The average particle-size of the RENaY zeolite is about 1-10 μm. This in combination with a particle density of 2.5 g/ml makes application of pure RENaY in a fluid bed reactor impossible; the small particles will be trailed by the gasstream and will pass the cyclones.

To solve these problems the RENaY zeolite particles can be embedded in a matrix. For example 5-25% RENaY can be embedded in low-alumina and from this mixture particles can be formed that can be used in a fluid bed reactor (average particle size about 60 μm). If we compare the performance of such a zeolite containing catalyst with an amorphous low-alumina catalyst we obtain the following picture.

	LA catalyst 100% LA	RENaY catalyst 10% RENaY 90% LA
relative activity for gasoil cracking	100	1000
selectivity at 75% conversion		
gas	15	10
gasoline	45	55
light cycle oil	10	7.5
coke	5	2.5

Compared with LA the zeolite catalyst has a much higher activity that allows much higher spacevelocities to obtain the same conversions. Furthermore the selectivity has drastically changed.
The selectivity for gasoline has been improved of the cost of the production of gas, light cycle oil and coke.
But also the composition of the gasoline has been changed [4]

Gasoline composition	LA catalyst	Zeolite catalyst
Paraffins	10	20
Olefins	45	15
Naphthenes	10	20
Aromatics	35	45

These differences can be explained by the relatively high rates of hydrogen transfer reactions in zeolites [10, 11].
Olefins, the primary cracking products, can be transformed to paraffins and aromatics according to the following reaction equation

$$4\ C_nH_{2n} \longrightarrow 3\ C_nH_{2n+2} + C_nH_{2n-6}$$

4 olefins   3 paraffins   aromatic

The relatively high rate of this reaction in zeolites can be understood in the following way:
carbenium ions can undergo two distinct reactions
i. they can be cracked by β-fission
ii. they can undergo a H-transfer reaction.
Schematically

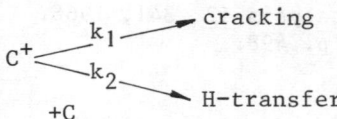

The rates of these reactions are given by

cracking reaction $\quad r_1 = k_1 [C^+] \quad$ monomolecular reaction
H transfer reaction $\quad r_2 = k_2 [C^+][C] \quad$ bimolecular reaction

The ratio between these two rates can be expressed as

$$\frac{\text{rate H-transfer reaction}}{\text{rate cracking reaction}} = \frac{k_2 [C^+][C]}{k_1 [C^+]} \approx k'[C]$$

Thus the rate of the H-transfer reaction is favoured by a high hydrocarbon concentration [C] at the surface of the catalyst. This especially is the case in the small pores of the zeolites.

REFERENCES

1. C. Okkerse in B.C. Linsen (ea) 'Physical and Chemical Aspects of Adsorbents and Catalysts', p. 213 Academic, New York 1970.
2. Akzo Chemie, brochure of Ketjen catalysts.
3. C.J. Planck and E.J. Rosinski, Chem. Eng. Progr. 63, 26, 1967.
4. S.C. Eastwood, C.J. Planck and P.B. Weisz, Proc. 8th World Petr. Congr. Moscow, 1971.
5. D.W. Breck, 'Zeolite Molecular Sieves', Wiley, New York, 1974.
6. J.V. Smith, Adv. Chem. Ser. 101, 171, 1971.
7. H.A. Benesi, J. Am. Chem. Soc. 78, 5490, 1956, and J. Phys. Chem. 61, 970, 1957.
8. L. Moscou, R. Moné, J. Catalysis, 30, 417, 1973.
9. J.W. Ward in J.A. Rabo (ea) 'Zeolite Chemistry and Catalysis' Amer. Chem. Soc., Washington, 1976.
10. C.L. Thomas and D.S. Barmby, J. Catalysis 12, 341, 1968.
11. P.B. Weisz, Chemtech, August 1973, p. 498.

BASIC FLUIDISATION

P.N. Rowe

Department of Chemical and Biochemical Engineering,
University College London

ABSTRACT. Introduction. Minimum fluidisation velocity. Experimental observation of bubbles. Particle movement caused by bubbles. Bubble size, coalescence and velocity. Gas flow through the bed. Two-phase theory. Heat and mass transfer. Physical considerations in designing a chemical reactor. A simple chemical reactor model.

1. INTRODUCTION

Imagine a dry powder such as fine sand or cracking catalyst resting in a vessel with a porous or perforated base as indicated in Fig. 1. If air is blown upwards through the bed of powder there will be a pressure drop across it, ΔP, which will increase with the gas velocity, U. At a critical velocity, U_{mf}, the pressure drop will equal the weight of the powder divided by the cross-sectional area of the bed. This is known as the minimum fluidisation velocity; i.e. at U_{mf}, $\Delta P = mg/A_R$. It is an experimental fact that thereafter the pressure drop remains constant as gas velocity increases, as is shown in Fig. 2. The entire bed is now supported by the gas stream rising through it. Particles are no longer resting on one another, inter-particle friction is largely eliminated and there is little resistance to relative motion so that the whole can be stirred or will flow like a liquid. Hence the name, "fluidised bed".

When and only when the supporting fluid is a gas the bed begins to boil or bubble once U_{mf} is exceeded and looks rather as indicated in Fig. 3. Increasing velocity increases the vigour of bubbling. The whole looks well mixed and turbulent but this

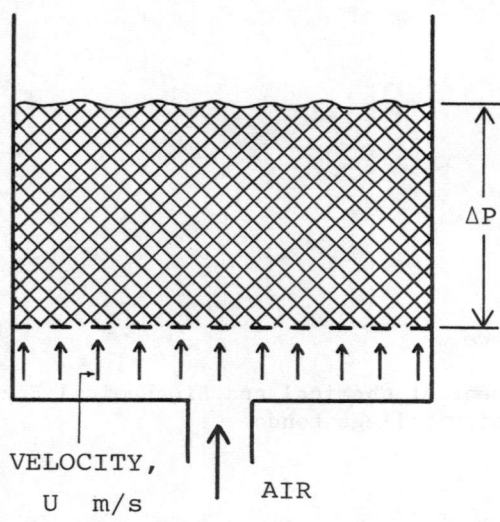

Fig. 1. Pressure drop across a fluidised bed.

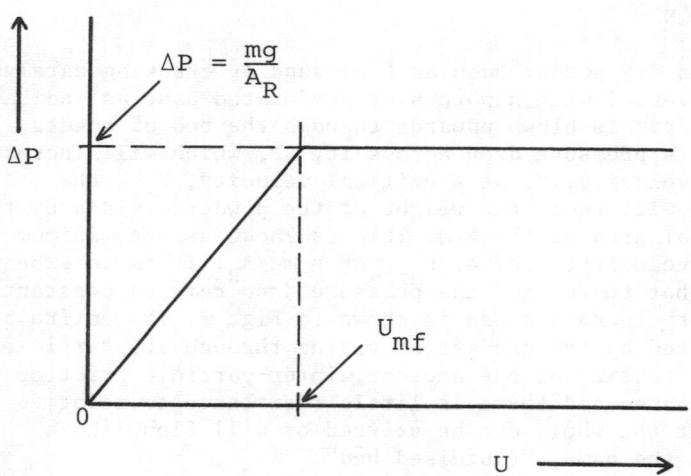

Fig. 2. Variation of pressure drop with gas velocity.

Basic fluidisation

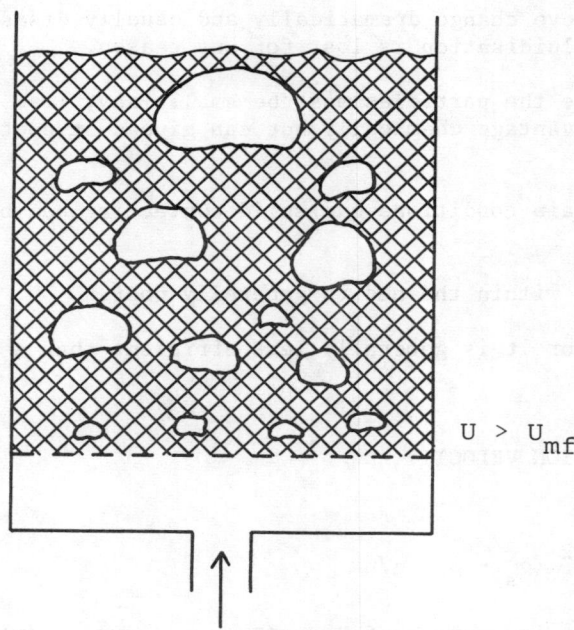

Fig. 3. Appearance of a bubbling fluidised bed.

appearance is misleading.

The fluidised bed is an attractive way of engineering gas/solid chemical reactions and has many advantages.

(1) Powder can be added to and removed from the bed continuously or periodically by allowing it to flow through pipes like a liquid.

(2) Temperature within the bed is very uniform even during vigorously energetic chemical reaction.

(3) The particles are generally well mixed and hence all take part in the reaction equally and composition is uniform.

(4) The pressure drop remains constant and is determined only by the bed height and powder density.

(5) Heat transfer between bed and walls (or other immersed surfaces) is very high so that temperature is easily controlled. This is in marked contrast to the packed bed where heat removal can be a serious problem.

There are some disadvantages:

(1) All the above change dramatically and usually disastrously if the state of fluidisation is lost for any reason.

(2) In practice the particles must be small (say, less than 1 mm) which is an advantage chemically but can give dust containment problems.

(3) Under certain conditions gas/solid contacting can be very poor.

(4) Gas mixing within the bed is generally poor.

(5) As a reactor it is generally less efficient than a plug flow reactor.

2. MINIMUM FLUIDISATION VELOCITY

Roughly,

$$U_{mf} = 0.0006 \, d_p^2 \, (\rho_s - \rho_f) \, g/\eta \tag{1}$$

This is only valid at low values of Re_{mf} ($Re_{mf} = U_{mf} \, d_p \, \rho_f/\eta$) and is not a very good predictor but it shows simply on what the velocity depends. With gases the fluid density tends to zero and for a given chemical reaction usually gas viscosity and solids density are fixed so that the only remaining variable is particle size. Minimum fluidisation velocity varies roughly as the square of this so it is particularly sensitive to the choice of granule size.

A better predictive equation is Leva's empirical formula (Ref. 1),

$$U_{mf} = 7.90 \times 10^{-3} \, d_p^{1.82} \, (\rho_s - \rho_f)^{0.94}/\eta^{0.88} \tag{2}$$

valid for $Re_{mf} < 10$. In both equations the constants refer to SI units.

This subject is discussed in greater detail by Richardson (Ref. 2).

3. EXPERIMENTAL OBSERVATION OF BUBBLES

The bubbles are responsible for many of the advantages and disadvantages of gas fluidised bed chemical reactors so it is necessary to understand their nature. Unlike, for example,

Basic fluidisation

bubbles of air in water, they cannot normally be seen in the interior of a bed of powder. Knowledge of them comes from either special laboratory experiments or from indirect observation (Ref. 3).

In most industrial situations the only practical means of observation is through probes. There are many different kinds - capacitance, heat transfer (thermistor), radiation adsorption, two-pronged light probes and others. All respond to changes in particle concentration around the tip. The signal from some refers to a not well defined volume and from others to more nearly a point value. They can be used to obtain information about the bubble concentration position, size and velocity. Generally considerable signal processing is necessary and the results depend on assumptions of various kinds.

The two-dimensional bed is very instructive in understanding the nature of the bubbles. It consists of two glass plates about 1 cm apart between which the fluidised powder is contained. It is effectively a thin slice of a normal bed. Fig. 4 shows the arrangement. When illuminated from behind the bubbles are seen in silhouette and their size, shape, number, position and velocity can be measured. Alternatively the bed can be viewed by reflected

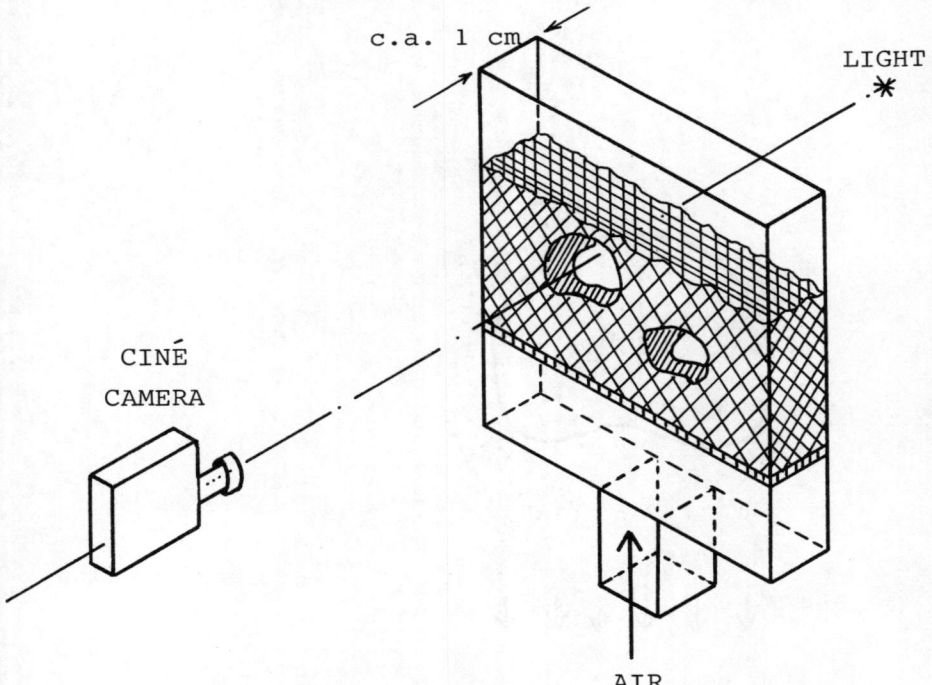

Fig. 4. Two-dimensional bed for experimental study.

light so that the movement of particles against the front face can be seen. Further, a dark tracer gas such as NO_2 can be introduced through the front face to reveal the patterns of gas flow within the bed.

Bubbles within the interior of a normal cylindrical or rectangular bed can be observed by X-ray ciné photography. This technique is limited to beds only a few tens of centimeters wide and materials of low atomic weight (Ref. 4).

Particle movement and the disturbance caused by bubbles can be seen by introducing coloured particles appropriately. One technique is to start with coloured layers and subsequently to section the disturbed bed (Ref. 5).

4. PARTICLE MOVEMENT CAUSED BY BUBBLES

Photographs of two-dimensional beds show that bubbles undisturbed by near neighbours and walls are roughly circular but with an indented base. They are essentially empty of particles and rise by particles flowing around them as indicated in Fig. 5

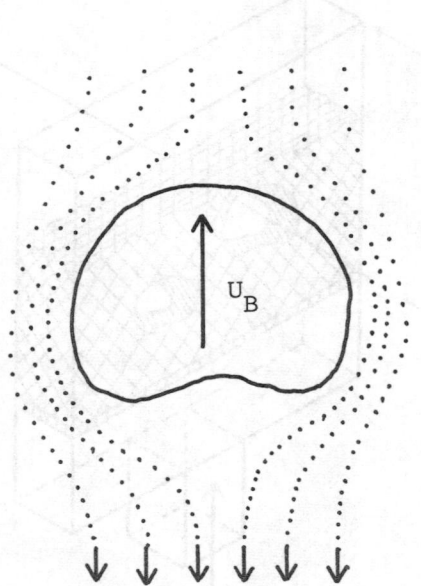

Fig. 5. Bubble rises by particles flowing round it.

Basic fluidisation

(Ref. 3). They behave rather like solid bodies moving through a true fluid in streamline flow. The apparent identation of the base in a wake of particles travelling with the bubble - rather like the attached vortices behind a slowly moving rowing boat in still water. The wake volume, V_w, averages one-third of the bubble volume, V_B, as shown in Fig. 6.

In consequence the overall disturbance caused by a single bubble is as follows. Imagine the bed divided into two layers, black particles below and white ones above. As a bubble crosses the colour interface it carries black particles across and eventually deposits them on the surface. It also draws a sharp spout of black particles behind solely as a result of the streamline motion. This is illustrated in Fig. 7.

Each bubble produces this basic displacement. When they are of different sizes and rising along different paths the result soon looks very confusing and rapidly approaches good mixing. However, it is not an entirely random process and it is misleading to describe its rate in terms of an eddy diffusion coefficient. The rate of particle mixing depends directly on the rate of bubbling.

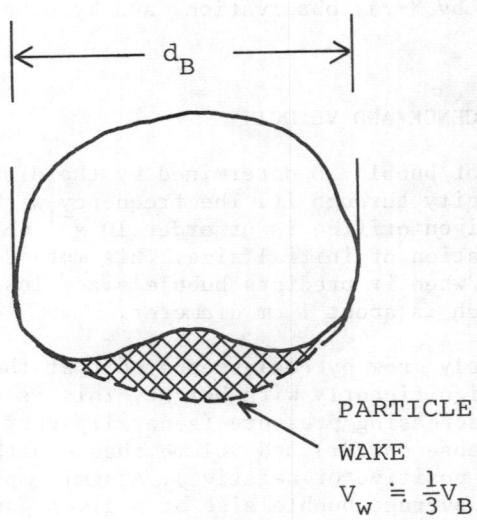

Fig. 6. Particle wake that travels with the bubble.

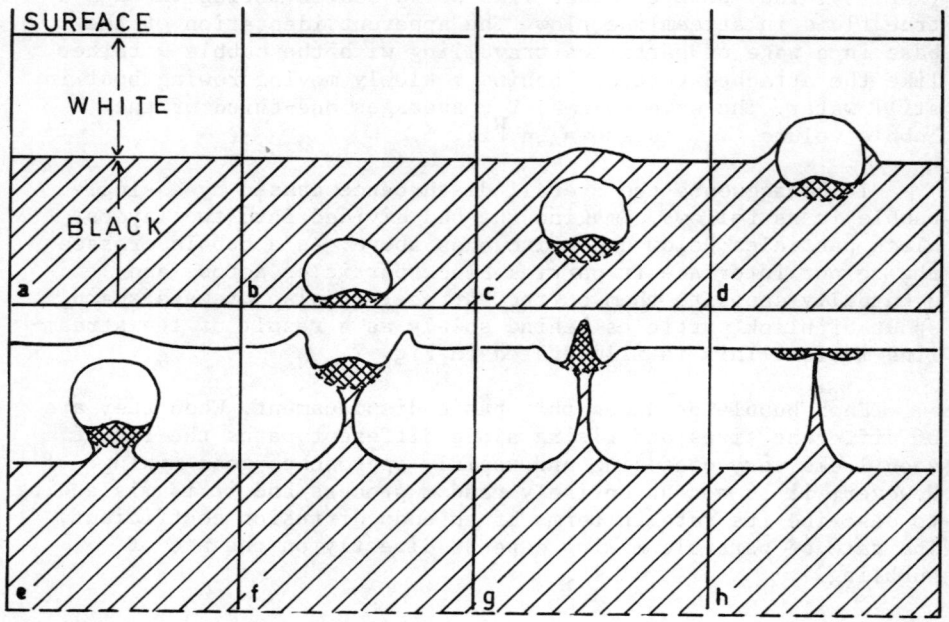

Fig. 7. Particle displacement caused by a bubble.

These observations have been confirmed in truly three-dimensional situations by X-ray observation and by other techniques.

5. BUBBLE SIZE, COALESCENCE AND VELOCITY

The initial size of bubble is determined by the distributor plate and the gas velocity through it. The frequency with which bubbles form above a given orifice is of order 10 s^{-1} which permits a rough calculation of initial size. This method of estimation breaks down when it predicts bubble sizes less than the stable minimum which is about 1 cm diameter.

Bubbles subsequently grow by coalescence so that the volume flow does not change significantly with height. This is ignoring volume increase with decreasing pressure (generally very small in all but deep beds of dense powder) and volume change with chemical reaction (which may be positive or negative). A semi-empirical equation that predicts average bubble size at a given gas flow rate and level in the bed is (Ref. 6)

$$\bar{d}_B = (U - U_{mf})^{\frac{1}{2}} (h + h_o)^{\frac{3}{4}}/g^{\frac{1}{4}} \tag{3}$$

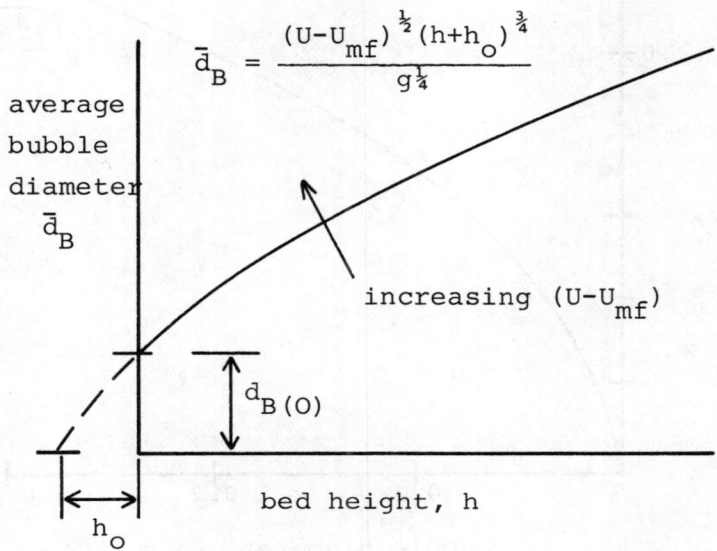

Fig. 8. Variation of bubble size with bed height.

As seen in Fig. 8, h_o is that imaginary distance beneath the distributor plate at which the initial bubble diameter, $d_{B(0)}$ would be zero.

A variety of bubble sizes exists at any level and their distribution is not normal but positively skewed for there is an excess of small ones (Ref. 7). This is because, although coalescence dominates, some splitting occurs and small daughters are shed.

Bubble rise velocity, U_B, varies with size according to (Ref. 8)

$$U_B = \sqrt{g \frac{d_B}{2}} \qquad (4)$$

as shown in Fig. 9. Since it varies with the square root of diameter (sixth root of volume) there is not large variation in velocity and a typical value is about 0.5 m/s. The velocity of individual bubbles varies quite a lot about the average value predicted by Equation (4) especially as it accelerates and decelerates during coalescence. Within a chemical reactor only average values are of interest.

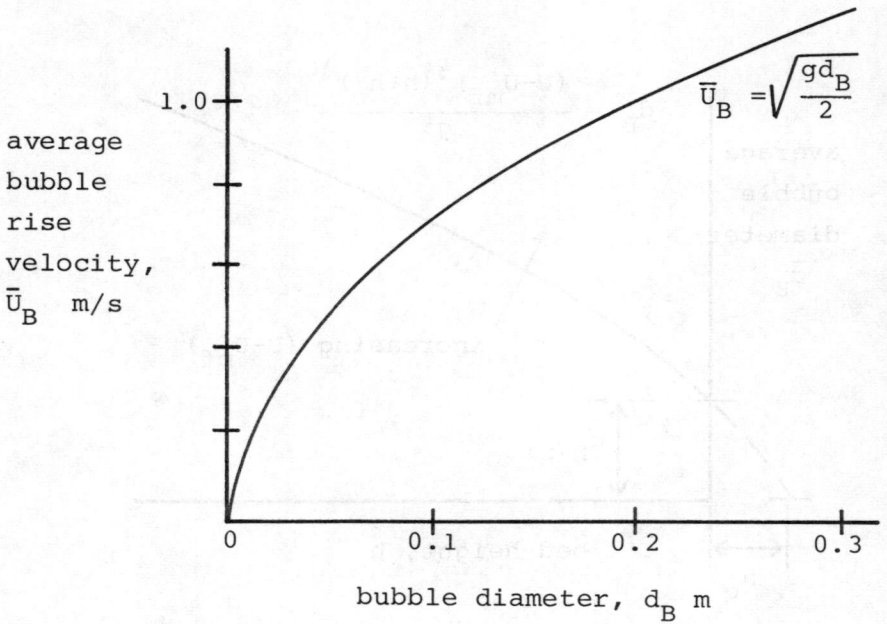

Fig. 9. Variation of bubble velocity with size.

6. GAS FLOW THROUGH THE BED

When tracer gas is introduced through the wall of a two-dimensional bed it is seen to follow a streamline path and not to disperse as might at first be expected. The path may be complex and vary with time but the thread retains its identity until it enters a bubble. Figure 10 illustrates a typical observation.

If the tracer thread is interrupted the head or tail can be followed to measure local interstitial gas velocity, U_i. From this it is observed that U_i remains at its value at the onset of fluidisation and does not increase when the total flow into the bed is increased. This observation has important consequences discussed in the next section. It should be noted here that gas velocities are conventionally expressed as superficial velocities, i.e. $U = F/A_R$ where F is the volumetric flow into a bed of cross-sectional area A_R. The interstitial velocity must take account of the space occupied by particles and thus

$$U_i = U_{mf}/\varepsilon_{mf} \tag{5}$$

where ε_{mf} is the voidage at minimum fluidisation.

Basic fluidisation

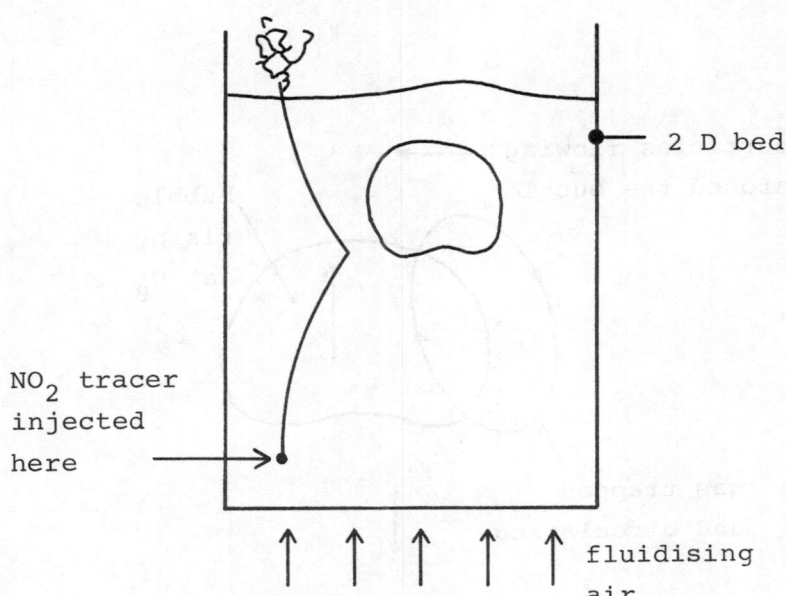

Fig. 10. Pattern of gas flow when $U_i > U_B$.

As indicated in Fig. 10, tracer gas is attracted towards a bubble and these can be thought of as regions of very high permeability in a volume of otherwise uniform permeability across which a potential exists - the pressure drop. The situation is analogous to the way magnetic lines of force converge towards a soft iron core, i.e. a region of high magnetic permeability. However, the fluidised bed situation is an unsteady one because the bubble is moving.

When the interstitial gas velocity, U_i, is greater than the bubble velocity, U_B, gas can catch up the bubble and enter it and within the open space some mixing occurs so that the tracer thread is lost. When $U_B > U_i$ an interesting and important situation develops. Gas must flow upwards through the bubble because there is a pressure difference and virtually no resistance. As it re-enters the powder above the bubble roof it finds itself in a rapidly moving medium that is flowing round the bubble. When the powder velocity (proportional to bubble velocity) is greater than the gas velocity (proportional to minimum fluidisation velocity) the gas cannot escape but is dragged round by the powder. Once it returns towards the base it is pushed back into the bubble by the pressure gradient. Thus, a body of gas is trapped and remains centred on the bubble whilst it circulates like a spherical vortex. This is illustrated in Fig. 11.

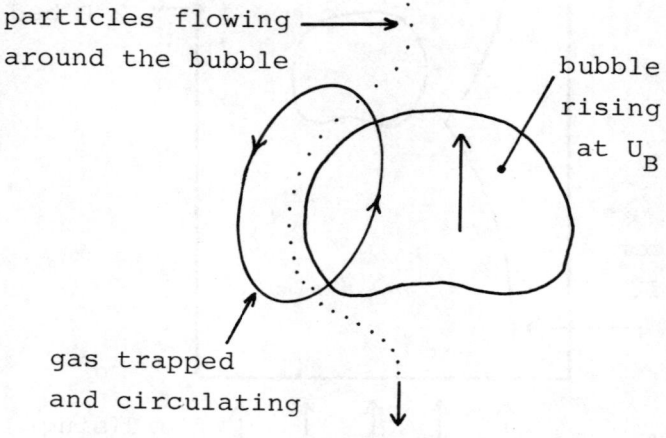

Fig. 11. How a cloud forms when $U_B > U_i$.

The gas centred on the bubble is called a cloud and there is necessarily an interface between it and other gas which, relative to the bubble, is moving downwards. Fig. 12 shows the form this takes, a shape that has been observed experimentally by forming NO_2 bubbles in a two-dimensional bed and also predicted by theory. A simple theory shows that for usual spherical bubbles the ratio of cloud to bubble diameter is given by (Ref. 8)

$$d_c/d_B = \left[(\alpha + 2)/(\alpha - 1)\right]^{1/3} \tag{6}$$

where $\alpha = U_B/U_i = U_B \varepsilon_{mf}/U_{mf}$. Equation (6) is plotted in Fig. 13. A semi-empirical equation that fits the data slightly better is (Ref. 9)

$$(V_C - V_B)/V_B = 1.17/(\alpha - 1) \tag{7}$$

where V_C and V_B are the respective volumes.

The important ratio α can vary widely amongst practical systems. For a given chemical process where the gas viscosity and the particle density is fixed, U_i varies approximately as d_p^2. Bubble velocity varies as $d_B^{\frac{1}{2}}$ and therefore

Basic fluidisation   193

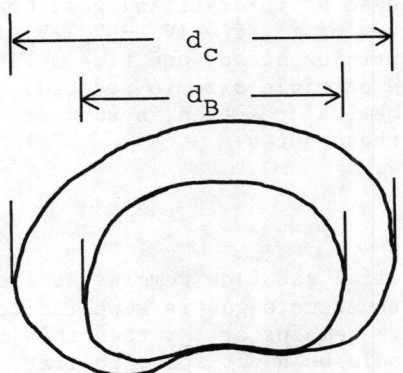

Fig. 12. Shape of cloud that forms.

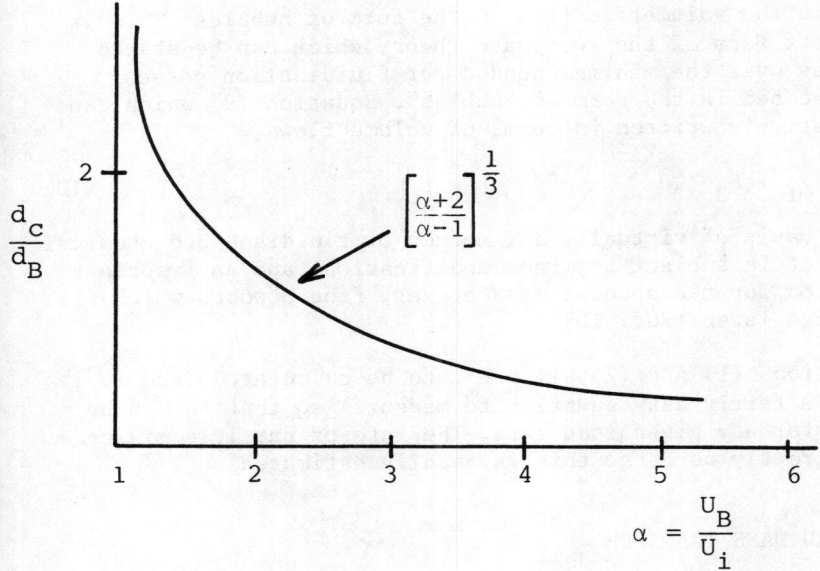

Fig. 13. Variation of cloud diameter with velocity ratio.

$$\alpha = \text{Constant} \times d_B^{1/2}/d_p^2 \tag{8}$$

In a relatively shallow bed bubble size may not vary very much so that to a first approximation α varies inversely with d_p^2.

By the mechanism described above the phase of cloud gas is largely isolated from the other phase of interstitial gas. Considering the former, only a fraction of it, $f = (V_C - V_B)/V_C$ is actually in contact with solid particles at any one time. The active proportion decreases as the particle size is reduced. This limits the overall rate of chemical reaction, a subject discussed in the last section of these notes.

7. TWO-PHASE THEORY

As stated above, the interstitial gas flow remains at its minimum fluidisation value at whatever rate gas is supplied to the bed. If the dense phase voidage remains at ε_{mf} then this must be so for if U_i increased there would be a net force to lift particles out of the bed. If the interstitial flow cannot increase then the excess flow must be carried by the bubbles and we can write

$$U = U_{mf} + \frac{F_B}{A_R} \tag{9}$$

where F_B is the volumetric flow in the form of bubbles. This is the simplest form of the two-phase theory which can be stated "excess gas over the minimum needed for fluidisation passes through the bed in the form of bubbles". Equation (9), which can be alternatively written in terms of volume flows,

$$F = F_{mf} + F_B \tag{10}$$

forms the basis of virtually all models of fluidised bed chemical reactors. It is subject to minor modifications and an important modification for the special case of very fine powders which will be discussed later (Ref. 10).

Equations (1) and (2) allow U_{mf} to be calculated (and it is generally a fairly easy quantity to measure) so that F_B can be estimated for any given feed rate. The rate of particle mixing depends directly on F_B so this is readily estimated

8. HEAT AND MASS TRANSFER

There are two distinctly different and unrelated heat transfer coefficients relevant to a fluidised bed. Heat may be

Basic fluidisation

transferred between the walls or other immersed surfaces and the bed for which the transfer coefficient $h_{B/W}$ is used. Also heat may be transferred between the entering gas and the particles for which $h_{G/P}$ is used. Normally mass transfer occurs only in the latter situation. The heat transfer coefficient is defined by

$$\emptyset = h A \Delta T \qquad (11)$$

where $\emptyset$ is the rate of heat flow under a temperature difference ΔT across a total surface area A.

Heat transfer to the walls increases dramatically once the bed is fluidised and bubbling. Fig. 14 shows how this changes with gas velocity (Ref. 11) and illustrates the low coefficient characteristic of packed beds. The maximum value of the coefficient is of order 100 w/m^{2}K and it is not usually necessary to know it accurately because overall heat transfer is limited by conditions outside the wall. The high coefficient arises because of particle mixing in the bed continuously bringing fresh particles to a wall of different temperature. It is the bubbles that cause this high heat transfer coefficient.

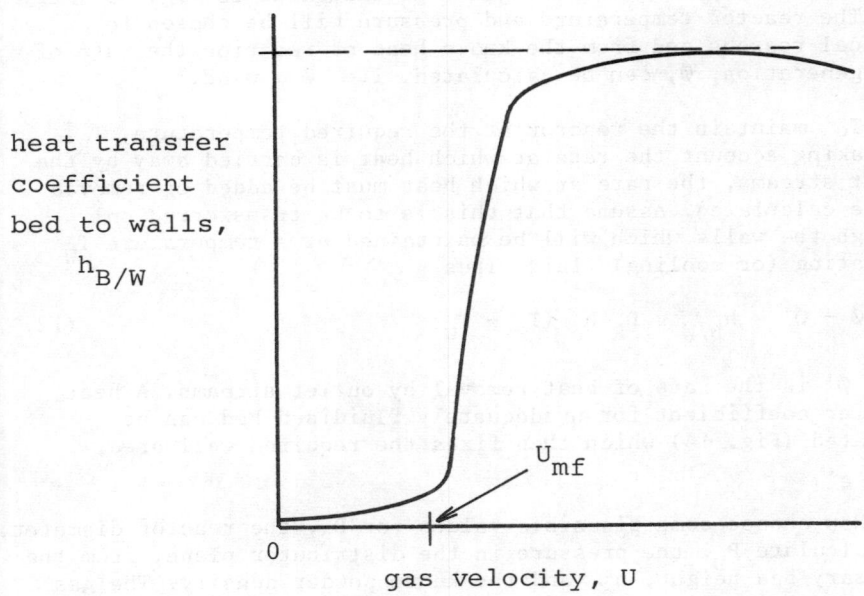

Fig. 14. Variation of heat transfer coefficient with gas velocity.

Heat and mass transfer between gas and particles is high only because of the large surface area available. The heat transfer coefficient on a particle basis is little different in a fluidised from a packed bed. A typical value corresponds to a Nusselt number, Nu ÷ 20 (Nu = $h_{G/P} \, d_p / \lambda$). Because of the large surface area presented by unit mass of powder and because of the relatively small volumetric heat capacity of gases, gas and particles reach temperature equilibrium very quickly, usually after travelling only one or two particle diameters into the bed. Because the transfer coefficients are similar, chemical rate constants obtained from laboratory experiments on small packed beds can be applied to the dense phase of a fluidised bed - but the local concentrations must be known, of course.

9. PHYSICAL CONSIDERATIONS IN DESIGNING A CHEMICAL REACTOR

Consider a gas/solid reaction that is to be engineered in a fluidised bed and assume that it is required to manufacture product continuously at a specified rate, w kg/s. From knowledge of the chemistry, stoichiometry and an assumed degree of conversion in the reactor, the flow rates into and out of the reactor of reactants, products and by-products can be written down. In general there will be a powder and a gas stream flowing in and gas and solids products flowing out as indicated in Fig. 15 (Ref. 12). The reactor temperature and pressure will be chosen for chemical reasons and from the known heat of reaction the rate of heat generation, $\emptyset$, can be calculated, i.e. $\emptyset \, \alpha \, w \, \Delta H$.

To maintain the reactor at the required temperature, T_R, and taking account the rate at which heat is carried away by the outlet streams, the rate at which heat must be added or removed can be calculated. Assume that this is to be transferred only through the walls which will be maintained at a temperature T_W by heating (or cooling) fluid. Thus

$$\emptyset - \emptyset' = h_{B/W} \, \pi \, D_R \, h_e \, (T_R - T_W) \tag{12}$$

where $\emptyset'$ is the rate of heat removal by outlet streams. A heat transfer coefficient for an adequately fluidised bed can be estimated (Fig. 14) which then fixes the required wall area, $\pi \, D_R \, h_e$.

Now choose some plausible values for D_R, the reactor diameter, and calculate P_D, the pressure in the distributor plane, from the necessary bed height, h_e, and the known powder density. The gas feed rate can now be calculated as a volume flow and hence the velocity, U_D. The same calculation can be done for the top of the bed taking into account volume change on reaction as well as pressure change.

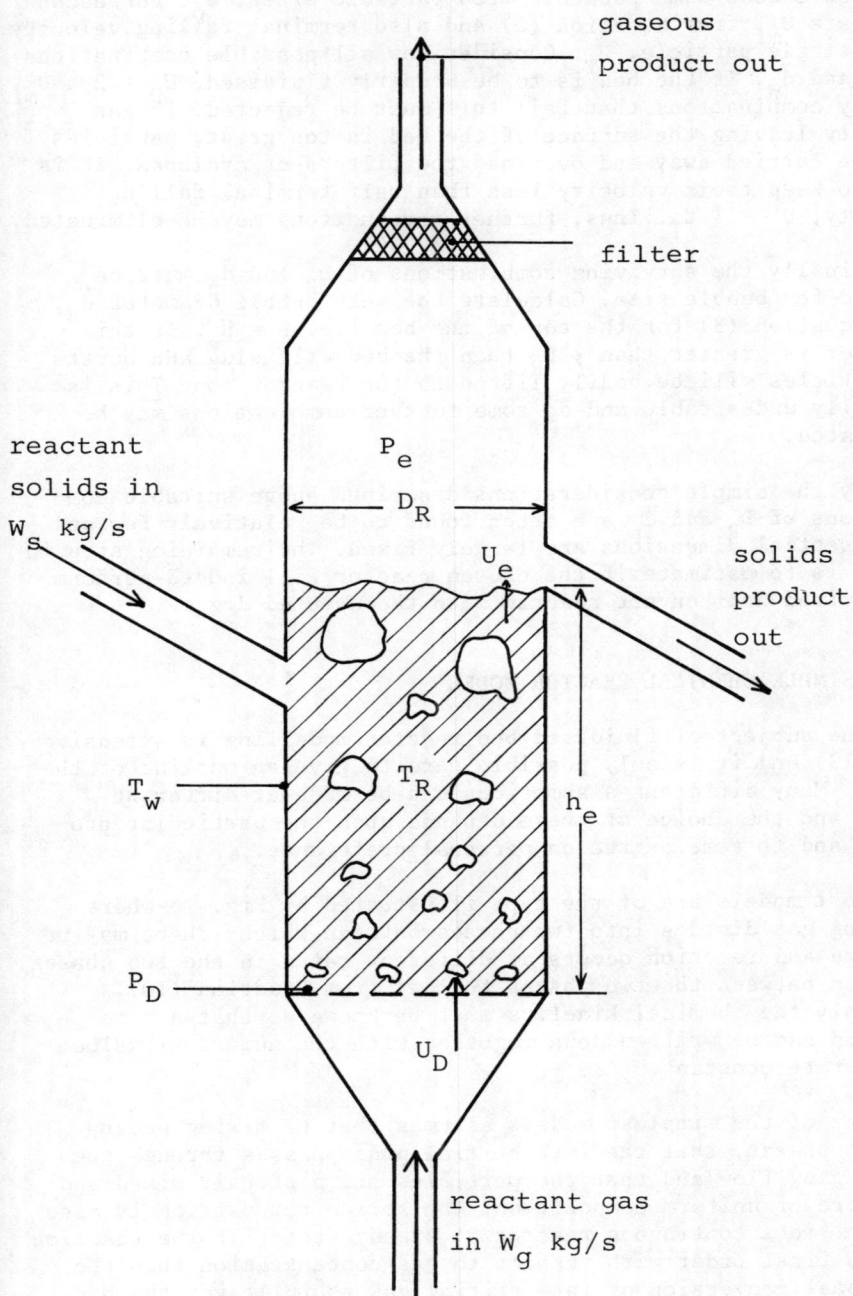

Fig. 15. Model of a gas/solids chemical reactor.

Now choose some possible mean particle sizes, $\bar{d}_p$. For each calculate U_{mf} from Equation (2) and also terminal falling velocity for a single particle, U_T. Consider now all possible combinations of D_R and $\bar{d}_p$. If the bed is to be properly fluidised, $U_D > 2 \times U_{mf}$ and any combinations that fail this must be rejected. If gas velocity leaving the surface of the bed is too great, particles will be carried away and overload the filters or cyclones. It is wise to keep their velocity less than half terminal falling velocity, $U_e < \tfrac{1}{2} U_T$. Thus, further combinations may be eliminated.

Finally the surviving combinations of D_R and $\bar{d}_p$ must be checked for bubble size. Calculate the mean bubble diameter, $\bar{d}_B$, from Equation (3) for the top of the bed i.e. $h = h_e$. If this diameter is greater than $\tfrac{1}{2} D_R$ then the bed will slug and bursts of particles will be bodily lifted to the reactor top. This is generally undesirable and so some further combinations may be eliminated.

By the simple considerations described above suitable combinations of D_R and $\bar{d}_p$ are often found to be relatively few and the essential dimensions are largely fixed. The remaining step in design is to estimate if the chosen reactor will indeed perform as required and convert reactants to the assumed degree.

10. A SIMPLE CHEMICAL REACTOR MODEL

The subject of fluidised bed reactor modelling is extensive (Ref. 13) and it is only possible here to give an outline of the method. Many different assumptions can be made at different points and the choice of these depends upon the particular process - and to some degree on personal preference.

Most models are of the type illustrated by Fig. 16 where entering gas divides into two phases between which there may be exchange and reaction occurs at different rates in the two phases. Division between the two phases follows from Equation (10). Obviously the chemical kinetics must be known so that a rate equation can be written down together with the numerical value of the rate constant.

One of the simplest models assumes that no mixing occurs between phases, that the interstitial phase passes through the bed in plug flow and that the particles are perfectly mixed and therefore of uniform composition. The solids composition is also constant in a continuous reactor at steady state. If the reaction is also first order with respect to gas concentration then the fractional conversion of interstitial gas as it leaves the bed surface will be given by

Basic fluidisation

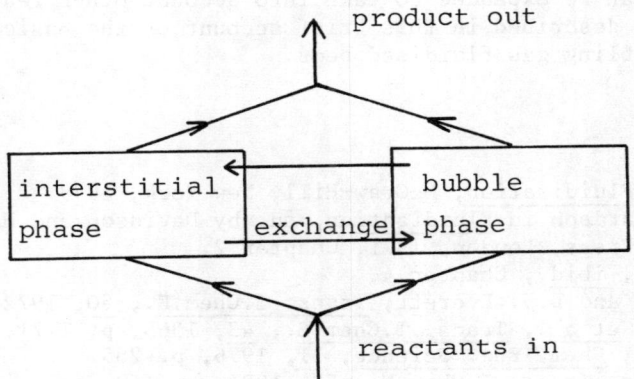

Fig. 16. Generalised two-phase chemical reactor model.

$$1-y_i = e^{-kt_r} \qquad (13)$$

where k is related to the rate constant and degree of solids conversion. The residence time, $t_r = h_e/U_{mf}$. If clouds form around the bubbles then a corresponding expression can be written for the cloud or bubble phase:

$$1-y_c = e^{-fk't'_r} \qquad (14)$$

where f is the fraction of cloud phase actually in contact with solids (Equation (7)) and the residence time, $t'_r = h_e/U_B$. The overall composition of gas leaving the reactor depends on the fractions in the two phases:

$$1-y = \frac{F_{mf}}{F} e^{-kt_r} + \frac{F_B}{F} e^{-fk't'_r} \qquad (15)$$

It should be noted that the cloud phase is far less efficient than the interstitial phase both because of restricted contact quantified by the fraction f and also by the reduced residence time when $U_B > U_i$. Both of these adverse effects increase as particle size decreases and, furthermore, the proportion flowing in the inefficient phase increases. From the point of view of gas/

solid contacting the bigger the particles the better - provided they are properly fluidised. The chemical rate constant generally increases as particle size is reduced but this is to little advantage if the contacting is poor. As in most engineering, a compromise is called for.

It will be evident to the imaginative student that this simple model can be expanded to take into account other features that have been described in this brief account of the basic physics of bubbling gas fluidised beds.

REFERENCES

1. M. Leva, Fluidisation, McGraw-Hill, New York, 1959.
2. J.F. Richardson in Fluidisation, ed. by Davidson and Harrison, Academic Press, London, 1971, Chapter 2.
3. P.N. Rowe, ibid., Chapter 4.
4. P.N. Rowe and D.J. Everett, Trans. I.Chem.E., 50, 1972, p. 42.
5. P.N. Rowe et al., Trans. I.Chem.E., 43, 1965, p. T271.
6. P.N. Rowe, Chem. Eng. Science, 31, 1976, p. 285.
7. J. Werther, Trans. I.Chem.E., 52, 1974, p. 149.
8. J.F. Davidson and D. Harrison, Fluidised Particles, C.U.P., Cambridge, 1963.
9. B.A. Partridge and P.N. Rowe, Trans. I.Chem.E., 44, 1966, p. T335.
10. P.N. Rowe, A.C.S. Symposium Series No. 65, 1978, p. 436.
11. J.S.M. Botterill, Fluid Bed Heat Transfer, Academic Press, London, 1975.
12. P.N. Rowe, Chem. and Ind., 17 June 1978, p. 424.
13. D. Kunii and O. Levenspiel, Fluidisation Engineering, John Wiley, New York, 1969.

LIST OF SYMBOLS

A	surface area for heat transfer	m^2
A_R	cross-sectional area of bed or reactor	m^2
D_R	diameter of bed or reactor	m
d_B	bubble diameter	m
$d_{B(0)}$	initial bubble diameter	m
d_c	cloud diameter	m
d_p	particle diameter	μm
F	volumetric flow rate into the bed	m^3/s
F_B	volumetric flow as bubbles	m^3/s

Basic fluidisation

F_{mf}	minimum fluidisation volumetric flow	m^3/s
f	fraction of gas cloud extending beyond the bubble $(1 - V_B/V_C)$	
g	acceleration of gravity	m/s^2
ΔH	heat of reaction	kJ/kg mole
h	heat transfer coefficient	$kW/m^2 K$
	height above distributor	m
h_e	total bed height	m
h_o	distributor bubble index (Equation (3))	m
$h_{B/W}$	heat transfer coefficient - bed to walls	$kW/m^2 K$
$h_{G/P}$	heat transfer coefficient - gas to particles	$kW/m^2 K$
k	first order rate constant	s^{-1}
m	total mass of fluidised particles	kg
Nu	Nusselt number $(h\, d_p/\lambda)$	
P_D	pressure just above the distributor	Pa
P_e	pressure just above the top of a bed of height h_e	Pa
ΔP	pressure drop across the bed $(P_D - P_e)$	Pa
Re_{mf}	Reynolds number at minimum fluidisation $(U_{mf}\, d_p\, \rho_f/\eta)$	
T_R	reactor temperature	°C
T_W	reactor wall temperature	°C
ΔT	temperature difference	K
t_r	residence time in reactor	s
U	superficial gas velocity through the bed	m/s
U_B	bubble rise velocity	m/s
U_D	gas velocity across distributor plane	m/s
U_e	gas velocity leaving the top of the bed	m/s
U_i	interstitial gas velocity	m/s
U_{mf}	minimum fluidisation velocity	m/s
	(all gas velocities are superficial except U_i)	
U_T	terminal falling velocity of a single particle	m/s
V_B	bubble volume	m^3

V_C	cloud volume	m^3
V_W	wake volume	m^3
w	feed rate to the reactor	kg/s
y	fraction of reactant gas converted	
y_c	fractional conversion in cloud phase	
y_i	fractional conversion in interstitial phase	
α	velocity ratio (U_B/U_i)	
ε_{mf}	powder voidage at U_{mf}	
η	gas viscosity	Ns/m^2
λ	thermal conductivity	kW/mK
ρ_f	fluid density	kg/m^3
ρ_s	solid density	kg/m^3
$\emptyset$	heat flow rate	kW

FLUIDISATION OF FINE POWDERS SUCH AS FCC

P.N. Rowe

Department of Chemical and Biochemical Engineering,
University College London

ABSTRACT. Difficulty of fluidising fine powders. Bed expansion without bubbling. Difficulty of measuring bed voidage. Division of gas flow between the two phases. Gas/solid contacting and reactor models. Operation at high gas velocity.

1. DIFFICULTY OF FLUIDISING FINE POWDERS

If the shape of particles is such that their dimensions in the three principal directions are similar, they are generally easy to fluidise provided their surfaces are not sticky. Plate-like and needle-like particles are often impossible to fluidise because the drag force changes greatly with orientation. In principle there is no upper limit to the size of particles that can be fluidised but since U_{mf} increases roughly as d_p^2 it is not generally practical to fluidise with a gas particles more than a few millimeters mean diameter.

Very small particles are often difficult to fluidise for different reasons. Difficulty increases with the ratio of surface to body forces, i.e. as d_p^{-1}. Fluidised with air at NTP most powders begin to show anomalies at $\bar{d}_p$ < 100 μm and when $\bar{d}_p$ < 50 μm many powders are impossible to fluidise normally. Surface forces, which may arise from electrostatic charge, capillarity when surface moisture is present or physico-chemical attraction, begin to dominate over the body forces (proportional to mass) as size is reduced. The causes are not well understood but the problems of fluidising fine powders are a matter of common experience.

It is difficult to distribute gas evenly through a bed of

fine powder when attempting to fluidise it. Gas will often tunnel a few channels through the bed and leave much of the rest undisturbed ("rat-holing"). Uniform fluidisation is difficult to maintain even at several multiples of U_{mf}. Reproducibility of the plot of pressure drop against gas velocity is poor and it is difficult to measure U_{mf}. Gentle fluidisation is often aided by mechanical means such as stirring or vibrating the bed. Sometimes the bed will behave reasonably well if it is first fluidised very vigorously and then the flow rate reduced to a small value. Beds of fine particles are generally operated at many multiples of U_{mf} (say, 10 to 100 X) but since U_{mf} is usually only a few millimeters per second the gas velocity may not be very great.

2. BED EXPANSION WITHOUT BUBBLING

Many fine powders exhibit a characteristic behaviour when first fluidised. Over a limited range of gas velocity they first expand uniformly without the formation of bubbles as if fluidised by a liquid. When a given mass of powder is placed in a suitable vessel and fluidised at increasing gas velocity, the bed height changes as shown in Fig. 1. When first fluidised it expands rapidly and smoothly. The effect is reproducible and observed

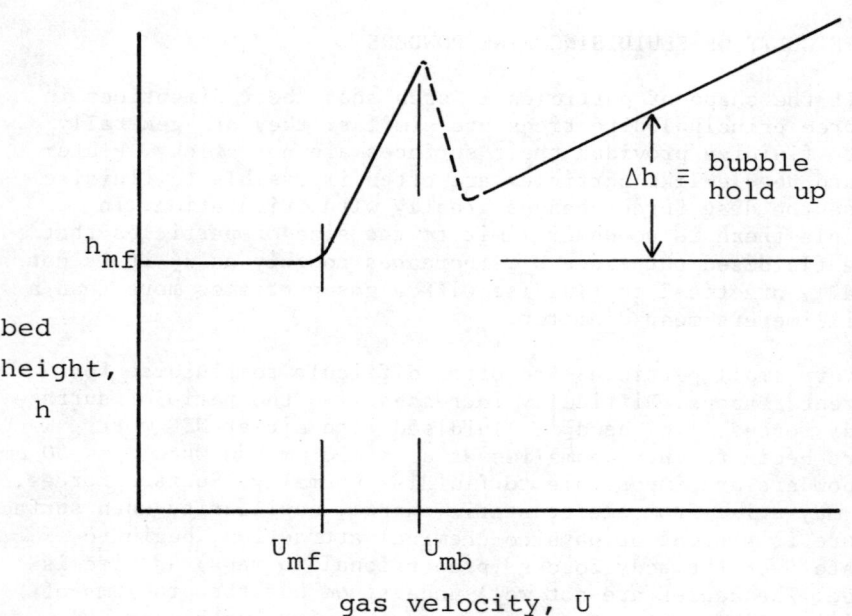

Fig. 1. Variation of bed height with gas velocity for fine powder.

Fluidisation of fine powders

with both increasing and decreasing flow. The height increase in this way can be considerable and a bed may double its volume.

At a particular velocity the bed appears to collapse and bubbles are first seen. The velocity at which this occurs is called minimum bubbling velocity, U_{mb}. It is not easy to measure reproducibly and experimental points in the region shown by the broken line in Fig. 1 are often scattered. Well beyond U_{mb} the bed bubbles normally and expansion is largely accounted for by the hold-up of bubbles within it.

The effect described above diminishes with increasing particle size and Fig. 2 illustrates the change. With, for example, catalyst grade alumina fluidised by air at NTP the effect is first observable at a mean particle diameter, $d_p \simeq 80$ μm. When $\bar{d}_p \gtrsim 40$ μm the powder becomes impossible to fluidise so the effect is confined to a narrow but important range.

The effect also depends on the range of particle sizes in the powder but this is imperfectly understood. The presence of very fine particles (< 20 μm) is probably an important factor. It is very difficult to prepare samples of close size range as size diminishes and so data from controlled experiments are very limited.

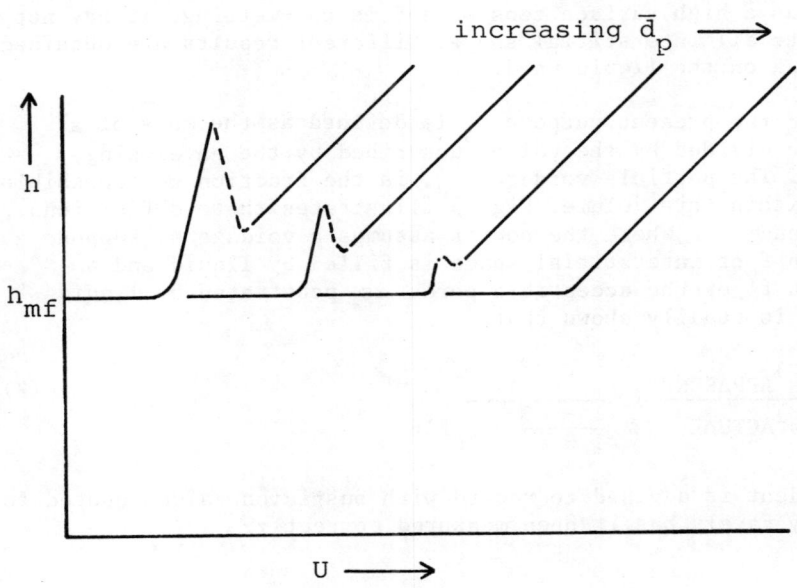

Fig. 2. Variation of height with velocity - effect of particle size.

The effect also depends on the nature of the material and apparently those qualities that effect the surface properties. Inorganic salts, for example, have been fluidised successfully down to $\bar{d}_p \risingdotseq 10$ μm. This subject is discussed further in Ref. 1.

3. DIFFICULTY OF MEASURING BED VOIDAGE

It is evident that if a bed expands uniformly more gas will be held up in the interstitial phase and, as this is the more effective one for gas/solid contacting, it is a matter of interest to measure it. The bulk density of a powder, ρ_{BULK}, is defined as mass divided by the volume it occupied and it is usually easy to measure this. ρ_{BULK} is related to the individual particle density, ρ_s, by

$$\rho_{BULK} = (1 - \varepsilon) \rho_s \qquad (1)$$

so the voidage, ε, is immediately known once the particle density is given.

With non-porous solid particles such as sand grains there is no difficulty in measuring density by the classical pycnometric method. When the particles are porous and have a coke-like structure (as do most catalyst particles) the problem is not so simple. In a pycnometer liquid may enter the particle pores or, if it has a high surface tension and is non-wetting, it may not penetrate all interstitial space. Different results are obtained depending on the liquid used.

For the present purpose ρ_s is defined as the mass of a particle divided by the volume described by the enveloping surface. The particle voidage, ε_p, is the fraction of accessible pores within this volume. Fig. 3 illustrates these definitions. In a pycnometer where the powder assumes a voidage ε, suppose a fraction f of interstitial space is filled by liquid and a fraction f' of the accessible pores is penetrated by liquid then it is readily shown that

$$\frac{(\rho_s)_{APPARENT}}{(\rho_s)_{ACTUAL}} = \frac{1}{\frac{1 - \varepsilon f}{1 - \varepsilon} - \varepsilon_p f'f} \qquad (2)$$

The student is advised to regard with suspicion values quoted for ρ_s. Very rarely has it been measured correctly.

4. DIVISION OF GAS FLOW BETWEEN THE TWO PHASES

For any bubbling fluidised bed we may write

Fluidisation of fine powders

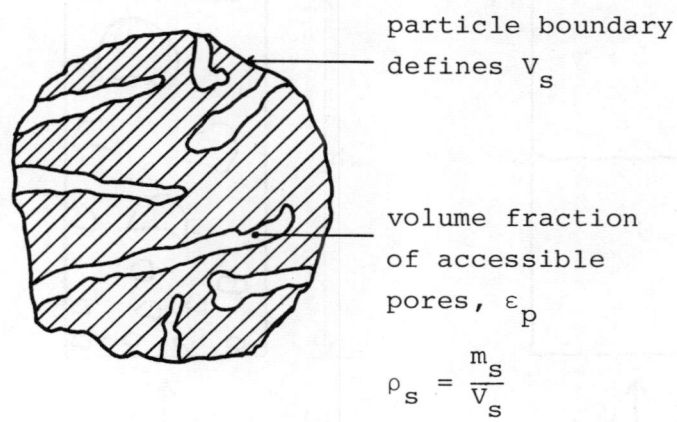

Fig. 3. Definition of particle density and voidage.

$$F = F_i + F_B \tag{3}$$

which merely states that the total volumetric flow into the bed, F, divides into an interstitial and a bubble flow. The bubble flow rate may alternatively be written

$$F_B = f_B \overline{U}_B A_R \tag{4}$$

where f_B is the fraction of bed volume occupied by bubbles and $\overline{U}_B$ their mean rise velocity. Combining (3) and (4) and dividing through by the bed cross-sectional area, A_R,

$$U = U_i + f_B \overline{U}_B \tag{5}$$

Consider a particle bed such as in Fig. 4 (a) filled with particles to a height h_{mf} at U_{mf}. From equation (1) and the definition of density,

$$h_{mf} A_R = \frac{\Sigma V_s}{(1 - \varepsilon_{mf})} \tag{6}$$

where ΣV_s is the sum of the individual particle volumes. Now imagine that the same mass of particles is fluidised at some velocity U and the bed height increases to h_e as in Fig. 4 (b).

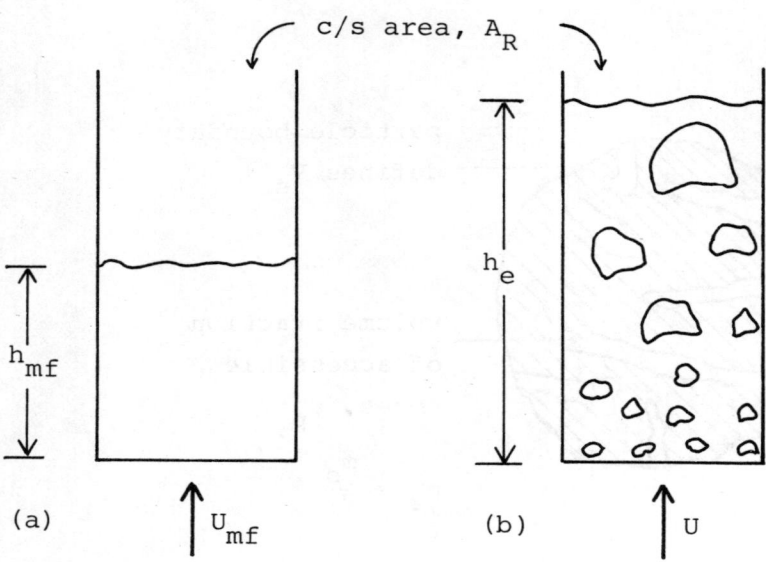

Fig. 4. Change of bed height with gas velocity.

The increase in volume is made up of the space occupied by bubbles in the bed and any change in the powder voidage which is assumed uniform in the dense phase. Corresponding to equation (6) we may write

$$h_e A_R = \Sigma V_B + \frac{\Sigma V_s}{(1 - \varepsilon)} \tag{7}$$

where ΣV_B is the total bubble volumetric hold-up. From the definition of f_B,

$$\Sigma V_B = f_B h_e A_R \tag{8}$$

Combining equations (6), (7) and (8) and remembering that ΣV_s is constant,

$$f_B = 1 - \frac{(1 - \varepsilon_{mf}) h_{mf}}{(1 - \varepsilon) h_e} \tag{9}$$

When the voidage remains constant at ε_{mf} then

$$f_B = 1 - (h_{mf}/h_e) \tag{10}$$

Fluidisation of fine powders

which simply states that the increase in bed height is a result only of bubbles within the bed. It is only necessary, therefore, to measure ε, ε_{mf}, h_e, h_{mf} and $\bar{U}_B$ to estimate the interstitial gas flow using Equations (5) and (9).

It is a fairly simple matter to measure h_e and h_{mf}. The former will fluctuate as bubbles leave the surface so some average must be taken. The fluctuations are not very large when A_R is of reasonable size and U not too great. $\bar{U}_B$ can be estimated but must be averaged over the full bed height as average bubble size changes. ε_{mf} follows directly from Equation (1) once the particle density is known. The remaining problem is to measure ε independently.

One method of measuring ε in the necessary dynamic condition is by using X-ray adsorption (Ref. 2). The bubbling bed is X-ray ciné-photographed when the bubbles are revealed. In regions of the bed that at the instant of photography are free of bubbles the film will be of optical density proportional to the density of powder along that path of the beam. Outside the fluidised bed is a wedge-shaped box containing the same powder as that fluidised and this box is included in the field of view. Fig. 5 illustrates the arrangement. The photographs appear as in Fig. 6 and are

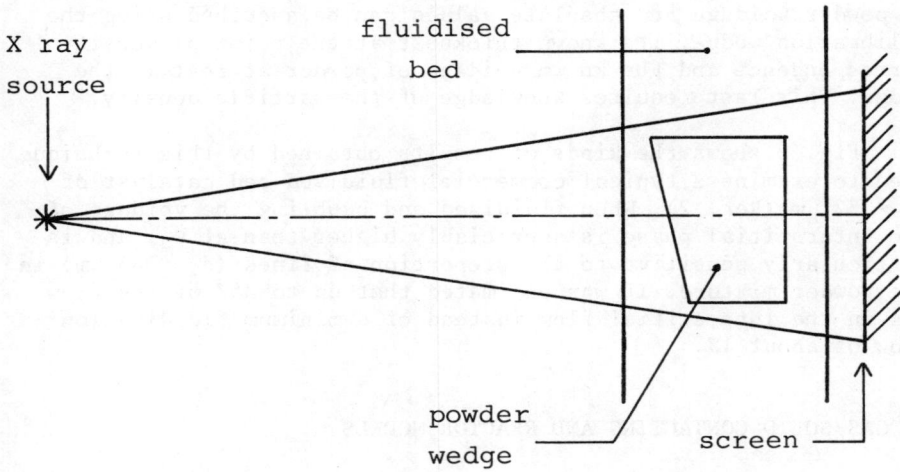

Fig. 5. X-ray method of measuring powder voidage.

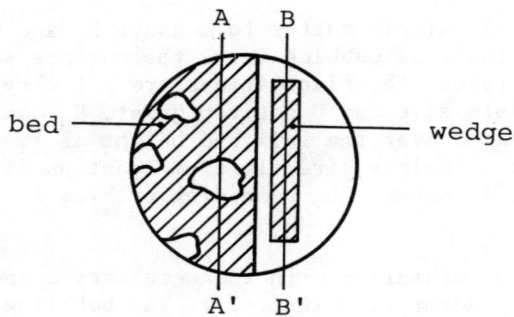

Fig. 6. The type of picture produced.

examined for optical density along lines such as AA' and BB' with results as in Fig. 7. Optical density is inversely proportional to powder voidage and absolute values can be ascribed using the calibration wedge, its known thickness at the point of density correspondence and the known voidage of powder at rest in the wedge. This last requires knowledge of the particle density.

Fig. 8 shows the kinds of results obtained by this technique used to examine a typical commercial fluidised bed catalyst of $\bar{d}_p$ = 52 µm (Ref. 2). When fluidised and bubbling the voidage of the interstitial phase is appreciably higher than at U_{mf} and is particularly sensitive to the proportion of fines (d_p < 45 µm) in the powder mixture. It was estimated that up to 35% of the flow was in the interstitial flow instead of a minimum fluidisation flow of about 1%.

5. GAS/SOLID CONTACTING AND REACTOR MODELS

An increased proportion of interstitial flow will certainly improve gas/solid contacting. Since this appears to increase as the particle size decreases below about 100 µm, it counteracts the deterioration in contacting expected from the reduction in

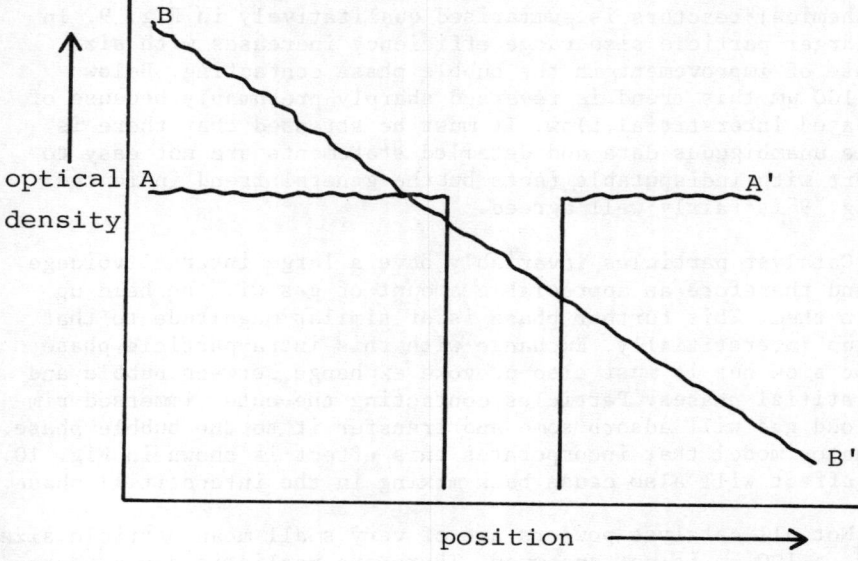

Fig. 7. Variation of optical density with position.

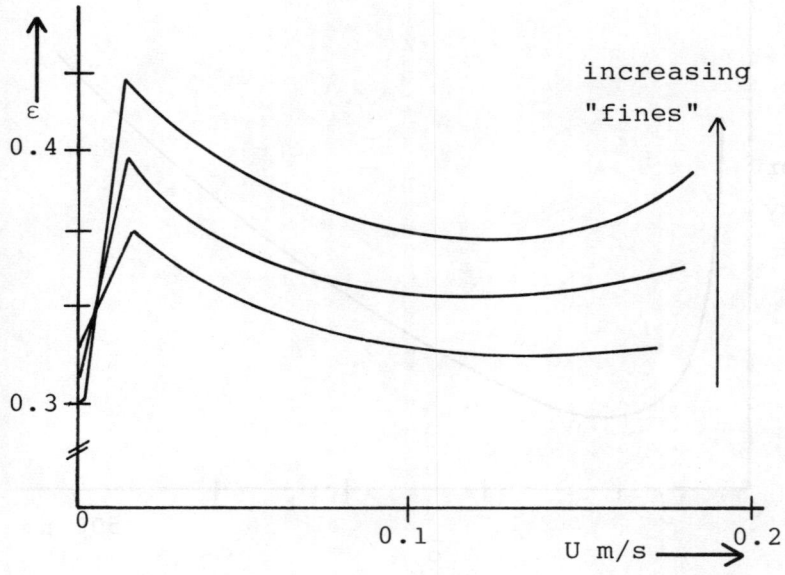

Fig. 8. Variation of voidage with gas velocity.

relative cloud size. Experience of the performance of fluidised bed chemical reactors is summarised qualitatively in Fig. 9. In the larger particle size range efficiency increases with size because of improvement in the bubble phase contacting. Below $\bar{d}_p \fallingdotseq 100$ μm this trend is reversed sharply presumably because of increased interstitial flow. It must be stressed that there is little unambiguous data and detailed statements are not easy to support with indisputable facts but the general trend indicated by Fig. 9 is fairly well agreed.

Catalyst particles invariably have a large internal voidage, ε_p, and therefore an appreciable amount of gas will be held up within them. This further phase is of similar magnitude to that held up interstitially. Exchange with this intra-particle phase may be slow but it must also provoke exchange between bubble and interstitial phases. Particles contacting the outer immersed rim of cloud gas will adsorb some and transfer it to the bubble phase. A reactor model that incorporates this effect is shown in Fig. 10. This effect will also cause back mixing in the interstitial phase.

Not all catalyst powders are of very small mean particle size and $\bar{d}_p > 100$ μm is not uncommon. There are realistic cases where

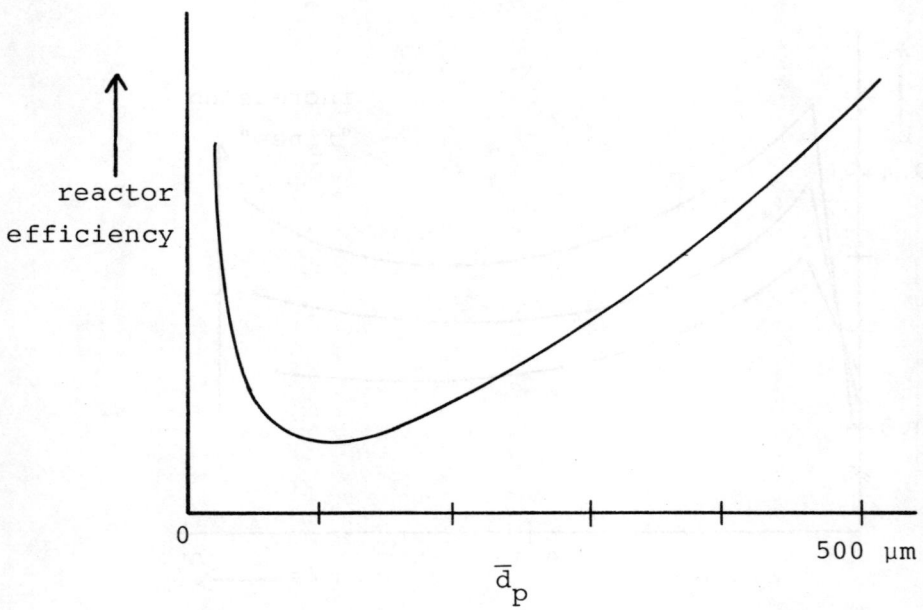

Fig. 9. Variation of reactor efficiency with particle size.

Fluidisation of fine powders 213

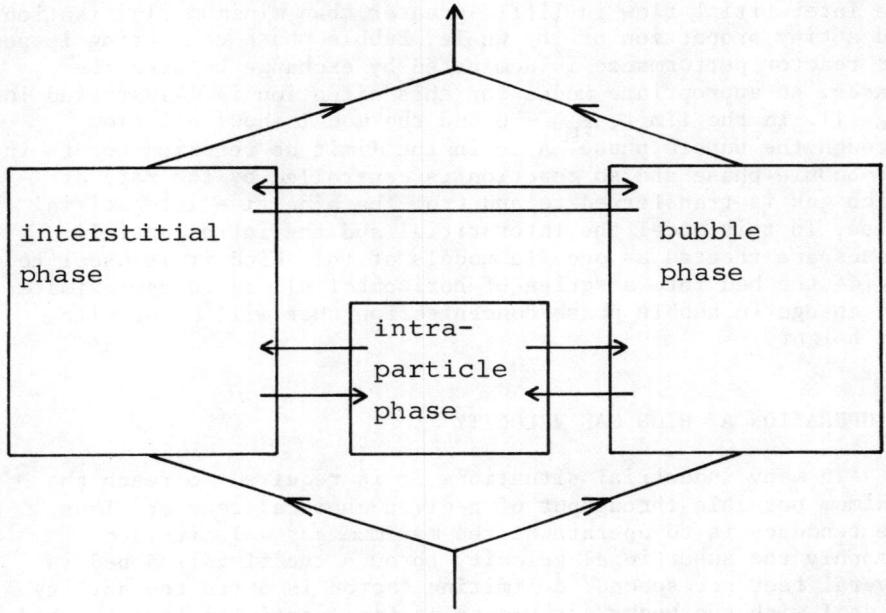

Fig. 10. Reactor model to include hold-up within the particle pores.

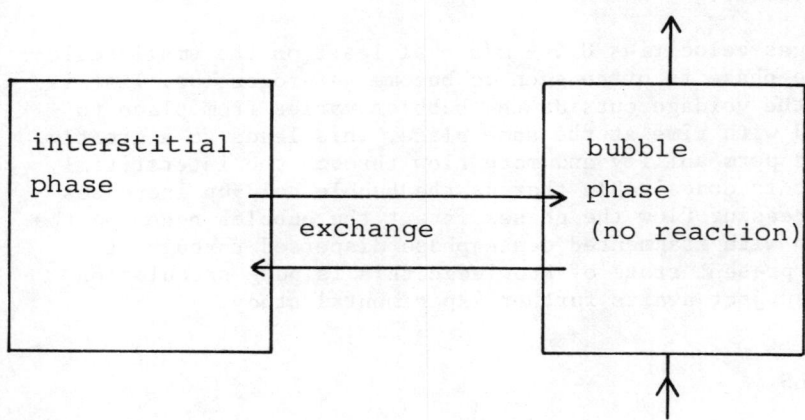

Fig. 11. Reactor model when little gas flows interstitially.

the interstitial flow is little greater than minimum fluidisation and a tiny proportion of the whole. Bubble phase contacting is poor and reactor performance is dominated by exchange between the phases. An appropriate model for this situation is illustrated in Fig. 11. In the limit, $F_{mf} \to 0$ and the model shows all flow through the bubble phase. Also in the limit no reaction occurs in the bubble phase and so reaction is controlled by the rate at which gas is transferred to and from the effective interstitial phase. In this model the interstitial and the intra-particle phases are treated as one. In models of this kind it is usual to divide the bed into a series of horizontal slices to approximate the change in bubble phase concentration that will occur with bed height.

6. OPERATION AT HIGH GAS VELOCITY

In many industrial situations it is required to reach the maximum possible throughput of a given chemical reactor. Thus, the tendency is to operate at the maximum gas velocity and commonly the superficial velocity through the fluidised bed is several feet per second. A limiting factor is often the ability to deal with the heavy carry-over of fine particles from the bed.

Experimental observation under these conditions is very difficult. Many bubbles in the bed form a confusing picture as they coalesce rapidly and are rarely of uniform shape. Observation on a small scale may be misleading for when large bubbles are present they can never be far from walls as they would be in a large vessel.

At gas velocities $U \gtrsim \tfrac{1}{2}$ m/s - at least on the small scale - the dense phase is often seen to become heterogeneous. That is to say, the voidage outside the bubbles varies from place to place and with time at the same place. This leads to a greatly increased permeability and more flow through the interstitial phase. It is conceivable that as the bubble hold-up increases with increasing flow the phases invert the bubbles becoming the continuum with fragmented dense phase dispersed through it. With the present state of knowledge this is pure speculation and the subject awaits further experimental study.

REFERENCES

1. J.F. Richardson in *Fluidisation*, ed. by Davidson and Harrison, Academic Press, New York, 1971, Chapter 2.
2. P.N. Rowe, L. Santoro and J.G. Yates, *Chem. Eng. Science*, 33, 1978, p. 133.

LIST OF SYMBOLS

Symbol	Description	Units
A_R	cross-sectional area of fluidised bed or reactor	m^2
d_p	particle diameter	μm
f	fraction of interstitial space filled by liquid in a pycnometer	
f'	fraction of particle pores filled by pycnometer liquid	
f_B	fraction of fluidised bed occupied by bubbles	
F	volumetric flow rate into the bed	m^3/s
F_B	volumetric flow as bubbles	m^3/s
F_i	interstitial volumetric flow	m^3/s
F_{mf}	minimum fluidisation volumetric flow	m^3/s
h	height above distributor	m
h_e	height to top (exit) of fluidised bed	m
h_{mf}	height of fluidised bed at U_{mf}	m
m_s	mass of a single particle	kg
U	superficial gas velocity into the bed	m/s
U_B	mean bubble rise velocity	m/s
U_i	mean velocity through the interstitial phase	m/s
U_{mb}	minimum bubbling velocity	m/s
U_{mf}	minimum fluidisation velocity	m/s
V_B	bubble volume	m^3
V_S	volume of a particle	m^3
ε	powder voidage	
ε_{mf}	powder voidage at U_{mf}	
ε_p	voidage within a particle	
ρ_{BULK}	bulk density of powder	kg/m^3
ρ_s	individual particle density	kg/m^3

CATCRACKER OPERATIONS
REACTION NETWORK AND KINETICS

H.S. van der Baan

Laboratory for Chemical Technology
Eindhoven University of Technology, The Netherlands

1. Product distribution

Depending on the feed composition and the process parameters the product distribution in catcrackers can vary widely.
In table 1 the product distributions are given for one type of feedstock but for varying process conditions. The latter have been chosen in such a way that the yield of respectively gasoil, gasoline and of butane and lighter have been maximized.

max	gasoil	gasoline	C_4^-
C_2^-	2.0	3.0	5.0
propane	0.5	1.5	3.0
propene	2.0	4.0	6.5
iso-butane	3.0	5.0	9.0
n-butane	0.5	1.0	2.0
butenes	3.0	7.0	10.0
gasoline	39.0	60.0	45.5
light cycle oil	38.0	7.0	7.0
heavy cycle oil	4.0	3.5	3.0
coke	7.5	7.5	8.5
loss	0.5	0.5	0.5

Table 1. Product distributions in % by wt. for a number of (rather extreme) operating conditions.

2. Reaction models

In principle the product distribution can be described if all the components of the feedstock were known and if for each component the reaction network would be known. How complex such an approach is follows from the work of Greensfelder, Voge and Good (1949) who composed a model for the catalytic cracking of hexadecane.

The model comprises the following rules:
1. Hexadecane forms carbenium ions by hydride abstraction from a secundary position. All secundary positions have the same probability;
2. Carbenium ions crack at the β position (in relation to the carbenium ion). The part with the carbenium ion forms an α olefine, the other part a new carbenium ion. All components have the same cracking rate constant. Products smaller than C_3 are not found;
3. The new carbenium ion isomerises to a secundary carbenium ion and is again subject to β scission. Fragments with 6 or less carbon atoms become either paraffins by hydride-ion abstraction from a hexadecane molecule or olefines by proton donation to a bigger olefin;
4. The olefines with 7 or more carbon atoms are for 50% protonated to carbenium ions.

(1) Thus the first reaction step can be, e.g.

$$C_{16}H_{34} + C_3^+H_7 \rightarrow C_4H_9 \cdot C^+H \cdot C_{11}H_{23} + C_3H_8$$

(in total 7 different carbenium ions can be formed).

(2) This carbenium ion splits into

$$C_3^+H_7 + C_{11}H_{23} \cdot CH = CH_2$$
or into
$$C_{10}^+H_{21} + C_4H_9CH = CH_2$$

(3) The carbenium ion $C_{10}^+H_{21}$ can isomerise into four different products, one of these being

$$C_2H_5 \cdot C^+H \ C_7H_{17}$$ that can split into

$$C_4H_8 \text{ and } C_6^+H_{13}.$$

The latter becomes hexane or hexene.

(4) The olefin $C_{11}H_{23}CH = CH_2$ formed in (2) is for 50% converted into the carbenium ion $C_{13}H_{27}$, that splits in a way comparable to that indicated for $C_{10}^+H_{21}$.

All reactions occuring according to this model are represented in figure 1. This figure is still somewhat simplified as it does not show the differences between olefins and carbenium ions for the C_6 and smaller fractions.

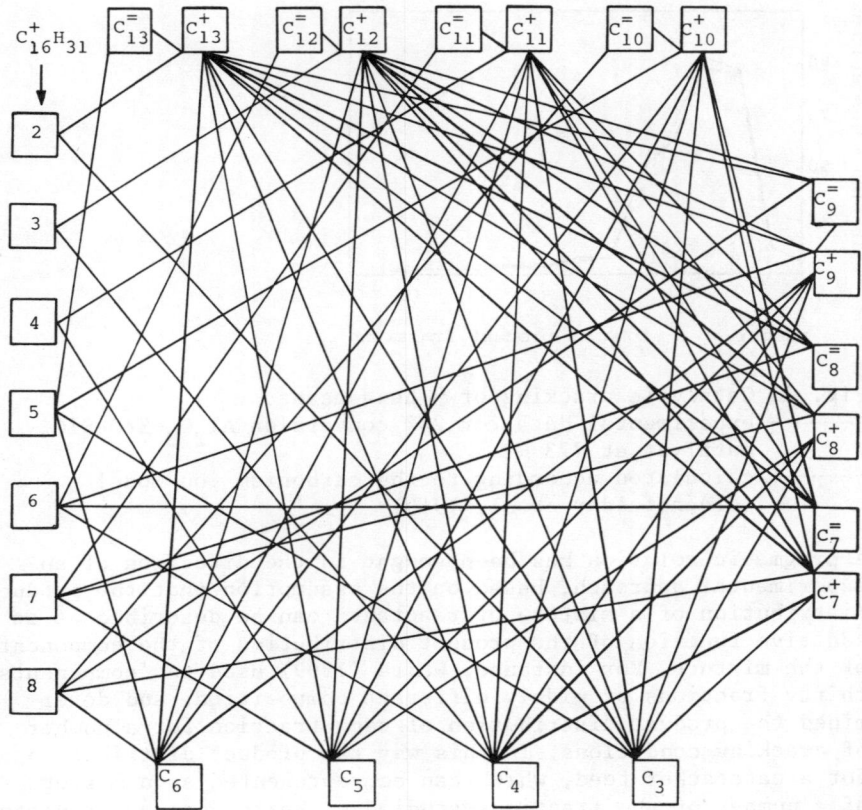

Fig. 1. Reaction network for the catalytic cracking of hexadecane. The squares in the left column indicate the place in the carbenium ion in the C_{16} molecule. The other squares indicate intermediates and end products. Lines indicating reaction emerge from the middle of a side of a square and lead to the corner of an other square.

With the simple assumptions that the rate constant for the formation of C_{16} carbenium ions is much smaller than all other rate constants and that the latter are all equal, a product distribution can be calculated that approaches the experimental value very near as shown in figure 2.
As however the feed of a catcracker consists of thousands of different species, and as in reality also other reactions than those assumed by Greensfelder et al take place, it will be clear that the complete reaction network describing all reactions occurring in a commercial installation will be too cumbersome to handle.

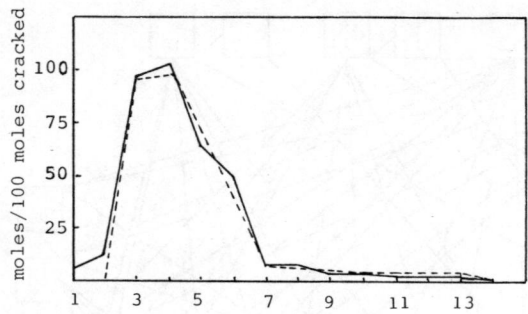

Fig. 2. Catalytic cracking of n hexadecane.
——— experimental data for 24% conversion $Al_2O_3-ZrO-SiO_2$ catalyst at 773 K
------ calculated according to the carbonium ion model (Greensfelder et al, 1949)

A pragmatic solution has been sought in the direction of an experimental approach, based on the assumption that the product distribution of a mixture of reactants can be described as an additive function of the product distribution of the components of the mixture. For instance, White (1969) used as 'components' thirty fractions of widely different composition, and determined the product distribution of each fraction for a number of cracking conditions. In this way the product distribution for a catcracker feed, which can be represented as a mixture of a number of the fractions studied by White, can be predicted. The method has not found universal application mainly because the performance of the laboratory reactor used by White performs in quite an other way as a commercial catcracking reactor.

A useful although very simple model is based on the consideration that the process conditions are generally choosen to maximize one of the following products:
1. <u>Gasoline</u>. This is the normal operation, provided an acceptable octane rating can be attained;
2. <u>Butane and lighter</u>. This is for the production of chemical feedstocks and LPG;
3. In the past catcrackers have also been used to lower the viscosity of the feed (visbreaking). See column gasoil in table 1.

Correspondingly Weekman and Nace (1968) developed the following model:

Catcracker operations

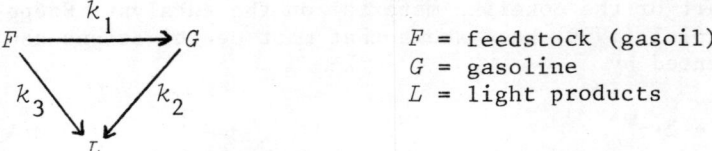

F = feedstock (gasoil)
G = gasoline
L = light products

Simplified reaction model for catcracking.

The wide applicability of this model has been proven in practice. The problem now is reduced to find expressions for the rate equations for the reactions of the model.

3. Factors influencing the rate equations of the simplified catcracker model.

Catalytic factors
In general a rate equation has the form:

$$-\frac{dA}{dt} = V_r \, k \, f \text{ (concentrations)} \qquad (1)$$

If (1) applies to a catalytic cracking reaction the rate constant k is proportional to the catalyst concentration and to the activity of the catalyst. In a dense phase reactor the catalyst concentration is more or less a constant; in the more modern riser reactor the catalyst concentration is a function of the cat-oil ratio, C.O.R. and the gasphase density.
Using the ideal gas law and indicating with w_F, w_G and w_L the weight fractions feed, gasoline and light products and with δ_G and δ_F the number of moles of gasoline respective light products formed from one mole of feed we obtain

$$\rho = \frac{P \, M_F}{R \, T} \cdot \frac{1}{w_F + w_G \delta_G + w_L \delta_L} \qquad (2)$$

where M_F = molecular weight of the feed (say 350) and δ_G and δ_L are about 3.5 and 9 respectively. Thus we find for the catalyst concentration in a riser reactor:

$$[cat] = (C.O.R.) = \frac{(COR) \cdot P \cdot M_F}{RT \, (w_F + w_G \delta_G + w_L \delta_L)} \qquad (3)$$

with the pressure P in pascals the value of R becomes 8.314 J/mol K

The catalyst activity is steadily decreasing by coke deposition, and after one pass through the reactor the catalyst has to be regenerated by just stripping off the volatile material adsorbed on the catalyst and thereafter by burning off the greater part of the cokelike material on the catalyst. Szépe and Levenspiel (1968) have shown that most deactivations can be represented by

$$-\frac{da}{dt} = k \cdot a^m \qquad (4)$$

As follows from data from Blanding (1953) and from Nace (1965) the equation

$$a(t) = A_c\, t^{-n} \qquad (5)$$

with A_c the activity after 1 second and $n \simeq 0.5$ describes the experimental results satisfactorily (see figure 3).
(Equation (4) follows from (5) by setting $m = \frac{n+1}{n}$).

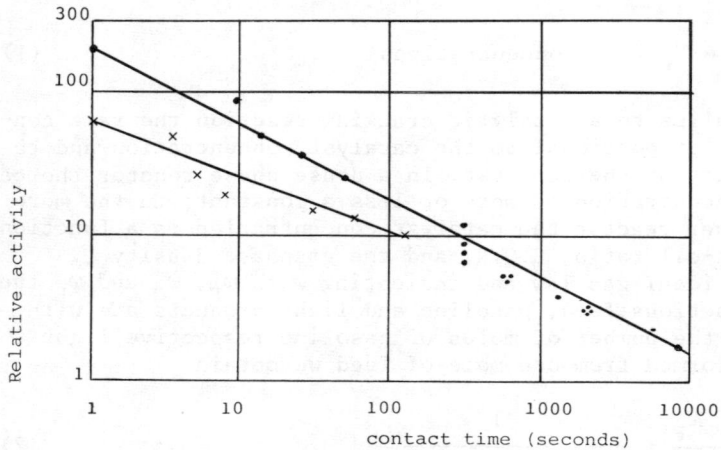

Fig. 3. Decrease in activity for cracking catalysts.
. Blanding (1953)
x Nace (1965)

For zeolite catalysts Gustafson (1972) has shown that the activity is best described by

$$a = a_o\, e^{-bt} \qquad (6)$$

which might indicate (see Tan and Fuller, 1970) that de deactivation is the result of an irreversible Langmuir-Hinshelwood adsorption.

Kinetic factors

Cracking reactions for a pure component are generally first order in the concentration of that component. When however a complete gasoil fraction is cracked a complication arises. Because the components that crack easiest are converted fastest the 'crackability' of the unconverted fraction decreases, i.e. the overall cracking rate constant decreases with conversion. If the effective cracking rate constant is proportional to the fraction unconverted:

$$k_{eff} = k_o \frac{w}{w_o} = k_o (1-x) \qquad (7)$$

in which w represents a weight and x the fraction converted. The gasoil cracking rate r then becomes:

$$r = k_{eff} \cdot \rho \cdot \underline{C} = k_o (1-x) \cdot \rho \cdot (1-x) \underline{C}_o$$

$$= k_o \underline{C}_o \rho (1-x)^2 \qquad (8)$$

For a truly second order reaction in component A we have

$$r_A = k C_A^2 = k (\rho \underline{C}_A)^2 = k \underline{C}_{A,o}^2 \rho^2 (1-x)^2 \qquad (9)$$

At constant volume (i.e. constant ρ), (8) and (9) are indistinguishable, but not at constant pressure. For the gasoline cracking, where we have a smaller number of reactants as in the gasoil fraction, a first order approximation is acceptable. The statement that the orders are approximately 2 and 1 makes a mathematical treatment of the kinetics with e.g. a Langmuir-Hinshelwood model superfluous.

Assuming that the catalyst deactivation is the same for the three reactions of the Weekman-Nace model, we can write:

$$- r_F = (k_1 + k_3) \rho \frac{\underline{C}_F^2}{\underline{C}_{F,o}} [cat] \, a(t) \qquad (10)$$

$$r_G = (k_1 \rho \frac{\underline{C}_F^2}{\underline{C}_{F,o}} - k_2 \rho \underline{C}_G) [cat] \, a(t) \qquad (11)$$

$$r_L = (k_3 \rho \frac{\underline{C}_F^2}{\underline{C}_{F,o}} + k_2 \rho \underline{C}_G) [cat] \, a(t) \qquad (12)$$

is shown in fig. 4

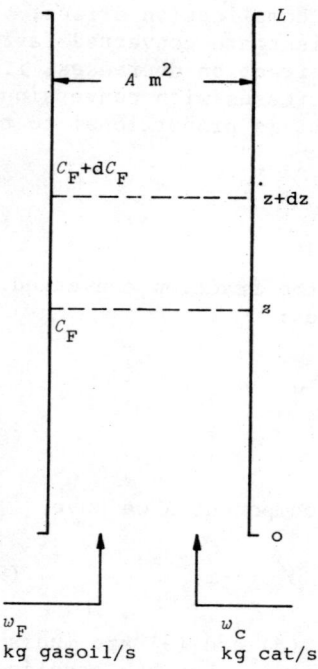

w_F
kg gasoil/s

w_c
kg cat/s

Fig. 4. Model of a catcracker riser reactor.

4. Application to the riser reactor

a. The massbalance
   For a differential volume element $A\,dz$ m^3 (see fig. 5) the massbalance for one second reads:

$$w_F\,\underline{C}_F\Big|_z = w_F\,\underline{C}_F\Big|_{z+dz} - r_F\,A\,dz + \frac{d}{dt}(\rho\,\underline{C}_F)A\,dz \qquad (13)$$

in case of plug flow.
In case axial dispersion has to be taken into account the terms:

$$-A\,D_{ax}\,\frac{d(\rho\,\underline{C}_F)}{dz}\Big|_z \quad \text{and} \quad -A\,D_{ax}\,\frac{d\,\rho\,\underline{C}_F}{dz}\Big|_{z+dz}$$

have to be added respectively to the right and left hand side of equation (13).

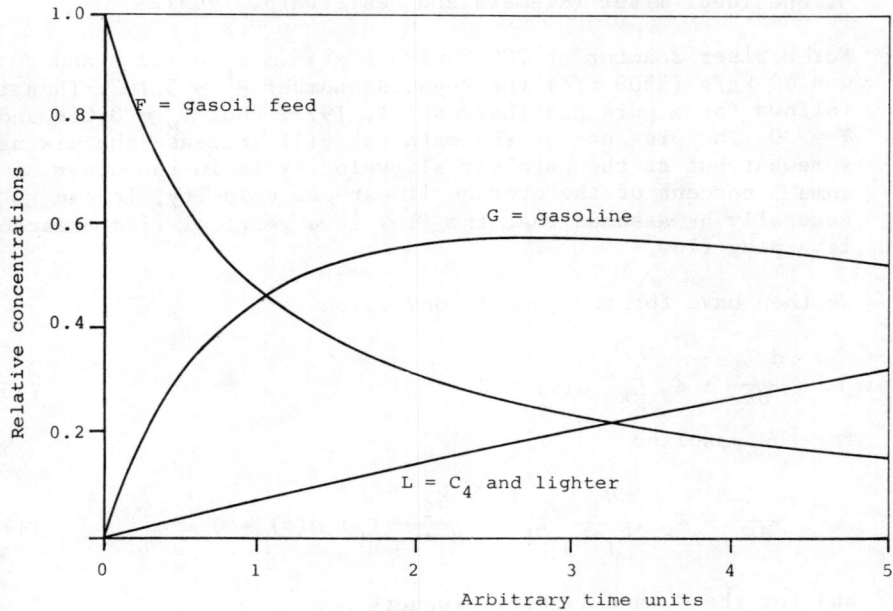

Fig. 5. Conversion pattern according to equations (10), (11) and (12). $k_1 = 1$, $k_2 = 0.1$, $k_3 = 0.1$.

In the steady state we then obtain

$$G_F \frac{d \underline{C}_F}{dz} - D_{ax} \frac{d^2 (\rho \underline{C}_F)}{dz^2} - r_F = 0 \qquad (14)$$

with $\lambda = z/L$, equation (10), $[\text{cat}] = (\text{C.O.R.}) \rho$,

$$\zeta_F = \underline{C}_F / \underline{C}_{F,o}, \quad K_F = (k_1 + k_3) \frac{L}{G_F} \underline{C}_{F_o} \rho^2 (\text{C.O.R.})$$

and $G_F = \rho \cdot u$ we obtain the dimensionless equation:

$$\frac{d \zeta_F}{d\lambda} - \frac{D_{ax}}{uL} \frac{1}{\rho} \frac{d^2 (\rho \zeta_F)}{d\lambda^2} + K_F \zeta_F^2 a(t) = 0 \qquad (15)$$

In this equation $\frac{ax}{uL}$ is the mass dispersion number N_M of the reactor.

For $N_M \to 0$ we have plug flow, for N_M small (< 0.1) the numbers N of ideal mixers in series that show the same behaviour as our reactor is $N = \frac{1}{2N_M}$ and for large N_M we have the equivalent of one ideal mixer (Kramers and Westerterp, 1963).

For a riser reactor at 780 K with $L = 15$ m $D_t = 1.2$ m and $w = 40$ kg/s (3500 t/d) the Reynolds number $R_e^t \sim 5.10^6$. Then it follows for a pure gas (Levenspiel, 1972) that $N_M = 0.016$ and $N = 30$. The presence of the catalyst will increase the mixing somewhat but as the catalyst slipvelocity is in the order of some 5 percent of the average linear gas velocity, it can generally be assumed that the flow in a vertical riser reactor is a plug flow.

We then have for the gasoil conversion

$$\frac{d\zeta_F}{d\lambda} + K_F \zeta_F^2 \, a(t) = 0 \qquad (16)$$

for the gasoline

$$\frac{d\zeta_G}{d\lambda} - K_F \left(\frac{k_1}{k_1+k_3} \zeta_F^2 - \frac{k_2}{k_1+k_3} \zeta_G\right) a(t) = 0 \qquad (17)$$

and for the C_4 and lighter products

$$\frac{d\zeta_L}{d\lambda} - K_F \left(\frac{k_3}{k_1+k_3} \zeta_F^2 + \frac{k_2}{k_1+k_3} \zeta_G\right) a(t) = 0 \qquad (18)$$

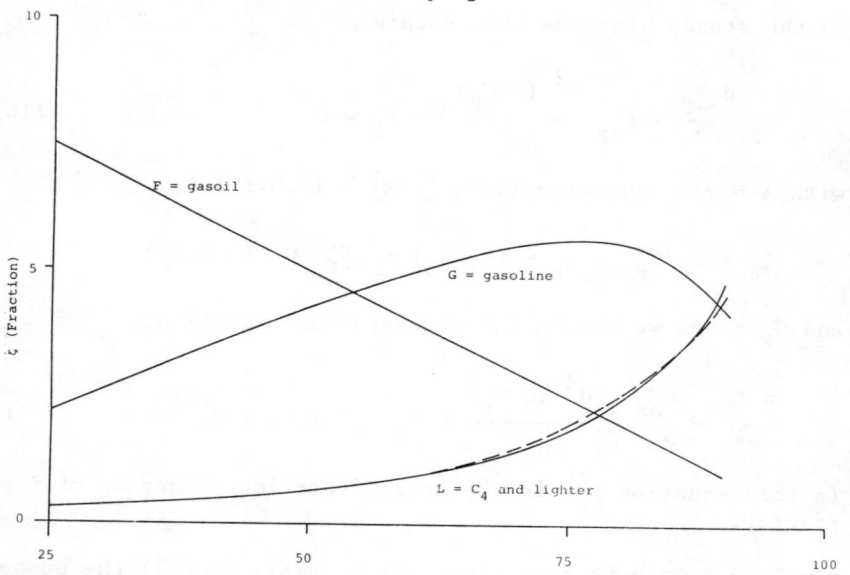

Fig. 6. The weight fraction of gasoil, gasoline and butane and lighter according to equations 10, 11 and 12 as a function of the conversion.
..... logarithmic approximation for fraction L.

Integration of these equations with the starting values:
$\lambda = 0 : \zeta = 1 \; \zeta_G = 0 \; \zeta_L = 0$ yields sets of curves of which fig. 6 is an example.

Further sophistication can be introduced by making a heat balance over the riser reactor, and correcting the rate constants and ρ for the change in temperature. Generally this type of fundamental approach is used only to develop useful correlations.

b. The practical approach

From fig. 6 we can see (dotted line) that the weight fraction C_4 and lighter can for a rather wide range of conversions be very well approximated by

$$\ln \zeta_L = a (1 - \zeta_F) + b \tag{19}$$

This is in agreement by the method described by Ewell and Gadmer (1978), who show that $\log \zeta_L$ and log (coke make) correlate linearly with conversion. This is shown in fig. 7 where also $\ln \zeta_F (\equiv \ln (1 - \text{conversion})$ is plotted as a function of the conversion.

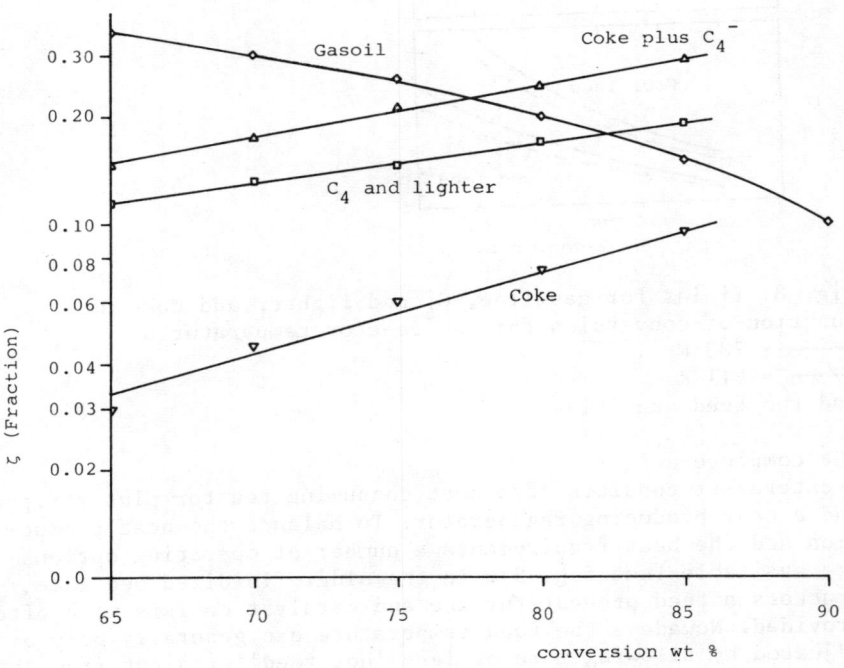

Fig. 7. Experimental conversion data on logarithmic scale.

The advantage of using conversion as the independent variable is that the effects of e.g. feed quality, reactor temperature and catalyst activity can to a large extent be lumped in the conversion parameter. This allows plots of the various yields against conversion to be made with other qualities as secundary parameters.

Fig. 8 gives an example of the type of relation used in this approach.

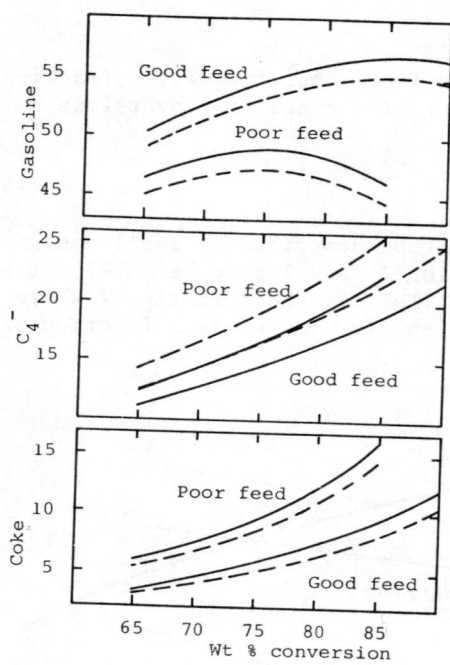

Fig. 8. Yields for gasoline, C_4 and lighter and coke as function of conversion for two reactor temperatures
———— = 783 K
- - - - - = 811 K
and two feed qualities.

5. The complete unit

A catcracker consists of a heat consuming reactor plus stripper and a heat producing regenerator. To balance the heat production and the heat requirements a number of operating options are available (see fig. 9). In the older fluidized bed catcrackers a feed preheat furnace and catalyst coolers were often provided. Nowadays the feed temperature can generally only be adjusted by allowing more or less 'hot feed' straight from the feed preparator (vacuum distillation) into the feed stream of the catcracker.

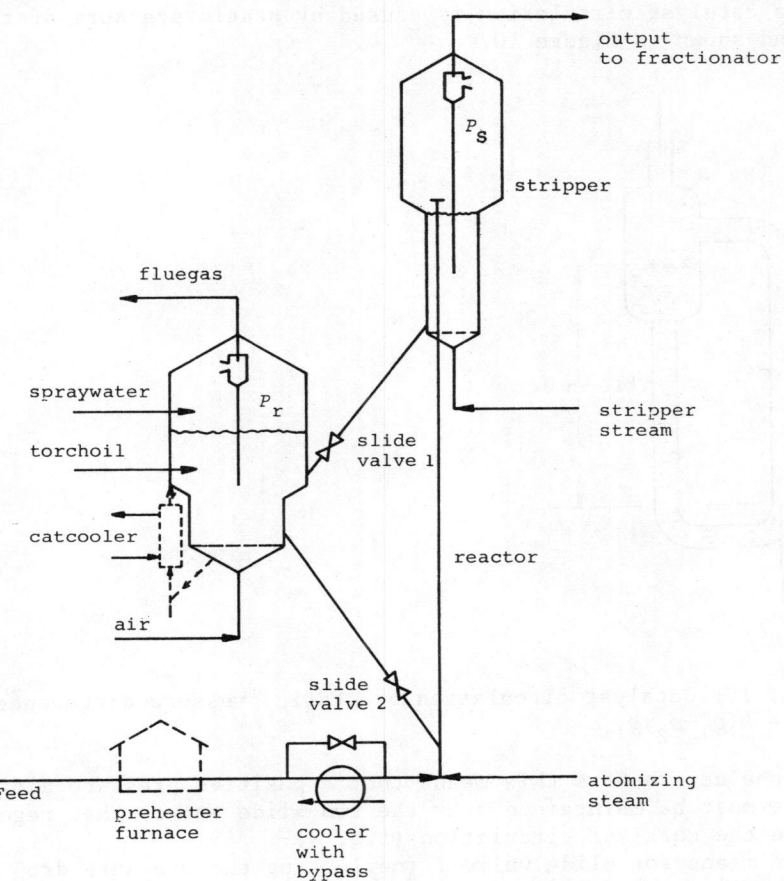

Fig. 9. Catcracker: heat balancing options.

More heat can be produced in the regenerator by reducing the stripping efficiency, adding torch oil or by decreasing the CO/CO_2 ratio in the flue gas. The catalyst can be cooled by adding spray water to the regenerator.

In case the capacity of the regenerator airblower is such that the coke burning capacity is the plant's bottleneck, a high CO/CO_2 ratio (say 0.7) will be choosen. If on the other hand enough air is available and the reactor has high heat requirements maximum CO combustion is advantageous. This results in high regenerator temperatures, which are also useful in reaching low coke on cat levels (say 0.10 % by wt) on the regenerated catalyst. Especially for the zeolitic catalysts this is advantageous, as this results in even greater activity and better selectivity. For the present day reactors this is a must since the residence time in the reactor is some two or three seconds only.

The catalyst circulation is caused by static pressure of the kind shown in figure 10.

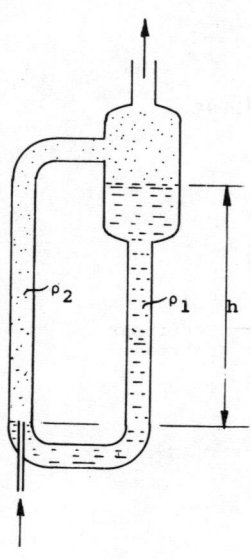

Fig. 10. Catalyst circulation by static pressure difference $\Delta p = h(p_1 - p_2)g$.

In the catcrackers this means that a positive pressure difference must be maintained over the two slide valves that regulate the catalyst circulation (fig. 9).
This means for slide valve 1 (neglecting the pressure drop from slide valve 1 to the cat level in the regenerator)

$$P_s + h_1 \rho_1 > P_r \tag{20}$$

and for slide valve 2

$$P_r + h_2 \rho_1 > P_s + h\,s\,\rho_2 \tag{21}$$

with P_s and P_r the stripper and regenerator pressures
h_1 the height between the cat levels in stripper and regenerator
h_2 the height between regenerator level and slide valve 2
h_3 the height of the riser reactor above slide valve 2
ρ_1 the density of the dense phase
ρ_2 the density of the dilute phase in the reactor
From this it follows that

$$P_s + h_1 \rho_1 > P_r > P_s - h_2 \rho_1 + h_3 \rho_2 \tag{22}$$

Catcracker operations

Due to the abrasive action of the circulating catalyst (some 700 kg/s ~ 60.000 t/d) no orifice flow measurement is possible. The cat circulation rate is calculated on basis of the air flow and a carbon hydrogen balance over the regenerator. The required data are
- carbon and hydrogen on spent and regenerated catalyst;
- flue gas composition;
- air flow from the airblower.

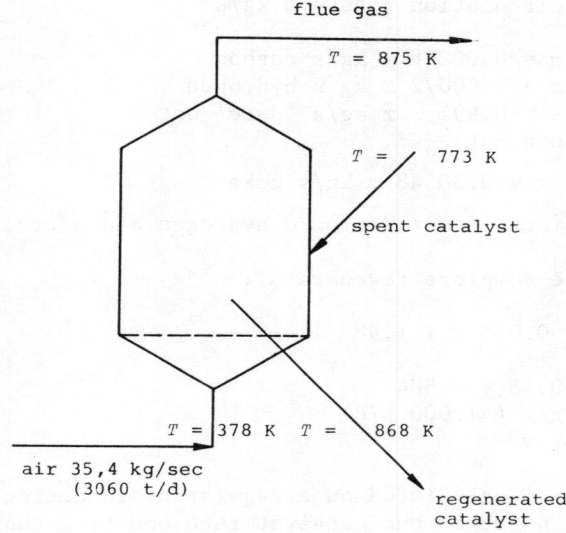

Fig. 11. Catcracker regenerator. Composition of (dry) flue gas (% by vol.): N_2 = 85.2, CO_2 = 8.0, CO = 6.5, O_2 = 0.3.
Coke on spent cat: 0.90 % by wt.
Coke on spent cat composition: 92 % by wt carbon
                                8 % by wt hydrogen

Example: For data see figure 11.
Air consists of 20.9 % by vol. O_2 and 79.1 % by vol. N_2.
Nitrogen balance:
In: $\dfrac{35.4}{79.1 \times 28 + 20.9 \times 32} \times 79.1 = 0.971$ k mol/s N_2
Out: 0.971 k mol/s N_2
  ← Mass flow of other components in flue gas.
Out CO_2: $\dfrac{8.0}{85.2} \times 0.971 = 0.091$ k mol/s CO_2

  CO : $\dfrac{6.5}{85.2} \times 0.971 = 0.074$ k mol/s CO

  O_2 : $\dfrac{0.3}{85.2} \times 0.971 = 0.003$ k mol/s O_2

Oxygen balance:

In: $\dfrac{35.4}{79.1 \times 28 + 20.9 \times 32} \times 20.9 = 0.257$ k mol/s

Out: in CO_2 0.091 k mol/s O_2
      in CO 0.037 k mol/s O_2
      in O_2 0.003 k mol/s O_2
      in H_2O (balance) 0.126 k mol/s O_2

C in flue gas : 0.165 k mol/s = 1.98 kg/s
H_2 in flue gas: 0.252 k mol/s = 0.504 kg/s
Say spent catalyst circulation rate = x kg/s
Catalyst balance:

In: $0.009 \times 0.92 \times x = 0.00828\ x$ kg/s carbon
     $0.009 \times 0.08 \times x = 0.00072\ x$ kg/s hydrogen
                          $0.991\ \ \ x$ kg/s 'pure' cat

Out: $0.991\ x$ kg/s pure cat

Coke: $\dfrac{0.55}{99.45} \times 0.991\ x = 0.00548\ x$ kg/s coke

Say spent coke consists of a fraction α hydrogen and a fraction $1-\alpha$ carbon.
We then have for the complete regenerator:
Carbon balance
$0.00828\ x = (1-\alpha)\ x\ 0.00548 + 1.98$
Hydrogen balance
$0.00072\ x = \alpha \cdot x \cdot 0.00548 + 0.504$
We find $x = 705.7$ kg/s (61.000 t/d)
      and $\alpha = 0.1$ % by wt

Such mass balances are calculated on a regular basis during operation in order to obtain the catalyst rate and from that the cat-oil ratio.
The heat transported from the regenerator is calculated from the cat and the dense bed regenerator temperature.
For design purposes this temperature is calculated from a heat balance of the regenerator. This is a rather complicated calculation as it not only requires data for coke on spent and regenerated catalyst, the stripper outlet temperature, the composition of the flue gas and for the external heat losses but, under afterburning conditions, also information on the interaction between the dense and the dilute part of the bed. Under severe afterburning the temperature difference between the dilute and the dense phase can increase to over 60 K and more than 25 percent of the total heat of combustion can be regenerated in the dilute phase.

References

Blanding, F.H., Ind. Eng. Chem., 45 (6), 1193 (1953)
Ewell, R.B. and Gadmer, G., Hydrocarbon Processing, 57 (4), p. 125 (1978)
Greensfelder, B.S., Voge, H.H. and Good, G.M., Ind. Eng. Chem., 41, 2573 (1949)
Gustafson, W.R., Ind. Eng. Chem. Process Develop., 11 (4), 507 (1972)
Gwyn, J.E., Advanc. in Chem. Ser., 109, 513-518 (1972) (1st Int. Symp. Chem. Reaction Eng.)
Kramers, H. and Westerterp, K.R., "Elements of Chemical Reactor Design and Operation", Amsterdam, p. 72 (1963)
Levenspiel, O., "Chemical Reactor Engineering", John Wiley and Sons, New York, p. 284 (1972)
Nace, D.M. (1965), see Weekman (1968), loc. cit., p. 92
Nace, D.M., Voltz, S.E. and Weekman, V.W., Ind. Eng. Chem., Proc. Des. Develop., 10, 530, 538 (1971)
van Swaay, W.P.M., Buurman, C. and van Breugel, J.W., Chem. Eng. Sci., 25, 1818 (1970)
Szépe, S. and Levenspiel, O., Proc. 4th European Symposium on Chemical Reaction Engineering, Brussel, p. 265 (1968) (Suppl. to Chemical Eng. Science)
Tan, C.H., Fuller, O.M., Can. J. Chem. Eng., 48, 174 (1970)
Voorhies, A., Ind. Eng. Chem., 37 (4), 318 (1945)
Weekman Jr., V.W., Ind. Eng. Chem. Proc. Des. Develop., 7, 90 (1968); 8, 385 (1969)
Weekman Jr., V.W. and Nace, D.M., A.I.Ch.E.J., 16, 397 (1970)
White, P.J., Hydrocarbon Processing, 47 (5), 103 (1968)

Part III

REFORMING OF HYDROCARBONS ON METALS AND ALLOYS

CHEMICAL BONDING

R. Prins

Laboratory for Inorganic Chemistry, Eindhoven
University of Technology, Eindhoven, The Netherlands

Whereas in the macroscopic world everything can be measured 'exactly', this is not possible in the microscopic world of the electrons that move around the nuclei in atoms and molecules. Because of the fact that it is impossible to exactly determine both place and momentum of an electron, we have to use a statistical rather than a deterministic description of the electrons in matter. The statistical information about the electrons is contained in the wave function Ψ and Ψ can be determined by solving the Schrödinger wave equation

$$H\Psi = E\Psi$$

The Hamilton operator H is easily obtained by a comparison with the classic expressions for kinetic and potential energy. Solving the Schrödinger differential equation then delivers both the wave function of the electrons and the energy that goes with it. We will illustrate this process of determining Ψ and E by solving the Schrödinger equation in a few examples. To refresh our memories we will start with the hydrogen atom and thereafter discuss the Molecular Orbital method as applied to the H_2^+, H_2 and other diatomic molecules. As an introduction to contributions by other lecturers we will end with a treatment of the bonding in metals and the bonding of an adsorbate like CO on a metal surface.

THE HYDROGEN ATOM [1]

```
   +e           -e
    •─────r─────•
    M           m
```

In classical mechanics the kinetic energy of the hydrogen atom would be equal to $\frac{p^2}{2\mu}$ with $\frac{1}{\mu} = \frac{1}{M} + \frac{1}{m}$. In quantum mechanics the corresponding Hamiltonian is

$$-\frac{\hbar^2}{2\mu}\left(\frac{\partial^2}{\partial x^2} + \frac{\partial^2}{\partial y^2} + \frac{\partial^2}{\partial z^2}\right) = -\frac{\hbar^2}{2\mu}\nabla^2$$

The potential energy is $V = -\frac{e^2}{r}$ so that $H = -\frac{\hbar^2}{2\mu}\nabla^2 - \frac{e^2}{r}$. Ψ and E can be found by solving the differential equation

$$-\frac{\hbar^2}{2\mu}\nabla^2\Psi - \frac{e^2}{r}\Psi = E\Psi$$

or

$$\frac{\hbar^2}{2\mu}\nabla^2\Psi + (E + \frac{e^2}{r})\Psi = 0$$

Using polar instead of cartesian coordinates we can write

$$\nabla^2 = \frac{1}{r^2}\frac{\partial}{\partial r}\left(r^2\frac{\partial}{\partial r}\right) + \frac{1}{r^2 \sin\theta}\frac{\partial}{\partial\theta}\left(\sin\theta\frac{\partial}{\partial\theta}\right) + \frac{1}{r^2\sin\theta}\frac{\partial^2}{\partial\phi^2}$$

Substitution gives a Hamilton operator that consists of three parts that are independent of each other, one depending only on r, another on θ and one on ϕ only. In such a case where the Hamiltonian consists of a sum of independent operators one can easily prove that the wave function Ψ will be a product of independent wave functions. In our case it means that

$$\Psi(r, \theta, \phi) = R(r).\Theta(\theta).\Phi(\phi) = R(r).Y(\theta, \phi)$$

By hard mathematical work, or by looking in textbooks, and after taking into consideration the boundary conditions one obtains the result that the wave function is dependent on the quantum numbers n, l and m_l, while the energy is a function of n only:

$$\Psi = R_{nl}\Theta_{lm_l}\Phi_{m_l} = R_{nl}Y_{lm_l} \text{ and } E = -\frac{\mu e^4}{2n^2\hbar^2}$$

Chemical bonding

The energy of the hydrogen atom is discrete and so are the wave functions. For l = 0 we have the so-called s functions, for l = 1 the p functions and for l = 2 the d functions. Mathematical expressions for these functions and drawings of their boundary surfaces are given in figure 1.

K Shell

$$n = 1, \; l = 0, \; m = 0: \quad \Psi_{1s} = \frac{1}{\sqrt{\pi}} \left(\frac{Z}{a_0}\right)^{3/2} e^{-Zr/a_0}$$

L Shell

$$n = 2, \; l = 0, \; m = 0: \quad \Psi_{2s} = \frac{1}{4\sqrt{2\pi}} \left(\frac{Z}{a_0}\right)^{3/2} \left(2 - \frac{Zr}{a_0}\right) e^{-Zr/2a_0}$$

$$n = 2, \; l = 1, \; m = 0: \quad \Psi_{2p_z} = \frac{1}{4\sqrt{2\pi}} \left(\frac{Z}{a_0}\right)^{3/2} \frac{Zr}{a_0} e^{-Zr/2a_0} \cos\theta$$

$$n = 2, \; l = 1, \; m = \pm 1: \quad \Psi_{2p_{x,y}} = \frac{1}{4\sqrt{2\pi}} \left(\frac{Z}{a_0}\right)^{3/2} \frac{Zr}{a_0} e^{-Zr/2a_0} \sin\theta \cdot \begin{matrix} \cos\phi \\ \sin\phi \end{matrix}$$

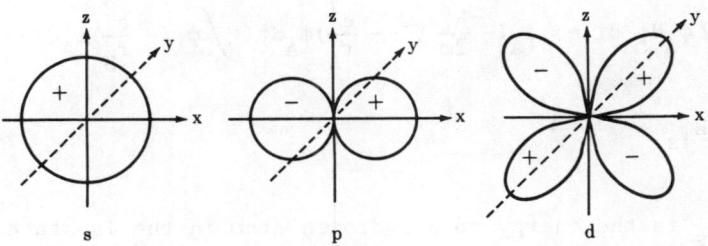

Fig. 1 Boundary surfaces and analytical expressions of atomic orbitals

A boundary surface is the contour inside which the chance of finding the electron is equal to a certain percentage. Usually boundary surfaces are drawn for a 95% chance and thus the percentage of the electronic charge that is outside the boundary surface is small.

THE H_2^+ ION [1]

$$H = -\frac{\hbar^2}{2m}\nabla^2 - \frac{e^2}{r_A} - \frac{e^2}{r_B} + \frac{e^2}{R}$$

Our task is to solve the equation $H\Psi = E\Psi$. This can be done in an exact way by using elliptical coordinates. The exercise is difficult, however, and the resulting wave function and energy are hard to analyse. Much more illustrative is a less exact approach in which an educated guess is made of the wave function, which is checked by substitution into the Schrödinger equation. So let us try if the 1s function of the ground state of the hydrogen atom is a reasonable wave function for the ground state of the H_2^+ ion. After all H_2^+ can be looked upon as a hydrogen atom perturbed by a positive charge. Let us therefore take Ψ equal to the 1s function on atom A: $\Psi = \phi_A$

$$H\Psi = E\Psi \qquad E = \int \Psi H \Psi d\tau$$

$$E = \int \phi_A H \phi_A d\tau = \int \phi_A (-\frac{\hbar^2}{2m}\nabla^2 - \frac{e^2}{r_A})\phi_A d\tau + \int \phi_A (-\frac{e^2}{r_B})\phi_A d\tau + \frac{e^2}{R}$$

$$E = E_{1s} + \alpha + \frac{e^2}{R}$$

where E_{1s} is the energy of a hydrogen atom in the 1s state. The integral α is not only negative but also $|\alpha| < \frac{e^2}{R}$, since simple electrostatics tell us that if the nucleus B is outside the electron cloud ϕ_A then $\alpha = \frac{-e^2}{R}$, while for B inside the electron cloud $|\alpha| < \frac{e^2}{R}$. Thus $E > E_{1s}$ and we have to conclude that the energy of the H_2^+ that is described by a 1s wave function is higher than that of a single H atom. This implies that there would be no chemical bonding between H and H^+ in the H_2^+ ion, in contrast to experimental observation. A hydrogen 1s function on one atom A therefore is not a good wave function for the ground state of the H_2^+ ion.

Instead of ϕ_A we could have taken ϕ_B and would have got the same result. But what happens if we take an equal mixture of ϕ_A and ϕ_B?

$$\Psi^+ = \frac{1}{(2 + 2S)^{\frac{1}{2}}} (\phi_A + \phi_B)$$

where $S = \int \phi_A \phi_B d\tau$ is the overlap integral which comes into play because of normalisation of the wave function. Using this wave function to describe the ground state of H_2^+ might improve the situation because it has more electron density in the internuclear region and may thus lead to a lower potential energy.

$$E^+ = \int \Psi H \Psi d\tau = \frac{1}{2(1 + S)} \int (\phi_A + \phi_B) H (\phi_A + \phi_B) d\tau = \frac{H_{AA} + H_{AB}}{1 + S}$$

$$H_{AA} = \int \phi_A (-\frac{\hbar^2}{2m} \nabla^2 - \frac{e^2}{r_A} - \frac{e^2}{r_B} + \frac{e^2}{R}) \phi_A d\tau =$$

$$= E_{1s} + \frac{e^2}{R} + \int \phi_A (-\frac{e^2}{r_B}) \phi_A d\tau = E_{1s} + \frac{e^2}{R} + \alpha$$

$$H_{AB} = \int \phi_A (-\frac{\hbar^2}{2m} \nabla^2 - \frac{e^2}{r_A} - \frac{e^2}{r_B} + \frac{e^2}{R}) \phi_B d\tau$$

$$= E_{1s} \cdot S + \frac{e^2}{R} \cdot S + \int \phi_A (-\frac{e^2}{r_A}) \phi_B d\tau = E_{1s} \cdot S + \frac{e^2}{R} \cdot S + \beta$$

$$E^+ = E_{1s} + \frac{e^2}{R} + \frac{\alpha + \beta}{1 + S}$$

Calculations show that because of the β integral E^+ can become lower than E_{1s} for a range of internuclear distances and that there is a bonding interaction between H and H^+ in the H_2^+ ion. In figure 2 it is seen that the energy passes through a minimum value E_{min} at a certain value of R. The energy stabilisation at this minimum is calculated to be 1.76 eV at an internuclear distance of 1.32 Å.

The experimental values are 2.79 eV at R = 1.06 Å and demonstrate that although the above wave function accounts for a major part of the bonding in H_2^+ it is not a very good wave function. By using a modified 1s function with e^{-Zr} instead of e^{-r} for ϕ_A and ϕ_B we can improve the situation already substantially. For such a wave function one gets E_{min} = 2.25 eV at R_{min} = 1.08 Å. The value of Z that corresponds to the stable H_2^+ ion is Z = 1.23. It tends to draw the electrons closer to the nuclei and in doing

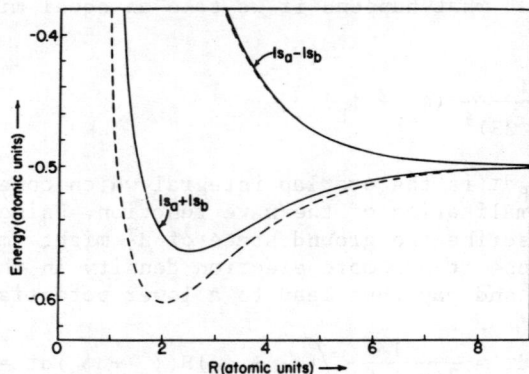

Fig. 2 Energies of the lowest bonding and antibonding states of H_2^+. The solid curves represent the energies of the molecular wave functions using hydrogen 1s atomic orbitals, while the dashed curve represents the true energy of the ground state.

so it shifts the electron cloud to some extent into the region between the nuclei. In this way the potential energy as well as the total energy is decreased.

These results show that a wave function of the type $(\phi_A + \phi_B)$ is able to account for most of the binding energy and suggests that such a wave function gives a reasonable description of the state of the electron. The calculations quoted above indicated that the bonding results from the so-called resonance integral β which comes into play because of the form chosen for the wave function. Contour curves given in figure 3 show that for this wave function the charge is concentrated more between the nuclei than would be expected by superposition of the component atomic orbital densities.

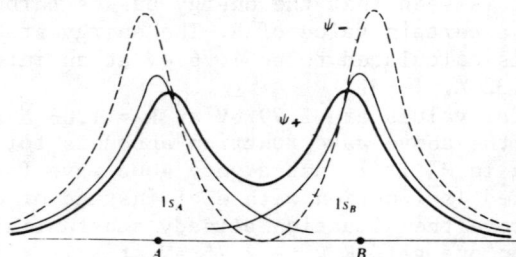

Fig. 3 The density ψ^2 along the nuclear axis of H_2^+ for the lowest bonding and antibonding wave functions.

The simultaneous attraction of the electron by the two nuclei concentrates the electron charge into the internuclear region thereby lowering the potential energy. This is the main reason for bonding in H_2^+.

The wave function used above is called bonding since it represents a bonding state. Another wave function Ψ^- that can be composed from ϕ_A and ϕ_B is the antibonding wave function

$$\Psi^- = \frac{1}{(2 - 2S)^{\frac{1}{2}}} (\phi_A - \phi_B)$$

Repeating the calculation presented above for the bonding state gives the following result for the antibonding state

$$E^- = E_{1s} + \frac{e^2}{R} + \frac{\alpha - \beta}{1 - S}$$

Remembering that all the bonding comes from β and that $\beta < 0$ it is clear that $E^- > E_{1s}$ for all internuclear distances and that Ψ^- represents a repulsive state of the H_2^+ ion. Contour curves show that in this wave function the charge is pushed away from the internuclear region and that actually there is a nodal plane in the electronic density midway between the nuclei (cf. figure 3). As a consequence there is no gain in potential energy that can overcome the nuclear repulsion when the nuclei in H_2^+ come together and an antibonding state results.

Summarising we can say that a very important reason for bonding in the ground state of the H_2^+ ion is the lowered potential energy of the electron in the internuclear region. This is well represented by a wave function of the type $\Psi = \frac{1}{\sqrt{N}}(\phi_A + \phi_B)$.

The original degeneracy of the two functions ϕ_A and ϕ_B is split up into a bonding and an antibonding combination. The energy of the bonding orbital is decreased and that of the antibonding orbital is increased relative to the energy of the atomic orbitals. Since the electron goes to the bonding orbital, the total energy of the system is lower than that of the individual components and the system H_2^+ is stabilised by bonding energy. Note that the energy decrease of Ψ^+ relative to ϕ_A or ϕ_B is always smaller than the energy increase of Ψ^- because of the difference in the normalisation factors $(1 + S)^{\frac{1}{2}}$ and $(1 - S)^{\frac{1}{2}}$, respectively. Often $S < 0.2$ and in that case we can neglect the overlap integral S in the numerator, resulting in a stabilisation of the bonding orbital by β and a destabilisation of the antibonding orbital also by β, relative to the energy of the atomic orbitals.

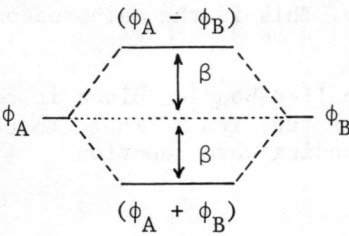

THE HYDROGEN MOLECULE [1]

The simplest molecular system is the hydrogen molecule. Its Hamiltonian is

$$H = -\frac{\hbar^2}{2m}\nabla_1^2 - \frac{\hbar^2}{2m}\nabla_2^2 - \frac{e^2}{r_{A1}} - \frac{e^2}{r_{B1}} - \frac{e^2}{r_{A2}} - \frac{e^2}{r_{B2}} + \frac{e^2}{R} + \frac{e^2}{r_{12}}$$

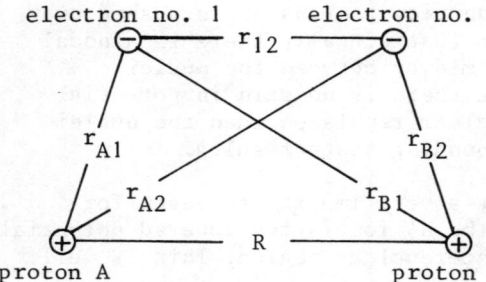

The Hamiltonian for the H_2 molecule leads to a complicated differential equation. The $\frac{e^2}{r_{12}}$ term, which represents the mutual repulsion of the two electrons, makes it impossible to separate the variables in any coordinate system and one therefore has to rely on approximate methods of solving the Schrödinger equation for the hydrogen molecule, as well as for all molecules containing two or more electrons.

A useful approximate wave function for the ground state of the hydrogen molecule is obtained when using the molecular orbital method. In this method it is realised that if the $\frac{e^2}{r_{12}}$ term in the Schrödinger equation is omitted, the resulting differential equation can be separated in the coordinates of the two electrons. The two resulting equations are those of H_2^+ molecular ions:

$$H = H(H_2^+, 1) + H(H_2^+, 2) - \frac{e^2}{R} + \frac{e^2}{r_{12}}$$

To a first approximation we can therefore write the wave function of H_2 as a product of H_2^+ wave functions and its ground state as

$$\Psi = \Psi(H_2^+, 1) \cdot \Psi(H_2^+, 2)$$

where $\Psi(H_2^+, i)$ is the wave function of the ground state of H_2^+. Knowing that each of the two occupied orbitals forms a strong bond in H_2^+ it is understandable that this wave function leads to a stable H_2 molecule. The electron repulsion weakens the bond somewhat but does not destroy it althogether.

Detailed calculations are most easily performed by using the linear combination of atomic orbitals (LCAO) approximation to the molecular orbitals:

$$\Psi(H_2^+) = \frac{1}{\sqrt{N}}(\phi_A + \phi_B)$$

It turns out that the energy-nuclear distance curve of H_2 is similar to that of H_2^+. Taking 1s functions for ϕ_A and ϕ_B the energy stabilisation of H_2 relative to the separate atoms is 2.68 eV at $R_{min} = 0.85$ Å. Introducing a scaling factor for the nuclear charge in $\phi \approx e^{-Zr}$ improves these values to 3.47 eV at 0.73 Å for $Z = 1.20$, whereas the experimental values are $E_{min} = 4.75$ eV at $R_{min} = 0.74$ Å.

Although the electron repulsion modifies the actual value of the bonding energy in H_2 relative to that in H_2^+, the description of the bonding is the same in both molecules: it is the increased electron density in the region between the nuclei that gives a decreased potential energy and a lowering of the total energy relative to that of the separate atoms.

DIATOMIC MOLECULES [1]

From the study of H_2^+ and H_2 we learned that, when bringing two similar atoms together, each atomic orbital ϕ_A or ϕ_B splits into two molecular orbitals whose LCAO forms are $\phi_A \pm \phi_B$. The energy difference between the bonding and antibonding molecular orbitals is about equal to twice the resonance integral β. Thus an atomic 2s orbital splits into $\phi_A(2s) \pm \phi_B(2s)$ and the atomic $2p_x$ orbitals (the x-axis is the main molecular axis) into $\phi_A(2p_x) \pm \phi_B(2p_x)$. Similarly the $2p_y$ orbitals give $\phi_A(2p_y) \pm \phi_B(2p_y)$. The general shape of the latter molecular orbitals is somewhat different from the others, though. For instance the bonding molecular orbital consists of two ribbon-like regions in which Ψ has different signs and the nodal plane of the $2p_y$ atomic orbital remains nodal plane of the molecular orbital. Such orbitals are called π orbitals, they play an important role in unsaturated hydrocarbons like ethylene and benzene. If we combine two $2p_z$ atomic orbitals instead of $2p_y$, we get the same molecular orbitals as before (and with the same energy), except that they are turned $90°$ around the molecular axis.

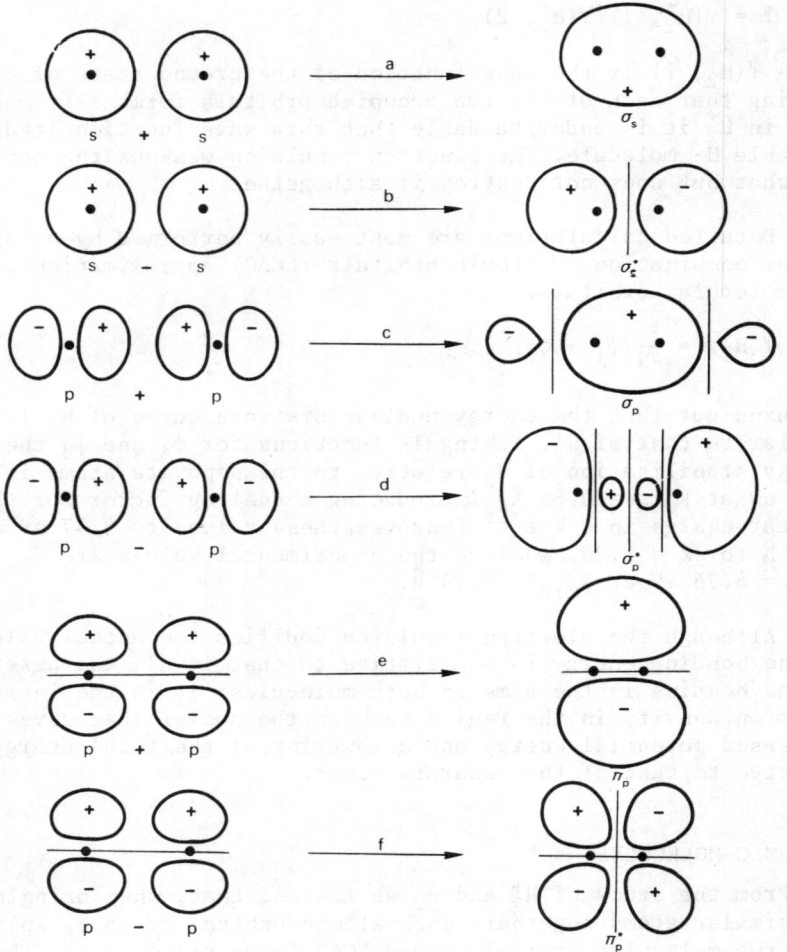

Fig. 4 Contours of molecular orbitals of diatomic molecules formed by linear combinations of atomic orbitals.

With these molecular orbitals we can draw the level scheme for the orbitals of homonuclear molecules as given in Figure 5. Molecular orbitals that have cylindrical symmetry around the molecular axis are indicated with the label σ, while orbitals with one nodal plane through that axis are called π. With the level diagram of fig. 5 we can qualitatively explain the stability or instability of diatomic molecules. In Li_2 there are two 2s electrons to be placed in this level scheme and in the ground state they will go to the σ molecular orbital. Since this

Chemical bonding

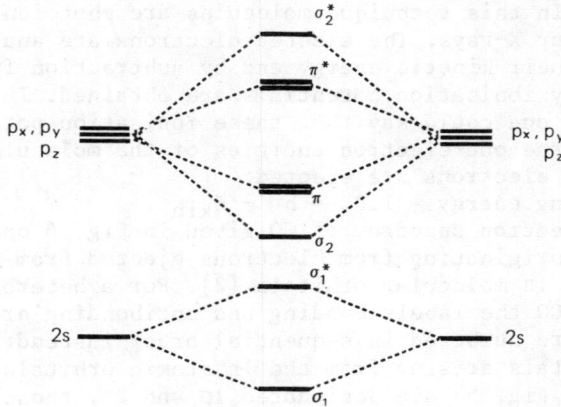

Fig. 5 Molecular orbital diagram of homonuclear diatomic molecules (neglecting s-p interactions)

is a bonding orbital there will be a net bonding energy, just like the H_2 molecule, and two lithium atoms may form a stable Li_2 molecule. In Be_2 we have to put two electrons in the σ orbital and the other two in the σ^* orbital. Remembering the fact that an antibonding orbital is increased in energy more than the corresponding bonding orbital is decreased relative to the energy of the atomic orbitals, we see that when both bonding and antibonding orbitals are equally filled there will be an increase in total energy relative to the energy of the constituting atoms. Two approaching berylium atoms thus repel each other, as will do two helium atoms and two neon atoms.

In a similar way it can be shown that the B_2, C_2 and N_2 molecules are stable. In this order the bond strength will increase due to the fact that an increasing number of electrons can be placed in bonding orbitals. Also O_2 and F_2 are stable diatomic molecules, but the bond strength in these molecules is weaker because of the fact that there are electrons in antibonding orbitals as well.

Heteronuclear diatomic molecules like CO and NO have a similar orbital level scheme and their properties can be well understood with the scheme given above. Thus in case of CO, which is isoelectronic with N_2, the $\sigma(2p)$ and $\pi(2p)$ orbitals are filled resulting in a large bond strength between the C and O atoms. In NO there is one electron more, which goes to the π^* orbital. The bond strength in NO is still appreciable and the unpaired electron explains the paramagnetism of the molecule.

Photoelectron spectroscopy enables us to check the orbital level scheme. In this technique molecules are photoionised with hard UV light or X-rays. The ejected electrons are analysed according to their kinetic energy and by subtraction from the ionising energy ionisation potentials are obtained. In molecular orbital theory one could say that these ionisation potentials correspond to the one-electron energies of the molecular orbitals from which the electrons are ejected:

electron binding energy = I.P. = $h\nu - E_{kin}$.

In the photoelectron spectrum of CO given in Fig. 6 one can indeed observe peaks originating from electrons ejected from the 3σ, 4σ, 5σ and 1π molecular orbitals [2]. For a heteronuclear molecule like CO the labels bonding and antibonding are not used. The orbitals are numbered in sequential order instead: molecular orbitals arising from the 1s atomic orbitals (not represented in Fig. 5) are designated 1σ and 2σ, those from the 2s atomic orbitals are designated 3σ and 4σ, and so on. Contrary to what might have been expected from Fig. 5 the ionization potential of the 1π electrons is larger than that of the 5σ electron. The reason for this is that our approach thus far has been to simple. By taking configuration interaction between 2s and 2p orbitals into account the reversal of the 1π and 5σ orbitals can be remedied [3].

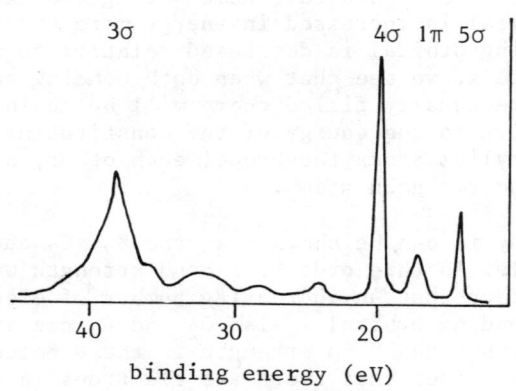

Fig. 6 Photoelectron spectrum of CO excited by Al Kα X-rays

Chemical bonding

METALS [1]

The essential point in the molecular orbital theory of molecules is that each electron moves in a potential field which extends over all atoms. By suitable averaging one can introduce a one-electron Hamiltonian and obtain the corresponding wave function and energy. The necessary condition for delocalisation of the molecular orbital is that the atomic orbitals on neighbouring atoms overlap significantly. In benzene, for example, each $2p_z$ atomic orbital (perpendicular to the molecular plane) overlaps equally with both neighbours and the resulting molecular orbitals extend over all six carbon atoms.

In metals one can follow the same procedure with infinite arrays of atoms. In a metal like lithium the overlap integral $S = 0.5$ and as a consequence the molecular orbitals are completely delocalised. These are the 'Bloch' wave functions. In case of a diatomic molecule we saw that by combining two atomic orbitals we obtained two molecular orbitals with energies $+\beta$ and $-\beta$ relative to those of the atomic orbitals. If we add a third atom we obtain three molecular orbitals with energies grouped around the energy of the constituting atomic orbital. This process may be continued, each successive addition of an atom adding one more energy level and at the same time altering those of the previous set slightly. For an infinite number of atoms we obtain a continuous band of energies around the energy of the original atomic orbital. The width of the energy band depends on the interaction between the atoms. If the interaction between an atom and its N neighbouring atoms is much larger than that with next nearest neighbours, the total width will be $2N\beta$, where β is the resonance integral between neighbouring atoms that we have already encountered in the H_2^+ ion. (Compare H_2 where the 'width of the 1s band' is about 2β and benzene where the 'width of the π band' is 4β).

The process of synthetizing an energy band from an atomic orbital level can be followed for any atomic orbital. The band width will vary with the internuclear distance R, since β will vary with R. As a consequence energy-internuclear distance diagrams may be obtained for the 3d and 4s bands, as exemplified in figure 7a.
Because of the fact that $\beta(3d)$ is much smaller than $\beta(4s)$, the width of the 4s band is much larger than that of the 3d band and the bands overlap at shorter interatomic distances. A section through this diagram at R equal to the experimental distance gives the experimental band widths. For metal theory it is furthermore important to know how the total number of allowed energies in a band varies with energy. This is given by the density of states $N(E)$, where $N(E)dE$ is the number of energy levels per unit cell which lie between E and $E + dE$. An example of a curve of $N(E)$ against E is given in figure 7b.

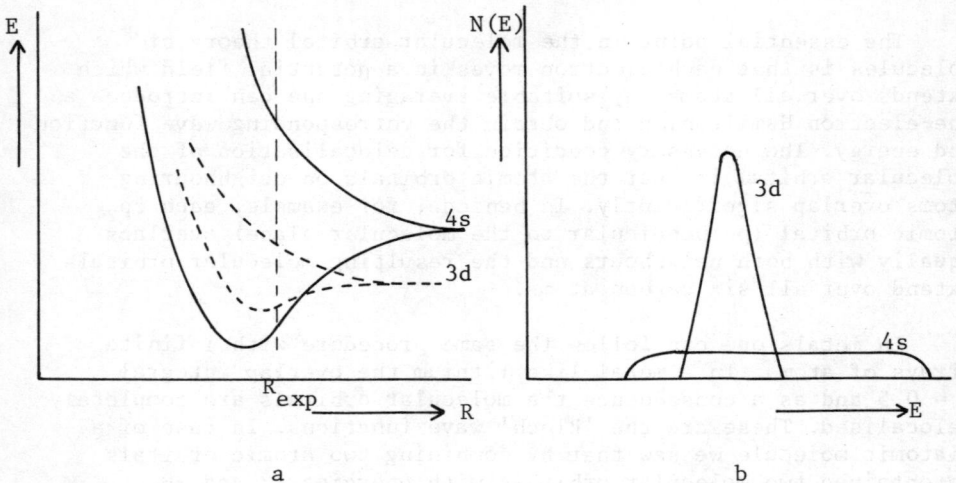

Fig. 7 a. Energy-internuclear distance diagram of the 3d and 4s levels in metals
b. Density of states as a function of energy for the 3d and 4s bands.

In principle density-of-states curves can be obtained from photoelectron spectra of the valence region of metals. In figure 8 the photoelectron spectra of the valence region of copper, silver and gold are presented. These spectra, measured with monochromatized Al Kα radiation, were obtained by Wertheim et al [4] by subtraction of the background due to inelastically scattered electrons and by enhancing the resolution to 0.2 eV with the aid of a deconvolution technique. In all three spectra the 4s band, with its relatively low density of states, can be clearly distinguished from the 3d band. The 3d band has a binding energy larger than the Fermi edge and is completely filled in these metals, whereas the 4s band is only half filled at the Fermi edge.

The dashed lines in figure 8 represent theoretical densities of states, with spin-orbit interaction included. The agreement between the location of the resolved features in theory and experiment is very good, but the intensities are quantitatively less accurately reproduced. Perfect agreement cannot be expected anyhow, because the intensities depend not only on the density of initial states, but also on transition matrix elements and on the density of final states. Other complicating factors have

been described in the literature [5]. With high energy radiation
most of these factors are, however, of minor importance and the
experimental X-ray photoelectron spectra may serve as a first
order estimate of the density of initial states.

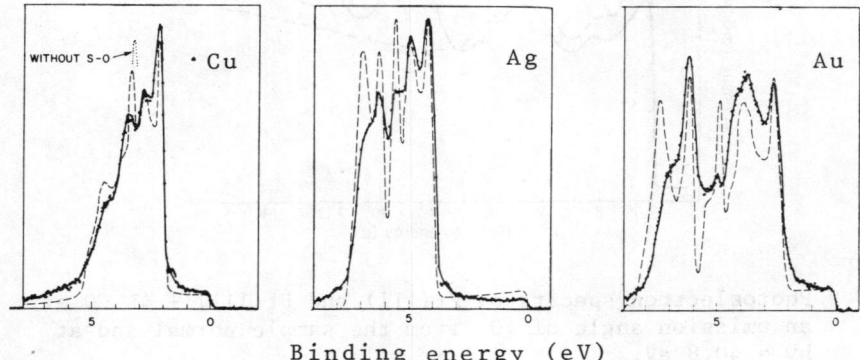

Binding energy (eV)

Fig. 8 X-ray photoelectron spectra of Cu, Ag and Au after subtraction
of the background and after deconvolution. Dashed lines represent
theoretical densities of states.

CHEMISORPTION OF CO ON METALS

For an interpretation of catalysis in terms of reactions
between adsorbates on a substrate, the energy changes that occur
on adsorption of molecules on a surface are of great importance.
Such energy changes have been studied lately with the aid of
photoelectron spectroscopy. The possibility of using (synchrotron)
radiation of varying energy and at different angles proved very
helpful. In view of their relevance to the presentations given
in subsequent chapters, we will here discuss some results for
the adsorption of CO on transition metals.

When CO chemisorbs on a transition metal, the photoelectron
peak nearest the Fermi edge always decreases in intensity more
strongly than other peaks, as for example demonstrated in the
photoelectron spectrum of CO adsorbed on Pt(111) in figure 9 [5].
This demonstrates that the states nearest the Fermi edge donate
electrons to the CO molecule on chemisorption. Calculations of
the band structure show that the peak nearest the edge arises
mostly from t_{2g} orbitals and the dramatic decrease in intensity
thus indicates the involvement of surface t_{2g} orbitals in the
metal-CO chemisorption bond.

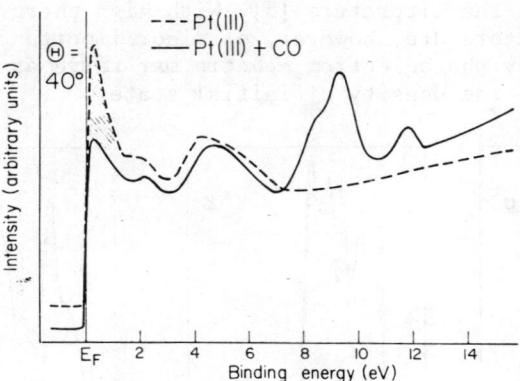

Fig. 9 Photoelectron spectra of Pt(111) and Pt(111) + 4L CO at an emission angle of 40° from the sample normal and at $h\nu$ = 40.8 eV.

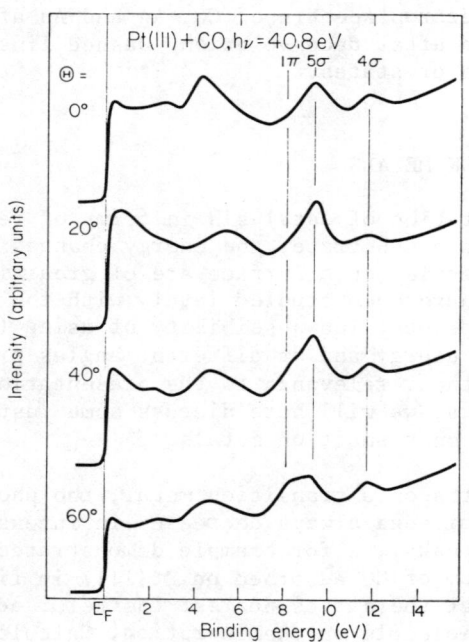

Fig. 10 Photoelectron spectra of Pt(111) + 4L CO at $h\nu$ = 40.8 eV at different emission angles from the sample normal.

Chemical bonding

Variation of the angle between the direction of the normal to the metal surface and the direction of photoemission shows that besides a peak around 11.7 eV binding energy, there are two other peaks at 8.1 and 9.3 eV, cf. figure 10. Since calculations of the intensity for a CO molecule perpendicular to the metal surface (see below) indicate that the probability for photoemission from the 1π orbital increases at high θ values, the peak at 8.1 eV is assigned to the 1π molecular orbital [5]. This means that on bonding to the metal surface the ordering of the 1π and 5σ orbitals is reversed from that of a gaseous CO molecule (cf. fig. 6). This reversal has also been observed for CO on Ni [6] and Pd [7] and was predicted in theoretical calculations by Bagus et al. [8]. In essence it arises from the fact that there is almost no interaction between the 1π molecular orbital* of CO and surface-metal t_{2g} orbitals, while there is a rather strong interaction between the CO 5σ orbital* and metal e_g orbitals, as demonstrated in a one-electron level scheme in figure 11.

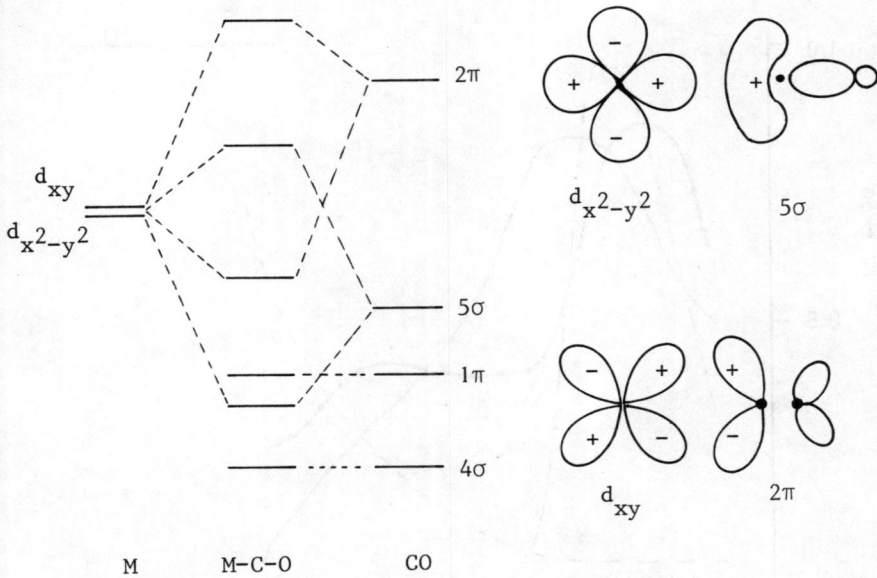

Fig. 11 Molecular orbital diagram of the interaction between CO and a transition-metal atom.

---

*For details on the composition and shape of the molecular orbitals of CO we refer to reference [9].

The so-called σ bonding interaction between M e_g and CO 5σ orbitals leads to a ligand-to-metal electron transfer. This interaction is further strengthened by a backflow of electrons via an interaction of M t_{2g} and CO 2π orbitals. It is this π backbonding which is responsible for the decrease in intensity of the photoelectron peak near the Fermi edge on adsorption of CO and also for the observed decrease in frequency of the infrared CO stretch band upon adsorption. This is because the electron flow away from the bonding 5σ CO orbital and towards the 2π antibonding orbital shifts electrons from a bonding to an antibonding orbital and thus weakens the CO bond. At the same time it strengthens the M-CO bond, however.

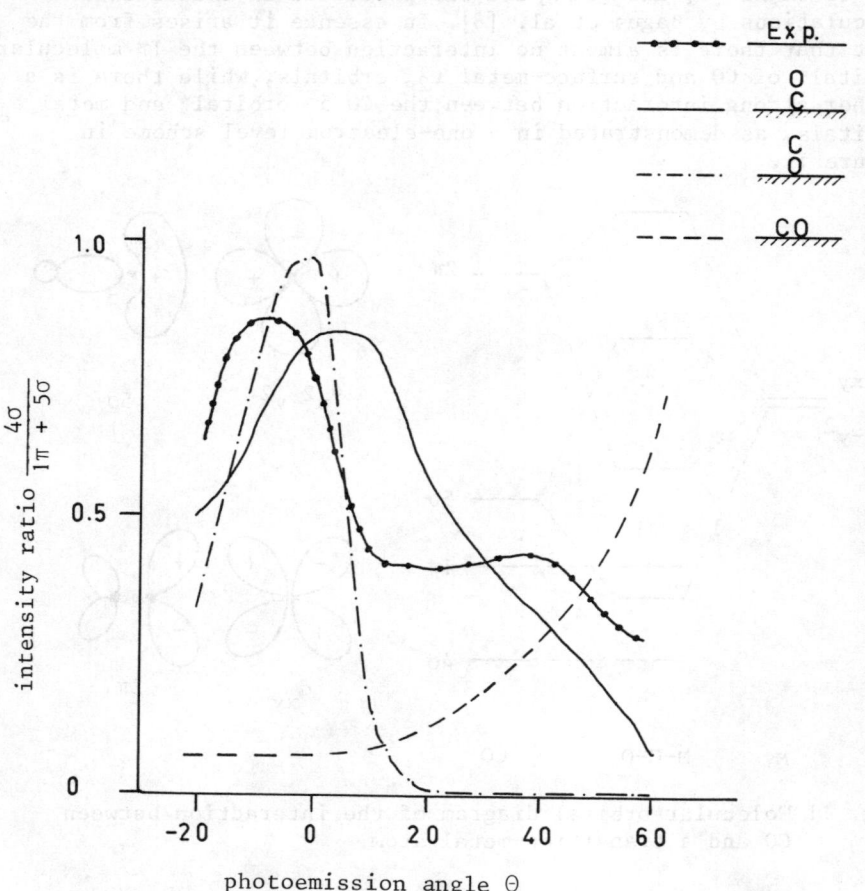

Fig. 12 Comparison of the experimental intensity ratio $4\sigma/(1\pi + 5\sigma)$ of photoelectron peaks for CO chemisorbed on Pt with theoretical ratios for three orientations of the CO molecule with respect to the metal surface.

Angular resolved spectra of CO on Pt(111) and Ni(111) not only display several characteristics of the electronic structure of CO on the metal, but also of the geometric structure. Figure 12 shows that a comparison of the experimental ratio of the intensity of the 4σ photoelectron peak and the sum of the 1π and 5σ peaks with theoretical intensity ratios for a CO molecule perpendicular to the surface, with either the C atom or the O atom bonded to the metal, and a CO molecule flat on the surface, strongly favors the configuration where CO stands up with the C atom bonded to the metal [5].

REFERENCES

1. Full details on chemical bonding theories can be found, among others, in the following textbooks:
   C.A. Coulson, 'Valence', Oxford Univ. Press, 2nd ed., London 1963
   W. Kauzmann, 'Quantum Chemistry', Acad. Press Inc., New York 1957
   H. Eyring, J. Walter and G.E. Kimball, 'Quantum Chemistry', Wiley, New York.

2. K. Siegbahn et al., 'ESCA Applied to Free Molecules', North-Holland Publ. Co., Amsterdam (1969), p. 76.

3. O. Henri-Rousseau and B. Boulil, J. Chem. Ed. 55, 571 (1978)
   K.F. Purcell and J.C. Kotz, 'Inorganic Chemistry', Saunders Comp., Philadelphia 1977, p. 141.

4. G.K. Wertheim, D.N.E. Buchanan, N.V. Smith, and M.M. Traum, Phys. Lett. 49A, 191 (1974).

5. D.A. Shirley, J. Stöhr, P.S. Wehner, R.S. Williams, and G. Apai, Phys. Scripta 16, 398 (1977).

6. P.M. Williams, P. Butcher, J. Wood, and K. Jacobi, Phys. Rev. B14, 3215 (1976).

7. D.R. Lloyd, C.M. Quin, and N.V. Richardson, Solid State Commun. 20, 409 (1976).

8. I.P. Batra and P.S. Bagus, Solid State Commun. 16, 1097 (1975);
   P.S. Bagus and K. Hermann, Solid State Commun. 20, 5 (1976).

9. R.S. Mulliken and W.C. Ermler, 'Diatomic Molecules, Results of ab-initio calculations', Acad. Press, New York, 1977.

BONDING IN AND ON METALS

V. Ponec

Gorlaeus Laboratoria, Rijksuniversiteit Leiden,
P.O. Box 9502, 2300 RA Leiden, The Netherlands

INTRODUCTION. It has always been considered as selfevident that an explanation of the catalytic behaviour of metals should finally be found in terms of the quantum theory of solids. We believe this is true, indeed, but because of the limitations which both the theory and experiments have at present, we are not able to do more than to rationalize some aspects of catalysis by metals and hope that in future these patches of knowledge will ultimately result in a full picture of catalytic and chemisorption phenomena on metals. Some of these patches will be mentioned in this lecture. In the year we commemorate the 200th anniversary of Berzelius who introduced the word catalysis into science, we can state there was some progress in catalysis since then, although some of us might find it too slow anyway. When speaking of catalysis we do not recall anymore terms like "vis vitalis" or "vis obscura" which make catalysed reactions possible, but we relate catalysis to chemisorption saying that the catalytic activity of solids (e.g. metals) is determined by properties of the complexes formed on the surface by chemisorption. Two properties of chemisorption complexes are particularly important in this respect:
1) the structure, determined a.o. by the number of solid state atoms involved and their geometrical arrangement when binding the complex;
2) the dissociation energy of the bonds in the complex as well as of the bonds with the metal. The dissociation energy is closely related to the reactivity of the chemical bonds and experience has taught us that in particular the metal-complex bond is decisive for the reactivity of the chemisorption complexes.
The following illustrates the last statement. Tanaka and Tamaru [1] observed that the adsorption heat of various gases is linearly correlated with $-\Delta H_f^o$, the formation heat of the highest stable oxide/

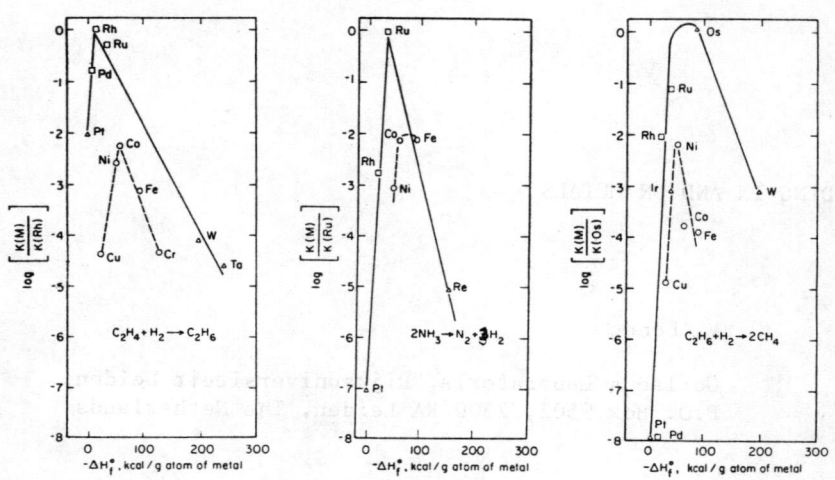

Fig. 1. The Tanaka-Tamaru correlation of catalytic activities of metals for the indicated reactions. The reaction rates (constants) for the most active metal are set equal to 1, as standards [2].

mole metal atom. When the activity in various <u>simple</u> reactions is plotted as a function of $-\Delta H_f^o$, a volcano-shaped correlation results (see Fig. 1 in Ref. 2). This is rationalized by the following assumption [3-6].

When the adsorption heat (metal-complex bond strength) is too low, the surface is scarcely covered by reactive species and these species are only marginally perturbed, i.e. activated. Both these factors cause a low rate of reaction. When on the other hand the heat is too high, the surface becomes covered by unreactive species, unwanted destructive side-reactions can take place and the reaction rate is again too low. Thus, there is always an optimum in the binding strength, characteristic for a given reaction (Sabatier [3] principle). Various authors [3-6] showed that this type of correlation is a very useful guiding principle for summarizing the data on metals and, sometimes, in predicting the behaviour of new materials. For the discussion in this lecture the most important message is that there is indeed an intimate <u>relation between the chemisorption bond strength</u> and <u>the reactivity</u> of chemisorption complexes, and through that, the metal activity in catalytic reactions. Let us, therefore, analyse the chemisorption bond strength a little more in details.

Chemical intuition supported by the ideas of quantum chemistry (think of the LCAO-MO approach, as mentioned in the lecture by Prins) tells us that there should be some relation between the bonds "in the metals" and "on the metals", between the A-M and M-M bonds in

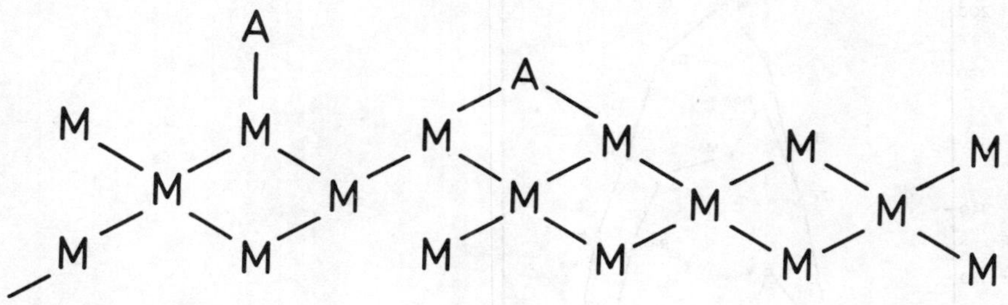

Fig. 2. Adsorption of atoms or molecular fragments on the metal M.

Fig. 2. Therefore, we turn our attention first to the questions of cohesion in metals.

Metal-metal bonds

The lecture by Prins shows that the metallic bond can be treated as a special case of chemical bonds. In an ideal metal with an ideal periodic potential, the valence electrons are completely delocalized (i.e. the probability to find them in all equivalent points of the lattice is the same) and their energy levels form a band, a system with a quasi-continuous distribution of energy levels. Instead of degeneracy of certain energy levels (as with hydrogen-like atoms) we talk about density of states, $N(E)$. This is the number of states belonging to the energy interval between E and $E + dE$. To characterize the binding strength in the metal we calculate the difference between the energy of all electrons participating in bonds in the state of free atoms (E_{at}) and in the metallic state; all related to one metal atom:

$$\Delta E_{cohesion} = Z \cdot E_{at} - \int_{-\infty}^{E_F} dE \cdot N(E) \cdot E \qquad (1)$$

Z stands for the number of valence electrons participating in the metallic bond and in both terms only occupied levels are considered.

We have seen with biatomic molecules that if <u>two</u> atomic orbitals combine in a molecular orbital, <u>two</u> atomic energy levels are replaced by <u>two</u> M.O. energy levels, one (bonding) under the A.O. levels, one (antibonding) above the A.O. levels. In the simplest LCAO-MO scheme placing two electrons ($Z = 1$) on the bonding levels leads to a stronger bond than only one electron and, on the other hand, the presence of one (or more-of two) antibonding electron destabilizes the chemical bond between atoms when compared with the optimal bond-

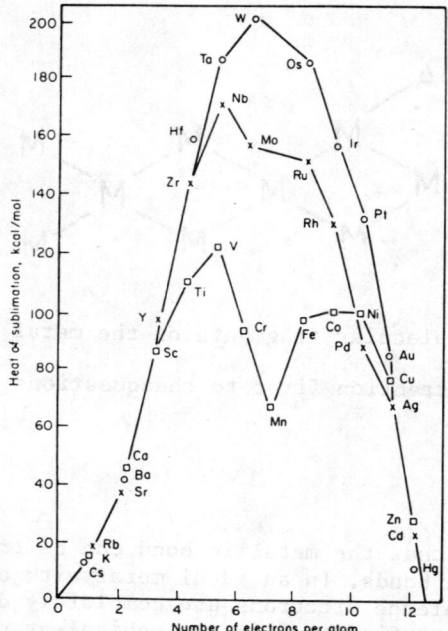

Fig. 3. Heat of sublimation of metals as a function of the number of [ns + (n-1)d] electrons [2].

ding. Essentially the same principles govern the strength of the metallic bond, the cohesion of metals. If "g" atomic orbitals are available per atom for the metallic bond and Z is the number of valence electrons available for the metallic bond, then it can be expected that the metallic bond strength tends to increase with Z for Z between 1 and g. When Z > g, the energy gain of the process "atomic electrons → metallic electrons" is always less than the maximum possible gain and the dependence of $E_{cohesion}$ on Z shows a maximum. This is indeed found experimentally, the melting points of metals or the sublimation (atomization) energies show a maximum as a function of Z, as in Fig. 3.

A word of caution is necessary. All this holds exactly only for a hypothetical model following the simple LCAO-MO rules. However, in reality, there are some complications. The position of atomic (in particular those of the d-orbitals) energy levels is also a function of Z as well as the extent of electron-electron repulsion. Going from left to right in the periods of the periodic system, we observe that, for example, the (n-1)d orbitals change their position in energy with respect to the ns orbitals. On the very left, the (n-1)d orbitals are above the ns level, but due to the bigger attraction of the d-electrons to the nucleus with higher positive change, the energy of the d-orbital undergoes a transition from above to under the corresponding ns levels when passing the period from the left to the right. At the same time, the d-orbitals become more contracted and offer less overlap (less binding strength) to their neighbours when the atom is placed in a metallic lattice. The energy of (n-1)d orbital electrons is also stronger than that of the ns orbital depending on the mutual repulsion of d-electrons. Fortunately, for our simplified picture, the three just-mentioned effects only strengthen but do not cancel the indicated correlation between the cohesion energy and the occupancy of MO levels, the latter being expressed by the parameter Z/g. More about these regularities will be found by the interested reader in the original literature[7-10] or in monographics [11-13].

Metal-adsorbate bonds

When an atom or a molecular fragment A is adsorbed on the surface of a metal, like in Fig. 2, the ideal periodicity of the potential is lost and as a consequence some electron(s) is (are) strongly localized in the immediate vicinity of the A-M bonds. Simultaneously, some of the M-M bonds in the neighbourhood can be weakened and the system reminds of a molecule A-M or AM_2 weakly bound to the rest of the metal M [15-16]. This is sometimes visualized as formation of "pseudomolecules" AM, AM_2 and the calculations performed on such hypothetical pseudomolecules show that many features of the whole system are already reproduced by these simplified models (adsorption heats, UPS/XPS spectra, etc.) [17-18]. Calculation of the chemisorption bond strength ΔE_{ads} comprises two terms:

$$\Delta E_{ads} = E_{cohesion} \text{ (without A)} - E_{cohesion} \text{ (with A)} \tag{2}$$

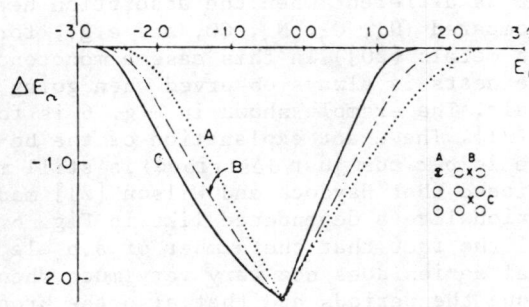

Fig. 4. Chemisorption binding energy plotted as a function of band filling for an unperturbed adsorbate level E_a near the band centre and perturbation energy V one quarter of the band width [16].

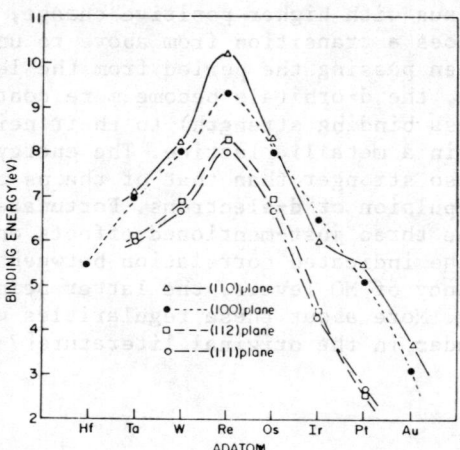

Fig. 5. Binding energy of the Period 6 metal atoms on the four low index planes of W.

However, a method of the so-called Green functions allows us to calculate ΔE_{ads} immediately without calculating explicitly [14-16] the two big terms which have to be abstracted from each other as in (2). Figure 4 represents the results of such model calculations for an atom with atomic energy level E_a [16]. For the purpose of these calculations the E_a level is placed near the band centre, i.e. near the energy of the atomic energy levels of those free atoms which constitute the metal. The perturbation energy V (due to A on M) is taken equal to one quarter of the band width. This result is to be compared with the measurements by Plummer and Rhodin [19] on the adsorption of various metals on the W tip. These results are in Fig. 5. However, the picture is different when the adsorption heats of simple gases are being compared (H_2, O_2, N_2, CO, see e.g., for the adsorption on transition metals [20]). In this case a monotonous decrease (no maximum) of the heats is always observed when going fr group IV to group VIIIC metals. The example shown in Fig. 6 is for oxygen on metal adsorption [21]. The exact explanation of the behaviour (monotonous decrease in periods just described) is still missing, but it should be mentioned that Haydock and Wilson [21] made the following attempt to rationalize a dependence like in Fig. 6.

The authors [21] recall the fact that the number of s,p electrons in the transition metal series does not vary very much when going from left to right along the periods and that also the broad s,p bands are similar for all the metals in question. Therefore, they assume that the interaction energy can be split into two terms

$$\Delta E_{ads} = \Delta E_{ads}^{S} \text{ (due to s,p orbitals)} + \Delta E_{ads}^{d} \text{ (d-orbitals)} \quad (3)$$

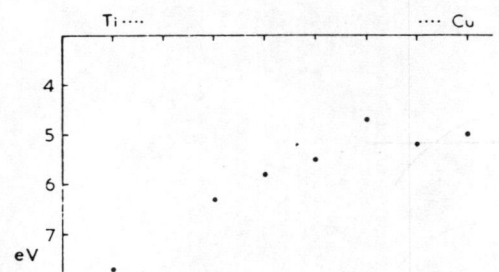

Fig. 6. Adsorption heat of O-atoms on the 3d transition metals (polycrystalline) [21].

from which the first one – according to them – is more or less constant for all the metals in the same period. The contribution of d-orbitals to the binding, the ΔE^d_{ads} term is then dependent on the position of ε with respect to the E_F (Fermi level) and the centre of the d-band; ε being the energy level formed after the interaction of the atomic orbital of the adsorbed atom and the s,p orbitals of the metal. When the ε level is near to E_F or slightly above it, $|\Delta E^d_{ads}|$ shows a low maximum for metals in the middle of the periods; when ε is considerably lower, e.g. under the d-band or, at least, under the centre of the d-band, the contribution $|\Delta E^d_{ads}|$ shows a monotonous decrease in periods when going from left to the right.

The latter behaviour is a consequence of a competition of two effects: the covalent interaction increases as the band shifts down in energy and ε gets by that nearer to the band centre (compare: interaction in an A-A biatomic molecule where the energy levels are equally high, is stronger than in an A-B molecule), but at the same time when the band shifts down with respect to E_F, the adsorbed atom-atomic orbital can interact strongly only with the unoccupied upper part of the band, i.e. with the antibonding states and successively with less and less of them. When going along the periods, e.g. from Ti to Cu, the d-band position and the occupancy (the part under the E_F level) change as shown in Fig. 7, reproduced from the paper by Haydock and Wilson [21]. All this seems to be a realistic rationalization of the data in Fig. 5, but we meet some difficulties with alloys. The initial adsorption heat of CO on Pd-Ag alloys is known [22] and it is independent of the band width and occupancy of the Pd d-band. A possible explanation here could be that among the group VIII metals the adsorption heats of CO do not vary strongly [20] anyway, so that the changes when going from Pd → Pd-Ag alloy might be within the experimental errors. However, more theoretical work is necessary in any case.

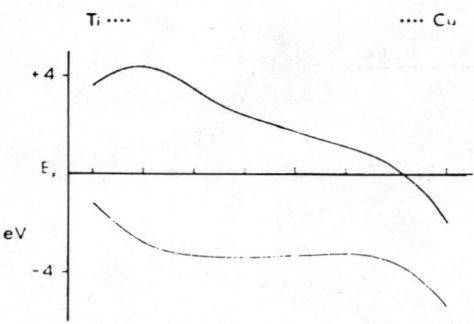

Fig. 7. Transition metal d-band characteristics [21]. The band width and its relative position with regard to the Fermi level E_F are shown. The trend is similar in other periods [21].

ALLOYS
Changes in the electronic structure of metals by alloying

At present a whole spectrum of various methods is available to determine the electronic structure of metals and to characterize it by appropriate parameters. The most valuable information was obtained by the ultraviolet and X-ray photoemission (UPS, XPS) spectroscopy. By measuring the distribution according to the energy of photoemitted electrons, essential features of the band structure are obtained; the physics of the process is such that the intensity of the photoemitted electrons I(E) is mainly determined by the density of states N(E) of the metal investigated. By this method it is mainly the structure of the $N(E)_d$ function of the d-electrons which is observed. There is already an extended literature information available (see e.g. ref. 23, 24) on this subject which can be summarized as follows, by means of fig. 8. All the metals of the transition metal series have a d-band close to the Fermi energy E_F; the IB and other non-transition metals have the d-band - if they have a valence d-band at all - several eV under the E_F level. When the group VIII metals are alloyed with Ib metals, both d-bands usually stay in their original position and only the size and shape of the bands change. In some cases (Ni-Cu) [23] the band width δ changes less than in others (Pd-Ag) [23]. When due to alloying the narrowed d-band completely falls under the E_F level, the alloy becomes diamagnetic (Pd-Ag). Only in strongly exothermic alloys, when intermetallic compounds are formed, d-bands seen by electron or X-ray spectroscopies shift to a lower energy; in many other cases of alloys the position of the d-band does not vary with alloy composition [25

Other methods bring information on the number of electrons in the orbitals of d- or s,p-character, respectively, [26] and this

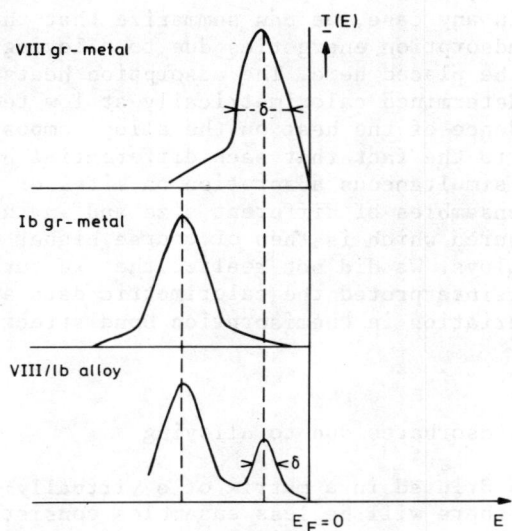

Fig. 8. Distribution I(E) of photoemitted electrons, schematically.
δ - band width; E_F - Fermi energy, an arbitrary zero level.

information confirms that in contrast to the assumptions of the old rigid band theory, there is very little transfer of electrons among the alloy components and also in the strongly exothermic alloys we can rather speak of sharing of electrons by components like in covalent bonds than of electron transfer.

Changes in the energetics of chemisorption by alloying

With simple gases (H_2, CO) adsorbed on metals, the thermal programmed desorption profiles usually show several peaks or peak-shoulders which are then ascribed to various chemisorption states. It was a crucial question whether and how much the position of the peaks and shoulders shift, due to alloying. Stephan investigated this problem and with the systems studied (Pt-Au/H_2, CO; Pd-Ag/CO; Ni-Cu/CO), there was no systematic shift of the peak position by alloying [27-29]. Similar results were obtained for Ni-Cu/H_2 by by Orlova et al. [30]. Christman and Ertl [22] found that also the initial isosteric adsorption heat of CO on Pd-Ag did not show any variations with alloying. However, different results were reported by Yu, Ling and Spicer [31], on Ni-Cu/CO systems. These authors found a slight systematic shift due to alloying with Cu to lower adsorption heats. However, it should be mentioned that these authors used alloys which systematically differed not only in their composition but also in their defect structure; the Ni-rich alloys were rougher than alloys rich in Cu, so that a part of the shifts found can be due to this factor. In any case, we can summarize that there are no dramatic changes in adsorption energetics due to alloying.

One more remark has to be placed here. The adsorption heats of H_2 and CO on alloys - when determined calorimetrically at low temperatures - revealed a dependence of the heat on the alloy composition[32-34]. Most likely due to the fact that each differential heat determination represented a simultaneous adsorption on sites of different quality, e.g. on ensembles of different size and in this way an average heat was measured which is then of course higher on group VIII metals than on alloys. We did not realize that in our earlier paper [35] and we misinterpreted the calorimetric data as bringing evidence for the variation in chemisorption bond strength by alloying.

Changes in the structure of adsorbates due to alloying

When an active metal is diluted in a matrix of a virtually inactive not-adsorbing metal, there will be less ensembles consisting of several active atoms and the fraction of isolated active atoms will increase. It is expected that this must suppress formation of those chemisorption complexes which require several active metal atoms to be formed and it will enhance the chances to form one-site complexes. There is evidence that this will happen to hydrocarbons (see lecture "Catalysis by Alloys"), but in this case the evidence is indirect, deduced from the catalytic data. However, there are also some chemisorption data which allow a straightforward explanation. When studying CO adsorption on Pd-Ag alloys by IR, Soma-Noto and Sachtler [36,37] discovered that alloying strongly suppresses the occurrence of the multicoordinated CO and favours the "on-top" CO (linear) adsorption. The suppression was so strong that there were doubts whether this could be explained at all by only dilution of Pd in Ag [38], because the literature available at that time did not confirm any strong segregation of Ag to the surface. However, when the Auger spectrometric analysis and data evaluation are properly performed as outlined briefly in this course in the lecture by Sachtler (see also the original literature, refs. 39-41), it appear that the IR data on CO adsorption are explained in a satisfying way by that extent of Ag segregation (that is, Pd dilution) as found by Auger spectrometry [42]. This is illustrated in Fig. 9, where the comparison of data is performed in the following way. Auger data are shown for vacuum and CO-treated surfaces (lowest and upmost curves) and the IR data are converted into the surface-bulk compositic curve by using statistical calculations derived by Sachtler [38]. Parameter λ is needed for these calculations, λ being a ratio of coverages and IR absorption coefficients of the two CO adsorbed complexes. Parameter λ is not known and, therefore, calculations were

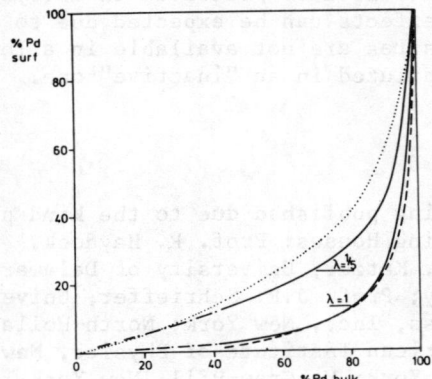

Fig. 9. Comparison of the surface composition (determined by Auger spectrometry) of Pd-Ag alloys sintered in vacuo (---) and after interaction with CO (...), with the calculated surface composition according to IR/CO adsorption data (—) and the theoretical model by Sachtler (see text for details).

performed for two extreme but reasonable estimates. We observe that the IR data can, indeed, be explained in terms of dilution, ensemble size, etc., without recalling some specific electronic structure effects.

The stretching frequency ν of the CO/IR band as well as $\Delta\Phi$, work function changes caused by adsorption are a measure of a.o. the backdonation of metal d-electrons into the π^*-CO orbital (see the lecture by Prins). It appears that ν is influenced by alloying and it is a continuous function of alloy composition. The changes in ν with alloying are always of the same character [43], irrespective whether the alloy is formed endo- or exothermically (one would expect different changes when speculating on the mutual ligand effects of alloy components), or whether there is a change in the number of group VIII metal d-electrons (Pd-Ag) or not (Ni-Cu), or whether the surface density of CO adsorbed molecules varies with alloying (group VIII/Ib metal alloys) or not (Ni-Co, Ni-Fe, etc.). We can thus exclude these factors from consideration and see that there is not very much left that can be suggested as an explanation. The only feature common to all alloys studied up to now by CO/IR is that always the band narrowing occurred for both alloy components. Band narrowing leading to an increase in the local density of states favours the electron transfer (direct and backdonation) and it is tempting to relate the changes in ν by alloying to this "collective" metal feature. However, it is not very likely that these small changes in CO behaviour would be detected by a catalytic reaction.

Summarizing we can state the following. The physical (UPS, XPS) as well as the chemisorption measurements show that very much of the individuality of the alloy components is also preserved in alloys; for catalysis the most pronounced effects can be expected due to the fact that big ensembles of active sites are not available in alloys where an active metal is strongly diluted in an "inactive" one.

ACKNOWLEDGEMENTS

The material reproduced is being published due to the kind permission of the authors and Publishing Houses: Prof. R. Haydock, University of Cambridge; Prof. J.R. Katzer, University of Delaware; Prof. T. Rhodin, Cornell University; Prof. J.R. Schrieffer, University of Pennsylvania; Academic Press, Inc., New York; North-Holland Publishing Company, Amsterdam; American Institute of Physics, New York; John Wiley & Sons, Inc., New York; Mc-Graw-Hill, New York.

REFERENCES

1. K. Tanaka and K. Tamaru, J. Catal. 2 (1963) 306.
2. B.C. Gates, J.R. Katzer and G.C.A. Schuit, Chemistry of Catalytic Processes, Mc-Graw-Hill, New York, 1979, chapter III.
3. P. Sabatier, Ber. Deutsche Chem. Ges. 44 (1911) 2001.
4. A.A. Balandin, Adv. Catal. 19 (1969) 1.
5. J. Fahrenfort, L.L. van Reijen and W.M.H. Sachtler, Z. Elektrochem. 64 (1960) 216.
6. V. Ponec, Z. Knor and S. Cerny, Disc. Faraday Soc. 41 (1966) 14
7. J. Friedel, J. Phys. 38 (1977) 697; 39 (1978) 651.
8. J. Friedel, Ann. Phys. 1 (1976) 257.
9. N.F. Mott, Report Progr. Phys. 25 (1962) 218.
10. W. Hume-Rothery and B.R. Coles, Adv. Phys. 3 (1954) 149.
11. N.F. Mott and H. Jones, Theory of Properties of Metals and Alloys, Oxford Univ. Press, 1936.
12. C. Kittel, Introduction to Solid State Physics, John Wiley and Sons, New York, 1976.
13. J.M. Zimmann, Principles of the Theory of Solids, Cambridge Univ. Press, 1972.
14. T.B. Grimley, Dynamic Aspects of Surface Physics, Proc. Int. School of Physics E. Fermi, 1974; Editrice Compositori Bologna Italy, p. 298 (and references therein).
15. J.R. Schrieffer, Dynamic Aspects of Surface Physics, Proc. Int School of Physics E. Fermi, 1974; Editrice Compositori Bologna Italy, p. 250.
16. J.B. Danese and J.R. Schrieffer, Int. J. Quant. Chem. 10 (1976) 289.
17. J.C. Slater and K.H. Johnson, Phys. Rev. B5 (1972) 844.
    K.H. Johnson and F.C. Smith Jr, Phys. Rev. B5 (1972) 831.
18. K.H. Johnson and R.P. Mesmer, J. Vac. Sci. Technol. 11 (1974) 236.

19. E.W. Plummer and T.N. Rhodin, J. Chem. Phys. 49 (1968) 3479.
20. L. Toyoshima and G.A. Somorjai, Catal. Rev. Sci. Eng. 19 (1) (1979) 105 (a complete, extended review on the subject).
21. R. Haydock and A.J. Wilson, Surface Sci. 82 (1979) 425.
22. K. Christman and G. Ertl, Surface Sci. 33 (1972) 254.
23. S. Hüfner, G.K. Wertheim and J.H. Wernick, Phys. Rev. B8 (1973) 4511.
24. V. Ponec, Electronic Structure and Reactivity of Metal Surfaces, eds. E.G. Derouane and A.A. Lucas (Proc. NATO-A.S.I., Namur, 1975), Plenum Press, 1976, p.537 (a review of results on catalytically interesting alloys).
25. K. Ichikawa, J. Phys. Soc. Japan 37 (1974) 377.
26. A. Wenger and S. Steinemann, Helv. Phys. Acta 47 (1974) 321.
27. J.J. Stephan, Thesis, Leiden, 1975.
28. J.J. Stephan, V. Ponec and W.M.H. Sachtler, Surface Sci. 47 (1975) 403.
29. J.C.M. Harberts, A.F. Bourgonje, J.J. Stephan and V. Ponec, J. Catal. 47 (1977) 92.
30. G.N. Orlova, I.T. Frolkina, V.M. Lebedev, Yu.A. Mischenko and A.I. Gel'bstein, Kin. Katal. 18 (1977) 980.
31. K.Y. Yu, D.T. Ling and W.E. Spicer, J. Catal. 44 (1976) 378.
32. L.S. Shield and W.W. Russell, J. Am. Chem. Soc. 64 (1960) 1592. (H_2).
33. T. Takeuchi, M. Sakaguchi, I. Miyoshi and T. Takabatake, Bull. Soc. Chem. Japan 35 (1962) 1320. (H_2).
34. M. Ralek, private communications, Techn. Univ. Berlin, 1976. (CO).
35. V. Ponec and W.M.H. Sachtler, Catalysis, Proc. Vth Int. Congr. Catal. Miami Beach, 1972; ed. J. Hightower, Elsevier Amsterdam, 1972, Vol. 1, p. 645.
36. Y. Soma-Noto and W.M.H. Sachtler, J. Catal. 32 (1974) 316.
37. M. Primet, M.V. Mathieu and W.M.H. Sachtler, J. Catal. 44 (1976) 324.
38. W.M.H. Sachtler, Catal. Rev. Sci. Eng. 14 (1976) 193.
39. F.J. Kuijers, Thesis, Leiden, 1978.
40. F.J. Kuijers, B.M. Tieman and V. Ponec, Surface Sci. 75 (1978) 657.
41. F.J. Kuijers and V. Ponec, Surface Sci. 68 (1977) 296.
42. F.J. Kuijers and V. Ponec, J. Catal., in print.
43. F.J.C.M. Toolenaar, D. Reinalda and V. Ponec, submitted to J. Catal..

SURFACE SCIENCE AND CATALYSIS ON METALS

G. Ertl

Institut für Physikalische Chemie,
Universität München, F.R. Germany

ABSTRACT. The development of various surface spectroscopic techniques allows a rather detailed characterization of elementary processes occurring at solid surfaces. After a short summary of the type of information necessary for the characterization of chemisorption on metals the application of these principles to two specific examples, viz. the oxidation of carbon monoxide at palladium and the ammonia synthesis at iron surfaces, is outlined.

1. INTRODUCTION

Based on the physical laws governing the interaction of electrons, photons, ions and neutral particles with solid matter a large variety of surface sensitive techniques has been developed [1]. These methods offer the possibility to get a more direct insight into the elementary steps involved in heterogeneous catalysis.

Since each method probes different aspects evidently only a combined use can yield a close picture. The applicability of these techniques is, however, frequently restricted to model systems ('surface science') which may be very far from the conditions applied in 'real' catalysis mainly in the following respects:

	'Surface Science'	Real catalysis
Surface composition	well-defined (clean surface)	undefined (promotors etc.)
Surface structure	single crystal	polycrystalline (active sites?)
Pressure	$\leq 10^{-4}$ Torr	$\gtrsim 10^{3}$ Torr

This contribution consists of two parts: At first it will be outlined, what kind of information on the surface chemistry of well-defined metal surfaces can be obtained and secondly, how this information may be transferred to the elucidation of the mechanism of catalytic reactions even under 'real' conditions.

2. CHEMISORPTION ON METALS [2]

A detailed characterization of chemisorbed systems comprises the following information:
- Elemental analysis by Auger electron spectroscopy (AES) or x-ray photoelectron spectroscopy (XPS)
- Identification of the molecular nature of the adsorbate: Ultraviolet photoelectron spectroscopy (UPS), vibrational spectroscopy, secondary ion mass spectroscopy (SIMS)
- Surface concentration (coverage): AES, XPS, thermal desorption spectroscopy (TDS), molecular beam techniques, low energy electron diffraction
- Electronic properties: UPS, XPS, electron loss spectroscopy (ELS), change of the work function ($\Delta\varphi$)
- Structure of the adsorbate layer: Low energy electron diffraction (LEED) with long-range order on single-crystal surfaces, angular resolved UPS, extended x-ray absorption fine structure (EXAFS).
  Types of overlayer structures: a) Lattice gas structures, b) Incoherent structures, c) surface reconstruction
- Adsorption energy: TDS, analysis of adsorption isotherms by means of the Clausius-Clapeyron equation
- Vibrations: Infrared and high resolution electron energy loss spectroscopy (HREELS)

Factors influencing the chemisorption bond:
- The nature of the substrate metal: So far no simple rules.
- The nature of the adsorbate: Strength of the bond
 $N > O > H > CO \gtrsim NO$, dissociative and non-dissociative chemisorption
- The local character of the chemisorption bond: Relations to complex chemistry, chemisorption on alloys ('ensemble' and 'ligand' effect)
- The crystallographic orientation of the surface:
 Bond strength may vary by about 10-20%, however kinetics by up to one order of magnitude ($\rightarrow$ 'structure sensitive' catalytic reactions)
- Monoatomic steps: Model for 'active' sites.
 Effects similar as with different crystal planes.
- Adsorption site: Variation of the bond strength across the unit cell of the substrate lattice (energy profiles), leading to either localized or delocalized adsorption. Consequences for surface diffusion.
- Interactions between adsorbed particles: These may be of direct (orbital overlap or dipole-dipole) or of indirect (via the metal valence electrons) nature and determine the mutual configuration of adsorbed particles, the maximum coverage, and the variation of the adsorption energy with coverage. In the case of two different types of adsorbed particles either the formation of a mixed phase (cooperative adsorption) or of separate islands (competitive adsorption) may be the consequence of the specific interactions.

Kinetics of surface processes:
- Kinetics of adsorption decribed by the sticking coefficient s through $dn_s/dt = s \cdot \dot{n}$, where n_s is the surface concentration of the adsorbate and $\dot{n}$ the rate of impingment from the gas phase. s depends usually in a complex manner on the coverage.
- Kinetics of desorption: $-dn_s/dt = k_d \cdot n_s^x$, where x is the reaction order for desorption, $k_d = \nu_d(n_s) \cdot \exp(-E_d^*(n_s)/RT)$ with ν_d preexponential and E_d^* activation energy for desorption
- Surface reactions: Usually no simple rate laws connecting the partial pressures of gaseous reactants with the rate of product formation, even if the mechanism of the surface reaction is simple. Therefore great care has to be taken with all attempts to conclude on the reaction mechanism

from the mere analysis of kinetic data.

3. THE OXIDATION OF CARBON MONOXIDE [3]

The reaction $CO + 1/2\ O_2 \rightarrow CO_2$ is easily catalyzed by the platinum group metals, the following dicussion will be restricted to palladium. The gap between surface science and real conditions as outlined in the introduction can be bridged according to the following observations:

a) Surface contaminants like S or C are cleaned-off by the reactants so that also a practical Pd catalyst exhibits an essentially clean surface. b) The catalytic activity is independent of the crystallographic structure of the surface, i.e. the reaction is 'structure-insensitive'. c) The stationary rate of CO_2 production is (under practically interesting conditions) given by $r = k p_{O_2}/p_{CO}$. That means that not the absolute pressure, but only the ratio of the partial pressures is decisive. As a consequence the conclusions reached for the microscopic processes occuring at a clean single-crystal surface at low pressures can directly be transferred to the conditions of real catalysis.

The adsorption of oxygen occurs dissociatively without any noticeable activation energy. The strength of the M-O bond is about 90 kcal/mole (i.e. about 30 kcal/mole smaller than the dissociation energy of O_2). With Pd(111) the initial sticking coefficient is $s_o = 0.4$ and the adsorbed oxygen atoms form an ordered 2x2 overlayer structure with a maximum coverage $\theta = 0.25$. This correspon to a minimum 0-0 separation of 5.5 Å which is rather large and accounts for the fact that additional uptake of CO may take place. CO is molecularly adsorbed with a sticking coefficient near unity and an adsorption energy of about 35 kcal/mole. Saturation of the adsorbed layers is characterized by the tendency of the CO molecules to form close-packed layers exhibiting an effective diameter of about 3 Å and leading to a saturation density of 1×10^{15} molecules/cm^2.

For the catalytic formation of CO_2 the following possible reaction steps may be formulated:

$$O_2 + * \xrightarrow{k_1} 2\,O_{ad} \tag{1}$$

$$CO + * \underset{}{\overset{k_2}{\rightleftharpoons}} CO_{ad} \tag{2}$$

$$CO_{ad} + O_{ad} \xrightarrow{k_3} CO_2 \tag{3}$$

$$CO + O_{ad} \xrightarrow{k_4} CO_2 \tag{4}$$

$$CO_{ad} + \tfrac{1}{2} O_2 \xrightarrow{k_5} CO_2 \tag{5}$$

* represents schematically an empty adsorption site, which however is not equal to identical groups of surface atoms for both adsorbates. The backward reaction of (1) can be neglected since desorption of oxygen takes place only above 800 K. CO_2 is so weakly adsorbed that it immediately leaves the surface above room temperature once it is formed, therefore no distinction is made between $CO_{2,ad}$ and $CO_{2,gas}$.

According to this sequence CO_2 formation might either take place through interaction between both reactants in their chemisorbed state (Langmuir-Hinshelwood reaction (3)), or by a process where only one reactant is chemisorbed with which the second interacts either by direct impact from the gas phase or from a weakly held physisorbed state (Eley-Rideal mechanism (4) or (5)).

That reaction (5) has to be ruled out may be demonstrated by a rather simple experiment: If a Pd surface is first saturated with CO and then exposed to gaseous O_2 no CO_2 formation at all is observed. This result appears to be rather plausible, since an essential role of the catalyst consists obviously in breaking the O-O bond by simultaneously forming two weaker M-O bonds. Another conclusion from such measurements is that preadsorbed CO inhibits the chemisorption of oxygen (although this would be energetically more favourable). A densely packed CO layer acts obviously like a carpet on the metal surface which prevents the intimate approach of the O_2 molecules.

Indeed the CO coverage has to be smaller than about 60% of its maximum value in order to enable dissociative oxygen adsorption as a prerequisite for CO_2 formation.

If on the other hand a surface saturated by O_{ad} is exposed to CO this molecule may be well adsorbed additionally. Depending on the surface concentrations of O_{ad} and CO_{ad} different types of

mutual configurations may occur which illustrate the complex nature of the LH-reaction (3). Therefore also no generally valid rate law $r = f(T, p_{CO}, p_{O_2})$ can be established.
A clear distinction between the LH-reaction (3) and the ER-reaction (4) is only possible on the basis of instationary measurements. Experiments with the molecular beam technique demonstrated that in fact the reaction proceeds exclusively via the LH-mechanism.

The approximate rate law $r = p_{O_2}/p_{CO}$, as established experimentally, may be verified from the reaction mechanism (1) to (3) as follows:

According to reaction (3)

$$r = \frac{d[CO_2]}{dt} \approx k_3 [O_{ad}] \cdot [CO_{ad}]$$

Under stationary conditions

$$\frac{d[O_{ad}]}{dt} \approx k_1 p_{O_2} \cdot [*] - k_3 [O_{ad}] \cdot [CO_{ad}] = 0$$

The CO adsorption-desorption equilibrium means that

$$k_2 \cdot p_{CO} [*] = k_{-2} \cdot [CO_{ad}],$$

leading to

$$[*] \approx \frac{k_{-2} [CO_{ad}]}{k_2 p_{CO}}$$

and

$$r \approx k_1 p_{O_2} \cdot [*] \approx k_1 \cdot \frac{k_{-2}}{k_2} \cdot [CO_{ad}] \frac{p_{O_2}}{p_{CO}}$$

At higher pressures $[CO_{ad}]$ will be large and approximately constant so that $r \approx k \cdot p_{O_2}/p_{CO}$.

4. THE SYNTHESIS OF AMMONIA [4]

Industrial production of NH_3 from the elements is performed with promoted iron catalysts at pressures above 100 atm and at temperatures around 400°C. The high pressures are needed for thermodynamic reasons and it is generally accepted that the reaction proceeds at a total pressure of 1 atm along the same mechanism. Even then the 'pressure gap' to surface science is very large. A similar situation is found with respect to the surface composition and structural factors.

a) Surface analysis of commercial catalysts shows strong enrichment of K_2O which acts as an 'electronic' promotor, as well as of other oxides which are 'structural' promotors and prevent the small particles from sintering. Under reaction conditions iron is reduced into the metallic state, whereas the other components remain oxidised. Model studies with clean Fe surfaces are therefore justified, whereby, however, the promotor effect of potassium has to be studied separately.

b) Investigations with different Fe single crystal surfaces revealed marked differences in the activity. That means the reaction is structure-sensitive which had already been concluded previously on the basis of measurements with varying size of small catalyst particles.

c) The most serious problem is offered by the 'pressure gap' which may be overcome along two lines:

i) The decomposition of NH_3 has to proceed through the same mechanism as the synthesis reaction and may be studied conveniently at low pressures.

ii) The synthesis reaction is performed at atmospheric pressure and the surface is analysed after subsequent evacuation.

The latter procedure requires of course first some knowledge on the thermal stability of the surface species in vacuo. It turns out that at 350°C N_{ad} will be stable (desorption of N_2 starts only above 450°C) so that the surface concentration of this species under reaction conditions (1 atm) may be determined after evacuation. Experiments with varying $H_2 : N_2$ partial pressure ratios revealed that with a stoichiometric mixture the stationary surface concentration of N_{ad} is very small. Analysis of these measurements leads to the following conclusions:

α) NH_3 formation proceeds over atomic rather than molecular nitrogen. β) The dissociative chemisorption of nitrogen is rate-limiting. γ) The reaction takes place on an essentially clean Fe surface (not on a nitride surface) even at high pressures. Keeping this in mind the individual reaction steps may now again be studied under ultrahigh vacuum conditions, and the main results are summarized as follows.

Adsorption of hydrogen:

- Dissociative non-activated adsorption (H_{ad})
- Adsorption energy 15-26 kcal/mole (E_{M-H} = 60-65 kcal/mole)
- Rapid adsorption with initial sticking coefficient $s_o \approx 0.2$
- Chemisorption bond essentially covalent (small dipole moment)
- Ordered overlayer structures on Fe(110)
- Very high surface mobility
- No significant differences between various crystal planes
- Reversible adsorption-desorption equilibrium will be established under reaction conditions.

Adsorption of nitrogen

$N_2 \rightleftharpoons N_{2,ad}$
- Rapid and almost reversible
- Adsorption energy < 10 kcal/mole

$N_2 \rightleftharpoons 2 N_{ad}$
(via $N_{2,ad}$)
- Very slow process with small activation energy (0-7 kcal/mole at zero coverage, depending on surface orientation)
- $s_o = 10^{-6}$ to 10^{-7}
- Plane specifity of kinetics
- Adsorption energy 50-58 kcal/mole ($E_{M-N} \approx 140$ kcal/mole)
- Chemisorption bond through N2p orbitals
- Structure: 'Surface nitrides' (similar to Fe_4N). Possible surface reconstruction.
- Bulk solubility and at high p_{NH_3}/p_{H_2} ratios formation of bulk nitrides
- Surface mobility smaller than of H_{ad}

Interaction with ammonia:

- Rapid and non-activated adsorption ($s_o \approx 0.2$)
- $E_{ad} \approx 10\text{-}15$ kcal/mole. Under reaction conditions rapid desorption (or decomposition)
- High dipole moment of the adsorbate complex (2.2 Debye)
- Bond formation through nitrogen lone electron pair. Ordered overlayers.
- Decomposition: $NH_{3,ad} \rightarrow N_{ad} + 3 H_{ad} \rightarrow \frac{1}{2}N_2 + \frac{3}{2}H_2$

via intermediates:
$$NH_{3,ad} \rightleftharpoons NH_{2,ad} + H_{ad}$$
$$NH_{2,ad} \rightleftharpoons NH_{ad} + H_{ad}$$
$$NH_{ad} \rightleftharpoons N_{ad} + H_{ad}$$

It is important to notice that direct spectroscopic evidence for the intermediates exists, which on the other hand are also passed during the synthesis reaction. Under stationary conditions of ammonia decomposition recombination and desorption of nitrogen is rate-limiting. Putting all this information together leads to the formulation of the reaction mechanism on iron surfaces

$$H_2 \rightleftharpoons 2 H_{ad}$$
$$N_2 \rightleftharpoons N_{2,ad} \rightleftharpoons 2 N_{ad}$$
$$N_{ad} + H_{ad} \rightleftharpoons NH_{ad}$$
$$NH_{ad} + H_{ad} \rightleftharpoons NH_{2,ad}$$
$$NH_{2,ad} + H_{ad} \rightleftharpoons NH_{3,ad} \rightleftharpoons NH_3$$

Under normal reaction conditions the step $N_{2,ad} \rightarrow 2 N_{ad}$ (which is structure-sensitive) is rate-limiting. Under conditions far away from the equilibrium (i.e. if the NH_3 partial pressure is negligible) the reaction rate is then simply given by $r = k \cdot p_{N_2}$ which is in agreement with general experience. A potential energy diagram illustrates how the catalysed reaction is favoured energetically over the homogeneous gas phase synthesis.

The last point concerns the role of the 'electronic' promotor: If a clean Fe surface is covered by small amounts of K the heat of adsorption of molecular nitrogen is increased by about 3 kcal/mole. As a consequence the surface concentration of $N_{2,ad}$ in the equilibrium $N_2 \rightleftharpoons N_{2,ad}$ is increased and the activation energy for the process $N_{2,ad} \rightarrow 2 N_{ad}$ is lowered. Both effects account for a net increase of the rate of dissociative nitrogen chemisorption by more than two orders of magnitude. The N atoms if once formed are diffusing away from the K-containing sites onto normal Fe adsorption sites where the further reaction proceeds.

REFERENCES

The following list contains a few general references where further information can be found.

1. a) L. Fiermans, J. Vennik and W. Dekeyser, eds., Electron and Ion Spectroscopy of Solids, Plenum Press, New York - London, 1978.
   b) R. Vanselow, ed., Chemistry and Physics of Solid Surfaces, CRC-Press, Cleveland, Ohio. Vol. I, 1977, Vol. II, 1979, Vol. III, in preparation.
   c) G. Ertl and J. Küppers, Low Energy Electrons and Surface Chemistry, Verlag Chemie, Weinheim, 1974.
   d) R. B. Anderson and P. T. Dawson, eds., Experimental Methods in Catalytic Research, Vol. III, Academic Press, New York, 1976.
2. a) M. W. Roberts and C. S. McKee, Chemistry of the Metal--Gas Interface, Clarendon Press, Oxford, 1978.
   b) G. Wedler, Chemisorption: An Experimental Approach, Butterworths, London, 1976.
   c) F. C. Tompkins, Chemisorption of Gases on Metals, Academic Press, New York, 1978.
   d) T. N. Rhodin and G. Ertl, eds., The Nature of the Surface Chemical Bond, North-Holland, Amsterdam, 1979.
3. T. Engel and G. Ertl, Adv. Catalysis 28, 1979, in press.
4. G. Ertl, Catalysis Reviews, in press.

SURFACE ELECTRON SPECTROSCOPY

R.F. Willis

Astronomy Division, Space Science Department, European Space Agency, ESTEC, 2200AG Noordwijk, The Netherlands*

1. PERSPECTIVE

We begin by asking "Why use electron spectroscopy to study surfaces?"

The reason becomes clear if we observe the mean free path (i.e. mean escape depth or mean penetration depth) of electrons in solids (Fig. 1). It can be seen that for a wide range of materials, the mean free path of electrons with kinetic energies in the range, 10 to 1000 eV, is of the order of 5 to 25 Å. Low energy electrons thus originate from (or sample) only the outermost atomic layers. This being the case, low energy electron spectroscopy is extremely sensitive to conditions right at the surface.

Great progress has been made over the past decade in our understanding of the atomic structure and of the electronic charge distribution at surfaces. A great variety of electron spectroscopic techniques have been developed, which are described in the literature under a bewildering array of acronyms: LEED (Low energy electron diffraction), EELS (electron energy loss spectroscopy), AES (Auger electron spectroscopy), FEED (field emission energy distributions), photoelectron spectroscopy (UPS and XPS covering the ultraviolet and x-ray regimes), INS (ion neutralization spectroscopy) etc. All have found application in the study of processes of importance to heterogeneous catalysis.

In these lectures, we will introduce the principles underlying these techniuqes with reference to only a few selected systems. The approach will be a heuristic one in which we begin by first asking the fundamental question: "What happens to the rotational

---

* Present address: Cavendish Laboratory, University of Cambridge, U.K.

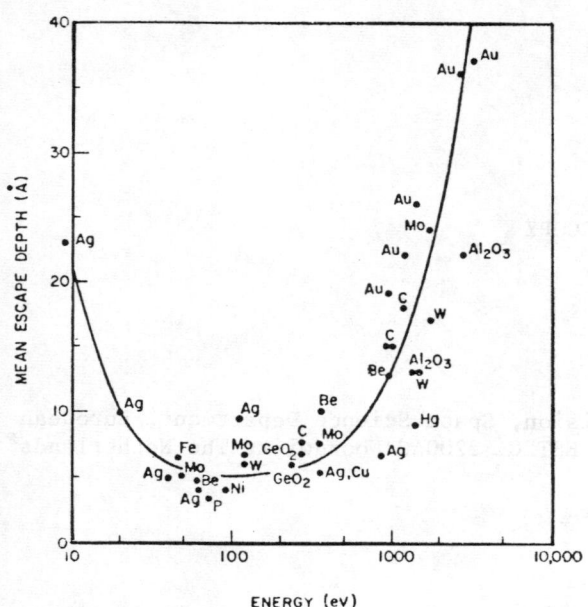

FIG. 1

Mean free path of excited electrons in solids.
From Rivière, Ref. 2.

and translational degrees of freedom of a molecule when it is adsorbed from the gas phase?" The answer is best considered from the viewpoint of adsorbate induced vibrational modes deriving from the now "frustrated" degrees of freedom of the gas phase species. Electron scattering spectroscopic methods for observing these modes viz. EELS, will be described (section 2). It will be shown that the molecular symmetry of the adsorbate together with the atoms in the substrate imposes certain selection rules which not only permit the "fingerprinting" of the adsorbed species but also help to determine its configuration and geometry at the surface.

We then proceed to ask the related question: "How do we observe the associated electronic states and changes in bonding?" Here the various electron emission techniques, such as UPS and INS, provide powerful tools. The problems of referencing adsorbates to their gas phase molecular levels, possible surface rehybridization, and the role of the substrate's electronic states will be addressed (section 3), mainly within the context of recent photoelectron spectroscopy studies.

Up to this point, we will have addressed mainly *inelastic* effects in surface electron spectroscopy, so that finally (section 4) we briefly touch upon those interference effects between the *elastically* scattered electrons, i.e., diffraction, which promises to yield further structural information. Whereas the now well-established tool of LEED requires films possessing long-range order on well-defined single crystal surfaces, which has long been a

Surface electron spectroscopy 283

limitation in relation to practical catalytic systems, recently observed "photoelectron diffraction" effects from adsorbates using synchrotron radiation (equivalent to Surface Extended X-Ray Absorption Fine Structure or SEXAFS) holds promise for studying highly disordered and small particulate materials.

The emphasis in these Lectures will thus be on the fundamental principles underlying the various electron spectroscopic techniques which can be used in the study of chemisorption processes. The list of general references provided [1] should be consulted for experimental details, as well as for the practical limitations inherent in the various techniques.

2. ADSORBATE VIBRATIONAL STATES

We consider first the <u>number</u> of adsorbate induced vibrational modes.

A molecule containing N atoms has 3N degrees of freedom in the gas phase, comprising 3 translational, 3 (2) rotational and 3N-6 (3N-5) vibrational modes (the brackets referring to linear species). Assuming that all translation and rotation is restricted when the molecule is adsorbed, we are now left with 3N vibrational modes, i.e., we have gained 6 (5) extra "frustrated" vibrational modes. To see how this comes about, consider the simple cases of a carbon monoxide molecule linearly bonded on-top of a metal atom and in a bridge site position (Fig. 2).

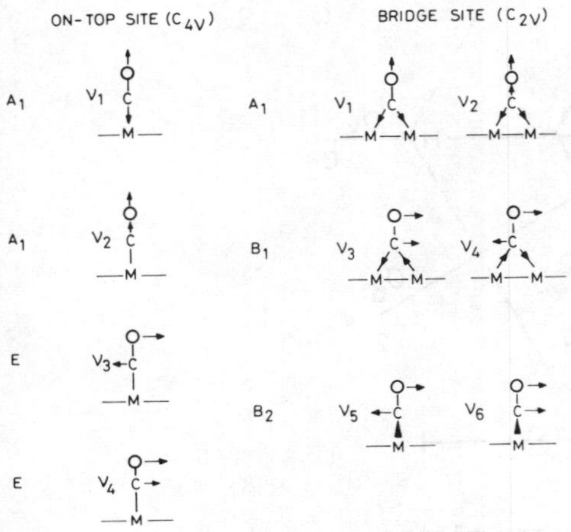

FIG. 2.

Illustrating the normal vibrational modes and symmetry representations of CO chemisorbed on a metal surface M: C_{4v} point group symmetry refers to bonding on top of a single metal atom in a four-fold rotation symmetry site; C_{2v} refers to a two-fold rotation symmetry bridge site.
(N.V. Richardson and A.M. Bradshaw, Ref. 3).

Referring to the C_{4v} point group symmetry of the CO adsorbed with its molecular axis perpendicular to (say) the (100) face of a cubic crystal in the on-top position, there are two stretching modes belonging to the totally symmetric A_1 representation, one of which is closely related to the free molecule C-O stretch and the other derives from the frustrated translational motion of the whole molecule against the surface (in the z direction), i.e., it becomes essentially a Ni-C stretching mode. The remaining 4 modes have bending character parallel to the surface and appear as degenerate pairs of E representation - the x and y frustrated translations and frustrated rotations about the x and y axes [3].

For the C_{2v} point group symmetry bridge position, again there are two A_1 stretching modes representing motion normal to the surface but now the degeneracies of the parallel modes are lifted, resulting in two modes vibrating in the plane of the bridge (B_1) and two modes perpendicular to this plane (B_2).

Thus, we can see that the number of possible localized vibrational modes is equal to 3N and independent of the adsorption site. However, the actual number likely to be observed in any spectrum will depend on the appropriate selection rules in relation to the symmetry of the surface site. Again, to see how such rules might operate, consider next the case of a highly symmetric molecule, such as benzene (Fig. 3).

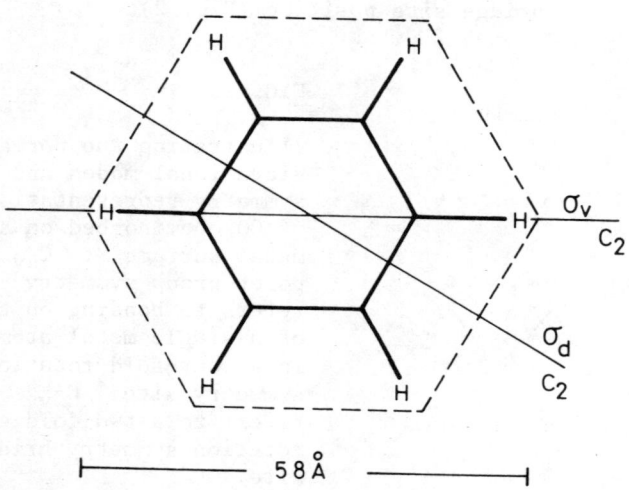

FIG. 3. Showing the mirror symmetry planes, σ_v and σ_d, and two-fold rotation axes, C_2, of the highly symmetric benzene molecule. The molecular dimension shown (5.8 Å) is that for atomic radii of the H atoms.

An isolated benzene molecule in the gas-phase has 20 fundamental vibrational frequencies, 10 of which are doubly degenerate. Of these, only 4 are infrared (IR) active, involving dipole moment changes either in the direction perpendicular to the aromatic ring (one a_{2u} mode) or parallel to this ring (three e_{1u} modes) (Ref. Table 1). Seven frequencies are Raman active because they cause a change in polarizability (two a_{1g} modes, one e_{1g} mode and four e_{2g} modes). These symmetry labels are appropriate to the D_{6h} point group and, because of the centre of symmetry, there are no coincidences between IR and Raman activities. In Table 1, the symmetry classification of the translational and rotational motions are also indicated. This is important since, as we have seen, fixing the molecule on a surface gives rise to six new vibrational modes, "frustrated" rotations and translations with symmetries $a_{2g}(R_z)$, $e_{1g}(R_xR_y)$, $a_{2u}(T_z)$ and $e_{1u}(T_xT_y)$ i.e., four possible new frequencies, of which e_{1g} is Raman active, a_{2u} and e_{1u} being IR active [3].

The question that now arises is that of how we go about observing these vibrational modes and what selection rules are applicable ? This will depend, to some extent, on the technique employed.

D_{6h} symmetry label	Frequency in cm^{-1} *	Optical Activity
a_{1g}	992, 3062	RAMAN
a_{2g}	1326, R_z	-
b_{1g}	---	-
b_{2g}	703, 993	-
e_{1g}	850, (Rx, Ry)	RAMAN
e_{2g}	606, 1178, 1595, 3048	RAMAN
a_{1u}	---	-
a_{2u}	671, T_z	INFRARED
b_{1u}	1010, 3060	-
b_{2u}	1160, 1310	-
e_{1u}	1035, 1485, 3080, (Tx, Ty)	INFRARED
e_{2u}	405, 970	-

* 1 milli-eV ≡ 8.066 cm^{-1}

TABLE 1 : Symmetries and frequencies of the vibrational modes of an oriented benzene molecule showing the symmetries of the "frustrated" rotations and translations.

2.1 The "Surface Normal Dipole Selection Rule"

The obvious technique to employ is that of infrared (IR) absorption spectroscopy and, indeed, this method has been widely used in the study of catalytic systems over the past 20 years, following pioneering work by Eischens and his collaborators in the fifties [4]. The infrared spectrum of CO chemisorbed on Cu(100) was the first reported application of infrared *reflection* absorption spectroscopy to metal single crystal surfaces [5]. This did not occur, however, until 1970, underlining the weak signals observable with this method. To date only the C-O stretch frequency ν_1, Fig. 1) has been observed. Moreover, it has become clear that IR suffers from the rather severe restriction that only those vibrations which produce a significant dipole change normal to the surface can be excited. The screening properties of metal surfaces at infrared frequencies effectively act in a manner as to produce an electric field intensity which is perpendicular to the mirror surface [6]. Vibrational modes, for which the matrix elements of dipole moments are parallel to the surface, appear abnormally weak, if observed at all, as is usually the case. A similar "selection rule" also applies in the case of electron energy loss spectroscopy (EELS), which has the advantage of greater surface sensitivity (though at much degraded resolution compared with IR). The origin of this "surface-screening selection rule" is a complex physical process [7], but one which may be readily visualized in terms of the production of an "image dipole" in a metal in response to an approaching point charge (Fig. 4).

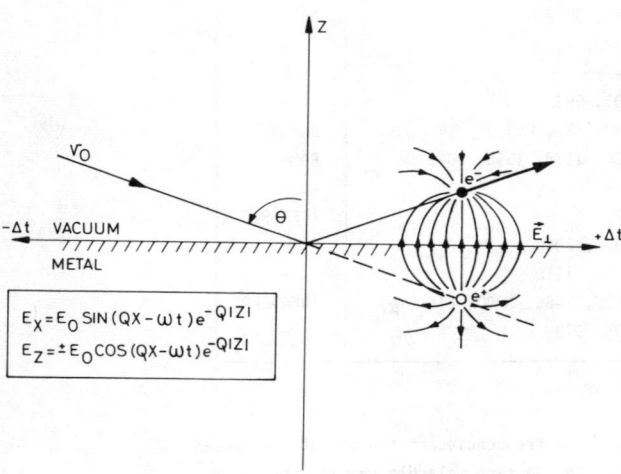

FIG. 4

Schematic illustration of the *long range* "image dipole" screening of an electron as it approaches, and is specularly reflected by a metal surface. The electron velocity $V_o = (2E_o/m_e)^{\frac{1}{2}}$ where E_o is the kinetic energy of the electron and Θ is the angle of incidence. The components of the surface plasmons wavefield is shown inset as a function of wavevector Q.

Consider a low energy electron incident on a mirror metal surface at an angle Θ and subsequently specularly reflected (e.g. as in a simple LEED experiment). The electron charge density in the solid responds to the Coulombic interaction of the approaching charge via electric fields produced by charge density oscillations at the surface – the *surface plasmons'* wave field (inset eqns., Fig. 4). These charge density oscillations have a continuum (almost) of wavelengths such that the longer wavelength oscillations extend far out (several hundred Angstroms) into the vacuum region: a distance $|Q_{sp}^{-1}|$ where $Q_{sp} = 2\pi/\lambda_{sp}$ is the wavevector of the surface plasmon wave. If we now regard the total time during which the electron interacts with this field Δt as being of the order of the period of the surface plasmon oscillations:

$$\Delta t = \frac{2}{V_e Q_{sp}} = \frac{2\pi}{\omega_{sp}} \qquad (1)$$

then maximum coupling occurs when (ignoring dimensionless factors the order of unity),

$$Q_{sp} \simeq \frac{\omega_{sp}}{V_e} \qquad (2)$$

i.e., when the parallel component of the electron velocity is equal to the phase velocity of the surface plasmon waves. (The factor 2 in Eqn. (1) accounts for time before and after reflection from the surface). For a surface adsorbate lattice, vibrational oscillations of wavevector $q_{//}$ will be excited when

$$\hat{q}_{//} = \hbar\hat{\omega}_L (2E_o)^{-\frac{1}{2}} \sin^{-1}\Theta \qquad (3)$$

where $E_o = \frac{1}{2}mV^2$ is the kinetic energy of the incident electron (Fig. 4) and Eqn. (3) is the equivalent relation to Eqn. (2) (expressed in atomic units). The condition $\hbar\omega_L \ll E_o$ implies that the scattering angle:

$$\Delta\Theta_s = \frac{\hbar\omega_L}{2E_o} \qquad (4)$$

is small and strongly directed about the specular beam direction. For example, the C–O stretching vibration in adsorbed CO produces a loss $\hbar\omega_L \simeq 0.25$ eV, which for an impact energy $E_o \simeq 5$ eV at an incident angle $\Theta \simeq 45°$ (ref. Fig. 4) gives an inelastic scattering cross section for the scattered electrons into a cone of half-angle $\Delta\Theta_s \lesssim 5°$ around the specularly reflected beam, and $\hat{q}_{//} \simeq 10^{-5}$ atomic units [8]. That is, $\hat{q}_{//}$ is much smaller than a reciprocal lattice vector and may be regarded as representative of a localized oscillator with negligible dispersion.

Effectively, the image potential (Fig. 4) results only from this polarization of the surface modes of the solid. For a simple metal these are just the surface plasmons and, for a dielectric, the surface optical phonons, surface excitations etc. The essential point is that the interaction is a *long range* one and since the wavelength of infrared radiation is long compared to atomic distances, absorption occurs via a similarly induced "dipole image field" as that for specularly scattered electrons. In this dipole theory, the relative electron energy loss intensity is given by [9] :

$$\frac{I_{loss}}{I_o} = \frac{4\pi m e^2}{h^2 E_o} |\mu_L|^2 \frac{n_s}{\cos \Theta} f(E_o, \Theta) \qquad (5)$$

where m is the electron mass, μ_L is the vibrational dipole matrix element and n_s is the number of molecules per unit surface area. $f(E_o, \Theta)$ is a function dependent on the scattering geometry [10]. For known surface coverages it is thus possible to determine the effective ionic charges of those dipoles vibrating normal to the surface from the relative intensities of the loss spectra [11]. For an harmonic oscillator:

$$\mu_\perp = \frac{\delta}{\delta z} \left(\frac{h}{2 M_r \omega_L} \right) \qquad (6)$$

where $\delta\mu/\delta z$ is the "dynamic effective charge" for displacements normal to the surface of a molecular complex of reduced mass M_r. For adsorbed CO, $\mu_\perp$ has been shown to be of the order 0.6e in good agreement with both the reflection IR measurements on adsorbates and the gas phase value [12].

EXAMPLE 1: CO on Ni(100) [12].
___

A recent measurement of the energy loss spectrum of CO chemisorbed on a Ni(100) surface is shown for a variety of surface structures in Fig. 5. The loss peak at 256 meV in the c(2x2) CO structure arises from a terminally bonded C-O stretch mode and that at 59 meV is due to the Ni-C stretch mode (ν_1 and ν_2 respectively, Fig. 2). With increasing coverage Θ, the film is compressed into a mixed phase corresponding to CO molecules chemisorbed on-top and in bridge sites simultaneously. Note that the parallel modes (ν_3 to ν_6, Fig. 2) are <u>not</u> observed.

A schematic of an electron energy loss apparatus [13] is shown in Fig. 6, together with a detailed scale drawing of the tandem cylindrical deflector type spectrometer used. The essential points are that the apparatus contains various facilities for characterizing the chemisorption system prior to EELS analysis. The latter requires that the incident and reflected electron beams are highly monochromatic (FWHH $\lesssim$ 10 meV) and collimated, necessitating a dispersion system of electrostatic deflector plates and appropriate optics. Design considerations have been reviewed by Roy and Carette [14].

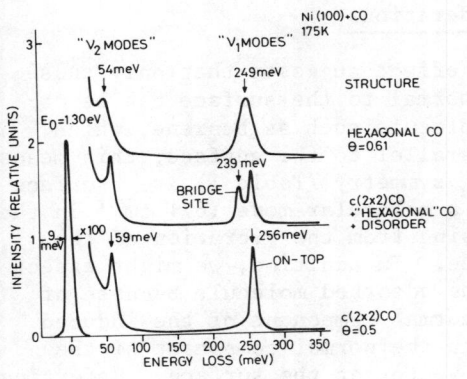

FIG. 5

Electron-energy-loss spectra of CO chemisorption on a Ni (100) surface showing the frequency variation of the ν_1 and ν_2 modes (z-motion, ref. Fig. 2) for different surface structures.
From Andersson, Ref. 12.

FIG. 6

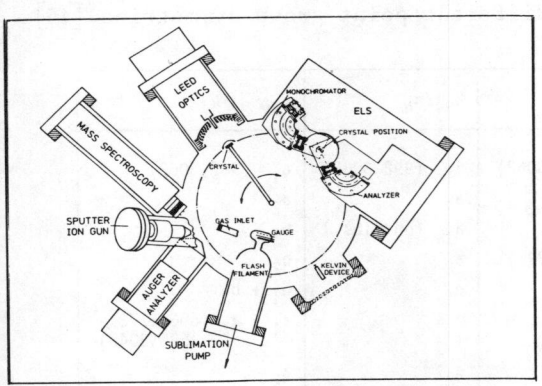

(a) Schematic of UHV apparatus incorporating an electron energy loss spectrometer (ELS) and other surface diagnostic techniques (Ref. 13).

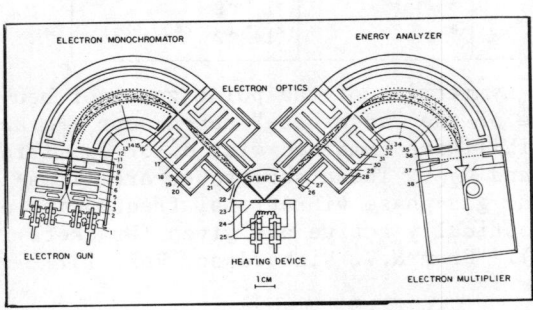

(b) Scale drawing of a tandem cylindrical deflector electron spectrometer (A. Adnot and J.D. Carette, Ref. 14) showing the electron optics appropriate for high resolving power vibrational spectroscopy.

2.2 Surface Site Symmetry Considerations

The image charge screening effect suggests that only those modes producing a dipole change normal to the surface can be observed. For a large symmetric molecule such as benzene, which adsorbs with its aromatic ring parallel to the surface, this means that only two bands, those of a_{2u} symmetry (Table 1) are "surface dipole active", corresponding to a molecular mode (671 cm^{-1} in the gas phase) and a new T_z mode arising from the vibration of the whole molecule against the surface. In addition, we might expect Raman active bands, induced in the adsorbed molecule because of its polarizability, involving a normal component of the induced molecular dipoles interacting with the normal component of the radiation-induced electric field vector at the surface. Referring to Table 1, this would indicate that the a_{1g} modes become surface active (bands at 3062 and 992 cm^{-1} in the isolated molecule) due to the α_{zz} component in the polarizability tensor.

These considerations relate, however, only to the symmetry of the oriented molecule. The number of actual modes likely to be observed will also be governed by the symmetry properties of the substrate's atomic site. This is shown in Table 2 and Fig. 7 for benzene occupying sites of different point group symmetries [15].

D_{6h}	C_{6v}		C_{3v}/σ_v		C_{3v}/σ_d		C_{2v}	
a_{1g}	a_1	(992,3062)	a_1	(992,3062)	a_1	(992,3062)	a_1	(992,3062)
a_{2g}	a_2		a_2		a_2		a_2	
b_{1g}	b_1		a_2		a_1	(no vibs.)	b_1	
b_{2g}	b_2		a_1	(703,995)	a_2		b_2	
e_{1g}	e_1		e		e		$b_1 + b_2$	
e_{2g}	e_2		e		e		$a_1 + a_2$	(606,1178, 1595,3048)
a_{1u}	a_2		a_2		a_2		a_2	
a_{2u}	a_1 (671,T_z)		a_1	(671, T_z)	a_1	(671, T_z)	a_1	(671, T_z)
b_{1u}	b_2		a_1	(1010,3060)	a_2		b_2	
b_{2u}	b_1		a_2		a_1	(1160,1310)	b_1	
e_{1u}	e_1		e		e		$b_1 + b_2$	
e_{2u}	e_2		e		e		$a_1 + a_2$	(400,985)

TABLE 2: Correlation of the symmetries of the point group of benzene D_{6h} with those of the "adsorbate" sub-groups C_{6v} and C_{3v} (with two possibilities combined with σ_v or σ_d mirror plane symmetries) and C_{2v}. These sub-groups are illustrated, Fig. 7. The gas-phase vibrational frequencies of benzene which are optically active are given (brackets in wave-numbers, cm^{-1}). From N.V. Richardson, Ref. 15.

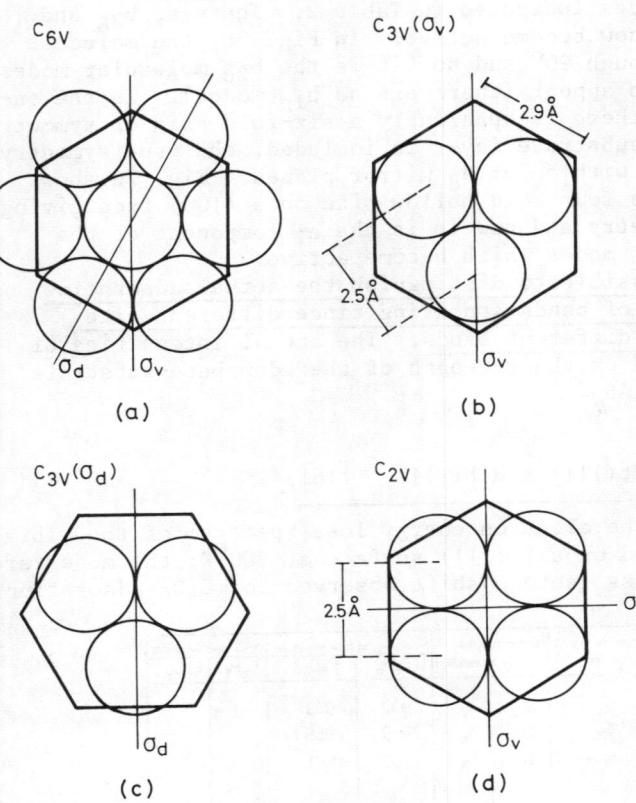

FIG. 7

Illustrating the different point group symmetries of benzene chemisorbed at different substrate sites: (a) 6-fold "on-top" site; (b) 3-fold "hollow" site with σ_v mirror symmetry; (c) 3-fold "hollow" site with σ_d mirror symmetry; and (d) 4-fold "hollow" site with σ_v <u>and</u> σ_d mirror symmetry planes. The metal atoms' spacing is appropriate to nickel, (a) - (c) corresponding to adsorption on the (111) surface and (d) on the (100) plane. From Richardson, Ref. 15.

Firstly, (weak) bonding along the direction of the surface normal ($AS_\perp$) removes: the mirror plane coincident with the aromatic ring; symmetry axes in this plane; improper axes normal to this plane; and the centre of symmetry. The point group changes from D_{6h} to C_{6v} (Fig. 7a), and the surface dipole active vibrations (changes in μ_z <u>and</u> α_{zz}) are now a_1 modes i.e., effectively the a_{1g} and a_{2g} active representations of D_{6h} symmetry correspond to a_1 in C_{6v} producing the equivalent four (IR and Raman active) bands described above. The extent to which these modes become "dipole active" will, of course, depend on the magnitude of the interaction with the surface, $AS_\perp$.

Also, localized bonding between the molecule and individual substrate atoms <u>in</u> the xy plane of the surface ($AS_{//}$ interaction) can introduce a_1 surface active modes. Such "local site symmetry" effects are illustrated for three possible orientations of the benzene molecule on the (111) face of a face-centred-cubic single crystal surface in Fig. 7. In Fig. 7b, the σ_v mirror planes of the isolated molecule (Fig. 3) are retained but the σ_d planes are lost producing C_{3v} point group symmetry and the <u>eight</u> possible surface active a_1 modes indicated in Table 2. That is, b_{2g} and b_{1u} molecular modes now become active. In Fig. 7c, the molecule has been rotated through $90°$ and now it is the b_{2u} molecular modes which are expected to appear (there are no b_{1g} modes). In the on-top site (Fig. 7a), there is apparently a six-fold axis of symmetry <u>but</u> when the second substrate layer is included, the true symmetry appears as C_{3v} again with σ_v or σ_d mirror planes. Fig. 7d shows benzene adsorbed in a four-fold hollow site on a (100) face giving C_{2v} point group symmetry and now it is the a_1 component of the molecular e_{2g} and e_{2u} modes which become active.

<u>Thus, it is possible to distinguish the actual adsorption site from the number of bands appearing since different site symmetries introduce different bands.</u> The actual intensities of the bands will depend on the strength of the adsorbate-substrate interactions $AS_\perp$ and $AS_{//}$.

EXAMPLE 2: C_6H_6 on Pt(111) and Ni(111) [16].

Fig. 8a shows the electron energy loss spectrum of the vibrations of C_6H_6 adsorbed on a Pt(111) surface at $300°K$; the modes are identified with $\sqrt{2}$ mass isotope shift observed for C_6D_6 adsorption.

C_{3v}	approx type of mode	symm species in free molecule	frequency in free molecule		Observed on Pt(111)	
			C_6H_6	C_6D_6	C_6H_6	C_6D_6
a) σ_v	ν_{CC}	a_{1g} ν_2	992	943		
	δ_{CC}	b_{1u} ν_6	1010	969		
	γ_{CC}	b_{2g} ν_8	703	601		
b) σ_v	ν_{CH}	a_{1g} ν_1	3062	2293		
	γ_{CH}	a_{2u} ν_4	673	497		
	ν_{CH}	b_{1u} ν_5	3068	2292		
$m_v = 4 \to 8$ A_1-modes	γ_{CH}	b_{2g} ν_7	995	827		
c) σ_d	ν_{CC}	a_{1g} ν_2	992	943		
	δ_{CC}	b_{2u} ν_9	1310	1286	<u>1420</u>	1350
	ν_{CPt}				{570 / 360}	{ — / 350}
d) σ_d	ν_{CH}	a_{1g} ν_1	3062	2293	3000	2240
	γ_{CH}	a_{2u} ν_4	673	497	{830 / 920}	{610 / 700}
$m = 2 \to 6$ A_1-modes	δ_{CH}	b_{2u} ν_{10}	1150	824	<u>1130</u>	800

TABLE 3 : EELS active normal modes of benzene adsorbed in the four possible sites of threefold symmetry on the (111) surface of Pt. All frequencies are in cm^{-1}. (Ref. 16).

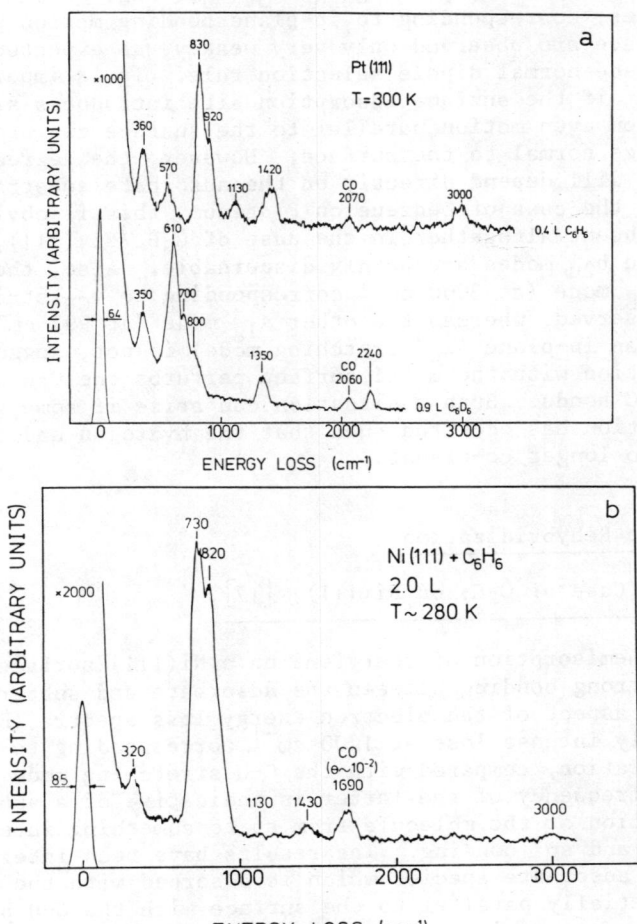

FIG. 8. (a) Electron energy loss spectra of the vibrations of C_6H_6 and C_6D_6 on Pt(111), submonolayer coverages at 300°K.

(b) EELS of C_6H_6 on Ni(111).
The CO contamination is about 1% of a monolayer.
ν (stretching modes) refers to motion $\perp$ surface;
δ and γ (in-plane and out-of-plane bending modes) refers to motion $\parallel$ surface (Ref. 16).

The spectrum has been assigned in terms of benzene occupying the three-fold hollow site, point group $C_{3v}(\sigma_d)$, Fig. 7c and Table 3d. The b_{2u} modes, corresponding to in-plane bending motion parallel to the surface are observed only very weakly, as expected in terms of the surface normal dipole selection rule. It is important to realize that if the surface adsorption site introduces a_1 type symmetry then even motion parallel to the surface can produce a dipole change normal to the surface. However, the degree to which this occurs will depend directly on the adsorbate/substrate interaction. In the case of benzene on Platinum, this is obviously weak, and absent altogether in the case of C_6H_6/Ni(111), Fig. 8b, in which the b_{2u} modes are hardly discernable. Also, the fact that only one a_{1g} mode (at 3000 cm^{-1} corresponding to ν_{CH} stretching mode) is observed, whereas the other a_{1g} mode (at 992 cm^{-1} corresponding to an in-plane δ_{CC} stretching mode) is not, suggests that the interaction with the metal surface perturbs the C-H bonds more than the C-C bonds. Such a situation can arise if some surface rehybridization has occurred such that the hydrogen and carbon atoms are no longer co-planar.

2.3 Surface Rehybridization

EXAMPLE 3: Case of C_2H_2 on Ni(111) [17].

The chemisorption of acetylene on a Ni(111) surface represent a case of strong bonding between the adsorbate and substrate. The significant aspect of the electron energy loss spectra (Fig. 9) is the extremely intense loss at 1200 cm^{-1} corresponding to the C-C stretch vibration, compared with the C-H stretching mode at 2910 cm^{-1}. The frequency of the latter is indicative of a substantial rehybridization of the molecule from sp to something intermediate between sp^2 and sp^3 bonding. The results have been interpreted in terms of an adsorbate species which is adsorbed with the C-C bond axis substantially parallel to the surface with the C-H bonds bent upwards. The strong intensity of the C-C stretch vibration (compared with C_6H_6, Fig. 8b) may then be explained by a mechanism involving charge transfer oscillations between the vibrating molecule and the surface[18].

This surface phase of adsorbed C_2H_2 has been found to be stable between 150°K and approx. 400°K. However, the reactivity of surfaces containing steps and kinks is seen by the dramatic differences observed between the flat Ni(111) surface (Fig. 9) and the spectra obtained from a stepped nickel surface (Fig. 10). The above intense C-C stretch mode is no longer observed, and a completely dehydrogenated metastable C ≡ C species (characterized by new losses corresponding to vibrations at 350 cm^{-1} and 2220 cm^{-1} Fig. 10a) is observed prior to complete fragmentation into carbon and hydrogen atoms. Fig. 10b shows a case in which only the "stripped" C ≡ C species exists, all of the C-H bonds being preferentially broken.

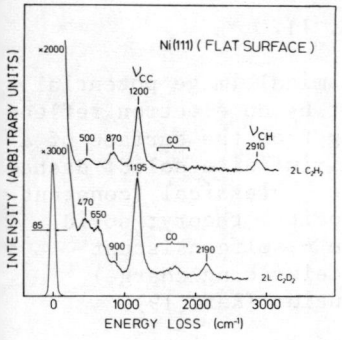

FIG. 9.

Vibration spectra of C_2H_2 and C_2D_2 on a <u>flat</u> Ni(111) surface.

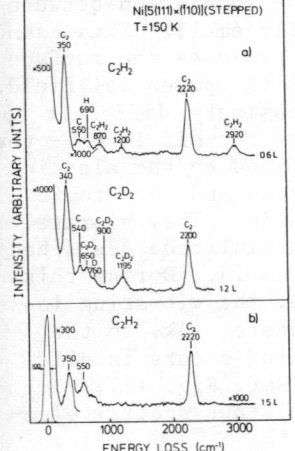

FIG. 10.

Vibration spectra of C_2H_2 and C_2D_2 on a <u>stepped</u> Ni $|5(111) \times (\bar{1}10)|$ surface; the convention refers to (111) terrace surfaces separated by steps oriented normal to the <$\bar{1}10$> crystallographic direction. The step atoms have a coordination number of 6 and 8 while those on the terraces have a coord. no = 9. (Ref. 17). While (a) represents a typical spectrum at 150°K, it is possible (b) to obtain only the completely dehydrogenated C_2 phase.

2.4 Short-range off-specular scattering: breakdown of the surface-dipole selection rule

We have seen that only vibrational modes with components of their dynamic dipoles acting normal to the surface are strongly excited under specular electron scattering conditions. The reason is the dipole electric field produced at the surface via the *long range* Coulombic interaction of the electron and the dielectric response of the solid. If we now consider this interaction in more detail, particularly the consequences of the *short range* ionic potential of the substrate lattice which effectively reflects the electron back towards vacuum, then we meet the problem illustrated

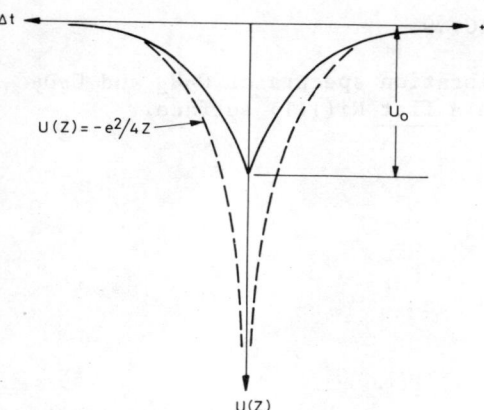

FIG. 11.

Dynamical image potential felt by an electron reflecting from the surface of a semi-infinite solid: dashed line - classical (constant velocity) theory; solid line - selfconsistent (accelerating charge) solution (Ref. 19).

in Fig. 11. That is, if we plot the strength of the dipole image potential, $U(z) = -e^2/4z$ as a function of the interaction time Δt, we see that $U(z)$ diverges rapidly as z^{-1} as the separation between the charge and its image becomes infinitesimally small. This cannot occur and the reason is that as the charge approaches the surface it accelerates into the increasing gradient of the potential field. The surface plasmon oscillations find it increasingly difficult to respond in time such that $U(z)$ reaches a finite cut-off level - the "inner potential" of the solid which is determined by the elastic scattering properties of the conduction electrons and the atomic ion-cores [19]. But the ion cores are not static. They vibrate about their mean lattice positions with a mean amplitude described by the Debye-Waller factor for the particular solid. During this short range part of the interaction, therefore, the vibrating ions (and adsorbed atoms) can impart a momentum transfer $|\Delta k|$ to the incident electrons such that "diffuse" scattering occurs in all directions about the specular beam direction (Ref. Fig. 12). Vibrational motion parallel to the surface will thus scatter electrons in off-specular directions.

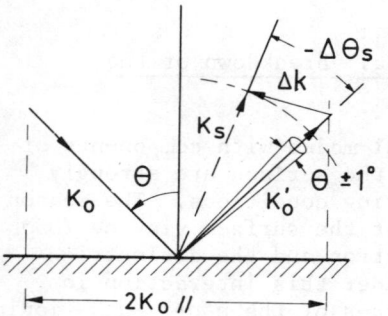

FIG. 12.

Illustrating the kinematics of "diffuse" inelastic electron scattering about the specular beam direction (full width $\pm$ 1° solid angle). Momentum transfer Δk can be furnished by atomic vibrations or lattice phonons.

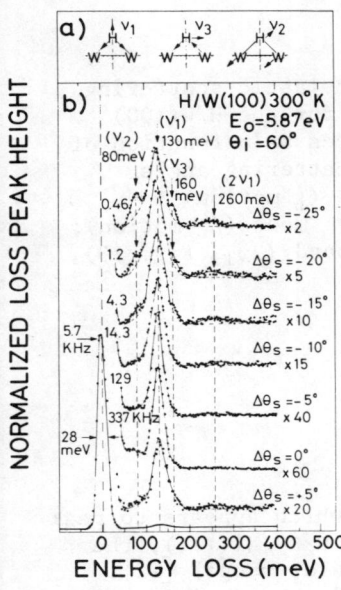

FIG. 13.

(a) Vibrational modes of H bridge-bonded between two W(100) substrate atoms.

(b) Associated electron energy loss spectra as a function of scattering angle $\Delta\Theta_s$ corresponding to a wave-vector range for surface excitations $\vec{q}_{\parallel}$ ranging from $0 \lesssim \vec{q}_{\parallel} \lesssim 0.4\ k_B$, where $k_B = \Gamma H$ surface Brillouin zone dimension (Ref. 20).

EXAMPLE 4: Angle-dependence of EELS, H on W(100) [20].

The consequences of off-specular scattering in EELS is illustrated for the case of a saturation coverage of atomic hydrogen adsorption on a tungsten (100) surface, Fig. 13. The hydrogen atoms chemisorb in bridge sites with C_{2v} point group symmetry giving rise to a p(1x1) LEED pattern and to 3 non-degenerate vibrational modes corresponding to "frustrated" translational motion in the x, y and z axes. The relative masses of the H and W atoms are such that we may consider only the motion of the very light adsorbate atoms against a "static" substrate lattice. We see, Fig. 13, that whereas only the symmetric stretch vibration ν_1 normal to the surface is seen in the specular scattering direction, with increasing off-specular scattering, $\Delta\Theta_s$, the asymmetric stretch ν_2 and wagging ν_3 modes (corresponding to in-plane and out-of-plane bridge site vibrations) appear, which represent motion parallel to the surface.

Thus, EELS offers the considerable advantage over IR spectroscopy of being able to detect all the vibrational modes, determined by the point group symmetry of the adsorption site, and to distinguish between modes vibrating normal from modes vibrating parallel to the surface.

This is seen clearly in the differential inelastic scattering cross-sections $(d\sigma/d\Omega dE)$ plotted as a function of scattering angle, Fig. 14. Whereas the symmetric stretch vibration shows a peak in intensity in the specular direction $\Delta\Theta_s = 0°$, and follows closely the elastic beam profile, as predicted by surface dipole theory,

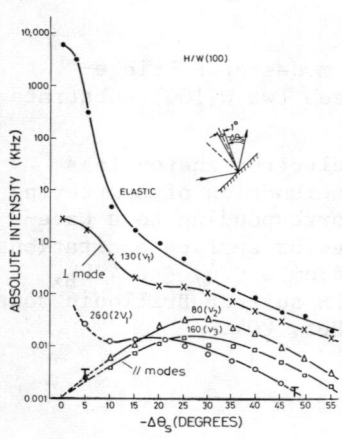

FIG. 14.

Differential inelastic scattering cross-sections for the H/W(100) vibrational modes as a function of off-specular scattering angle $\Delta\Theta_s$: $\times$-ν_1 mode ($\perp$ motion) Δ-ν_2 mode and ▯ - ν_3 mode (// motion); 0-combination band ($2\nu_1$, Fig. 13), Ref. 21.

the parallel vibrations have an intensity which appears to peak off-specular [21]. (In other cases, e.g. C_2H_2/Ni(111), the distribution remains flat over a large angular range) [22]. The absolute intensity of a broad combination band of parallel and perpendicular vibrational modes ($2\nu_1$, Fig. 13) shows the "mixed" behaviour expected.

To understand this off-specular distribution in vibrational intensity it is useful to write the intensity at any point in the Brillouin zone of the surface lattice in the form |23|:

$$I(\Delta k) = e^{-2M} I^{(o)}(\Delta k) + 2Me^{-2M} I^{(1)}(\Delta k) + (1 - e^{-2M} - 2Me^{-2M}) I^{(m)}(\Delta k) \quad (6)$$

or simply

$$I(\Delta k), T) = I^{(o)}(\Delta k, T) + I^{(1)}(\Delta k, T) + I^{(m)}(\Delta k, T) \quad (7)$$

where $2M = \langle(\Delta \underline{k} \cdot \underline{u})^2\rangle$ is the Debye-Waller factor written in terms of the change of momentum $\Delta \underline{k}$ and the mean atomic vibrational amplitude $\underline{u}$. Eqn. (6-7) simply expresses the fact that the first term, the zero phonon intensity of the lattice vibrations, removes intensity from the elastically scattered beams and the second and third terms represent the one- and multiphonon intensities respectively which redistribute intensity throughout the Brillouin zone as diffuse scattering. 2M is essentially the ratio of the *integrated* intensity in the one-phonon scattering to that in the zero-phonon scattering. Expanding the displacements in Eqn. (6) in terms of the normal modes of a lattice yields (in the high temperature limit), the one phonon intensity

$$I^{(1)}(\Delta k, T) = \frac{d\sigma}{d\Omega} = \frac{2Me^{-2M}}{|q^n|} \quad (8)$$

where $1 \leq n \leq 2$. This function must go to zero at the center of the zone since for a finite crystal there are no phonons with wave vector $|\bar{q}| = 0$.

That this behaviour should also hold for adsorbate modes, would appear to be confirmed by the H/W(100) results for the parallel modes (Fig. 14). However, in other adsorbate systems, the degree to which vibrational intensity in the thermal diffuse scattering is distributed in one phonon (or single inelastic scattering) intensity will depend strongly on the effective Debye-Waller factor of the adsorbate/substrate lattice. The lower the effective Debye temperature (or the higher the temperature for a given system), the more the inelastic intensity will be distributed evenly throughout the Brillouin zone due to the multiphonon component. This is more likely to be the case for the larger more complex hydrocarbon molecules having a wide variety of particularly (soft) vibrational modes [21].

This temperature dependence of the integrated zero-, one- and multiphonon terms in Eqn. (6/7) is illustrated as a function of the magnitude of $|2M|$ in Fig. 15a. Also shown (Fig. 15b) is the mean square vibrational amplitudes of the hydrogen vibrational modes on W(100) as a function of temperature and frequency [21,24].

FIG. 15.

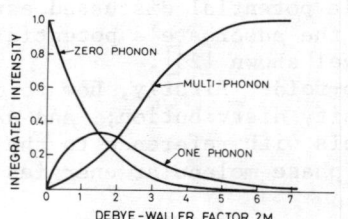

(a) The integrated intensity in the zero, one and multiphonon scattering as a function of the exponent of the Debye-Waller factor, $2M = <(\Delta k \cdot u)^2>$ Ref. 23.

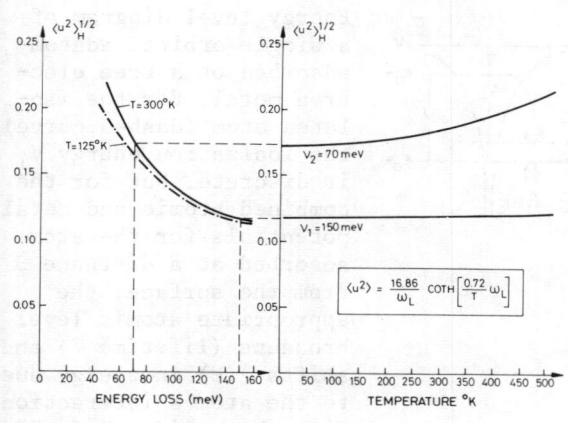

(b) The mean square vibrational amplitudes of the adsorbate vibrations of H chemisorbed on W(100). Ref. 24.

3. ADSORBATE ELECTRONIC STATES

We turn our attention now to the question of the nature of adsorbate bonding in relation to the electronic charge distribution induced at the surface.

First consider the simplest case of a single orbital adatom (e.g. hydrogen) chemisorbing on a free electron substrate (the simplest model for a metal). The overlap of the electronic wavefunctions permits charge to tunnel freely back and forth between the adsorbate and the metal surface. The lifetime of an electron on the atom is thereby reduced leading to a broadening of the original adatom energy level. The repulsive Coulombic interaction between the adsorbate's electronic charge and the metal's electrons produces the same kind of complex dynamical screening and polarization of the surface electron charge density as that described for an approaching point electronic change (Fig. 4). This in turn modifies the potential at the surface leading to a (self-consistent shift in energy of the original "atomic" level, as indicated schematically, Fig. 16. Here the metal is depicted by a simple potential well filled to a depth up to the Fermi energy E_F with the vacuum level at E_v and a work function $(E_v - E_F) = \phi$. The surface potential barrier (the shape of which is determined largely by the surface image dipole potential discussed earlier, Fig. 11) is strongly perturbed by the adsorbate's potential producing the broadened adsorbate level shown [25].

The problem therefore is two-fold: firstly, how to observe this adsorbate-induced charge density distribution; and secondly, how to identify the adsorbate levels with reference to the original unperturbed metal surface and gas phase molecular energies ?

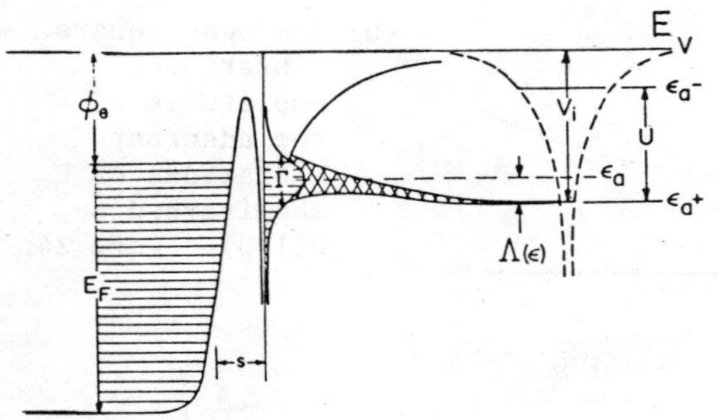

FIG. 16.

Energy level diagram of a single orbital adatom adsorbed on a free electron metal. For the isolated atom (dashed curve), the ionization energy V_i is discrete, but for the combined atomic and metal potentials for the atom adsorbed at a distance S from the surface, the appropriate atomic level broadens (lifetime Γ) and shifts $\Lambda(\epsilon)$ in energy due to the atom's interaction with the solid. Ref. 25.

3.1 Observation

There are principally two direct excitation *electron emission* methods which have been employed: field electron emission and photoelectron spectroscopy. The energy distribution of electrons emitted in an accelerating field $V_F = eF_z$ and of photoelectrons emitted following irradiation with ultraviolet light (UPS) are illustrated schematically in Fig. 17 (Irradiation with X-rays, i.e. XPS, may also be used but, conventionally, it is usually referred to in connection with deep atomic core-level effects).

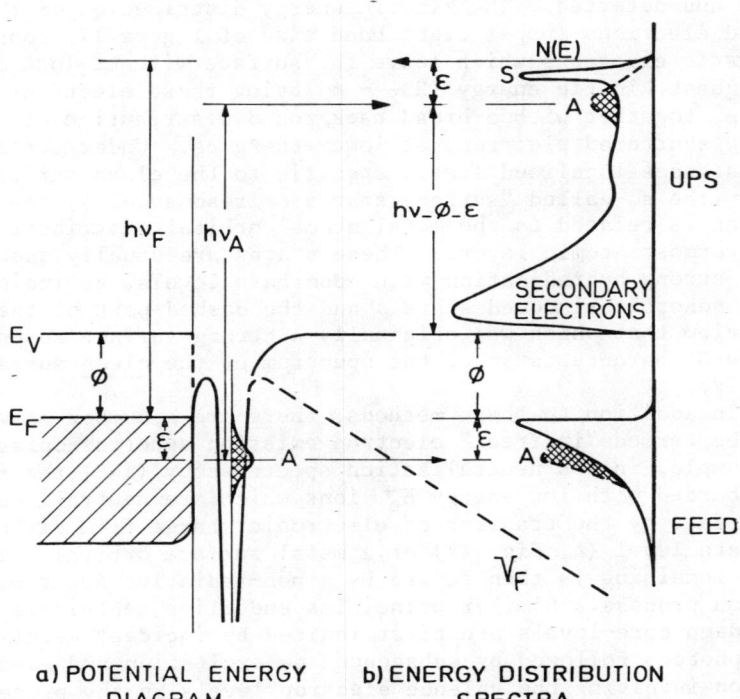

a) POTENTIAL ENERGY DIAGRAM b) ENERGY DISTRIBUTION SPECTRA

FIG. 17. Potential energy diagram (a) of a metal with an adsorbate atom A on its surface showing the origin of (b) the field emission energy distribution (FEED) and the photoelectron emission spectrum (UPS) under the stimulus of an applied electric field V_F or ultra-violet light of energy h_ν respectively. Both distributions reveal features associated with clean surface resonances S and adsorbate-induced states A, the latter often suppressing the former (dashed spectrum).

In a strongly accelerating field (dashed line), the potential barrier E_v is lowered allowing electrons in the occupied states at the surface (shaded) to tunnel out into vacuum and be detected as a field emission energy distribution (FEED). The disadvantage of this technique is that the tunneling probability decreases exponentially below the Fermi level cutoff, due to the increasing width of the wedge-shaped tunnelling barrier, so that only those adsorbate-induced states within one eV or so of E_F can be studied [26].

A much more widely used technique is that of ultraviolet photoemission. Here photons of energy $h\nu$ excite electrons into unoccupied states *above* E_F from which they are then emitted into vacuum and detected. The kinetic energy distribution of the emitted electrons (upper right hand side of figure 17) consists of *elastic* electrons which leave the surface without loss of energy the highest kinetic energy ($h\nu_F - \phi$) being those electrons excited from E_F, together with a broad background distribution of *inelastically* scattered electrons at lower energies. Under certain circumstances localized states specific to the clean surface S may appear (the so-called "surface states or resonances"), the origin of which is related to the metal atoms' orbital distribution in the outermost atomic layers. These states are usually quenched due to strong hybridization with adsorbate levels, as indicated by the adsorbate-induced state A and the dashed part of the spectrum below that which was originally a strong surface resonance feature S characteristic of the spectrum of the <u>clean</u> surface. (Fig. 17).

In addition to these methods, there are a number of what might be termed "indirect" electron emission spectroscopies. For example, in ion neutralization spectroscopy (INS) the surface is bombarded with low energy H_e^+ ions which are neutralized at the surface by the transfer of electronic charge from either an adsorbate level (A, Fig. 17) or a metal surface orbital. The "hole" remaining is then filled by a non-radiative Auger electron emission process. Similar principles underline techniques in which deep core-levels are first ionized by incident electrons or X-ray photons followed by subsequent de-excitation and electron emission involving the valence electron levels in the process. Auger electron spectroscopy (AES) derives from electron excitation and Appearance potential spectroscopy (APS) derives from the use of X-ray radiation [27].

These methods are "indirect" in the sense that a two-electron excitation/de-excitation process is involved requiring a complex deconvolution of the data in order to derive the adsorbate electronic structure. This is more of an interpretative limitation than is the case with photoelectron spectroscopy, and for this reason, the latter is more widely used as a probe of surface electronic structure.

3.2 Identification.

The question of the unequivocal identification and subsequent interpretation of adsorbate-induced features is a complex one. Consider, for example, the case of hydrogen chemisorbed on a transition metal surface. In addition to the "screening" effects described above, and usually a property of the s-electrons, there are strong localized bonding effects arising from the more tightly bound d-electrons with their associated directed orbitals at the surface which form bonds with the adatom. This is illustrated in Fig. 18 for hydrogen chemisorption on W(100).

EXAMPLE 5: Case of Strong Chemisorption, H/W(100) [28].

In Fig. 18a we show the results of a linear-combination of atomic-orbitals (LCAO) calculation of the two-dimensional distribution of the prominent surface states/resonances of a bare W(100) surface together with that of an isolated sheet of H atoms (dashed curve) for two principal symmetry directions. The bands are labelled with the principal orbital content in the LCAO eigenvector i.e., those W orbitals which have the largest amplitude in the surface. These orbitals hydridize strongly with both the bonding and antibonding orbitals of the H(1s) levels to produce the complex "band structure" shown in Fig. 18b. The states whose principal orbital content is hydrogen 1s (labelled H in Fig. 18b) produce not one (as suggested by the simple model depicted in figure 16) but several adsorbate-induced levels.

Angle-resolved photoelectron spectroscopy has been successfully employed in the identification of these bands, principally by measuring the dispersion of the states as a function of angle relative to the surface normal direction i.e., energy E as a function of parallel wavevector $\vec{k}_{//}$ [28]. In many cases, however, it has proved difficult to resolve the different states. This problem can be seen particularly for the case of many perturbed electronic levels in weakly chemisorbed hydrocarbon molecules.

EXAMPLE 6: Case of Weak Chemisorption, C_6H_6/Ni(111) [29].

In Fig. 19, we present UPS data from Demuth and Eastman for chemisorbed benzene on Ni(111). In panel (a), the clean metal spectrum is compared with that obtained after an exposure to 2.4 Langmuirs (1L $\equiv 10^{-6}$ Torr sec^{-1}) of benzene (dashed curve). The adsorbate-induced effects are emphasized by taking the *difference spectrum*, panel (b). The most obvious differences are: 1) the suppression of the metal d-states located within a few eV of E_F; and 2) the appearance of adsorbate levels, which correlate closely with the envelope of the gas phase molecular levels (panel d) [30], seen more clearly by comparison with a condensed film (panel c). The important aspect of these results is that by rather arbitrarily

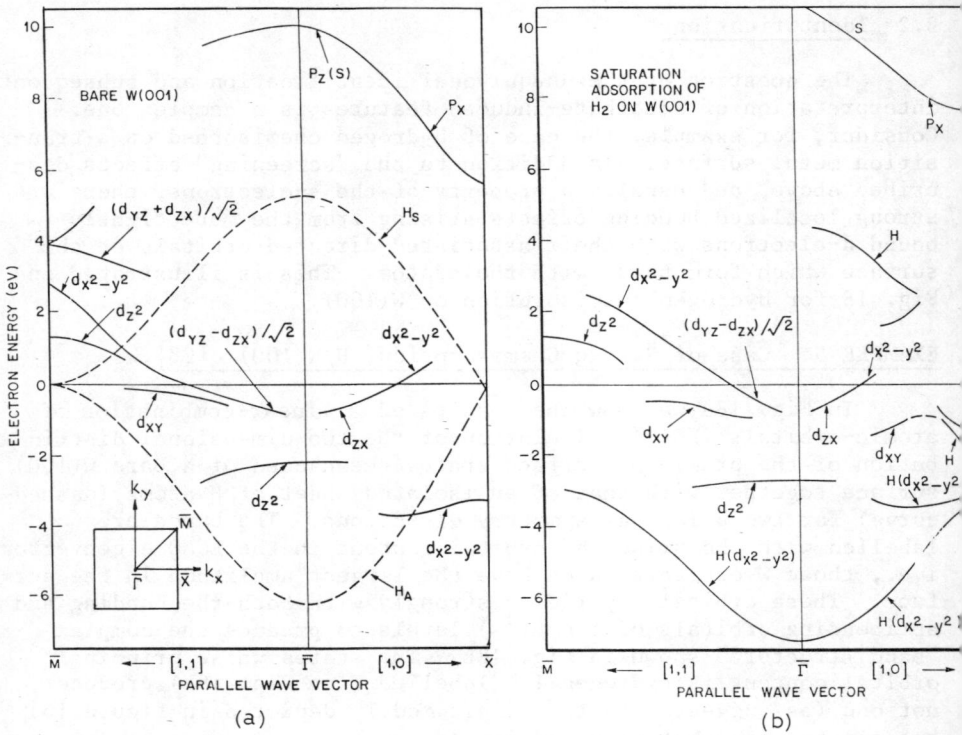

FIG. 18. Two dimensional energy band structure for (a) surface states/resonances deriving from W orbitals with large amplitudes in the <u>clean</u> surface and (b) for saturation coverage of bridge-bonded H atoms. The dashed curve (Fig. a) is the "band structure" of an isolated sheet of H atoms which then interact (hybridize) with the clean surface bands (full lines). N.V. Smith and L.F. Mattheiss, Ref. 28).

referencing the gas phase ionization levels of the weakly perturbed σ-bonded molecular orbitals with those of the adsorbed phase, the largest shift (relative to the other "non-bonding" orbitals) is seen to occur for the π orbitals. These are just the molecular orbitals which one might expect to be perturbed the most for a benzene molecule lying flat on the surface.

The magnitude of the π-orbital shift is a measure of the (weak bonding) interaction with the surface. However, a more detailed analysis of these results is complicated by the fact that, in addition to such <u>bonding shifts</u>, in photoelectron emission there are additional <u>relaxation shifts</u>. The latter arise from the screening energy of the nearby metal electron gas in response to the sudden creation of a hole left behind by the departing electron. Such effects can complicate the identification of the parent gas phase levels.

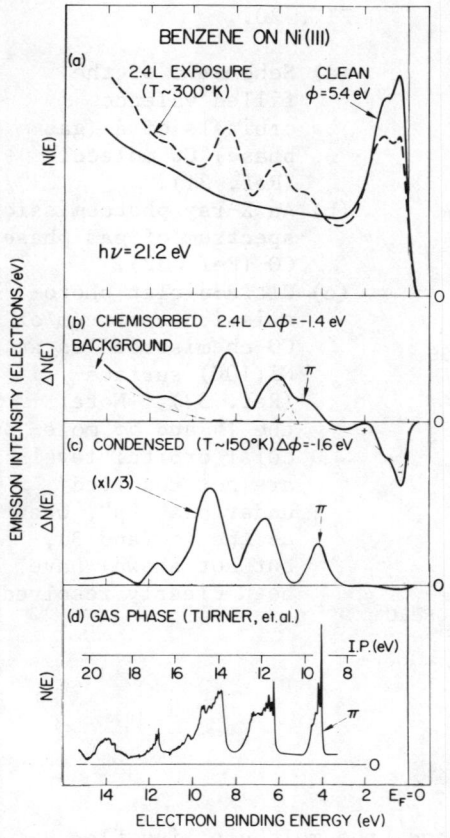

FIG. 19.

(a) UPS spectra, N(E) plotted as a function of electron binding energy relative to the Fermi level of clean nickel (solid curve) and a chemisorbed layer of benzene (dashed curve).
(b) Adsorbate-induced difference in emission $\Delta N(E)$ for chemisorbed benzene, and
(c) $\Delta N(E)$ for a condensed film, compared with
(d) gas phase photoelectron spectrum (Ref. 30).

EXAMPLE 7: Chemisorption of CO/Ni(100).

This problem is particularly acute for the case of CO chemisorption, which provides something of a model test of existing theory. The reason becomes apparent if we examine Fig. 20, in which we show: a) a sketch of the filled valence orbitals of the free gaseous CO molecule [31]; b) an X-ray photoemission spectrum of gas phase CO in which the various orbitals are identified [32]; and c) the energy distribution curve of CO chemisorbed on Ni(100) [33].

The first question to arise is to whether or not we can determine if the CO molecule is adsorbed with its molecular axis oriented normal to the surface or lies down by inspection of the photoelectron distribution ? If the CO forms a chemisorption bond via carbon-metal bonding (which is most likely in view of the chemistry of the transition metal carbonyls), then we would expect the 5σ bond to be most strongly shifted in energy relative to the

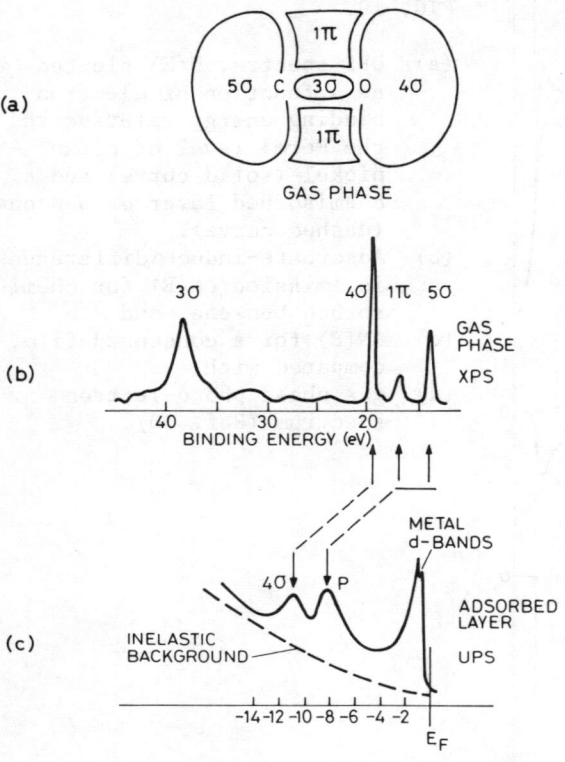

FIG. 20.

(a) Schematic of the filled valence orbitals of a (gas phase) CO molecule (Ref. 31).
(b) An X-ray photoemission spectrum of gas phase CO (Ref. 32).
(c) Ultra-violet photoemission spectrum of CO chemisorbed on a Ni(100) surface (Ref. 33). Note: the 1π and 5σ molecular orbital levels are not resolved under peak "p", whereas the 4σ (and 3σ, but not shown) have been clearly resolved.

other orbitals. If, on the other hand, the molecule lies flat, we might expect the 1π orbitals to be most strongly affected. Unfortunately, inspection of the UPS spectrum (Fig. 20c) shows only two adsorbate peaks, the 5σ and 1π orbitals being the unresolved peak "p".

Metal-carbonyl cluster calculations [34] for terminally bonded CO suggest that the 5σ level suffers a strong bonding energy shift and ends up just below the 1π level. This turns out to be the case when we examine the polarization dependence of photoelectron spectroscopy in relation to the symmetry properties of the adsorbate's electronic wavefunctions.

3.3 Polarization Dependence and Symmetry.

Angular selection of the photoemitted current implies a definite dipole selection rule operating on the coupling of the initial electron state $|i\rangle$, from which excitation occurs, to some final state $|f\rangle$, i.e. the relative symmetries of the quantum states can be measured using polarized light. Since different adsorbate states have different symmetries, it then becomes possible to distuinguish between photoexcitation from these states.

The basis for this can be seen by consideration of the nature of the dipole operator $\vec{p}$ in the optical transition matrix element. That is, the interaction of an electron with an electromagnetic wave is described by a vector potential $\vec{A}$ and, for the simplest case in which $\vec{A}$ remains a purely transverse field (the so-called "local approximation"), the energy and angle-resolved photocurrent can be written in the Fermi golden rule form:

$$\frac{d^2j}{dE_f d\Omega} \sim \sqrt{E_f} \sum_k \left| <f_k|\vec{A}\cdot\vec{p}|i> \right|^2 \delta(E_f - E_i - \hbar\omega) \tag{9}$$

treating the final state $<f_k|$ as a state in the vacuum continuum carrying current through a solid angle $d\Omega$ into the detector, the momentum of the outgoing electron being $\hbar\vec{k}$. This dipole approximation, Eqn. (9), implies a strict selection rule between states contained in a mirror symmetry plane, defined in terms of the plane of incidence of the light and the emission direction. This is illustrated in Fig. 21.

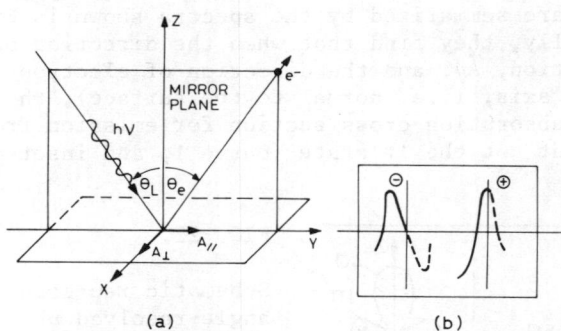

FIG. 21. Schematic representation of a) geometry of polarization dependence of the electric field vector $\vec{A}$ for photoelectron emission in a mirror symmetry plane normal to a surface, and b) illustrating even (+) and odd (-) parity of a (wave) function under reflection in this mirror plane. The dipole transition matrix element $|<f|\vec{A}\cdot\vec{p}|i>|$ preserves parity between the coupled initial and final states depending on the orientation of $\vec{A}$ with respect to the (mirror) symmetry plane.

Consider a mirror symmetry plane M normal to an emitting surface. If we keep the emission direction limited to this plane, the "dipole" matrix element $<f|\vec{A}\cdot\vec{p}|i>$ has to remain invariant under reflection in this plane if parity of the relevant wavefunctions is to be conserved. Now, if the light is polarized such that $\vec{A}$ is normal to this mirror plane $\vec{A}_\perp$ then the matrix element changes sign upon reflection and can be assigned odd parity. This being the case, only quantum states having <u>different</u> parity in Eqn. (9) with respect to emission in this plane can couple in order to preserve reflection invariance. The opposite situation is found for the polarization vector within the plane, $\vec{A}_{//}$; now only states of the <u>same</u> parity are coupled. <u>An important observation is that the final state in an angle resolved photoemission experiment always has even parity for emission in a mirror plane.</u> The reason is that the emitted electron is a plane wave state at the detector with its momentum vector lying in this plane; as such, the wave fronts are orthogonal to the mirror plane so that the parity of $|f\rangle$ is even [35].

For the vector field at some arbitrary direction relative to the emission direction, the components A_x, A_y and A_z can lead to complex "interference effects" since the transition matrix element in Equation (9) contains the square of the sum of such terms.

Similar considerations lead Allyn et al. [33] to perform an angle-resolved polarization dependent experiment on CO adsorbed on Ni(100) in attempt to unravel the adsorbate energy levels, Fig. 20c. The results are summarized by the spectra shown in Fig. 22.

Basically, they find that when the direction of polarization of the radiation, $\vec{A}_{//}$, and the direction of electron emission is along the CO axis, (i.e. normal to the surface), they observed a peak in the absorption cross section for emission from the 4σ and 5σ states, but <u>not</u> the 1π state (curve 1, and inset diagram).

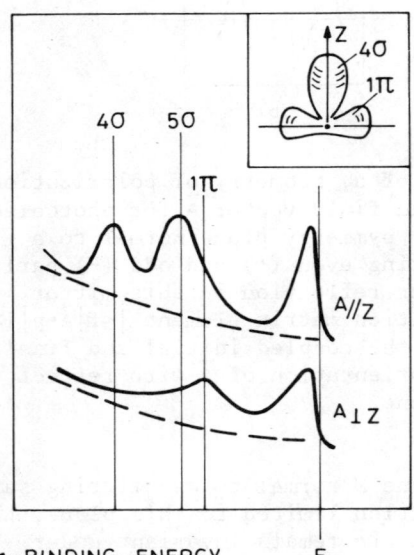

FIG. 22.

Schematic representation of the angle-resolved photoemission results of Allyn et al. (Ref. 33): $\vec{A}_{//}Z$ represents the spectrum obtained for the electric field vector $\vec{A}$ component along the molecular CO axis (i.e. angle of incidence 45°) and emission normal to the surface. $\vec{A}\perp Z$ refers to A normal to the CO axis and emission at 45°. The insert is a sketch of the differential cross sections for photoionization from the "unperturbed" 4σ and 1π orbitals of an isolated, but oriented, CO molecule, Ref. 36.

The 1π state couples mainly to $A_\perp$ (perpendicular to the molecular axis) and emits mostly at large angles from the molecular axis (i.e. curve 2) [36]. The important aspect of these results is: 1) the 1π and 5σ states are now clearly distinguishable, the 5σ level lying below the 1π level as predicted by the cluster calculations [34], and 2) the CO molecule is bonded via its 5σ bond to the surface, which provides independent evidence to the EELS data, Fig. 5, that the CO molecular axis is substantially perpendicular.

This example thus illustrates that the angle and polarization dependence of the photoelectron intensities provide a sensitive means for determining the bonding configuration of adsorbed molecules.

4. ADSORBATE STRUCTURE

4.1 Photoelectron Diffraction.

A resonance in the sigma orbitals' photoionization cross-sections for polarization parallel ($\vec{A}_\parallel$) the molecular CO axis, described above [33] (section 3c), is due to scattering of the excited electrons between the carbon and oxygen atoms. Its strong directional dependence is therefore not wholly surprising if we view it in terms of the excited electron wave undergoing coherent scattering at the oxygen atom i.e. essentially a "photoelectron diffraction effect". This has recently been demonstrated to be the case in angle-resolved X-ray photoemission from the C 1s core-level of CO adsorbed on Ni(100) [37].

Fig. 23a is a schematic illustration of this "*intra*-molecular diffraction effect", showing the extremely forward-peaked nature of the C 1s emission intensity profile, i.e., along the perpendicular oriented molecular CO axis. The extremely forward-peaked single scattering nature of electron atom scattering at XPS energies of $\sim 10^3$ eV is well described by the kinematical model shown in Fig. 23a. The polar angle measurements (Fig. 23b) confirm the up-right orientation of CO chemisorbed on a Ni(100) surface, thereby providing a direct method for determining adsorbate configuration relative to the surface.

Similar measurements of the azimuthal anisotropies in the XPS intensities of deep core levels from atoms adsorbed on single crystal surfaces have also been observed [38] and found to be well described by the single-scattering (i.e. kinematical) model, so that their local structural geometry in relation to neighbouring atomic positions may be deduced.

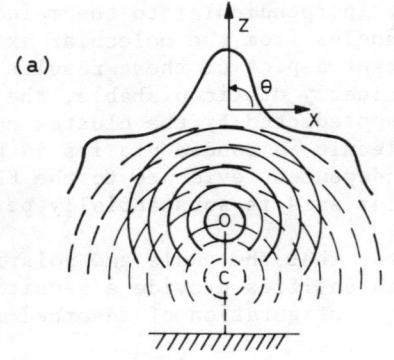

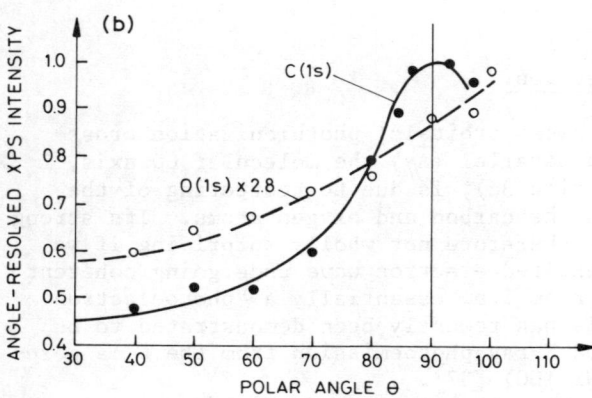

FIG. 23.

(a) Schematic of *intra*-molecular photoelectron diffraction of carbon 1s core-level electrons from chemisorbed CO on Ni(100) showing the kinematical (i.e. single elastic scattering) photoemission intensity expected, and

(b) the polar-angle dependence observed for the absolute intensities of the O1s (dashed) and C1s (full line). From Ref. 37.

4.2 Photoelectron SEXAFS.

A complete explanation of the above "photoelectron diffraction effects" in terms of adsorbate geometry requires detailed scattering calculations. A simpler kind of diffraction effect in photoemission intensity measurements arises if one measures the total (*angle-integrated*) current as a function of incident photon energy. Associated with the photoionization of each core level threshold, modulations in the total current vs photon energy curves can be observed, which arise from processes in which the photoelectron wave is scattered by one of the nearest neighbour atoms directly back into the emitting atom. Subsequent interference occurs between the excited electron's wave and the backscattered waves, as shown in Fig. 24a. Here we have an example of an "*inter*-molecular photoelectron diffraction" process.

Processes of this kind are responsible for the surface-sensitive extended X-ray absorption fine structure (EXAFS) observed at energies extending above each X-ray absorption threshold [39]. If we assume only one kind of neighbouring atom at a distance

$\vec{r}$ from the emitting atom, the $\vec{k}$-periodic factor in the absorption coefficient modulation is given by:-

$$\Delta\alpha(\vec{k}) = A(\vec{k}) \sin [2\vec{k}r + \delta(\vec{k})] \qquad (10)$$

where $\vec{k}$ is the magnitude of the photoelectron wavevector and $\delta(k)$ is the <u>total</u> scattering phase shift. $A(k)$ is an amplitude function which takes into account the probability for electron backscattering and the vibrational fluctuations of the surrounding atoms. The argument of the EXAFS sinusoid contains the interbond distance information.

EXAMPLE 8: InSe Photoelectron SEXAFS [40].
___

This is illustrated by the nature of the oscillations observed in the photoemission intensity associated with the photoionization threshold of the 3d-atomic level of Se in the layered crystal InSe, Fig. 24b. (The Se atoms may be regarded as an "adsorbate monolayer" on an In substrate, as in Fig. 24a). Two series of peaks are observed: series (1) arises from Se → In → Se backscattering; and series (2) arises from Se ⇄ Se intra-monolayer scattering. The wave vector of the photoemitted electron $k \gg |\vec{r}|^{-1}$ is zero at the actual absorption threshold. A least-squares fitting routine is used, together with a variable amplitude function $A(k)$, to determine the <u>total</u> phase shift function $\delta(\vec{k})$ from the data (Fig. 24b) i.e., the "best fit" oscillatory profile with background removed. The important point is that the observed structure in $\Delta\alpha(k)$ provides direct information about $\vec{r}$ <u>without</u> requiring knowledge of the individual k-dependent phase shifts of the outgoing d-wave and the contribution of large numbers of backscattered l waves ($l \geq 0$), as required in the analysis of the *angle-dependent* photoelectron diffraction process (section 4b). (The phase shift problem between scattered partial waves is illustrated, Fig. 24c).

Thus, the relative simplicity in the application of Eqn. 10 is one of the main reasons for the widespread interest in applications of the EXAFS technique. Photoemission intensity measurements (Fig. 24) offer the possibility of applying the technique to surface-specific problems i.e., SEXAFS. An important aspect is that the scattering is <u>local</u> and does not require well-ordered monolayer films or surfaces. This is the case for both photoelectron diffraction processes described above. Hence, they are applicable to disordered films and particulate systems [41].

5. CONCLUDING REMARKS

Surface electron spectroscopy now covers a very wide field of interdisciplinary activities encompassing very many techniques which have provided a wealth of information on surface and mole-

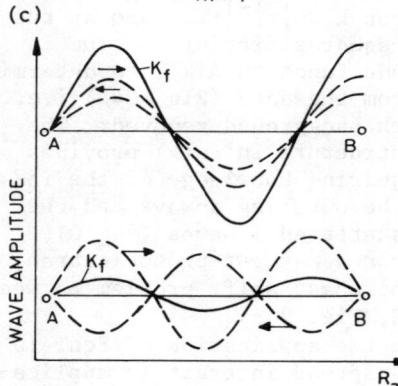

FIG. 24.

(a) Showing *inter*-molecular interference between an electron wave excited on an adsorbate atom (filled circles) and wavelets back-scattered from adjacent and substrate atoms (open circles).
(b) The resultant total (angle integrated) photoemission (SEXAFS) oscillations observed following photoionization of the Se(3d) (and In(4d)) levels in indium selenide, Ref. 40.
(c) Illustrating the phase dependence of the interference between waves scattered between adjacent atoms A and B on the amplitude of the resulting wave field representing electron emission from the atom A.

cular electron states and excitation processes [1]. These methods involve excitation of electrons by photons, ions, electrons, high-level electric fields etc. The methods we have chosen to discuss in these lectures have evolved only within the past few years and are particularly important for the study of adsorbate systems for two principle reasons: 1) they provide fundamental *spectroscopic* information on the vibrational and electronic states, and 2) they provide *structural* information on the geometry and structure of the adsorbate-substrate system. This is prerequisite for any understanding of systems which can be used to model our ideas pertaining to more complex (and practical) catalytic systems.

The photoelectron diffraction SEXAFS technique holds particular promise in this area. To date, there have been few such measurements on adsorbate systems since the photon frequency must be swept over a continuum range of energy using synchrotron radiation as the light source. As such sources become more available, application of this sort will grow. In the meantime, surface electron spectroscopy will continue to develop and deepen our understanding of fundamental chemisorption processes. This author feels encouraged to continue his efforts in this field and the hope is that he may have stimulated (some) members of his audience likewise !

1. GENERAL REFERENCE SOURCES:

- *Low Energy Electrons and Surface Chemistry*, G. Ertl and L. Kuppers (Verlag Chemie, 1974).
- *Electronic Structure and Reactivity of Metal Surfaces*, ed. E.G. Derouane and A.A. Lucas (Plenum, 1976).
- *Treatise on Solid State Chemistry*, (Vol. 6, Surfaces), ed. N.B. Hannay (Plenum, 1976).
- *Electron Spectroscopy: Theory, Techniques and Applications*, (4 Vols.), ed. C.R. Brundle and A.D. Baker (Academic, 1977).
- *Electron Spectroscopy for Surface Analysis*, ed. H. Ibach (Springer-Verlag, 1977).
- *Electron and Ion Spectroscopy of Solids*, ed. L. Fiermans, J. Vennik, and W. Dekeyser (Plenum, 1978).
- *Photoemission and the Electronic Properties of Surfaces*, ed. B. Feuerbacher, B. Fitton and R.F. Willis (Wiley, 1978).
- *Infrared Spectra of Adsorbed Species*, L.H. Little (Academic, 1966).

REFERENCES:

2. J.C. Rivière, Contemp. Phys. $\underline{14}$, 513 (1973).
3. This approach was first introduced by: N.V. Richardson and A.M. Bradshaw, in Proc. Int. Conf on *Vibrations in Adsorbed Layers*, Jülich, KFA-Jülich Publication (1978).
4. R.P. Eischens, S.A. Francis and W.A. Pliskin, J. Phys. Chem. $\underline{60}$ (1956) 194.
5. M.A. Chesters, J. Pritchard and M.L. Sims, Chem. Commun. (1970) 1454.
6. S.A. Francis and A.H. Ellison, J. Opt. Soc. Amer. $\underline{49}$, 131 (1959); R.G. Greenler, J. Chem. Phys. $\underline{50}$, 1963 (1969).
7. A.A. Lucas and M. Sunjić, Progr. Surface Sci., $\underline{2}$, 75 (1972); D.L. Mills, Progr. Surface Sci. $\underline{8}$, 143 (1977).
8. D.M. Newns, Phys. Letters $\underline{60A}$, 461 (1977); B.N.J. Persson, Sol.State Commun. $\underline{24}$, 573 (1977).
9. E. Evans and D.L. Mills, Phys. Rev. $\underline{B5}$, 4126 (1972).

10. F. Delanaye, A.A. Lucas and G.D. Mahan, Surf. Sci. 70, 629 (1978).
11. H. Ibach, Surf. Sci. 66, 56 (1977).
12. S. Andersson, Proc. 2nd European Conf. Surf. Sci., Cambridge, U.K., 1979, to be published.
13. H. Froitzhiem, in *Electron Spectroscopy for Surface Analysis* (Springer Verlag, 1977), p. 205.
14. D. Roy and J.D. Carette, in *Electron Spectroscopy for Surface Analysis* (Springer Verlag, 1977), p. 13.
15. N.V. Richardson, Surf. Sci. to be published.
16. S. Lehwald, H. Ibach and J.E. Demuth, Surf. Sci. 78, 577 (1978).
17. S. Lehwald, W. Erley, H. Ibach and H. Wagner, Chem. Phys. Letters, in press.
18. C. Backx, R.F. Willis, B. Feuerbacher and B. Fitton, Surf. Sci. 68, 516 (1977).
19. R. Ray and G.D. Mahan, Phys. Letters 42A, 301 (1972).
20. W. Ho, R.F. Willis and E.W. Plummer, Phys. Rev. Letters 40, 1463 (1978).
21. R.F. Willis, to be published.
22. S. Lehwald and H. Ibach, to be published.
23. R.L. Dennis and M.B. Webb, J. Vac. Sci. and Technol. 10, 192 (1973); M.B. Webb and M.G. Lagally, Sol. State Phys. 28, 301 (1973).
24. N.V. Richardson, private communication.
25. J.W. Gadzuk, Surf. Sci. 6, 133 (1967); *ibid* 43, 44 (1974).
26. J.W. Gadzuk, and E.W. Plummer, Rev. Mod. Phys. 45, 487 (1973).
27. For a review see: E.G. McRae and H.D. Hagstrum in *Treatise on Solid State Chemistry, Vol. 6A, "Surfaces"*, Chap. 2, ed. N.B. Hannay (Plenum, 1976).
28. Theory: N.V. Smith and L.F. Mattheiss, Phys. Rev. Letters 37, 1494 (1976).
    Experimental: B. Feuerbacher and R.F. Willis, Phys. Rev. Letters 36, 1339 (1976).
29. J.E. Demuth and D.E. Eastman, Proc. 2nd Int. Conf. on Solid Surfaces, 1974, Jap. J. Appl. Phys. Suppl. 2, Pt. 2, 1974.
30. D.W. Turner, C. Baker, A.D. Baker, and C.R. Brundle, *Molecular Photoelectron Spectroscopy* (Interscience, 1970).
31. W.E. Spicer, in *Electron and Ion Spectroscopy of Solids*, Chap. 2, ed. L. Fiermans, J. Vennik and W. Dekeyser (Plenum, 1978).
32. U. Gelius, E. Basilier, S. Svensson, T. Bergmark and K. Siegbahn, J. Elec. Spec. 2, 405 (1973).
33. C.L. Allyn, T. Gustafsson and E.W. Plummer, Chem. Phys. Letters 47, 127 (1977).
34. I.P. Batra and P.S. Bagus, Solid State Commun. 16, 1097 (1975).
35. J. Hermanson, Solid State Commun. 22, 9 (1977).
36. E.W. Plummer, Proc. 7th Intern. Vac. Congr. and 3rd Intern. Conf. Solid Surfaces, Vienna, p. 647 (F. Berger and Söhne, Vienna 1977).

37. L.G. Petersson, S. Kono, N.F.T. Hall, C.S. Fadley and J.B. Pendry, Phys. Rev. Letters $\underline{42}$, 1545 (1979).
38. S. Kono, C.S. Fadley, N.F.T. Hall and Z. Hussain, Phys. Rev. Letters $\underline{41}$, 117 (1978); *ibid* 41, 1831 (1978).
39. E.A. Stern, Phys. Rev. $\underline{10B}$, 3027 (1974).
40. G. Margaritondo and N.G. Stoffel, Phys. Rev. Letters $\underline{42}$, 1567 (1979).
41. For example: D.E. Sayers, E.A. Stern and F.W. Lytle, Phys. Rev. Letters $\underline{35}$, 584 (1975).

ACKNOWLEDGEMENT:

I would like to thank those authors who were kind enough to allow me to include references to their unpublished, as well as, published work. In particular, S. Andersson, H. Ibach and N.V. Richardson.

SURFACE COMPOSITION OF BINARY ALLOYS*

W.M.H. Sachtler

Koninklijke/Shell-Laboratorium, Amsterdam
(Shell Research B.V.)

ABSTRACT. The specific catalytic properties of bimetallic catalysts are correlated with the surface composition of the alloy particles. Only in exceptional cases are the compositions equal for surface and bulk of an alloy; more frequently one component will be enriched in the surface. The physical and chemical phenomena responsible for this segregation are described in a qualitative manner. Some advices are given how the surface composition can be estimated also for supported industrial catalysts, systems for which at present theoretical formulas are inadequate, while experimental techniques are faced with serious problems.

1. INTRODUCTION

In recent years a tendency has become noticeable in industrial processes, such as catalytic reforming, to replace the monometal catalysts by preparations which contain, besides an oxidic and - in some cases - acidic support, a combination of two or more metals. The attractiveness of this new generation of catalysts consists in their superior stability, as evidenced by a lower rate of decline of the catalytic performance, and selectivity. In particular undesired side-reactions, such as hydrogenolysis and formation of "coke", are often suppressed on such catalysts, resulting in higher selectivities for the desired processes of (de-)hydrogenation, isomerization and (dehydro-)cyclization [1, 2,3]. Since the same phenomena have been observed experimentally with unsupported binary alloys of the same metals [4], it seems

---

* Some parts of this paper have been taken from Ref. [15], with kind permission of the publisher.

justified to conclude that the above characteristics of industrial bi- or multi-metallic catalysts are to a considerable extent caused by the formation of alloy particles in these catalysts.

A rationalization of the typical catalytic performance of bimetallic catalysts requires a sound knowledge of the surface composition of the alloy particles. This statement is related to three basic phenomena, at some time regarded as axioms, which we shall briefly mention first.

1.1. The axiom of the individual surface atom [5]

As shown schematically in Fig. 1, an atom X adsorbed on the surface of an alloy exposing atoms A and B, can form a number of complexes. When confining ourselves to those complexes where X forms a bond with one surface atom only, we assume that the bond X-A is different from the bond X-B. This assumption might appear trivial, but is at variance with an older hypothesis that described chemisorption and catalysis in terms of "collective" parameters of the adsorbing solid, in particular the concentration of quasi-free electrons, or the filling of the d-band. These parameters had been used successfully to describe cooperative phenomena such as ferromagnetism or electric conductivity,

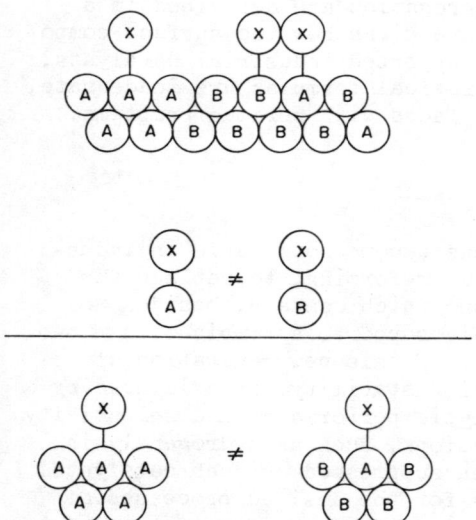

Fig. 1. Chemisorption complexes formed by the "on top adsorption" of an X atom on the surface of a binary alloy AB_X, illustrating the "individual surface atom approach" and the "ligand effect" concepts (schematic).

phenomena which by their very nature are not even definable for an isolated atom. Assuming that such collective parameters provide an exclusive description of chemisorption and catalysis, there was no reason to consider chemical bonds with individual surface atoms. This "collective" approach to catalysis has, however, been abandoned in recent years for a number of reasons; these include (1) the progress made in the physics of alloys, which disproves the "rigid band model" [6], (2) characterizations of adsorption complexes on metals showing the often highly localized nature of chemisorption bonds [7], and (3) results obtained on alloys, which showed, for instance, that a CO molecule or a hydrogen atom adsorbing on a PdAg surface clearly distinguishes between Pd atoms and Ag atoms [5].

1.2. The ligand effect [2,8]

In Fig. 1. two of the three X atoms shown are bonded to an A atom. In one case this A atom is surrounded by other A atoms in and below the surface, while in the other case all the atoms adjacent to this A atom are B atoms. Bearing in mind that in metal-organic complexes the bond between the metal atom and a given ligand is significantly influenced by the other ligands (they might be electron-donating or electron-accepting e.g.) one must expect that also for chemisorption complexes the characteristics describing the chemisorption bond should be influenced by the nature of the surrounding atoms. This "ligand effect" provides a second reason why a good understanding of alloy catalysis should be based on a knowledge of the composition of the surface; it, moreover, illustrates the desirability of knowing the concentration "depth profile", i.e. the composition of the first few subsurface layers.

1.3. The ensemble effect [2,8]

Many physical data collected over the last twenty years have shown that a given gas can form a number of discernible chemisorption complexes. Even on a well-defined crystal surface a molecule can form a distinct number of complexes. Often they show very similar heats of adsorption; by consequence, different complexes will, in equilibrium, coexist on the same surface. Each of them can be the entrance port to a catalytic reaction, so the occurrence of different reactions simultaneously proceeding on the same catalyst can be ascribed to the existence of these different complexes.

In simple cases these complexes differ by the number of surface atoms involved. This is schematically illustrated in Fig. 2. A multi-atomic molecule can be visualized to be adsorbed on one surface atom, forming, for instance a π-complex or - in the case of oxygen - a superoxide ion [9,10]. The molecule can dissociate, each atom forming a localized bond with one surface atom, or a "bridging bond" with two surface atoms. If all the surface atoms are chemically identical, each of these complexes can be formed, provided that the thermodynamic and kinetic conditions are fulfilled, which then determine the relative concentrations of each of the coexisting complexes. For an alloy surface, however, the availability of the required "ensembles" of surface atoms becomes an additional prerequisite. If, in the simplest possible case, the surface consists of two kinds of atoms, only one of which is capable of forming any bonds with the adsorbate considered, the concentrations of each complex will, of course, decrease with the concentration of the active alloy component in the surface, but this decrease will be most pronounced for the complex requiring the largest ensembles. If the atoms in a diluted alloy are distributed statistically, the concentration of pairs will decrease with the second, but that of four-membered ensembles with the fourth power of the surface concentration and this can result in a dramatic change of the catalytic selectivity.

1.4. The surface free energy

While the above considerations explain the interest of the catalytic chemist in the surface composition of an alloy, the question can be raised why this cannot simply be equalized with the known overall composition of the alloy components. Indeed one can imagine that an alloy single crystal when broken at liquid-

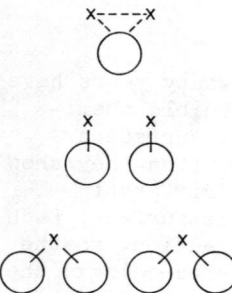

Fig. 2. Chemisorption complexes formed by the adsorption of the molecule X_2 on the surface of a metal, illustrating that complexes can differ in the size of the required "ensembles" of surface atoms (schematic).

hydrogen temperature under ultra-high vacuum, will expose new
surface facets with a composition equal to the interior of the
original crystal.

An alloy surface, acting as catalyst will, however, undergo
a huge number of adsorption and desorption events. The formation
and rupture of chemical bonds, being accompanied by corresponding
changes in the interactions between the surface atom and its
neighbours in and below the surface, is known to facilitate
surface reconstruction [11]. Therefore, the surface composition
of a working catalyst in the steady state is necessarily the
composition corresponding to the lowest free energy of the system.
This system includes, of course, besides the metal atoms, the
ambient medium.

More than a century ago it was shown by Gibbs [12] that the
free energy of any condensed phase - solid or liquid- includes a
term characteristic of the surface and, therefore, changes upon
changing the surface area, for instance by cutting a solid into
smaller particles. In the modern chemical approach it is conceptually convenient to associate the surface (free) energy with the
existence of atoms which are coordinatively unsaturated. In a
simple idealization, the energy of atomization, E_{at}, i.e. the
energy necessary to transform one gramatom of a surface-less
metal into a monoatomic gas is put equal to the product of the
number of bonds between closest neighbours in the metal crystal,
and the energy E per bond:

$$E_{at} = \frac{m+l}{2} E \qquad (1)$$

where $(m+l)$ is the coordination number. We can then characterize
the surface atoms by the number of "dangling bonds" q, which is
simply equal to the number of missing neighbours. If l is the
number of neighbours inside the plane and m is the number of the
neighbours off-plane for an atom in the interior of the crystal,
the number of dangling bonds of a surface atom in this plane is
simply

$$q = \tfrac{1}{2} m \qquad (2)$$

If, furthermore the surface free energy is interpreted in the
usual way as being due to a surface enthalpy and an entropy term,
one can associate the surface enthalpy with the product of qE and
the number of atoms in the surface. Minimization of this term for
a crystal of given size then permits one to understand the crystal
shape, which is a compromise of the tendency to minimize the
surface area and to minimize the number of surface atoms with
large values of q (Gibbs-Wulff principle [13].

For an alloy and, more generally, any mixture of more than one component, it is essential that surface enthalpy and surface free energy become functions of the surface composition. If we consider, again, a crystal which we cut in two pieces at such a low temperature that no surface reconstruction occurs, the composition will initially be equal to the composition of the interior. In general, this system is not in a state of lowest free energy, since reconstructions can be carried out which enrich the surface with the component lowering the surface free energy. This might be the component with lowest E and, consequently, lowest E_{at}. This enrichment of the surface will, however, results in a change of the composition of the interior. If the interior is of infinite volume, all the atoms have equal size and the alloy is an ideal solid solution, it will suffice to consider the entropy change due to the concentration difference of surface and interior. This is done in the so-called "broken-bond model". Since these conditions are, however, seldom fulfilled, the energy effects of changing the composition of the interior must be considered. In particular with ordered alloys forming Daltonide phases where each atom sort forms a well-defined sublattice, a change in composition of the interior is only possible by putting atoms in the wrong sublattice, thus increasing the energy of the system. Further, it must be realized that in the presence of a gas which is capable of forming strong bonds with one of the alloy components only, there is a driving force for enriching the surface with that particular element. Similarly, the interface between the alloy particle and the solid support may segregate one alloy component. It is therefore clear that the equilibrium composition of an alloy surface depends on a great number of factors. Quantitative predictions are hampered by the lack of numerical data of the parameters and functions involved. A quantitative theory for the surface composition of alloys in equilibrium should be based on a knowledge of:

    surface enthalpy
    surface entropy
    dependence of either of these on
        composition
        temperature
        crystal face
        particle size
        gas atmosphere
        interaction with support.

At the present time such a comprehensive theory does not exist. Recent review articles [2,14,15,16,17,18] give the present state of theories, based on simplifying models which focus on one phenomenon but disregard others. In the present paper these models are described in a qualitative way in order to enable the catalytic chemist to make a reasonable guess for a novel bimetallic catalyst as to what kind of surface segregation might be expected. We shall also briefly mention which methods can be used to obtain information on the surface composition of alloys in the laboratory. We shall make use of the following definitions:

x = composition of element (1) in the interior
y = composition of element (1) in the surface

$$\frac{y}{1-y} \left(\frac{x}{1-x}\right)^{-1} \equiv \chi \tag{3}$$

χ is called the "surface enrichment factor"

Applying Boltzmann's law to this:

$$-RT \ln \chi = \Delta F_a \tag{4}$$

defines the "Free energy of Gibbs adsorption", which is not necessarily independent of x and y.

2. THEORETICAL AND PHENOMENOLOGICAL APPROACHES

2.1. Broken-bond approximation

The broken-bond model approximates the unknown ΔF_a in Equation (4), by the difference in heat of atomization and usually all lattice energies are approximated by first neighbour interactions.

For a semi-infinite regular solution [19,20,21,22] the following expression is derived, if enrichment is assumed to occur in the outer plane only:

$$-kT \ln \chi = \frac{m}{4}(E_2 - E_1) + \alpha \{(2\ell + m)x - 2\ell y - \frac{m}{2}\} \tag{5}$$

where: E_1, E_2 are bond energies between nearest neighbours of components 1 and 2, respectively,
$E_{1\,2}$ = bond energy between nearest neighbours 1 and 2,
α = $E_{1\,2} - \frac{1}{2}(E_1 + E_2)$,
ℓ = number of nearest neighbour bonds per atom in the plane parallel to surface plane,
m = number of neighbour bonds per atom in the bulk outside the plane parallel to the surface.

Expressions accounting for enrichment in more than the outermost layer can be found in [20,23-26].

To apply Equation (4) to an actual alloy one has to derive the bond energies and α from experimental values. If Q denotes the heat of formation of the alloy, α can be determined from:

$$\alpha = \frac{1}{\ell + m} Q \tag{6}$$

The bond energies E_1 and E_2 can be derived from the atomization energies by Equation (1).

We would prefer values of E derived from the surface free energies. Since, however, experimental information on surface free energies is very limited, the question of the proportionality factor between E_1 and E_{at} is still not completely solved [27,28]. This is important since small changes in the proportionality factor influence the enrichment calculated strongly.

Formula (5) shows that the enrichment factor depends on the bond energy difference of the two components, the heat of formation and surface unsaturation.

Two limiting cases are interesting: If metals are chemically different, but have equal heats of sublimation, enrichment is governed by the last term, which becomes zero for $x = y = \frac{1}{2}$. This is the "symmetrical" case, where 1 lowers the surface energy of 2 and vice versa. In the surface one finds enrichment of the minority component, but no enrichment for a 50:50 alloy.

The second limiting case is that of the ideal solution where $\alpha = 0$. The component of lower heat of atomization will be found enriched on the surface and the degree of enrichment is highest on the higher index faces and therefore on edges and corners. The surface composition changes with the composition of the interior and the enrichment factor is constant. Surface entropy differences are neglected in this model. According to Kelly [17] this neglect does not lead to serious deviations.

A qualitative experimental verification of these phenomena was given in 1969 by Bouwman et al. in a study of Pd-Ag alloy films [29] which were chosen because the phase diagram for Pd-Ag unlike Ni-Cu and Pt-Au has no miscibility gap. The changes in work function of Pd-Ag films after chemisorption of carbon monoxide showed that the Pd concentration of the surface increases with that of the bulk (see Fig. 3), in contrast to other alloy systems (vide infra) where in a wide region the surface concentration is independent of the overall composition. The work function change due to CO was much smaller for alloys than for Pd, suggesting that the alloy surfaces after equilibration became enriched with silver.

Surface composition of binary alloys

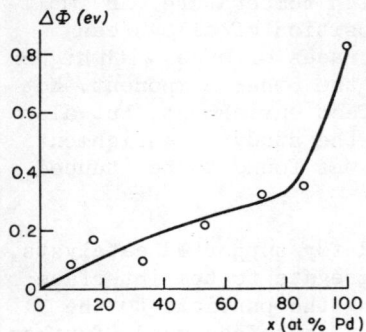

Fig. 3. Increase in work function $\Delta\phi$ of Ag-Pd films, caused by CO chemisorption, plotted against overall composition x [29].

2.2. Influence of gas atmosphere

If gas phase molecules are selectively chemisorbed by atoms of element 1, this will lower the chemical potential of component 1. So chemisorption provides a driving force to enrich the surface with the element forming the strongest adsorption bonds. This phenomenon has been called chemisorption-induced surface segregation [29]. In the case of Pd-Ag alloys, e.g., CO molecules are selectively chemisorbed on Pd atoms, so these atoms segregate to the surface. This is shown in Fig. 4. The phenomenon is found

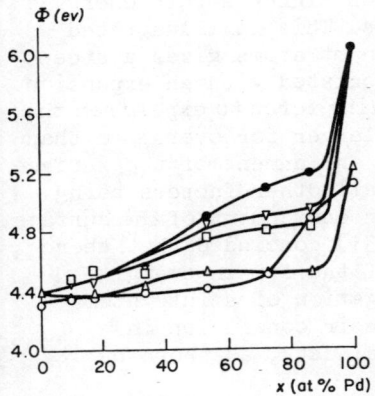

Fig. 4. Work function of Ag-Pd films, plotted against overall composition [29]; o freshly evaporated film, △ equilibrated film, □ immediately after admission of CO ($p = 10^{-4}$ Torr), ▽ after 16 h exposure to 10^{-4} Torr CO at 293 K, ● after 16 h exposure to 10^{-4} Torr CO at 273 K.

to be reversible upon desorbing CO at high temperature. An implication is that data on the surface composition of alloys can easily be falsified by the presence of traces of a gas with higher affinity to one alloy component than to the other component. Not only impurities of the gas phase can affect enrichment, but also impurities stemming from the sample. In the study of enrichment of Au-Cu alloys strong enrichment of Cu was found to be induced by segregation of sulphur [30].

A similar effect can be anticipated for supported catalysts, where one component of the alloy may segregate to the interface and become effectively a "glue" attaching the particle to the support. Analysis of the particle support interface may be further complicated by lattice mismatch between the metal particle and, typically, oxide support.

2.3. Size effects

Superimposed on the electronic effects determining the heat of formation of alloys are size effects causing elastic strain. All other things being equal, a lattice can relax by precipitating impurities of different size on grain boundaries. These, incidentally, can be many atomic layers thick. There are, however, two reasons for assuming that strain effects are not really symmetrical but oversized atoms are preferentially segregated at the surface.

(1) Oversized atoms implanted in an ideal crystal lattice will cause a compression leading to a much higher strain energy than that caused by undersized atoms. This is illustrated schematically in Fig. 5. Compression of atoms gives a steep rise in energy while the energy associated with an expansion is smaller. By consequence, the driving force to expel from the lattice atoms of deviating size is larger for oversized than for undersized atoms. The resulting enrichment of the surface will therefore be asymmetric, i.e. all other factors being equal, lattice strain will favour an enrichment of the surface with oversized atoms. Tsai et al. [31] concluded that there is good agreement with experiment if the strain energy relaxation is only applied to segregation of solute atoms that are larger than the solvent. Their conclusion is supported by a wealth of experimental data, as shown by Seah [18].

(2) With equal coordinative unsaturation per surface atom, the energy of the surface per unit surface area can be higher if a given surface area is occupied by many small atoms than if the same area is filled with a smaller number of larger atoms. This explains, for instance, why Au has a lower surface energy

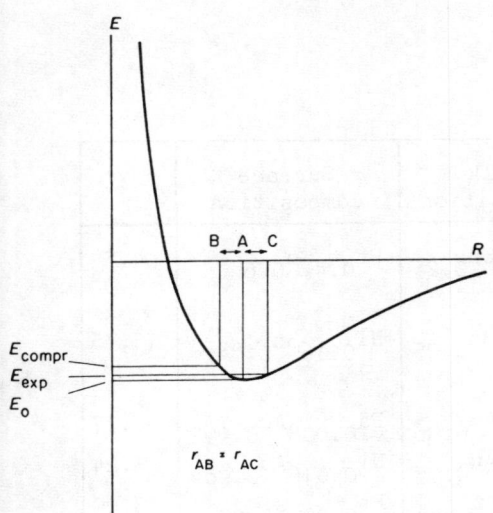

Fig. 5. Changes in potential energy upon compression or extension of relative atomic distance.

than Cu, although Au has the higher heat of atomization [30,32]. Miedema [27] has used this principle to explain heat in copper-gold alloys gold enrichment occurs over a broad concentration range [33,34]. Fewer Au atoms are needed than Cu atoms to occupy the surface of a copper-gold alloy. In this case the enrichment is accompanied by a large rearrangement of surface atom positions.

The size effects might explain the much larger enrichment factor found recently for Ni-Sn alloys than for Pt-Sn alloys (Table I) [35]. While the difference in sublimation energy between Sn and Pt is larger than that between Sn and Ni (E_{Sn} = 12 kcal/gat, E_{Pt} = 20 kcal/gat, E_{Ni} = 17 kcal/gat), the average atomic surface difference between Ni and Sn is much larger than that between Pt and Sn (a_{Sn} = 4.38 Å^2, a_{Pt} = 4.33 Å^2, a_{Ni} = 3.89 Å^2). The χ factor of 15.75 in Table I, therefore might reflect a large strain effect in Ni_3Sn.

McLean [36] has computed surface segregation on the basis that only strain energy release provides the driving force, using for ΔF_a in Equation (4):

$$\frac{-24 \pi KG r_0 r_1 (r_0 - r_1)^2}{3 K r_1 + 4 G r_0} \tag{7}$$

TABLE I

ENRICHMENT RATIO χ

Alloy	Method	Bulk composition	Surface composition	χ
Pt_3Sn	CO_{ads} [a)] LEIS	$Pt_{0.75}Sn_{0.25}$	$Pt_{0.4}Sn_{0.6}$	4.5
Ni_3Sn	CO_{ads} LEIS	$Ni_{0.75}Sn_{0.25}$	$Ni_{0.16}Sn_{0.84}$	15.75
sol.sol.PdAu	LEIS	$Pd_{0.04}Au_{0.96}$	$Pd_{0.008}Au_{0.992}$	5.17
sol.sol.NiAu	LEIS	$Ni_{0.05}Au_{0.95}$	$Ni_{0.016}Au_{0.984}$	3.24
sol.sol.PtAu	CO_{ads}	$Pt_{0.14}Au_{0.86}$	$Pt_{0.05}Au_{0.95}$	3.09

a) Ref. [40], all the other [35].

where K is the solute bulk modulus, G is the solvent shear modulus and r_0 and r_1 are the appropriate radii for solvent and solute atoms in their pure states, respectively. A second treatment [37] use the volumes of the elements in their own pure form instead of a final radius and is therefore somewhat more convenient:

$$E_e = \frac{2}{3} \frac{G_m K_i}{3 K_i + 4 G_m} \frac{(V_i - V_m)^2}{V_m} \tag{8}$$

where V_m and V_i are the atomic volumes of the solvent and solute, respectively.

These elastic strain models should be used cautiously, however, since they assume a dilute solution so that solute-solute effects can be neglected. While the bulk metal in a segregation experiment may indeed be described as a dilute solution, a surface which is almost fully covered by solute clearly cannot. Further, keeping lattice registry with the bulk requires that the solute atoms be strained to the same size as the solute and/or a defect structure be introduced near the surface to accommodate the misfit. While such an analysis is beyond the present scope, it is evident that the strain energy released for a concentrated surface will be less than what is estimated from these simple elastic calculations.

Burton and Machlin [38] tested the prediction of the elastic strain theory and they found a very poor correlation with experiment. Part of this poor agreement may result from the fact that Equation (7) also contributes to the heat of formation of the alloy and the contribution of strain energy release in the heat of formation should be properly included in the calculation.

On the other hand, it is probably better to include the driving force to enrichment calculated according to the broken-bond model into ΔF_a as well. This has been done by Wynblatt and Ku [39]. They find excellent agreement with their experimental data, if they also include changes in partial entropy.

2.4. Two-phase systems

Endothermic alloys have a miscibility gap. The Pt-Au and Ni-Au alloys are examples. At temperatures where the miscibility for these alloys is limited, however, homogenous alloys are formed only for a restricted range of concentrations. Surface enrichment is determined by the laws discussed in previous sections. A very different situation arises if the relative concentrations of the components of a binary system e.g. are within the miscibility gap. Establishment of equilibrium results in a two-phase system. Each crystallite consists of a kernel enveloped in a skin of a different alloy. The concentrations of kernel and skin and their composition can be derived from the phase diagram using the lever rule. Which of the two phases will form the outer envelope is determined by the surface energy of that phase. This, so-called cherry model [8] describes surface enrichment as the presence of three concentric spheres:

the kernel = alloy phase with higher surface energy
the flesh = alloy phase with lower surface energy
the skin   = surface layers enriched with respect to "flesh" because of any of the phenomena described in the preceding sections.

The implication of this arrangement is that the surface composition remains constant over almost the full range of immiscibility. Changes in overall composition affect primarily the relative sizes of flesh and kernel.

Chemisorption of a gas will induce a surface segregation as in the case of monophasic alloys. For the Pt-Au situation this is illustrated in Figure 6, where the hydrogen adsorption per unit surface area of Pt-Au films after sintering is plotted versus the alloy composition. As the surface area was determined from the physical adsorption of xenon, the ratio of H_{ads}/Xe_{ads} is plotted versus the overall composition.

2.5. Ordered alloys

So far, the depth profile has not been considered. Intuitively one might expect a smooth transition from the highly enriched surface layer to the interior. This is indeed the case for solid solutions that are endothermic [20].

In the case of ordered exothermic alloys [19] we showed by a simple theoretical calculation that enrichment should occur by "inversion" of the concentration gradient, i.e. exchange of atoms in the surface layer with atoms in the layer second to the surface. The two outer layers, taken together, have the same stoichiometry as the interior.

This result follows by considering only short range order. The simple chemical principle which is valid here can be expressed as the tendency, for ordered exothermic alloys, of each atom of a given element to attract preferentially atoms of the other element in its vicinity. It has, therefore, been speculated that the inversion might be repeated several times i.e. atom layers enriched with element 1 should alternate with atom layers

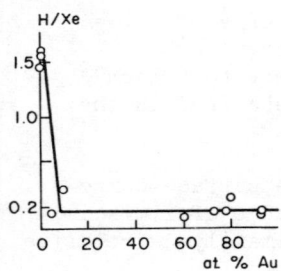

Fig. 6. Surface composition of Pt-Au films, derived from the selective chemisorption of hydrogen on the surface platinum atoms and plotted against overall composition (Ref. [42]).

$$H/Xe = \frac{\text{number of chemisorbed H atoms}}{\text{number of adsorbed Xe atoms}}$$

enriched with element 2, the concentration profile being thus similar to a damped oscillation, but experimental confirmation for such a multiple inversion is still lacking.

Electron spectroscopy methods, such as Auger electron spectroscopy (AES) and X-ray photoelectron spectroscopy (XPS) have often been used to obtain information on the surface composition of alloys. However, in these methods subsurface atoms contribute significantly to the measured signal, and this leads to serious errors, in particular if the ln χ of the outermost and following layers are opposite in sign.

The depth profile of concentration can, however, be estimated by combining methods with different contributions of subsurface layers. Chemisorptive titration by selective adsorption of hydrogen or carbon monoxide, and low energy ion scattering provide information on the outermost layers, while the contribution of subsurface layers to electron spectroscopy increases with increasing escape depth of the electron whose energy is recorded. This type of titration has been done with the compounds Pt_3Sn and Ni_3Sn, systems for which there is no doubt that the surface is enriched by tin (Table I) [36]. Chemisorptive titration with CO shows very strong enrichment [40]. AES, which contains a non-negligible contribution from subsurface layers, shows less enrichment with Sn, and XPS, characteristic for a still larger escape depth of the signal, a still smaller enrichment. It was found that no consistency can be obtained for a smooth profile, but consistency was obtained for inversion [41] assuming that the calculated outermost layer was better in agreement with chemisorption data. The analysis of the experimental data showed, however, inversion between layers deeper in the solid than predicted by the model, which considers only interactions between nearest neighbours.

2.6. Guesstimates and experimental determinations

While further refinement of the models sketched is a challenging task for surface science, the catalytic chemist will often be obliged to make a reasonable estimate of the surface composition of a particular bimetallic catalyst for which precise data are not available in the literature. Table II gives a schematic survey of the phenomena described above, so that the enrichment factor can be estimated from handbook data. Segregation will, of course, be large if the broken-bond model, the elastic strain model and the chemisorption induced surface segregation predict enrichment of the surface with the same component. If the various driving forces counteract each other, however, it will be of help to know their relative strengths. It is the author's experience, that in most catalytic systems where a gas is strongly chemisorbed to one alloy component, the chemisorption-induced surface segregation predominates, in other words, this phenomenon decides which alloy component is concentrated in the surface of a working catalyst. This is particularly important when sulphur is part of the ambient atmosphere.

A more quantitative formula has been proposed by Seah [18], who limited himself to the case of binary, diluted, monophasic solid solutions, disregarding chemisorption effects, metal/support interactions and composition changes in subsurface layers. He arrived at a semi-emperical formula of the type:

$$RT \ln \chi = A\, T_{m,1} - T_{m,2} + BQ + C\, V_2(V_1 - V_2)^2 \tag{9}$$

where $T_{m,1}$ and $T_{m,2}$ are the absolute melting temperatures of solute and solvent, V_1 and V_2 are their atomic volumes and A, B, C are numerical coefficients, with the proviso that C is part equal to 0 when the atomic volume of the solvent is larger than that of the solute.

1. QUINTO, SUNDARAM, ROBERTSON; [47]
   ERTL, KÜPPERS (1971, AES 800 eV) [48]
2. TAKASU, SHIMIZU (1973, AES 800 eV) [49]
3. WATANABE, HASHIBA, YAMASHINA (1976, AES 100 eV) [50]
4. VAN DER PLANK, SACHTLER (1968, H_2 ads.) [51]
5. BRONGERSMA, BUCK (1975, LEIS) [52]
   KUIJERS, PONEC (1977, AES DEPTH PROFILE) [53]
   HELMS (1975, AES 100 eV) [54]

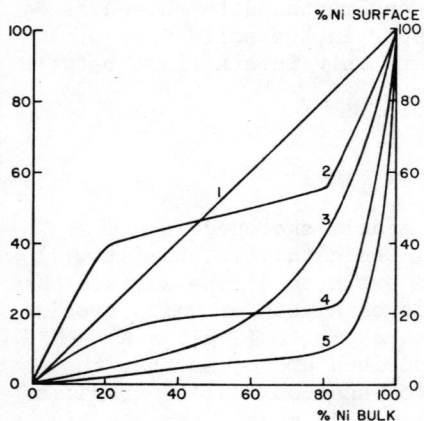

Fig. 7. Literature data of the surface composition of Cu-Ni alloys
ordinate: $(\frac{Ni}{Ni + Cu})_{surface}$; abscissa: $(\frac{Ni}{Ni + Cu})_{overall}$.

Where the a priori estimate is unsatisfactory, the catalytic chemist will wish to measure the surface composition experimentally. This is not easy even for well defined films and single crystals, as is illustrated by the data published by various authors on the most studied system, i.e. Cu-Ni. The data in Fig. 7 show that older Auger data did not reveal any surface segregation of copper, while the work function results, the data obtained by low energy ion scattering and the modern Auger data which correct for the large contribution of subsurface layers leave little doubt that enrichment with Cu is very pronounced.

For highly diluted and supported catalysts the only method easily applicable is chemisorptive titration, which requires, however, two gases, one counting all metal surface atoms, while the other method should count specifically one atom sort only. The method has, however, a number of problems. If chemisorption is dissociative, as in the case of hydrogen, it is not obvious whether isolated Pt atoms, e.g. completely surrounded by Au atoms, will be able to adsorb hydrogen atoms. For gases which can be adsorbed in two or more different states, the relative population of these states may be different for surfaces consisting of the adsorbing atoms only and for surfaces where these atoms are interdispersed between other atoms. A typical example are the "multi-site" or "bridged" complexes, for instance of CO and Pd [43]. Even atoms pairs of the active and the inactive element may be considered as potential sites for a bridged complex [44]. Further, the "ligand effect" can alter the partial heat of adsorption on the active metal and hence also lead to a variable degree of coverage of these atoms depending on their environment in and below the surface. Finally, we have mentioned already the phenomenon of "chemisorption induced surface segregation" which leads to an enrichment of the surface with the element of highest affinity towards the adsorbate and thus alters the surface composition.

The chemisorptive titration methods when applied in full awareness of these complications to estimate the possible error of the final values can, however, provide very relevant results with simple and inexpensive equipment. For an extensive discussion of this technique we refer to [2], where we presented a critical analysis. With respect to the other methods used we refer to Refs. [15] en [45]. For the present paper a brief and schematic review is given in Table II.

3. CONCLUSIONS

The above description of the phenomena contributing to the segregation of one element in the surface of alloy particles of supported and unsupported catalysts is intended to help the

TABLE II

GUESSTIMATES OF χ

1) $\chi > 1$	for component with low heat of sublimation (broken-bond, cherry models)
2) $\chi > 1$	for solute if $(r_0 - r_1)^2 \gg 0$ (elastic strain)
3) $\chi > 1$	for component with <u>highest</u> chemical affinity to ambient gas or <u>lowest</u> affinity to supporting solid
4) $\chi \sim 1$	for Daltonide phases or if above effects counteract each other

catalytic chemist for an orientation, applicable to the system he has under study. For more quantitative information it will often be desirable to study an unsupported model catalyst alongside with the supported system. The unsupported preparation will then have to be used as catalyst for the reaction under study in order to ascertain that it is indeed an alloy which is responsible for the performance of the supported bimetal system. If this assumption appears justified care must still be taken to define where the supported and the unsupported system are comparable.
An oxidizing regeneration procedure e.g. might lead to the irreversible formation of a mixed oxide comprising the support and one of the metal components. It is also possible that alloy systems with a wide miscibility gap form solid solutions when present as small clusters. Another complication to be considered when dealing with technical catalysts is the non-uniformity of composition of individual alloy particles [46]. While the overall composition may be A_xB_y, some particles may be richer in either component. Also the existence of unalloyed particles of A and/or B can be important for the catalytic behaviour of the system. It is, therefore, useful also to consider the miscibility of the compounds formed during catalyst preparation. If, for instance, both elements were brought to the surface as carbonates by an impregnation technique and this was followed by calcination and reduction steps, it is essential to consider whether the carbonates and oxides of A and B form solid solutions or separate phases [46].

In conclusion it can be stated that the surface composition of alloys can be estimated qualitatively by applying well-known principles of thermodynamics and solid state chemistry.

REFERENCES

1. J.H. Sinfelt, Catal. Rev. - Sci. Eng., 9, 147 (1974).
2. R.A. van Santen and W.M.H. Sachtler, Adv. Catalysis, 26, 69 (1977).
3. V. Ponec, Catal. Rev. - Sci. Eng., 11, 1 (1975).
4. F.M. Dautzenberg, J.N. Helle, P. Biloen and W.M.H. Sachtler, J. Catal. (in press).
5. W.M.H. Sachtler and P. van der Plank, Surf. Sci., 18, 162 (1969).
6. N.D. Lang and H. Ehrenreich, Phys. Rev., 168, 605 (1968).
7. W.M.H. Sachtler, Angew. Chemie, 80, 155 (1968).
8. W.M.H. Sachtler, Vide, 164, 67 (1973).
9. L. Imre, Ber. Bunsenges. Phys. Chemie, 74, 220 (1970).
10. P.A. Kilty, N.C. Rol and W.M.H. Sachtler, Proc. Vth Int. Congress Catalysis 1972, J.W. Hightown (Ed.), North-Holl. Publishing company Amsterdam, 1973.
11. A.A. Holscher and W.M.H. Sachtler, Disc. Faraday Soc., 41, 29 (1966); W.M.H. Sachtler, Angew. Chemie, Int. Ed., 7, 668 (1968).
12. J.W. Gibbs, Trans. Connecticut Acad. Sci., 3, 108 (1975/76).
13. G. Wulff, Z. Kristallographie, 34, 449 (1901).
14. J.J. Burton and R.S. Polizzotti, Surf. Sci., 66, 1 (1977).
15. W.M.H. Sachtler and R.A. van Santen, Application Surf. Sci., 3, 121 (1979).
16. V. Ponec, Surf. Sci., 80, 352 (1979).
17. M. Kelly, J. Catalysis, 57, 113 (1979).
18. M.P. Seah, J. Catalysis, 57, 450 (1979).
19. R.A. van Santen and W.M.H. Sachtler, J. Catal., 33, 202 (1974).
20. F.L. Williams and D. Nason, Surf. Sci., 45, 377 (1974).
21. E.A. Guggenheim, Trans. Faraday Soc., 41, 150 (1945).
22. R. Defay, I. Prigogine, A. Bellemans and D.H. Everett, Surface Tension and Adsorption, Longmans, Green, New York, 1966).
23. J.J. Burton, E. Hyman and D. Fedak, J. Catal., 37, 106 (1975).
24. R. Defay and I. Prigogine, Trans. Faraday Soc., 46, 199 (1950).
25. R. Bouwman, L.H. Toneman, M.A.M. Boersma and R.A. van Santen, Surf. Sci., 59, 72 (1976).
26. J.L. Meyering, Acta Metall., 14, 251 (1966).
27. A.R. Miedema, Z. Metallkunde, 69, 455 (1978).
28. H.H. Brongersma, M.J. Sparnaay and T.M. Buck, Surf. Sci., 71, 657 (1978).
29. R. Bouwman, G.J.M. Lippits and W.M.H. Sachtler, J. Catal., 25, 300 (1972).

30. R.A. van Santen, L.H. Toneman and R. Bouwman, Surf. Sci., 47, 64 (1974).
31. N.H. Tsai, G.M. Pound and F.F. Abraham, J. Catal., 50, 200 (1977).
32. J.W. Taylor, J. Inst. Metals, 86, 456 (1958).
33. J.M. McDavid, S.C. Fain, Surf. Sci., 52, 61 (1975).
34. W. Losch and J. Kirschner, 7th Int. Vacuum Congress & 3rd Int. Conference on Solid Surfaces, ed. by L. Dobrozensky, F. Rüdenauer, F.P. Viehböck, A. Breth, Vienna, 1977, p. 823.
35. P. Biloen, R. Bouwman, R.A. van Santen and H.H. Brongersma, Appl. Surf. Sci., 2, 532 (1979).
36. D. McLean, Grain Boundaries in Metals, Oxford Univ. Press, London, 1957.
37. J.D. Eshelby, in: Solid State Physics - Advances, Vol. 3, p. 79, ed. by F. Seitz and D. Turnbull, Academic Press, New York, 1956.
38. J.J. Burton and E.S. Machlin, Phys. Rev. Letters, 37, 1433 (1976).
39. P. Wynblatt and R.C. Ku, Surf. Sci., 65, 511 (1977).
40. H. Verbeek and W.M.H. Sachtler, J. Catal., 42, 257 (1976).
41. R. Bouwman and P. Biloen, Surf. Sci., 41, 348 (1974).
42. F.J. Kuijers, R.P. Dessing and W.M.H. Sachtler, J. Catal., 33, 316 (1974).
43. Y. Soma-Noto and W.M.H. Sachtler, J. Catal., 32, 315 (1974).
44. H.C. de Jongste, F.J. Kuijers and V. Ponec, in: Proc. VIth Int. Congress Catalysis, Paper B 31, London, 1976.
45. H.H. Brongersma, F. Meijer and H.W. Werner, Philips Techn. Tijdschrift, 34, 367 (1974).
46. S.D. Robertson, S.C. Kloet and W.M.H. Sachtler, J. Catal., 39, 234 (1975).
47. D.T. Quinto, V.S. Sundaram and W.D. Robertson, Surf. Sci., 28, 504 (1971).
48. G. Ertl and J. Küppers, Surf. Sci. 24, 104 (1971).
49. Y. Takasu and H. Shimizu, J. Catal., 2, 479 (1973).
50. K. Watanabe, K. Hashiba and T. Yamashina, Surf. Sci., 61, 483 (1976).
51. P. van der Plank and W.M.H. Sachtler, Surf. Sci., 18, 62 (1968).
52. H.H. Brongersma and T.M. Buck, Surf. Sci., 53, 649 (1973).
53. F.J. Kuijers and V. Ponec, Surf. Sci., 68, 294 (1977).
54. C.R. Helms, J. Catal., 36, 114 (1975).

CATALYSIS BY METALS AND ALLOYS
REFORMING OF HYDROCARBONS AND SOME OTHER REACTIONS

H.C. de Jongste and V. Ponec

Gorlaeus Laboratoria, Rijksuniversiteit Leiden,
P.O. Box 9502, 2300 RA Leiden, The Netherlands

Let us start this chapter by summarizing once more what has already been documented in the foregoing lectures.
1) Alloying causes rather moderate changes in the electronic structure of alloy components.
2) Alloying causes rather moderate changes – if any at all – in the adsorption energetics (adsorption bond strength).
3) Alloying may cause quite considerable changes in the structure of chemisorption complexes. Dilution strengthened by surface segregation (see lecture by Sachtler on this point) frequently makes the changes quite pronounced.

Conclusions 1 – 3 form the starting point of the discussion which follows.

A) Reaction intermediates on pure metals; the types of bonding to the metal surface.

A lot of valuable information has been obtained by the use of isotopic exchange reactions (mainly with D_2) of various compounds, the way pioneered by Kemball [1,2], Burwell [3,4], Rooney [5], Gault [6] and others [7]. Let us reproduce in telegraphese the main results of these experiments.
1. The mere existence of the exchange of "H" for "D" may be safely considered as evidence for the C-H, O-H, N-H, etc. dissociation and the formation of "α"-species:

$$CH_2-(CH_2)_n-CH_3$$
$$|$$
$$*$$
(1)

$$\begin{array}{ccc} R\diagdown_O & R\diagdown_{NH} & \\ | & | & \\ * & * & \end{array}$$

(α–)

2. In some cases, depending on the metal and reaction conditions, molecules stay on the surface so long that they manage to exchange more than one H during one sojourn on the surface (multiple exchange). Since the completely saturated forms are easily desorbable, evidently some intermediates are held to the surface by multiple bonds:

$$CH_4 \diagup \atop \diagdown_{-H} \atop CH_3 \atop | \atop * \xrightarrow{-H} CH_2 \atop \| \atop * \xrightarrow{D} CH_2D \atop | \atop * \xrightarrow{-H} CHD \atop \| \atop * \xrightarrow{D} CHD_2 \atop | \atop * \longrightarrow etc. \quad (2)$$

(αα)

On Ni, Rh and some other similar metals[7] an assumption of a multiple bond intermediate explains how the exchange jumps from one set of hydrogen (one side) to another (other side of the ring). In these cases the literature speaks of "αα"-species.

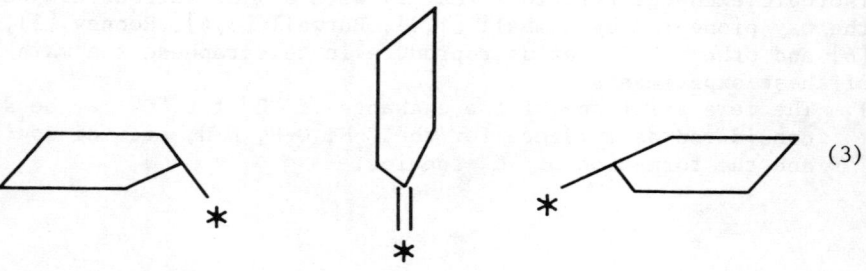

(3)

It must be remarked that the propensity of metals to form a multiple bond (αα) is not eliminated even by alloying a group VIII metal with very exothermically reacting components like Al [8]. The rate of formation of simple α-species is even not slowed down (T.O.N. per active atom) by alloying group VIII- with group Ib-metals [9].

3. The multiple exchange of ethane with deuterium takes place much more readily than that of methane so that there may be another multiple bond intermediate responsible for that. It has been suggested that its structure is like this [1-7]:

$$H_2\overset{}{C}-\overset{}{C}H_2 \quad \text{or} \quad CH_2 \ddagger CH_2 \qquad (4)$$
$$(\alpha\beta) \qquad\qquad (\pi-)$$

The fact that ethane hydrogenolysis proceeds on some metals quite easily, but at temperatures rather high for π-complexed species being still densely populated on the surface, indicates that the (αβ) intermediate is a possible and very likely intermediate of hydrogenolysis on metals. Experiments with adamantanes and similar molecules showed that (αβ) species are only formed when the two hydrogens involved are in a cis-eclipse form [3,4].

4. Other multiple bound species, αγ, αδ, αε, can exist as well, but evidently their formation is accompanied by a higher activation energy, a deeper dehydrogenation of adsorbed species, etc., so that these species if present are formed only at higher temperatures. For example, it has been shown that a quarternary carbon is an obstacle for the multiple (α,β)-like exchange; a molecule such as [3,4] (expected maxima in product distribution are indicated):

$$\underset{d_5}{\nearrow}CH_3CH_2 - \underset{\underset{CH_3}{|}}{\overset{\overset{CH_3}{|}}{C}} - CH_2CH_2CH_3\underset{d_7}{\searrow}$$

does not show after exchange with D_2 any combination of numbers of D-atoms which could indicate that the exchange can easily go over from the left to the right side of a molecule demonstrating that the multiple exchange is not propagated easily by a αγ-intermediate. In other words: αβ-exchange is "stopped" at the quarternary C. In a similar way a heteroatom such as in dialkyl ethers or dialkylamines limits the multiple exchange to one side of a molecule. There are many indications that the αγ-intermediate is indeed not formed easily at low temperatures [1-7]. However, at high temperatures and with metals which do not cat-

$$\begin{array}{c} \text{C} \quad \text{C} \\ \diagdown\!\!\text{C}\!\!\diagup \\ \text{C} \quad \text{C} \\ \| \quad \mid \\ * \quad * \end{array} \qquad (\alpha\alpha\gamma-) \tag{5}$$

alyse hydrogenolysis so vigorously so that the molecule can survive that treatment (Pt), some multiple exchange in αγ-positions of neopentane may take place and at the same time isomerisation and hydrogenolysis of this molecule is also possible [1-7].
This has led to the suggestion [10] that an intermediate like the one shown here is operating. Closing and breaking of the bands in the indicated way leads to isomerisation. When the metal is dehydrogenating the adsorbed species too vigorously, hydrogenolysis is strongly prevailing, which is most likely the case for Ni, Rh and similar metals. The relation between the degree of dehydrogenation and selectivity for hydrogenolysis has been found for cyclopropane but most likely it is of a quite general character [11]. It has also been found that on Pt the 2,2,3,3-tetramethyl butane is preferentially split in the middle of the molecule, because the main products are the isobutane molecules [12] and very little methane is found. It is then tempting to assume that at a certain stage of the reaction, αδ species are formed like

$$\begin{array}{cc} \begin{array}{c} \text{C} \quad \text{C} \\ \mid \quad \mid \\ \text{C}-\text{C}-\text{C}-\text{C} \\ \mid \quad \mid \\ \text{C} \quad \text{C} \\ \mid \quad \mid \\ * \quad * \end{array} \quad (\alpha\delta-) & \begin{array}{c} \text{C} \quad \text{C} \\ \mid \quad \mid \\ \text{C}-\text{C}-\text{C}-\text{C} \\ \mid \quad \mid \\ \text{C} \quad \text{C} \\ \diagdown\!\diagup \\ * \end{array} \end{array} \tag{6}$$

Kazansky et al. [13], Pines et al. [14] and others showed that polysubstituted alkanes which cannot form 1,6-intermediates produce a lot of substituted cyclopentanes. The 1,5-ring closure is a good indication for the existence of a 1,5(αε)-intermediate on the surface or of its more dehydrogenated alternative forms (αα,ε   αα,εε or ααα,εεε etc., see below).

5. To explain the low temperature exchange of cyclopentane on metal like Pd which does not easily form αα-bonds, an intermediate has been suggested bound simultaneously to several metal atoms [3,4]

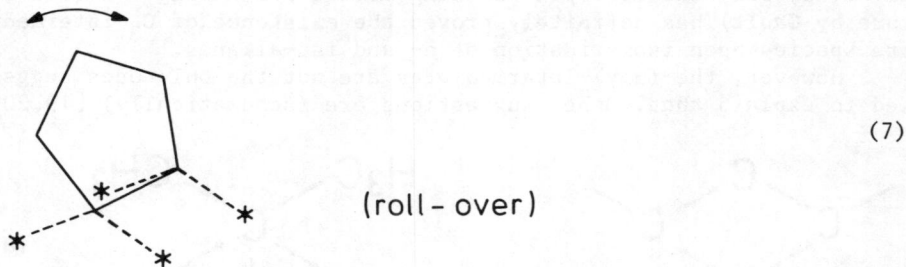

(7)

(roll-over)

The multiple bonding, a big ensemble of equivalent sites required for that intermediate explained nicely the results obtained with Pd-Ni alloys [15]. On the other hand, the confidence in the other alternative explanation of the same exchange results by the π-allylic intermediates is nowadays much less than it used to be [16].

6. It should be stressed that it is not clear how many hydrogen atoms are lost at a given temperature, i.e. how many multiple bonds are formed when the intermediates are formed bound to the surface through their $\alpha\gamma$, $\alpha\delta$ or $\alpha\varepsilon$ positions.

B) Reaction intermediates of hydrogenolysis, dehydrocyclisation and isomerisation

There are several suggestions on the mechanism of these reactions in the literature, and for example, the terms "bond shift" and "cyclic" intermediates are often used in relation to it. However, this terminology which has its good historical reasons does not characterize well the difference among the intermediates, as we shall see below. In each mechanism suggested, there is some bond-shift and at a certain stage of the reaction all intermediates have the form of a ring, i.e. they are essentially all cyclic. Therefore, a terminology is preferable which indicates how many C atoms are involved (and become equivalent) upon formation of a surface complex. In this way we discern C_3, C_5 and C_6 intermediates [17].

C_3-intermediates

We have already seen one of the intermediates of this group, the $\alpha\alpha\gamma$ species suggested in the literature [10] for the mechanism of isomerisation and hydrogenolysis of neopentane. The same intermediate can, of course, also play a role in isomerisation and hydrogenolysis of other alkanes. While with neopentane isomerisation it is almost self-evident that one or another C_3-intermediate must be involved, only the isotopic labeling method [18] using ^{13}C (see lecture by Gault) has definitely proved the existence of C_3-intermediate species upon isomerisation of n- and iso-alkanes.

However, the (ααγ)-intermediates are not the only ones suggested to explain that. Other suggestions are (schematically) [19,20]:

(8)

abstraction of 2H- abstraction of 1H-

The metallocyclobutane [19] intermediate (left) has been suggested on the strength of analogy with similar inorganic complexes. It is expected that it should react as e.g. propylene does upon metathesis that is, by forming a carbene and a π-complexed olefine, both attached to the same (*) site. Upon hydrogenolysis these two fragments would just desorb after partial hydrogenation. Upon isomerisation the olefinic fragment has to rotate around the normal to the π-bond and reestablish a hydrocarbon with the original number of C-atoms. However, a free rotation around a π-bond or a free π → σ → π transition does not seem to be confirmed yet for Pt complexes (like Zeise salt), although it is known for group VI metal complexes. The pseudo metal-carbenium ion (right) has been suggested [20] in order to explain the isomerisation of cage molecules like (3,3,2)-bicyclooctane and similar ones, which under the conditions of running isomerisatio do not exchange more than one H-atom.

We believe it is premature to try to decide on the basis of the existing experimental material which of the intermediates in (8) can be excluded from considerations on the mechanism. At the first glanc it seems that the metallocyclobutane can better explain hydrogenolysis whereever it can take place on one site (W, Mo and similar metals might be examples of the best analogy with metathesis catalysts) while the pseudocarbenium ion seems to be better suited for isomerisation on very diluted Pt alloys.

C_5-intermediates in dehydrocyclisation

There are several suggestions [19,21,22] in the literature (for references, see also the lecture by Gault) how to close a 5-ring, closure can take place when a bond shifts from one to another carbon one carbon (at least) being bound by a multiple metal-carbon bond. The main features of intermediates are expressed by the following structures:

```
      C              C              C
    /   \          /   \          /   \
   C     C        C     C        C     C
   |     |        |     |        |     |       Class I
   C     C        C     C        C     C
   ‖     |        ‖     ‖        ⫴     ⫴
   *     *        *     *        *     *                        (9)

      C              C              C
    /   \          /   \          /   \
   C     C        C     C        C     C
   |     |        |     |        |     |       Class II
   C     C        C     C        C     C
    \\ //          \\ //          \  /
      *              *              * ⤳ C
```

There are some experimental indications or reasons for any of the intermediates. A principle of "minimum hypothesis" favours the least dehydrogenated species ($\alpha\alpha\epsilon$) while the very high and negative orders for hydrogen pressure sometimes found, seem to favour the more dehydrogenated forms. It has been found that well dispersed metals differ in the product pattern of the MCP-hydrogenolysis from massive flat surfaces and this has led to the suggestion of class I and class II types of intermediates, i.e. those requiring a big or small ensemble of sites (in extreme: one- and two-site ensembles).

It is our opinion also in this case that the literature data do not allow a definite choice of really operating intermediates and none of the intermediates shown can be completely disregarded at the moment. Namely, a great deal of necessary information is still missing. The number of H-atoms split off is not known, neither is the exact number of atoms in the ensemble required for the reaction. The metallic site (*) can also be a whole valley inbetween several metal atoms (and not just a single atom), so that the metal carbon bond would be actually delocalized and could not be visualized as a classical double or triple bond to one metal atom, etc.

C_5-intermediates of isomerisation

While the necessity of the C_3-species as intermediates seems to be almost selfevident (consider isomerisation of neopentane, neohexane!), in some cases it is quite difficult to prove that intermediates like those shown above for the 1,5-dehydrocyclisation are also operating upon isomerisation of those hydrocarbons which allow a 1,5-ring closure like n-hexane methylpentanes, etc. The definite evidence for C_5-intermediates has been delivered by ingenious experiments designed and performed by Gault et al. [18,19]. The principle of the method is illustrated by the way of the following example. Consider 2-methylpentane labeled by ^{13}C in the proper position and let us follow its conversion:

(10)

In the mechanistic studies, the products of isomerisation and other reactions are first separated by GLC and then the position of the label in different molecules is determined by analysis of the mass-spectra fragments [18]. The scheme above indicates the role of the respective intermediates and it is obvious that the ratio of the two 3MP's with ^{13}C in different positions is a measure of the contribution to the total isomerisation of the C_5-intermediates, the existence of which is simultaneously proved by that.

C_6-intermediates

Earlier literature often speculated (see e.g. refs 13,14) that aromatisation proceeds via a 1,5-ring closure followed by ring-enlargement, the latter running through C_3-intermediates. Upon such a reaction a 1,6-intermediate would also be formed when the ring is just being enlarged. However, this idea is generally doubted now, because the 1,5- and 1,6-ring closures are usually not influenced in parallel, when a catalyst is modified or the reaction condition varied. Moreover, it has been shown that in all cases where a direct 1,6-ring closure is possible, this closure proceeds preferentially to the alternative pathway comprising 1,5-ring closure and ring-enlargement. In principle, the 1,6-ring closure could proceed via intermediates like in (9), but the pronounced differences in behaviour of 1,5- and 1,6-ring closures indicate that some other intermediates are not unlikely. Indeed, there is some evidence already available [23,24] that this 1,6-ring closure may proceed through

Catalysis by alloys. Reforming of hydrocarbons 345

dehydrogenation via dienes (e.g. hexadiene) and trienes (hexatriene) which form aromatic rings readily, even thermally, without a catalyst.

C) Alloys, classification of reactions

The most complete information on various reactions on the same series of alloys is available for the Ni-Cu alloy system. Moreover, the surface composition of these alloys is also reasonably well known [25-27]: alloys with about 5-85% Cu in bulk have almost a constant surface composition varying between 15 and about 5% Ni. Chemisorption indicates a little higher Ni content in the surface than AES and LEISS. Nevertheless, it is in any case a safe statement to say that if each Ni atom could catalyse a given reaction, the reaction rate per total alloy surface area should not decrease by alloying (5-85% bulk Cu) with more than a factor of 10-20. When this factor is exceeded, evidently the given reaction requires an ensemble of several Ni atoms and these are more suppressed in their number than by a factor of 10-20. Another possibility is that the given reaction can proceed on either a big ensemble as a fast reaction or a single site as a slow reaction (slower by a factor higher than - say - 20). The variety of reactions studied on Ni-Cu alloys can be subdivided into two subgroups [28,29].

I. Hydrogenation of C=C, C≡C, C=O and similar (cyclopropane) bonds, dehydrogenation, hydrocarbon-deuterium exchange reactions. For these reactions the activity is slightly increased or it drops by alloying, but then never more than by a factor of 10-20.
II. Hydrogenolysis, isomerisation, dehydrocyclisation, coke deposition (all these reactions comprise a -C-C-C bond rearrangement and breaking), disproportionation of alkylamines, ether formation from alcohols on metals, methanation and Fischer-Tropsch reactions; all reactions where at least one C-C, C-N or C-O bond is being broken in the adsorbed state of the reacting molecules. Evidently, upon fission or rearrangement

$$\begin{array}{c} C - C \to C - C \\ | \quad | \quad || \quad || \\ * \quad * \quad * \quad * \end{array} \qquad (11)$$

several metal atoms are simultaneously involved (= big ensemble required), one or more sites mediating upon that, a multiple bond between the chemisorbed fragments and the metal surface. Obviously, those conditions are not fulfilled when an active (group VIII) metal such as Ni is alloyed with an inactive (much less active) Ib-metal, etc. and as a consequence the activity drops by alloying more than a factor of 10-20. If one-site alternatives were operating on Ni alloys, these had to be slower by a factor more than 20, than the multi-site pathways. By the absence of the fast group II-reactions (like coke-deposition) the activity in the group I-reactions may increase sometimes above the value for the pure group VIII metal.

D) Pt alloys, an example of subtile variations in the mechanism due to alloying

When studying Pt-Au alloys [30], it has been found that isomerisation and dehydrocyclisation were suppressed by alloying in the region of 0-15% Pt in Au, more than would correspond to the surface composition of these alloys. According to measurements of CO adsorption on Pt-Au alloys [31] on about 15% (bulk) Pt-Au film alloys, the surface composition was about equal to the bulk composition; according to the LEISS experiments [32] on Pt-Au powders, the Pt concentration could be about a factor of 2 lower than that. However, the overall rate was about 300 times lower than on Pt of the same surfac area [33]. Although there is such a decrease in the overall rate, there is always some isomerisation persisting upon alloying and strong dilution of Pt [30]. These findings can be rationalized as followed. On pure Pt various mechanisms run parallel through C_3- and C_5-intermediates which can both be present in their "one- and "two" site variants. One can expect that by diluting Pt in Au the multisit variants are selectively suppressed and because the overall rate drops so strongly we conclude that the one-site variants have lower rates than the two-site alternatives.

This fundamental idea is supported by the results obtained with Pt-Cu alloys [34,35]. With these alloys again the overall activity drops quite considerably, but even very Pt-diluted alloys keep some activity in isomerisation. Along the series of alloys the contribution of C_3- and C_5-intermediates varies as well. Pure Pt shows 40-50 of the C_3 contribution, alloys with about 50% Pt show more than 90% of the C_5 contribution and the alloys with only a few bulk % Pt show again an increase (up to 40%) of the C_3-intermediate contribution. At the same time, the character of the MCP splitting reveals the sam changes like when going from a massive flat Pt surface to well dispersed Pt, i.e. like going from a situation where many Pt atoms can be involved simultaneously to a situation where a limited number of Pt or single Pt atoms must do the job. Therefore, the results of ^{13}C labeled 2MP were explained as follows: on pure Pt the multisite variants prevail; on alloys with about 40% Pt and lower one-site variants prevail and at the same time the relative contribution of C_5-intermediates goes through a maximum.

E) Rate and selectivity determining steps

The classical networks of consecutive steps, adsorption – surface reaction – desorption, the usual basis for treatment of kinetics by means of the Langmuir-Hinshelwood type of equations, usually comprise one rate determining step, other steps being assumed to be in equilibrium. However, we have seen that the network of reaction comprises not only consecutive but also many parallel reactions. A certain product of one reaction - isomerisation - can be simultaneously formed by at least four different pathways through one- and two-site C_3- and C_5-intermediates. In this situation it seems extremely difficult to formulate kinetic equations which are not only formally in agreement with the data, but also correspond with reality. It is also obvious that the rate and selectivity determining steps need not always be identical.

An interesting example of a system of simultaneous reactions is the Fischer-Tropsch synthesis. With regard to this reaction it has been proved [36,37] that

i) the CO dissociation on metals active in Fischer-Tropsch synthesis is possible and carbon formed in this way is certainly a most likely intermediate of methanation and of Fischer-Tropsch synthesis as well (the latter first derived from experiments with alloys [38] and later definitely confirmed by isotopic labeling [39]). It has also been shown that CO dissociation takes place on active metals at much lower temperatures than CO disproportionation ($2CO \rightarrow C + CO_2$) which is probably not - as such - an intermediate step of the synthesis [40].

ii) CH_x is a proved intermediate of hydrocarbon chain growth [39]. Yamamoto [40] and Young and Whiteside [42] have shown that this chain growth reaction can proceed also on an isolated single active site. On the other hand, Araki [36] proved that CO dissociation needs an ensemble of several active sites, while hydrogenation (partial or complete) again does not have the same requirements.

In this way we come to the conclusion that for these particular reactions the rate and selectivity determining steps have different requirements with regard to the ensemble of active sites necessary for the reaction.

F) Conclusions

We have seen from several examples how alloying changes the metal selectivity. Up to now the main operating factor has always been identified as the availibity of active sites - the size of available ensembles of active sites. There are, surprisingly enough, no pieces of direct evidence that subtle changes in the electronic structure sometimes occurring (see lecture on "Bonding in and on metals") play indeed a role in governing the alloy selectivity. The electronic structure factor in catalysis by alloys is represented by the fact that for certain reactions or chemisorption, a certain electronic structure of metal atoms is necessary and the given metal cannot be fully substituted in its function by another metal.

REFERENCES:

1. C. Kemball, Advances Catal. 11 (1959) 37.
2. C. Kemball, Catal. Rev. 5 (1971) 33.
3. J.R. Burwell, Accounts Chem. Res. 2 (1969) 289.
4. J.R.J. Burwell, Catal. Rev. 7 (1) (1972) 25.
5. J.J. Rooney, Chemistry in Britain, 1966, 242.
6. F.G. Gault, J.J. Rooney and C. Kemball, J. Catal. 1 (1962) 255.
7. G.C. Bond, Catalysis by Metals, Acad. Press, London, 1962
   (a complete evaluated review of results up to 1962 is presented)
8. B. van Keulen, W.R. Wichers, and V. Ponec, Kin. Katal. Letters, in print.
9. V. Ponec, and W.M.H. Sachtler, J. Catal. 24, (1972) 250.
10. J.R. Anderson, and N.R. Avery, J. Catal. 5 (1966) 446.
    J.R. Anderson, Adv. Catal. 23 (1973) 1.
11. R. Merta, and V. Ponec, IV Int. Congr. Catal. Moscow, 1968, ed. Akademiai Kiado, Budapest, Vol. 2 (1971) p. 53.
12. G. Leclerq, L. Leclerq, and R. Maurel, Lecture on the Symposium 'Hydrocarbon Reactions', Brussels, March 1977. Proceedings will be published in Bull. Soc. Chim. Belg., 1979.
13. B.A. Kazanskii, Kin. Katal. 8 (1967) 977.
    J.V. Kalechits, ibid. 8 (1967) 1114.
14. H. Pines, and C.T. Goetschel, J. Org. Chem. 30 (1965) 3530, 3546.
    H. Pines, and J.W. Dembinski, ibid. 30 (1965) 3537.
15. J.L. Vlasveld, and V. Ponec, J. Catal. 44 (1976) 352.
16. H.A. Quinn, J.H. Graham, M.A. McKervey, and J.J. Rooney, J. Catal. 22 (1971) 35.
    J.J. Rooney, J. Catal. 22 (1971) 35.
17. H.C. de Jongste, and V. Ponec, Lecture on the Symposium 'Hydrocarbon Reactions', Brussels, March 1979. Proceedings will be published in Bull. Soc. Chim. Belg. Nov. 1979.
18. C. Corolleur, S. Corolleur, and F.G. Gault, J. Catal. 24 (1972) 385.
    C. Corolleur, D. Tomanova, and F.G. Gault, J. Catal. 24 (1972) 401.
19. F.G. Gault, Lecture on the symposium 'Hydrocarbon Reactions' Brussels, March 1979. Proceedings will be published in Bull. Soc. Chim. Belg. 1979.
20. M.A. McKervey, J.J. Rooney, and N.G. Samman, J. Catal. 30 (1973) 330.
21. J.H. Muller, and F.G. Gault, J. Catal. 24 (1972) 361.
22. Y. Barron, G. Maire, J.M. Muller, and F.G. Gault, J. Catal. 5 (1966) 428.
23. P. Tetenyi, L. Guczi, and Z. Paal, Acta Chim. Acad. Sci. Hungary, Vol. 83 (11) (1974) 37.
24. W.L. Callender, S.G. Brandenberger, and W.K. Meerbott, Catalysis, Proc. V Int. Congr. Catal. Miami Beach, 1972, ed. J. Hightower, Elsevier Amsterdam, 1973, p. 1265.

B.A. Kazanskii, G.V. Isagulyants, M.I. Rozengart,
Yu.G. Dubinsky, and L.I. Kovalenko, ibid., p. 1277 and the
discussion on these papers.
25. P.E.C. Franken, and V. Ponec, J. Catal. 42 (1976) 398.
26. J.C.M. Harberts, A.F. Bourgonje, J.J. Stephan, and V. Ponec,
J. Catal. 47 (1977) 92.
27. F.J. Kuijers, and V. Ponec, Surface Sci. 68 (1977) 299.
28. V. Ponec, Int. J. Quant. Chem. 12 (2) (1977) 1.
29. V. Ponec, Surface Sci. 80 (1979) 352.
30. J.R.H. van Schaik, R.P. Dessing, and V. Ponec, J. Catal. 38 (1975) 251.
31. J.J. Stephan, V. Ponec, and W.M.H. Sachtler, Surface Sci. 47 (1975) 403.
32. P. Biloen, R. Bouwman, and R.A. van Santen, Proc. 7th Int. Vac. Congr. and 3rd Int. Conf. Solid Surfaces, Vienna 1977, ed. R. Dobrozemsky et al., Vol 2, p. 1401.
33. H.C. de Jongste, F.J. Kuijers, and V. Ponec, Proc. Symp. Scientific Basis of Preparation of Heterogeneous Catalysts, eds. B. Delmon et al., Elsevier Amsterdam 1976, p. 207.
34. H.C. de Jongste, F.J. Kuijers, and V. Ponec, Proc. VI Int. Congr. Catal. London 1976, eds. G.C. Bond, P.B. Wells, F.C. Tompkins, Vol. 2, 915, paper B31.
35. H.C. de Jongste, V. Ponec, and F.G. Gault, submitted to J. Catal.
36. M. Araki, and V. Ponec, J. Catal. 44 (1976) 439.
37. W.A.A. van Barneveld, and V. Ponec, Ind. Eng. Chem., in print.
38. W.A.A. van Barneveld, and V. Ponec, J. Catal. 51 (1978) 426.
39. P. Biloen, J. Helle, and W.M.H. Sachtler, J. Catal. 58 (1979) 95.
40. H. Wise, Stanford Res. Inst., private communication.
41. T.J. Yamamoto, J. Chem. Soc. Chem. Commun. 1003 (1978).
42. C.B. Young, and G.M. Whitesides, J. Am. Chem. Soc. 100 (1978) 5808.

SKELETAL ISOMERIZATION OF HYDROCARBONS ON METALS

F. GARIN and F.G. GAULT

Laboratoire de Catalyse, Université Louis Pasteur,
67008 Strasbourg, France.

This paper was partially written by Prof. Dr. F.G. Gault
before his sudden death 4.8.1979.

ABSTRACT. Metallocyclobutanes and metallocarbenes are proposed as precursors in bond-shift and cyclic-type isomerization of hydrocarbons on metals. The proposed mechanisms account for a number of experimental facts - Selectivity, structural effects, Kinetic data - which could not be explained previously.

This paper is devoted to skeletal rearrangements, an important class of reactions catalysed by metal surfaces which has no or very little counterpart in homogeneous catalysis.
Pioneering work by the soviet school of catalysis [1-4] showed that many reactions such as ring cleavage, dehydrocyclisation, and aromatization, involving the rupture and formation of carbon-carbon bonds, took place on metal surfaces. However it was long believed that skeletal isomerization on supported metal catalysts necessarily occurred according to a "bifunctional" mechanism which includes three consecutive steps :
- dehydrogenation on the metal,
- isomerization of the resulting olefin on the support by a carbenium ion mechanism
- and re-hydrogenation of the isomerized olefin [5], figure 1.

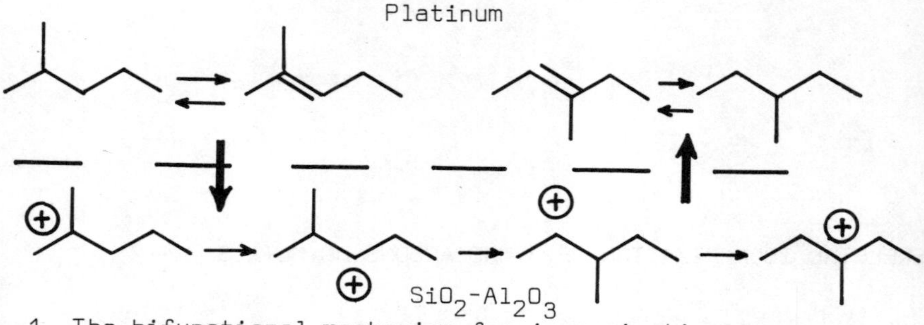

Fig. 1. The bifunctional mechanism for isomerization.

However, the occurrence of skeletal isomerization on platinum or palladium films and on supported platinum under such conditions (200-300°C) that the support (glass or alumina) is catalytically inactive [6-7-8] clearly demonstrates that the metal itself catalyses this reaction.

A. SKELETAL ISOMERIZATION

I. Bond-shift and cyclic "mechanisms"

As soon as it was found that skeletal isomerization of hydrocarbons could occur on metal surfaces, two distinct reaction mechanisms, bond-shift and cyclic, have been characterized.

The bond-shift mechanism, [10] responsible for the isomerization of small molecules, corresponds to the simple displacement of carbon-carbon bond. While two metals, platinum and palladium, may promote isomerization of isobutane to n-butane, platinum and, at a much smaller extent, iridium catalyse the isomerization of neopentane to isopentane.

Fig. 2. Bond-Shift mechanism.

For larger molecules, methylpentanes, normal hexane but not 2,3-dimethylbutane, a second type of mechanism may occur, involving consecutive 1-5 ring closure and ring opening. This "cyclic" mechanism [8] which also accounts for hydrogenolysis of cyclopentane molecule and 1-5 dehydrocyclization of alkanes is best represented in figure 3 where C represents a series of adsorbed intermediates, some of which having a cyclopentane structure.

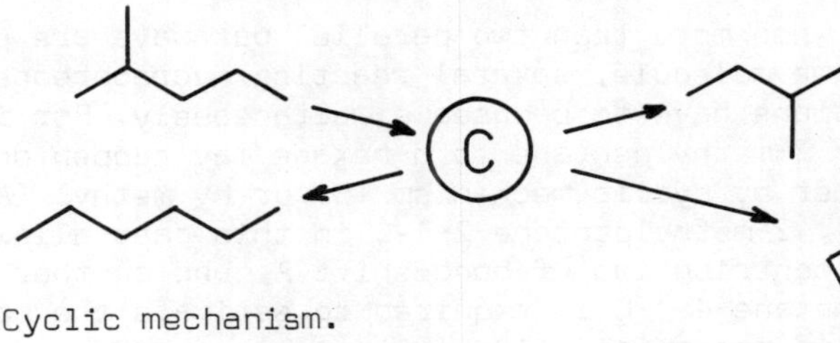

Fig. 3. Cyclic mechanism.

The cyclic mechanism, in the case of <u>highly dispersed catalysts</u>, was demonstrated by comparing the initial product distribution in methylcyclopentane hydrogenolysis and methylpentane or hexane isomerization [8-9] . For instance, the ratios 3-methylpentane over n-hexane, extrapolated to zero conversion, are the same in 2-methylpentane isomerization and in methylcyclopentane hydrogenolysis. Since cyclic-type isomerization involves first C-C bond formation and then C-C bond rupture, one does not expect hydrogenolysis of alkane to occur by this process. In contrast, as suggested very early [11] , if bond-shift isomerization involves primarily C-C bond rupture and then C-C bond reformation, a common intermediate should exist, leading to both the isomerization and the hydrogenolysis product.

II. Distinction between cyclic and bond-shift isomerization : ^{13}C-tracer experiments [12-13]

While in the case of very small molecules, bond-shift is the only possible isomerization reaction, in the case of larger molecules, with five carbons at least in the chain, cyclic mechanism widely predominates only in the case of extremely dispersed catalyst. On most catalysts - films, bulk metals, supported catalysts of medium and low dispersion - both isomerization mechanisms, bond-shift and cyclic compete, and the first task is to evaluate the contribution of each of them. The best way to do it is to label the reacting hydrocarbon with carbon 13. For instance isomerization of 2-methylpentane-2-^{13}C yields two different 3-methylpentane isomers, 3-^{13}C and 2-^{13}C, according to whether the reaction mechanism is cyclic or bond shift (Figure 4).

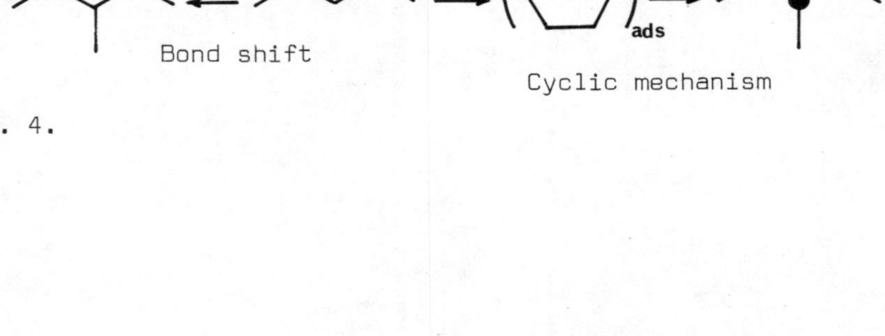

Bond shift   Cyclic mechanism

Fig. 4.

Most generally, when more than two parallel pathways are possible to obtain the same molecule, several reacting hydrocarbons labeled in various positions have to be used simultaneously. For instance isomerization of 2-methylpentane to n-hexane may happen on 3 different ways either by cyclic mechanism (C) or by methyl (A) or propyl shift (B). 2-methylpentane-2-^{13}C in this case allows only to estimate the contribution of bond-shift A, and another molecule, 2-methylpentane-4-^{13}C is required to estimate the contribution of the cyclic mechanism.(Figure 5).

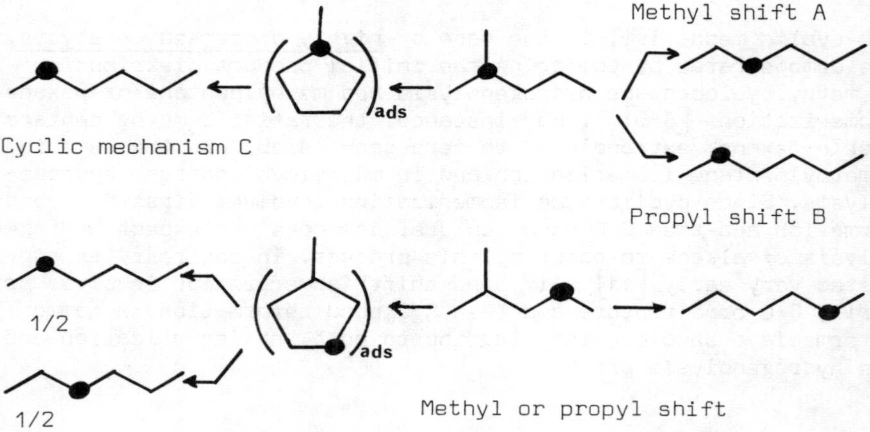

Fig. 5.

The use of labeled molecules allows also to determine the amounts of self-isomerized molecules, i.e. of molecules obtained by isomerization but having the same structure as the reactant. Formation of such molecules may contribute for an important part to the overall reaction and cannot be recognized by using non-labeled molecules. For instance, on a Pt catalyst of low dispersion, self isomerized 2,3-dimethylpentane is the major reaction product [12], and may be traced only by labeling the reactant on carbon 2. (Figure 6).

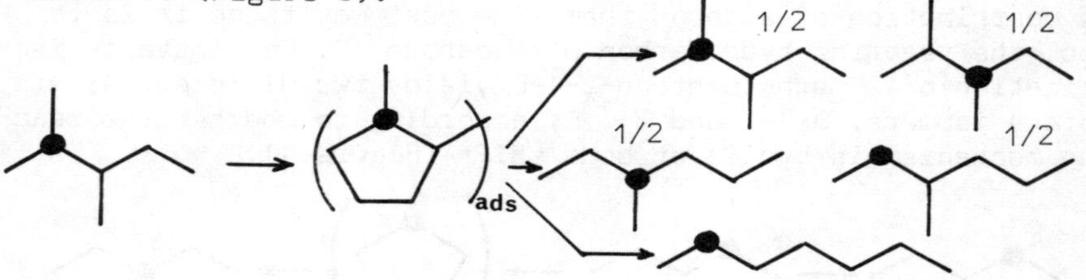

Fig. 6.

Similarly methyl shift in isopentane is a reaction ten times faster than chain lengthening and can be observed only with the appropriate labeled hydrocarbon [15] .(Figure 7).

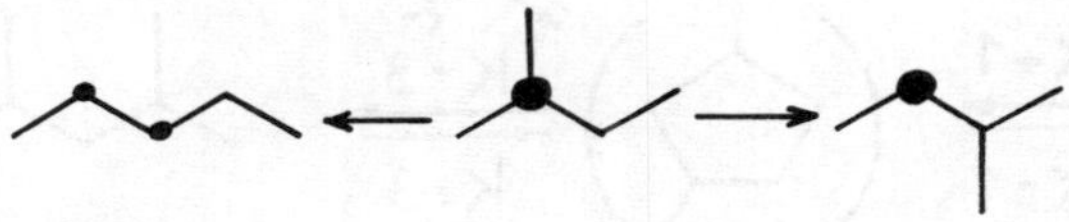

Fig. 7.

On the other hand for n-pentane on a highly dispersed Pt-Al_2O_3 catalyst, the major reaction is self-isomerization of n-pentane, by a cyclic mechanism and is traced as a label-scrambling reaction when using n-pentane-2-^{13}C as a reactant. (Figure 8).

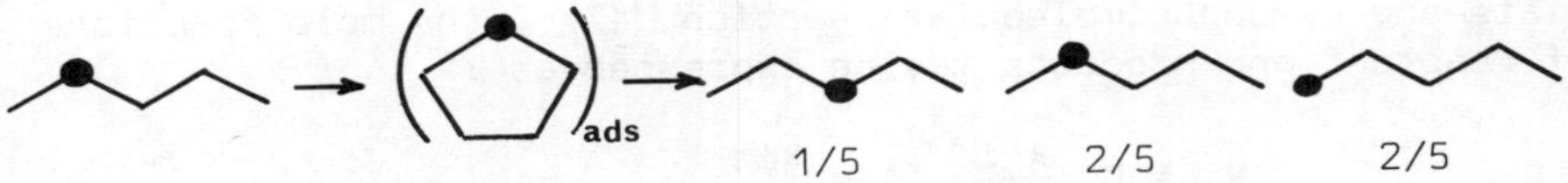

Fig. 8.

During the past decades, isomerization of hexanes or pentanes have very often been taken as a test reaction to study particle size or alloying effect. In these studies, the product distribution and especially the percentage of cyclic molecules (cyclopentane or methylcyclopentane) in the reaction products have often been used to estimate the contribution of the cyclic mechanism. Such a procedure may be misleading. First, self isomerization reactions, which may be predominant, are overlooked when not using labeled molecules. Secondly, the percentage of cyclic products is highly dependent upon experimental conditions, [15] increasing with decreasing hydrogen pressure or increasing hydrocarbon pressure or temperature. Comparison between two catalysts are only valid then when using strictly identical experimental conditions. Moreover, a third less obvious and more fondamental reason to discard the product distribution approach results from the mechanism itself : as shown for 3-methylhexane isomerization [16] , the ratio ρ between cyclic and acyclic products varies widely with conversion, even at very small conversion when the distribution of the acyclic isomers remains practically constant : this can be easily accounted for by assuming a simplified model where C represent a single cyclic intermediate. (Figure 9).

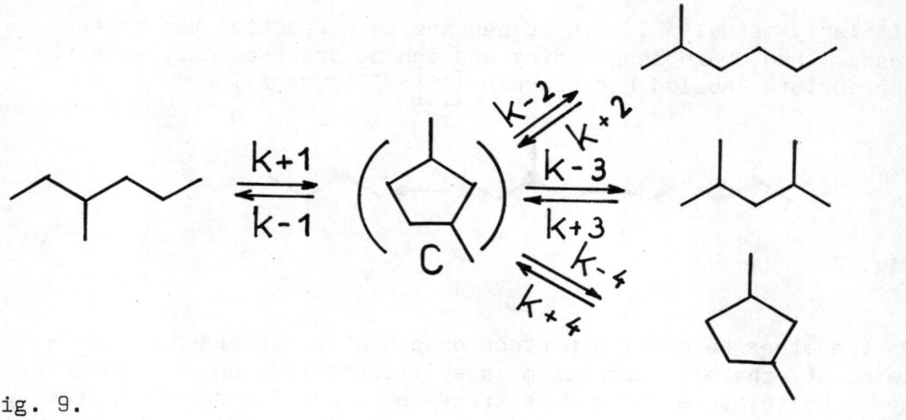

Fig. 9.

Assuming further steady state concentration C for this intermediate and pseudounimolecular reaction [17], the mole fractions of reactant and products may be expressed as :

$$x_i = C \frac{k_{-i}}{k_i} (1-e^{-k_i t})$$

the ratio between the amounts of cyclic and acyclic products is

$$\rho = \frac{k_{-4}/k_4 \, (1-e^{-k_4 t})}{\sum_{2}^{3} k_{-i}/k_i \, (1-e^{-k_i t})}$$

and r between the amounts of the two isomers are :

$$r = \frac{k_{-3}/k_3 \, (1-e^{-k_3 t})}{k_{-2}/k_2 \, (1-e^{-k_2 t})}$$

Since the "adsorption" rate constant k_4 is always much larger (by 2 to 3 orders of magnitude), than the dehydrocyclization rates constants k_2 and k_3, even at very small conversion where $1-e^{-k_2 t}$ and $1-e^{-k_3 t}$ may be approximated to $k_2 t$ and $k_3 t$, $1-e^{-k_4 t}$ is already close to unity.
Therefore, while the selectivity factor r remains constant, equal to k_2/k_3, the ratio ρ decreases sharply with contact time. The only correct way then to compare the contribution of the cyclic mechanism for various catalysts is to determine the percentage of cyclic products at various contact times and to extrapolate at zero conversion. Even in this case, such a comparison is valid,

only for a given set of experimental conditions and may be misleading, since the desorption rate constant k-4 may vary relative to the hydrogenolysis rate constant, k-2 and k-3 from catalyst to catalyst.

From the above discussion, it ensues that the tracer techniques are the most suitable to estimate the contributions of the various mechanisms. However this method provides information only if skeletal rearrangement and not adsorption-desorption is rate-determining.

B. REACTION MECHANISMS

I. General

By tracing the displacement of the various carbon atoms during the isomerization process, carbon 13 labeling allows a rough classification of the reactions into two groups, bond-shift and cyclic type reactions. It does not provide however any information concerning their detailed mechanism. In order to achieve a complete description, on a molecular scale, of skeletal isomerization, three different approaches have been used and very often combined. The first one consists in changing the structure of the reacting hydrocarbon. Complicating the molecule provokes drastic changes in the selectivity and also in the relative contributions of the various mechanisms. In the second approach, the usual kinetic parameters (orders vs hydrogen and hydrocarbon, apparent activation energy) are determined for the various parallel pathways including hydrogenolysis. Large differences between some of the parameters, Ea for instance, allow to distinguish between different reaction mechanisms and, conversely, finding identical kinetic parameters for several reactions points out for common reaction intermediate and interrelated mechanisms. A third method to gain information on reaction mechanisms consists in systematically changing the structure of the catalysts either by modifying the metal particle size or by alloying with a non-active metal. Lastly as an ultimate refinement of the molecular description of the mechanism, reference is made, as far as possible, to corresponding reactions which have been characterized or at least are supposed to occur in organometallic chemistry.

REACTIONS	E_a	ORDER VS HYDROGEN
2-methylbutane → n-pentane (labeled)	71.4 ± 1.5	− 3.4 ± 0.1
n-pentane → isopentane	55.3 ± 1.5	− 1.8 ± 0.1
isopentane → n-pentane	54.3 ± 1.5	− 2.3 ± 0.15
2-methylbutane → 2-methylbutane (labeled)	45.3 ± 1.5	− 1.9 ± 0.2
neopentane → isopentane	45 ± 3	− 1.65 ± 0.15
2,2-dimethylbutane → isobutane + C_2H_6	45 ± 3	− 1.65 ± 0.2
n-hexane → n-butane + C_2H_6	44.5 ± 3	− 1.3 ± 0.2
n-pentane → n-butane + CH_4	38.5 ± 3	− 0.6 ± 0.2
isopentane → n-butane + CH_4	35 ± 3	− 0.7 ± 0.1

P_{H2} = 760 torrs P_{HC} = 5 torrs   T = 260–290°C

Table 1. Isomerization of C_5 hydrocarbons: orders vs hydrogen. and activation energies.

II. Existence of two bond-shift mechanisms

The kinetic parameters - apparent activation energies, orders vs hydrogen and hydrocarbon - have been determined for all the contact reactions, isomerization and hydrogenolysis, of n-pentane and isopentane over a Pt-Al_2O_3 catalyst of low dispersion (10 % Pt, d = 90Å) [15]. N-pentane-2-^{13}C and 2-methylbutane-2-^{13}C were used to recognize cyclic-type isomerization and methyl shift in isopentane. As shown in table 1, the reactions may be classified into four groups, according to their apparent activation energies. The reaction with the highest activation energie is the cyclic-type isomerization of n-pentane which yields a completely scrambled molecule. The second group includes isomerization of isopentane to n-pentane and the reverse reaction. The reactions of group III are isomerization of 2-methylbutane-2-^{13}C to 2 methylbutane-3-^{13}C, isomerization of isopentane to neopentane, and hydrogenolysis of internal C-C bonds in isopentane and n-pentane. Lastly, demethylations are the less activated reactions. Remarkable in this table is the splitting of the bond-shift reactions into two groups, with activation energies differing by 10 kcal/mole, and the fact that one of these groups (III) is associated with hydrogenolysis reactions, while the other one (groupe II) is not. Interesting also is the distinction, among the hydrocracking reactions, between internal fission and demethylation. Mechanisms could account for the reactions of group III : Anderson's one involving as precursor an $\alpha\alpha\gamma$ triadsorbed species and as transient state a π complex of the Dewar type attached to the surface by two carbon metal bonds [10-11] (Figure 10),

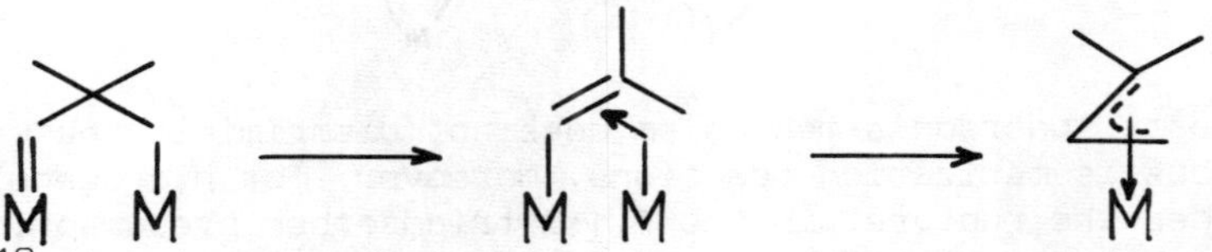

Fig. 10.

and the mechanism involving an adsorbed metallocyclobutane [19] similar to the trimethylene di σ complex of platinum, a complex that is formed readily by reacting cyclopropane and chloroplatinic acid [18] and has no analogue with any other metal (Figure 11).

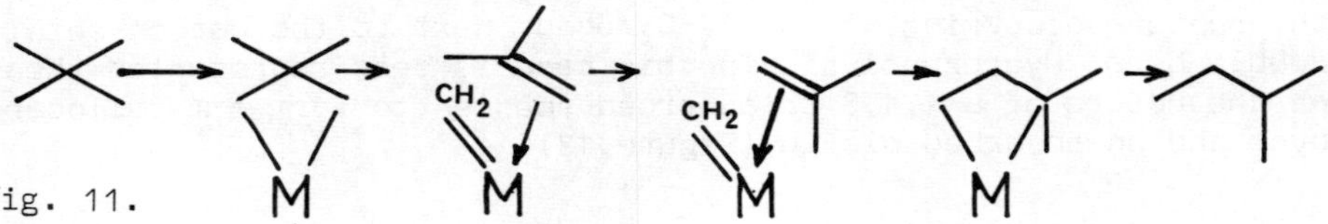

Fig. 11.

That all these reactions are on palladium either negligible (internal fission, methyl-shift) or inexistent (neopentane isomerization) is a first argument in favour of the metallocyclobutane dismutation mechanism. Moreover, when one tries to apply this mechanism to the isomerization of isopentane to n-pentane, (figure 12) adsorbed ethylidene is formed instead of adsorbed methylene like in methyl shift and neopentane formation, and it is well known, from carbene and metallocarbene chemistry [20, 21] that substituted metallocarbenes are rapidly isomerized by hydrogen shift to adsorbed olefins. The reformation of the metallocyclobutane cannot occur then and bond-shift isomerization is replaced by hydrogenolysis of the internal C-C bond. One explains then at the same time why internal fission has the same activation energy as methyl shift and why chain lenghtening does not take place by the metallocyclobutane mechanism and requires a more activated mechanism.

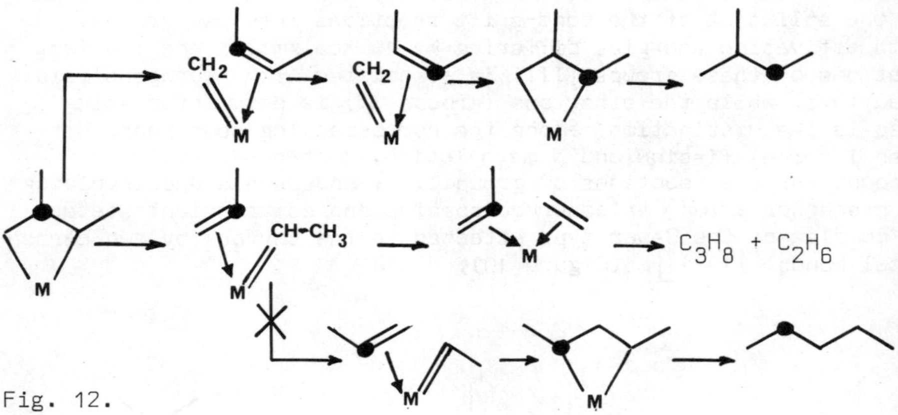

Fig. 12.

In contrast, Anderson's mechanism does not discriminate between the various isomerization reactions. Moreover, for hydrogenolysis, it provides the rupture, in the 1,1,3-triadsorbed precursor, of the C_2-C_3 bond, next to the simple carbon-metal bond. However, in a recent investigation of the hydrogenolysis of a number of hydrocarbons on platinum catalysts, Leclercq, Leclercq and Maurel showed that the C-C bonds in β position to a tertiary carbon atom were preferentially ruptured [22] . As pointed out by Leclercq, this result is best explained by a mechanism involving a 1,1,3 triadsorbed species, the precursor in Anderson's mechanism, but with the rupture occurring at the C_1-C_2 bond, next to the carbon-metal double-bond. Hydrogenolysis in this case is best represented then as the rupture of a 1,1,3 triadsorbed species to form a metallocarbyne and an adsorbed olefin (Figure 13).

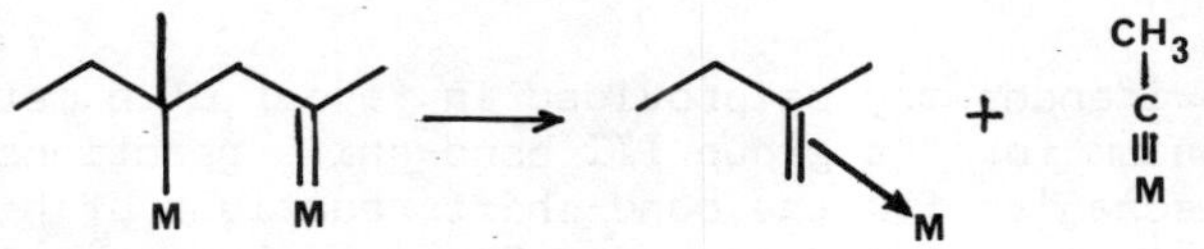

Fig. 13.

Isomerization of ^{13}C-labeled C_7 hydrocarbons-2,3 dimethylpentane [14], 2-methylhexane [23], 3-methylhexane [16] - on platinum provides further evidences for the existence of two bond-shift mechanisms and also for the metallocyclobutane mechanism.

Comparison between 2,3-dimethylpentane and 2-methylhexane isomerization on one hand, [14][23], and 2-methylpentane and 3-methylpentane isomerization on the other hand [24], shows that the contributions of bond-shift reactions are decreased in the case of C_7 hydrocarbons, but not at the same extent for methyl-shift and chain-lengthening. In the case of 2,3-dimethylpentane, when comparing with the homologous reactions of 2-methylpentane and 3-methylpentane, the contribution of bond-shift relative to cyclic mechanism is divided by 1.5 for chain lengthening and by a factor of 4 for methyl displacement.

All these results, and similar ones, are readily interpreted by assuming the existence of two bond-shift mechanisms. The first one which accounts for methyl displacement, may be identified to the metallocyclobutane mechanism responsible for the group III reactions of isopentane and n-pentane. It is very sensitive to alkyl substitution. The second one, which accounts, for chain lengthening, is the same as the mechanism of higher activation energy responsible for n-pentane isopentane isomerization [15] [19]. This latter mechanism seems relatively insensitive to hydrocarbon structure.

The sharp decrease of the contribution of the metallocyclobutane mechanism when increasing substitution should probably be connected with an increase of the energy barrier for rotating (or displacing) the π-adsorbed olefin before C-C bond reformation [25]. Since the metallocyclobutane mechanism requires <u>rotation of the π-adsorbed olefin</u>, replacement in this species of hydrogen and methyl by more bulky substituents would increase the rotation barrier and decrease significantly the reaction rate. One could explain on this way the decreasing contribution of the bond-shift mechanism in isomerization when passing from pentanes to hexanes and to heptanes [16]. A further proof for this mechanism is the occurrence at very low hydrogen pressure of homologation on platinum catalysts.

This reaction, consisting in the addition of one carbon unit at once, takes place at a very small extent because the metallocyclobutane reformation is much faster than the required methylene migration [26].

While some good evidences may be provided in favour of a metallocyclobutane mechanism for the group III bond-shift reactions of isopentane, the mechanism for the bond-shift reactions of higher activation energy is not elucidated yet. Since no hydrogenolysis reactions belong to group II one should favour for these reactions either the cyclopropane mechanism (Figure 14) [27] in which stabilization of the intermediate by methyl substituents explains the predominance of ethyl shift (path B) over methyl shift (path A) in isomerization of isopentane to n-pentane [15]

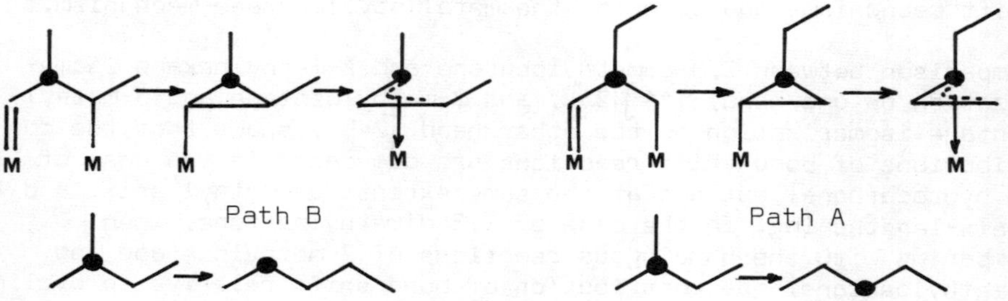

Fig. 14.

or better a concerted mechanism such as Rooney's mechanism in its original form [28] where the transient species involved three center orbitals like in the non classical carbenium ion and simultaneous π-bonding to the metal (Figure 15).

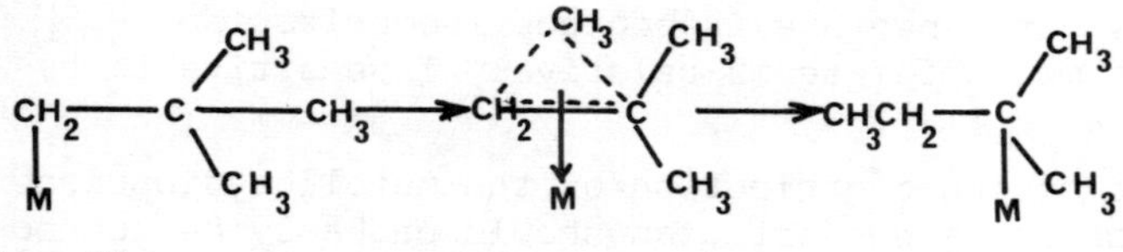

Fig. 15.

As a matter of fact, none of the above mechanisms account in their original form for the differences of behaviour between Pt and Pd in skeletal rearrangement. As noted before, while isobutane is isomerized on both platinum and palladium, platinum but not palladium catalyses the isomerization of neopentane to isopentane [10].

On the other hand on platinum of very high dispersion, methyl shift in isopentane has the same activation energy as n-pentane ⟶ isopentane interconversion, isomerization to neopentane

Skeletal isomerization of hydrocarbons

becomes a very minor reaction and hydrogenolysis of the internal C-C bonds have the same kinetic parameters as demethylation [19]. Clearly, all the reactions that belong to group III on the catalyst of low dispersion completely disappear and the only bond-shift mechanism left is the one with the higher activation energy, since this mechanism does not allow rearrangement at a quaternary carbon atom.

The easiest way to explain these differences is to assume that in the case of Pd (or mechanism for the group II) the precursor species is attached to the metal by three consecutive carbon atoms. With this asumption a mechanism has been proposed for bond-shift (and ring enlargement) on Pd involving, as a precursor, a π-allylic species easily formed on this metal, and as the rate-determining step, the rupture of a carbon-carbon bond by hydrogen attack to form a metallocarbene and a π-adsorbed olefin [29].

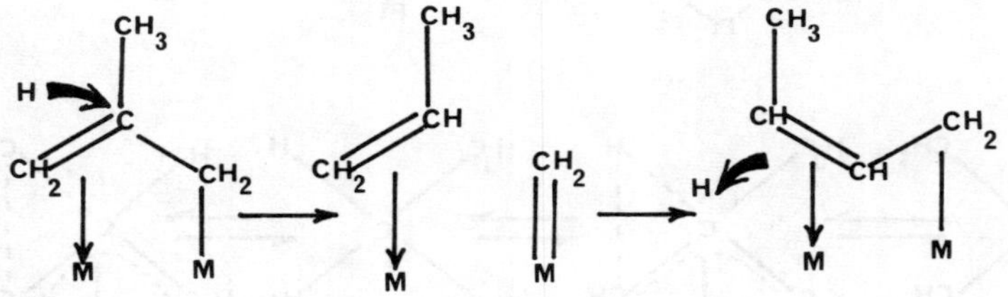

Fig. 16.

Note that the first and the last steps in figure 16 should obviously be decomposed into two, each of which are well-known reactions in organometallic chemistry : in figure 17, the attack of the π-allylic species to form a metallocyclobutane has its analogue in nucleophilic attack of a π-allylic complex by hydride ions and the dismutation of metallocyclobutane into metallocarbene and adsorbed olefin is the first step in the now well established mechanism of metathesis, as proposed by Chauvin and Herrisson [30].

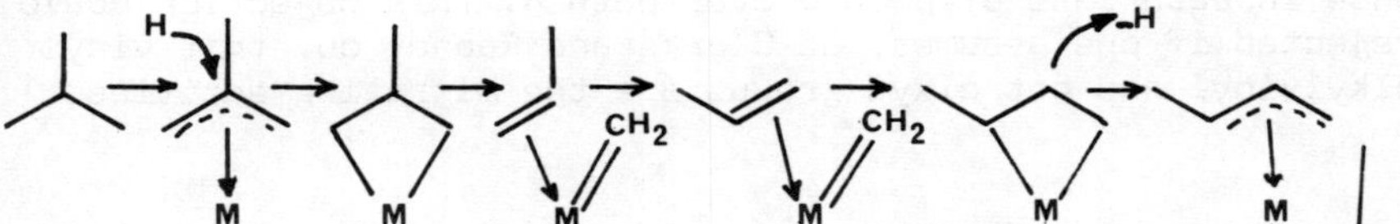

Fig. 17.

A last attempt to account for the difference of behaviour between Pd and Pt in bond-shift isomerization has been presented in a recent review by Clarke and Rooney. This new mechanism, which is also based on the ability of platinum and not palladium to form a metallo-cyclobutane, is derived from early Rooney's mechanism by replacing the σ-alkyl precursor by a metallocyclobutane and in the transient state π-olefinic bonding by π-allylic bonding. Like in the previous mechanism it is assumed that on platinum the metallocyclobutane is directly formed, while in the case of Pd, it would result from a 1,2 hydrogen shift via transient species of π-allylic character.

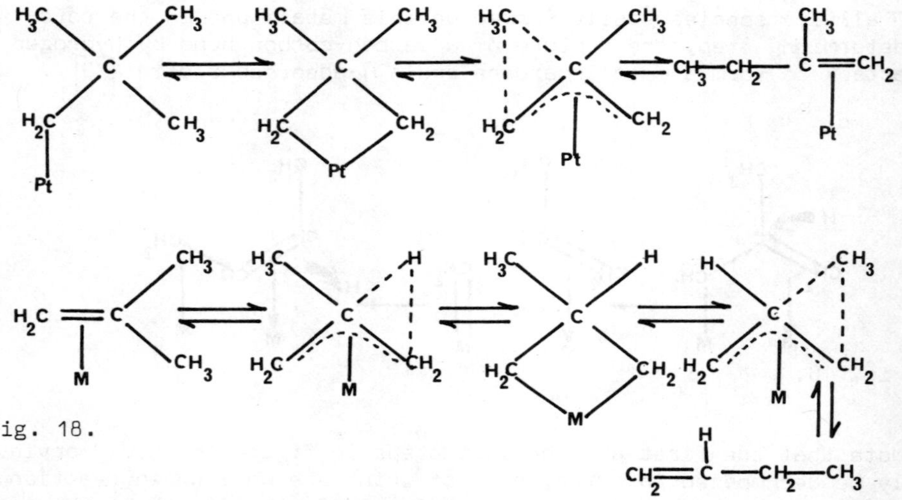

Fig. 18.

From all these proposed mechanisms, the first one (cyclopropane mechanism [27]) could not account for the isomerization of a series of 2-methylalkanes to n-alkane, for the increase, when increasing the chain length, of methyl shift (A) relative to alkyl shift (B). In contrast, Rooney's mechanism would explain this progressive increase, if one assumes that the rate of formation of the transient species with three center orbitals decreases with increasing the size of the migrating alkyl group. In its original form however Rooney's mechanism does not explain the predominance in each case of path B over path A. This objection could be rejected if one assumes, as Clarke and Rooney do, that vinyl or alkylvinyl and not alkyl groups are the migrating entities.

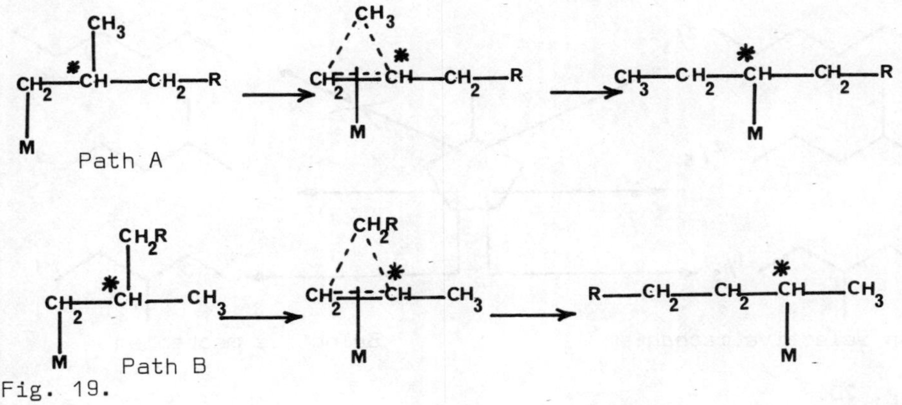

Skeletal isomerization of hydrocarbons

[Figure 19: Reaction schemes showing Path A and Path B with CH₃ groups, asterisks, and M (metal) labels]

Fig. 19.

III. Cyclic mechanisms

Isomerization according to a cyclic mechanism includes several consecutive steps.
1. Dehydrocyclization to a cyclopentane intermediate, 2. interconversions between adsorbed cyclopentane intermediates, 3. ring opening and desorption of the acyclic product. Steps 2, which allow the ring to be ruptured at a different position from where it has been formed, are most probably the ones which account for multiple HD exchange of hydrocarbons. They are very fast reactions, while step 1 and the reverse step 3 are rate-determining for alkane dehydrocyclization (or isomerization) and cyclopentane hydrogenolysis respectively. We shall therefore discuss in turn these two reactions, reviewing the most significant experimental data, in order to derive a reasonable reaction mechanism.

1. Hydrogenolysis of cyclopentane hydrocarbons on platinum

In a study of the hydrogenolysis of methylcyclopentane and 1,3-dimethylcyclopentane on a series of $Pt-Al_2O_3$ catalysts with different metal loading (from 0.2 to 20 %), it was found that the distribution of the reaction products changed substantially with changing the percentage of platinum on the carrier. An almost selective rupture of the CH_2-CH_2 bonds was found on the most concentrated catalysts (more than 2 % of Pt) [33], this mechanism does not allow however the rupture of quaternary secondary or quaternary tertiary C-C bonds [1]. On the less loaded catalyst (less than 1 %) the chances to rupture the five C-C bonds of the ring were approximately equal [9] [33]. This effect is interpreted as a particle size effect [31] - [32] - [9] - [35].

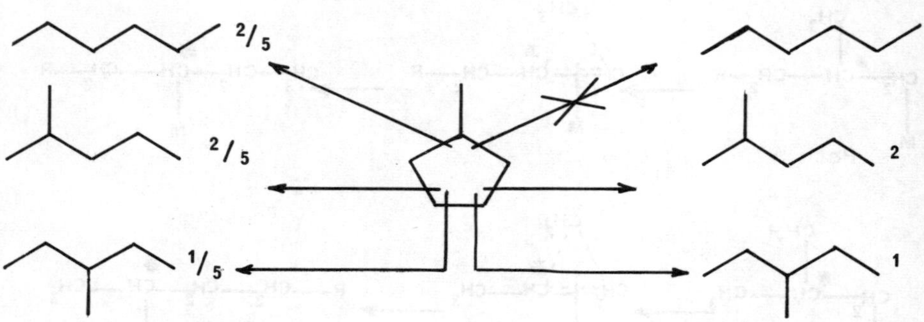

Non selective mechanism   Selective mechanism

Fig. 20.

A more careful study of methylcyclopentane hydrogenolysis, made with the two catalysts of extreme dispersions, showed that on 0.2 % $Pt-Al_2O_3$ the product distribution did not vary with temperature between 250 and 320°C, while it did significantly on the catalyst of low dispersion. On this 10 % $Pt-Al_2O_3$ catalyst all the observed distributions appeared as linear combinations of two limit distributions, one of which included only 2-methylpentane and 3-methylpentane and corresponded then to a completely selective hydrogenolysis of the bisecondary C-C bonds. The second limit distribution contained n-hexane, but was different from the one obtained on 0.2 % $Pt-Al_2O_3$. It was proposed then that hydrogenolysis of cyclopentanes could occur according to three distinct mechanisms : one, the non-selective mechanism A occurring on highly dispersed catalysts, corresponds to an equal chance of breaking any C-C bonds. Two other mechanisms, the selective one B and the partially selective one C, involving or not selective rupture of CH_2-CH_2 bonds, compete on catalysts of low dispersion. While mechanism C could not be isolated, on account of the fast consecutive isomerization of acyclic products that occur, above 320°C, mechanism B was associated with the methylpentanes distribution obtained at the lowest temperature (220°C). The distinction between two reaction mechanisms on the catalyst of low dispersion might appear formal, and the variations of the product distributions with temperature, although not their expression as linear combination, could be interpreted by a temperature dependent steric interaction between surface and adsorbate. However it was found that hydrogenolysis of methylcyclopentane on iridium supported catalysts was completely selective, whatever the temperature and whatever the dispersion of the metal on the carrier [36] . Mechanism B therefore, like mechanism A is real, and the temperature dependency of the product distribution on 10 % $Pt-Al_2O_3$ shows also the existence of a third mechanism C.

Skeletal isomerization of hydrocarbons

Precursor species in hydrogenolysis

Mechanism B, associated, with the selective rupture of CH_2-CH_2 bonds, obviously involves $\alpha\alpha\beta\beta$ tetraadsorbed species - 1,2 dicarbenes -. On the other hand, a suitable precursor species for mechanism A is the $\alpha\beta$ diadsorbed species-adsorbed-π-olefin.

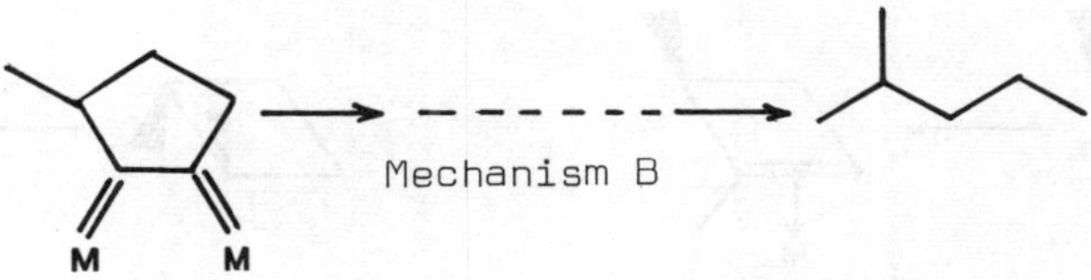

Fig. 21.

Such a species would account for the equal rupture of bisecondary, secondary-tertiary and ditertiary C-C bonds and for the inertness of quaternary-secondary and quaternary-tertiary C-C bonds.

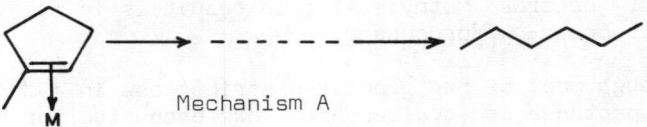

Mechanism A

Fig. 22.

Several additional facts could be presented in favour of the precursors species presented in figures 21 and 22.
In the case of 1,2-dimethylcyclobutanes, cis trans isomerization occurs at a moderate rate on platinum at low temperature and is negligible on nickel and rhodium. It is therefore possible to obtain the initial hydrogenolysis distributions for both cis and trans isomers; n-Hexane obtained by rupture of the C_1-C_2 bonds is formed in much larger amounts from the cis than the trans isomers, as one would expect since π-adsorbed olefin is known to be formed by cis elimination of two hydrogen atoms in $\alpha\beta$ position [37] [31].

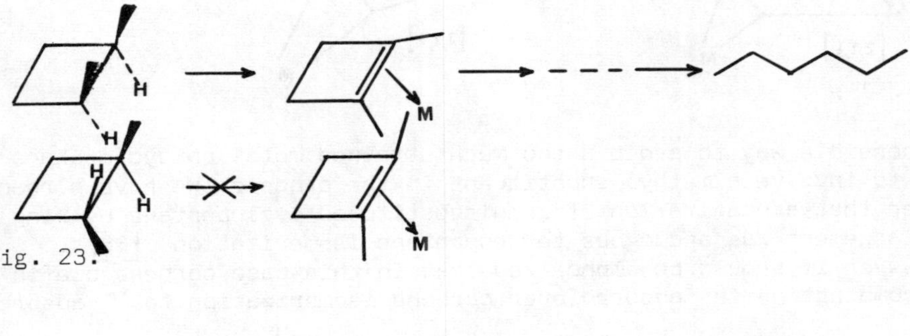

Fig. 23.

On account of steric interaction between the surface and the methyl groups, adsorption of the tertiary secondary C-C bonds sould be much less easy in the trans than in the cis isomer : that could account for the smaller amount of 3-methylpentane that is obtained from the trans isomer.

Fig. 24.

In the experiments on films, unlike the ones on 0.2 % Pt-Al_2O_3 made at higher temperature, part of the 2,3 dimethyl butane is formed by mechanism B. The contribution of this selective hydrogenolysis increases when changing the metal in the same order Pd < Pt < Ni < Rh as multiple exchange of methane. Since multiple exchange of methane is usually taken as a criterion for the ability of a metal to form adsorbed methylene, this result is in fair agreement with the proposed $\alpha\alpha\beta\beta$ precursor.

Although most of the product distributions in hydrogenolysis of cyclopentanes and cyclobutanes, are accounted for either by mechanism A or by mechanism B, a number of anomalies remain, which, we believe, should be attributed to a third mechanism C analogous to the bond-shift metallocyclobutane mechanism. Since this mechanism accounts not only for bond-shift but also for C-C bond rupture, hydrogenolysis of cyclic hydrocarbons might also occur accordingly. However, metallobicyclic compounds of the [3,1,1] and [2,1,1] series might be unfavoured on account of strain. (Figure 25).

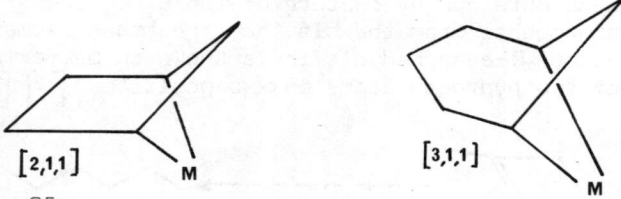

Fig. 25.

A possible way to avoid a too much strained metallobicycloalkane is to involve a methyl substituent in the process. We have already seen that aromatization of gemdisubstituted cyclopentane by ring enlargement was analogous to neopentane isomerization [38] . However it should be emphasized that in this case carbene olefin recombination is favoured over carbene isomerization to π-adsorbed olefin, since xylenes and not 2,4-dimethylhexane are obtained in significant amounts [38].

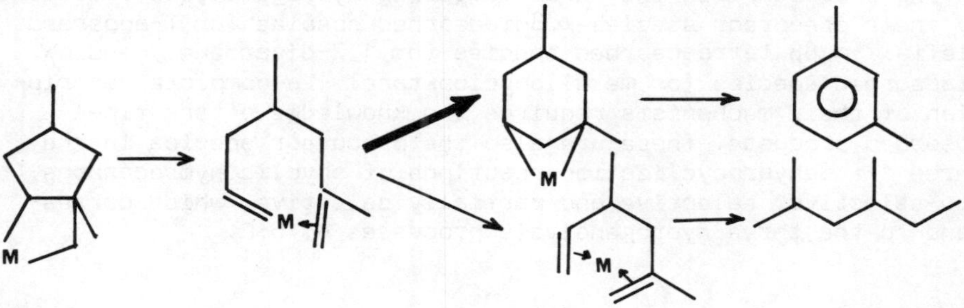

Fig. 26.

Another example of probable intervention of the metallocyclobutane mechanism C is the hydrogenolysis of the C_1-C_2 bond in trans 1,2-dimethylcycloalkane. The formation of small but significant amounts of n-hexane from <u>trans</u> 1,2-dimethylcyclobutane cannot be explained by mechanism A or B but readily by mechanism C.

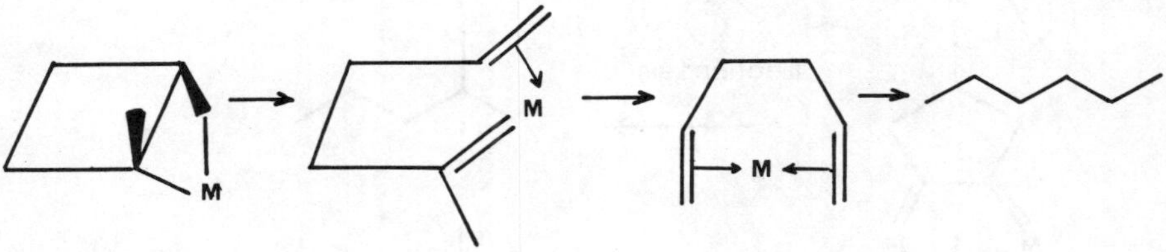

Fig. 27.

Abnormal also, according to mechanisms A and B, are the large amounts of n-heptane obtained from 1,2-dimethylcyclopentane on 10 % Pt-Al_2O_3. On this catalyst of low dispersion, according to the amounts of 3-methylhexane obtained by rupture of the tertiary-secondary bonds, the selective mechanism B largely predominates over the non-selective mechanism A, and n-heptane should be obtained in very small amounts. Most of the n-heptane then, we believe, is formed by a metallocyclobutane mechanism identical to the one represented in figure 27.

Lastly, the n-hexane which appears at the higher temperature on the catalyst of low dispersion in methylcyclopentane hydrogenolysis and was attributed to the so called "partially selective" mechanism C could also be accounted for by the metallocyclobutane mechanism.

2. Dehydrocyclization of alkanes

Having characterized the three competing hydrogenolysis reactions by their precursor species - $\alpha\beta$ diadsorbed species (or π-adosrbed olefin) - $\alpha\alpha\beta\beta$ tetraadsorbed species (or 1,2-dicarbene), and $\alpha\gamma$ diadsorbed species (or metallocyclobutane) - a complete description of their mechanisms requires the knowledge of the final adsorbed products. These are also the precursor species in the three 1-5 dehydrocyclization reactions of acyclic hydrocarbons, non-selective, selective and partially selective, which correspond to the three hydrogenolysis processes A, B, C.

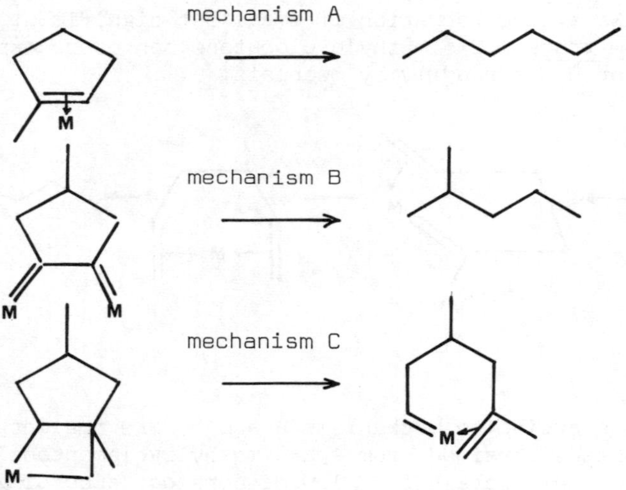

For mechanism C, identical to the bond-shift mechanism already discussed, [15] the precursor species is a 1,1,4,5-tetraadsorbed species (1,1-carbene 4,5-π-adsorbed olefin) and dehydrocyclization consists in a carbene-olefin addition.

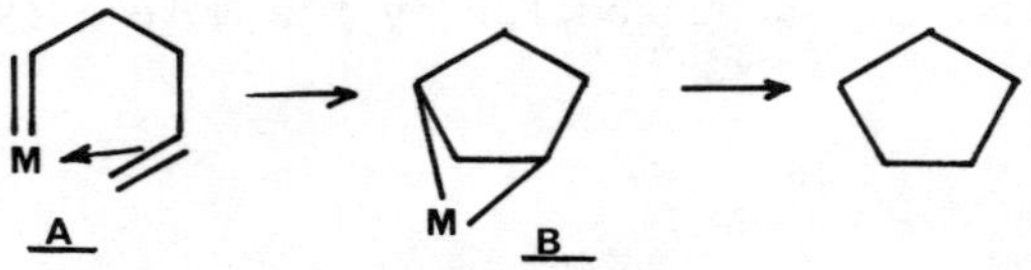

Fig. 28.

Skeletal isomerization of hydrocarbons

As already pointed out this mechanism as such is rather unfavourable on account of the strain in the metallobicycloalkane B, except if B is a transition complex between the diadsorbed carbene-olefin and a π-allylic complex C.

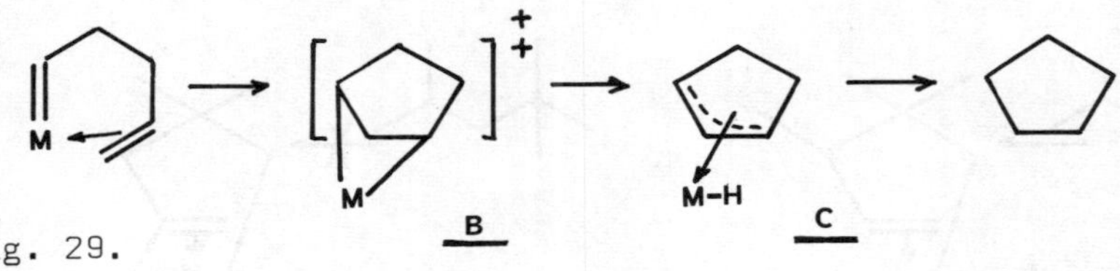

Fig. 29.

Mechanism C for dehydrocyclization (and for hydrogenolysis) might therefore be important for these metals such as Pd which favour the formation of π-allylic species. [39-40]

It should be outlined that the mechanism in figure 29 has some analogy with a mechanism proposed by Shephard and Rooney to explain the isomerization of o-ethyltoluene to propylbenzene on platinum [41].

A similar mechanism, in a slightly different version, was also proposed to account for cyclic type isomerization of hexanes and non-selective hydrogenolysis of methylcyclopentane, (figure 30) [42].

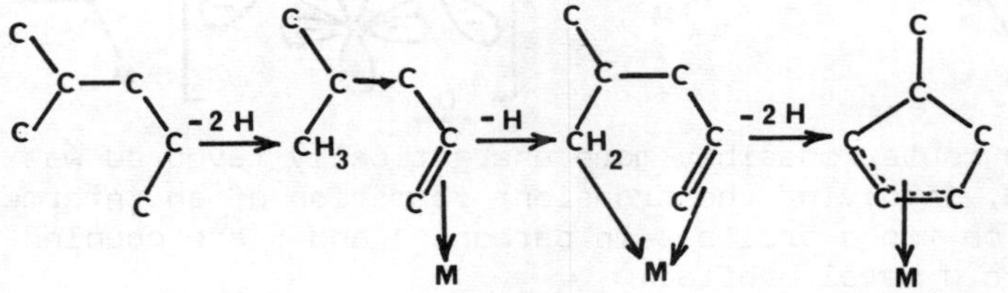

Fig. 30.

The common feature in figures 28, 29 and 30 was that they require preliminary formation of a π-adsorbed olefin. Therefore a molecule like 2,2,4,4-tetramethylpentane I, which cannot dehydrogenates without skeletal rearrangement should not dehydrocyclize to 1,1,3,3-tetramethylcyclopentane.

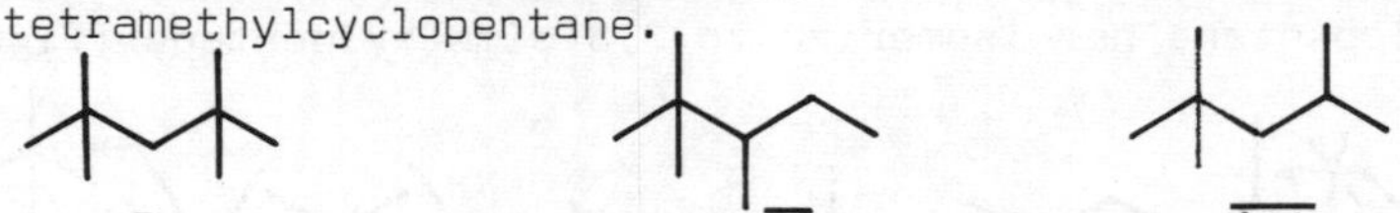

I   II   III

This is the case indeed on palladium [29] : at 300°C on Pd film, the relative rates of dehydrocyclization of I, 2,2,3-trimethyl (II) and 2,2,4-trimethylpentane (III) are respectively $<10^{-2}$, 2.6

and 0.18. Olefin formation is clearly required then for dehydrocyclization and steric hindrance by olefin substitution drastically decreases the reaction rate, as it was already found for 1-6 ring closure.

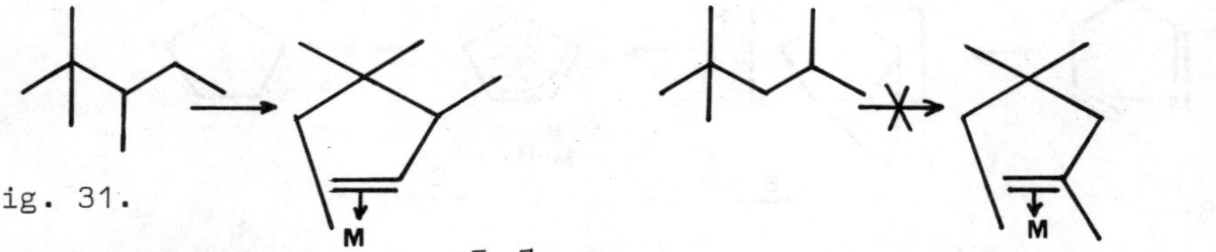

Fig. 31.

On platinum by contrast, [29] all three compounds I, II and III dehydrocyclise with very similar rates (1/1, 2/1.2 respectively), which shows that dehydrocyclization does not require olefin formation and that the precursor species is attached to two carbon atoms only in 1,5 position. The metallocyclobutane mechanism (or the alkene/alkyl insertion mechanism) should be rejected in this case, and a mechanism involving carbene insertion in a σ metal-carbon bond was proposed to account for the results [43].

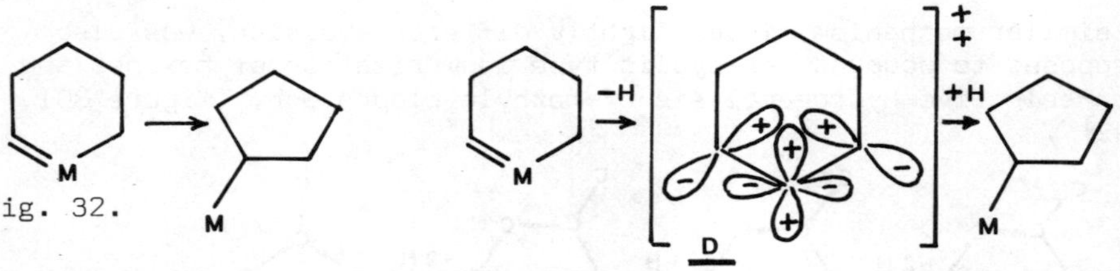

Fig. 32.

An alternative route, possibly more energetically favoured was also suggested, involving the transient formation of an intermediate D in which two p orbitals in carbons 1 and 5 are coupled together with a d metal orbital.

Although the carbene-alkyl insertion mechanism explains the identical dehydrocyclization rates of all three substituted pentanes I, II, III on Pt, it does not account for the selectivity in hydrogenolysis and 1,5 ring closure. Quaternary-secondary C-C bond cannot be ruptured in hydrogenolysis of cyclopentanes and conversely cyclization involving a tertiary carbon atom does not occur: for instance 2-methylhexane does not dehydrocyclise to gemdimethylcyclopentane nor isomerize to 3,3-dimethylpentane (figure 33).

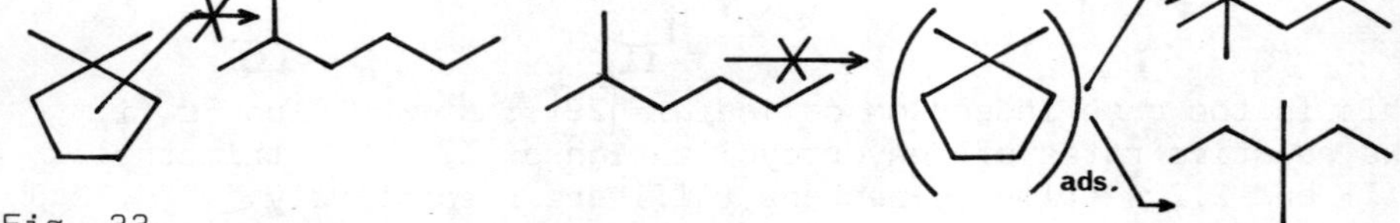

Fig. 33.

1-5 ring closure therefore only takes place between methylenic
or methylic carbon atoms in 1-5 position, which is not accounted
for by the mechanism in figure 32. Dimetallocarbenes then are obviously the precursor species in this reaction and mechanism A
therefore is best described as a dicarbene recombination forming
an adsorbed olefin ; a reaction which is well-known in carbene
and metallocarbene chemistry.

Fig. 34.

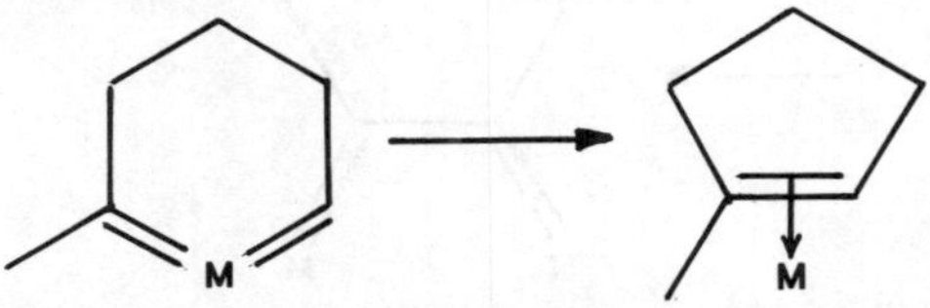

On the other hand the selective dehydrocyclization (mechanism B)
which does not allow formation of tertiary secondary C-C bond
should only involve two methylic carbon atoms in 1-5 position.
Although the reverse reaction, selective hydrogenolysis, could be
observed on platinum catalysts of low dispersion at 220°C, selective dehydrocyclisation is noticeable only at higher temperature
where it competes on 10 % Pt-Al$_2$O$_3$ with an other process, mechanism C. Fortunately metal catalysts were found recently - Ir-
Al$_2$O$_3$ or Ir-SiO$_2$ - which catalyse selectively at 160°C the cyclic
type isomerization of 2-methylpentane to 3-methylpentane [36].

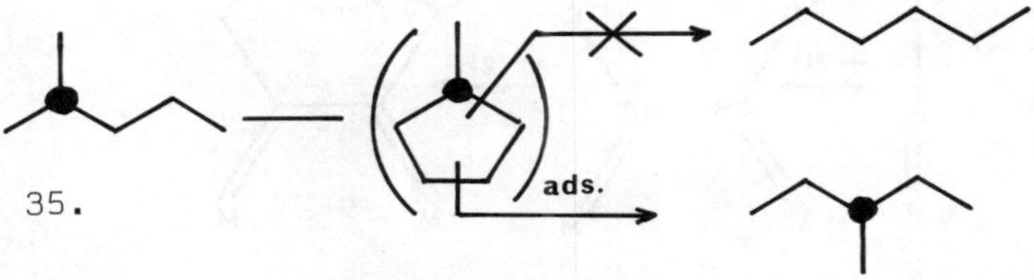

Fig. 35.

The interconversion between ethylpentane and 3-methylhexane also
takes place on the same catalysts while 2-methylhexane and n-hexane
yields only hydrogenolysis products.

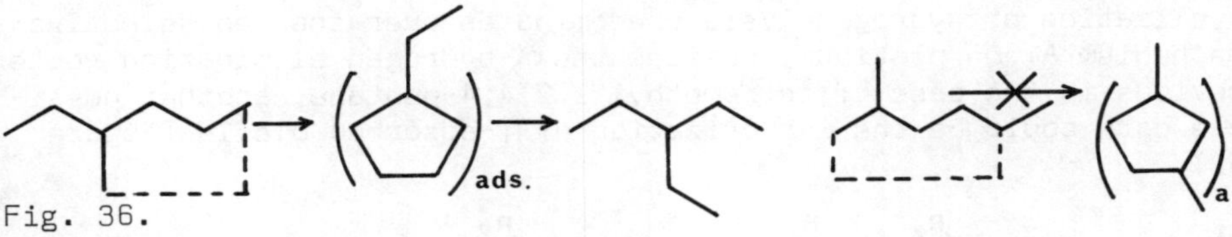

Fig. 36.

1-5 ring closure in this case takes place only between two primary carbon-atoms and the precursor species in mechanism B could be
a metallodicarbyne E . Such a species E , on account of the lineari-
ty of the M≡C-C-bonds involves necessarily two metal atoms instead
of one, which is in fair agreement with the fact that mechanism B

on platinum unlike mechanism A takes place only on very large metal crystallites (vide infra). Selective dehydrocyclization of type B is best pictured then as a dicarbyne recombination (figure 37) involving two metal atoms.

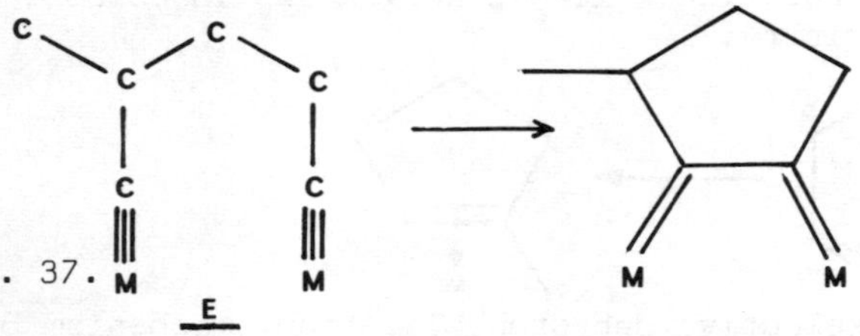

Fig. 37.

Although metallocarbyne has been isolated, dicarbyne has yet no analogue in coordination chemistry. One could object that the dicarbyne recombination in figure 37 is made difficult by the distance between the two carbon atoms 1 and 5, equal to the intermetallic distance. To overcome this difficulty, one could propose that dicarbyne recombination involve two consecutive steps, with hydrogen addition and removal, and the transient formation of a dicarbene intermediate.

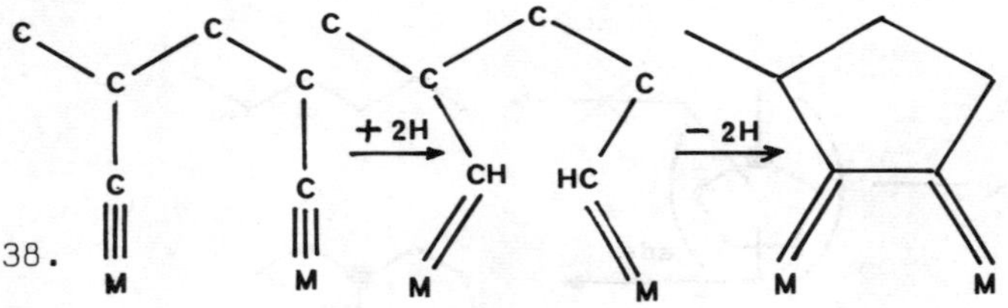

Fig. 38.

A common precursor for both dicarbene and dicarbyne mechanisms is adsorbed metallocarbene. This species may be formed by α-hydrogen elimination [44] and that is most probably the case on iridium. On this metal, further dehydrogenation to metallocarbyne is also easy and that explains why the selective mechanism B of dehydrocyclization or hydrogenolysis predominates over the non-selective mechanism A. On platinum, besides the α-hydrogen elimination route, obvious in the case of tetramethyl 2,2,4,4-pentane, another possible path could be the isomerization of π-adsorbed olefin (figure 39).

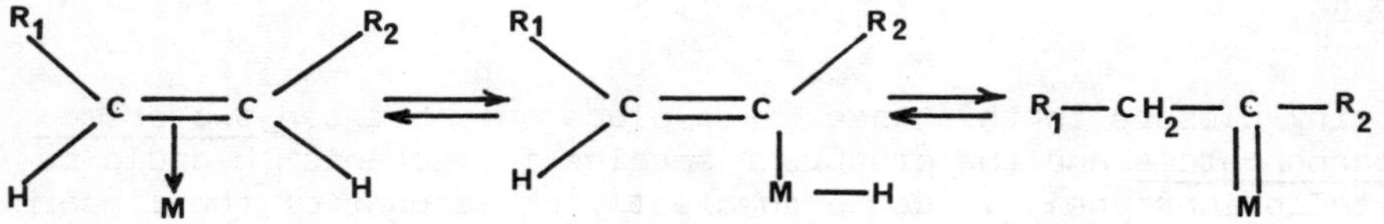

Fig. 39.

The equilibrium in figure 39 is of course shifted much to the left and the more especially as the π-adsorbed olefin is more substituted. Nevertheless the concentration of metallocarbenes by reaction 39 could be raised high enough, at elevated temperature, to permit the formation of dicarbene or dicarbyne and the occurrence of dehydrocyclization.

C. AROMATIZATION

Conflicting results are presented in the literature in favour of two mechanisms of aromatization on platinum : direct 1-6 ring closure [45-46] and 1-5 ring closure followed by ring enlargement [2] - [47] . The 1-6 ring closure mechanism is evidenced by the initial formation of benzene from n-hexane and not from methylpentanes on highly supported platinum catalyst and by the aromatic product distributions obtained from a number of methylheptanes and dimethylhexanes. On the other hand the identical distributions of para- and metaxylenes obtained from 2,2,4-trimethylpentane and 1,1,3-trimethylcyclopentane is definitely in favour of a 1-5 ring closure - ring enlargement mechanism. Even in experiments done with non geminated acyclic hydrocarbons, formation of 1,2,3-trimethyl and 1,2,4-trimethylbenzene from 2-methyloctane, of meta and paraxylenes from 3,4-dimethylhexane cannot be accounted for by simple 1-6 dehydrocyclization.

Davis suggests to explain such abnormal products the occurrence of skeletal isomerization before 1-6 ring closure. However on platinum films aromatics are formed as initial products from methyl substituted pentanes (2 and 3 methyl, 2,2,4-and 2,2,3-trimethylpentanes) and from substituted cyclopentanes (1,1,3- and 1,1,2-trimethyl, 1,1-dimethyl, methyl- and ethylcyclopentanes). The 1-6 ring closure mechanism is not the only one then and an additional mechanism also takes place, the importance of which depends upon the structure of the reacting hydrocarbons and on the experimental conditions. This mechanism in the case of gemdisubstituted cyclopentane is best represented, we believe, by figure 26, and by figure 40 in the case of gemdisubstituted pentanes.

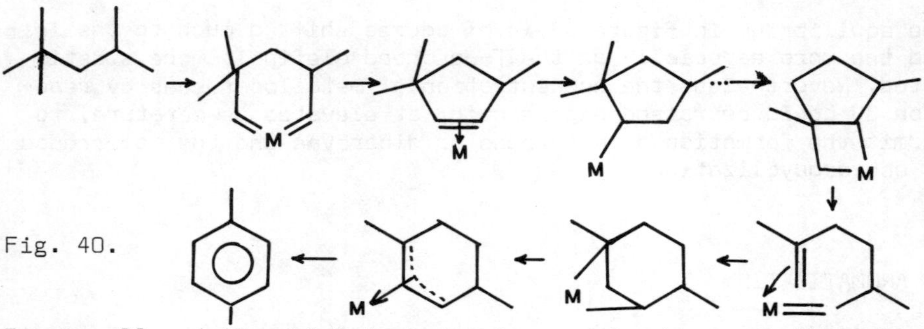

Fig. 40.

Figures 26 and 40, which include ring opening by a metallocyclobutane mechanism followed by carbene-olefin addition account for a number of experimental facts.

1) That gemdisubstituted cyclopentanes on platinum films produce much more aromatics initially than methyl and ethylcyclopentane is readily explained by methyl stabilization of the π-adsorbed olefin in the carbene olefinic species, although some other factor might also intervene such as relative rates of ring hydrogenolysis by competing mechanisms A and B.

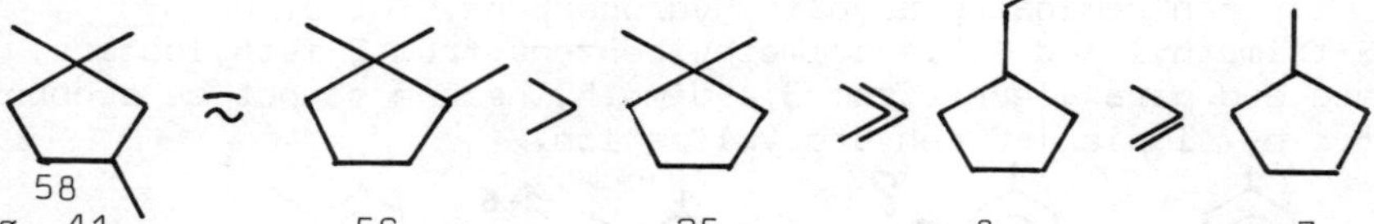

Fig. 41.   58   ≈   56   >   25   ≫   8   >   7

2) In 1,2-dimethylcyclopentane hydrogenolysis, the amount of toluene parallels the amount of n-heptane, formed by the metallocyclobutane mechanism C. Both products increase with decreasing hydrogen pressure. Moreover under the same experimental conditions toluene is not formed when mechanism C does no occur like in 1,3-dimethylcyclopentane and ethylcyclopentane.

The mechanism presented in figure 26, 40 and 41 has the advantage not only to explain the aromatization of methylsubstituted cyclopentane and pentane hydrocarbons, but also to give a simple representation of 1-5 ring closure.

Another advantage of the above mechanism (figures 26, 40) is to consider that the abnormal multistep mechanism does not involve two steps - 1-5 ring closure + ring enlargement - but three - 1-5 ring closure, ring opening, 1-6 ring closure - In such a succession of events, already proposed by Herrington and Rideal [48] for aromatization of substituted pentanes, ring opening by

Skeletal isomerization of hydrocarbons

π-adsorbed-olefin mechanism A ("non-selective mechanism"), provided the resulting dicarbene produces, by hydrogen shift, a carbene-olefin species suitable for aromatization. For instance, in the abnormal aromatization of n-heptane via adsorbed ethylcyclopentane, transient formation of adsorbed n-heptane but not of adsorbed 3-methylhexane could initiate 1-6 ring closure. (Figure 42).

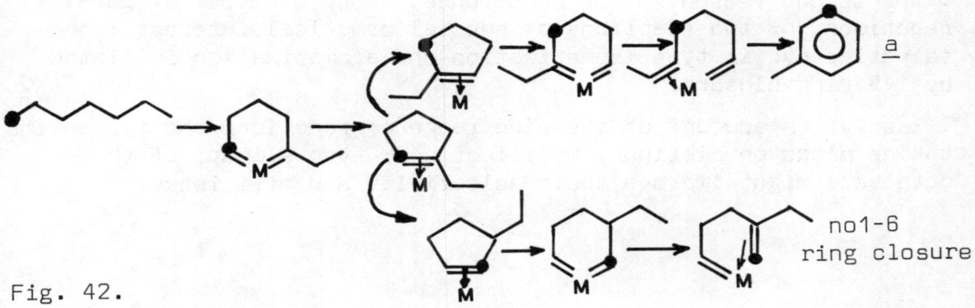

Fig. 42.

One could explain on this way the characteristic scrambling of the label during aromatization of n-heptane 1-^{13}C. The possible scrambling reactions producing meta-labelled toluene, are, besides reactions of type a involving adsorbed ethylcyclopentane and adsorbed n-heptane, reactions of type b involving adsorbed 1,2-dimethylcyclopentane and adsorbed 3-methylhexane.[49]

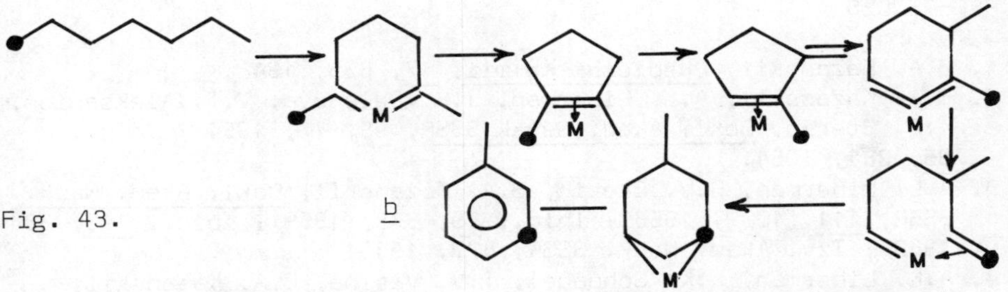

Fig. 43.

It should be noted that half of the toluene obtained by reactions a and b is methyl-labelled. As observed, while a mechanism involving cyclic type isomerization and consecutive aromatization of the isomers by 1-6 ring closure would produce less than 50 % of methyl labelled toluene.

Therefore in isomerization of acyclic hydrocarbons as in aromatization and hydrogenolysis, metallocarbenes and metallocarbynes play an important role.

CONCLUSION

Platinum is by far the best metal for the catalytic reforming of hydrocarbons. There are three reasons for that :

. the first one is the very high selectivity of platinum catalysts in isomerization. Hydrogenolysis is much less important than on any other metal, so that the yields are high.

. the second reason is the occurrence, among a number of parallel reactions, of two reactions of special practical interest : non-selective cyclic type isomerization and aromatization initiated by 1-5 ring closure.

. lastly, on account of the wide range of reactions and mechanisms taking place on platinum, modifications, even slight, of the catalysts might improve their selectivity and efficiency.

ACKNOWLEDGMENT

F.G. thanks Pr. D. Cornet for helpful discussions and Dr. L. Hilaire, Dr. G. Maire and Dr. V. Amir-Ebrahimi for their aid in the preparation of this text.

REFERENCES

1. B.A. Kazanskii, Ouspiekhi Khimii, 17, 655, 1948.
2. B.A. Kazanskii, A.L. Liberman, J.F. Bulanova, V.T. Aleksanian, K.E. Sterin, Dokl. Akad. NaukK SSSR, 95, 77, 1954 ; ibid., 95, 281, 1954.
3. A.L. Liberman, O.V. Brazin, B.A. Kazanskii, Dokl. Akad. Nauk. SSSR, 111, 1039, 1956 ; ibid., 29, 578, 1959 ; ibid., 148, 338 1963 ; Izv. Akad. Nauk. SSSR, 879, 1959.
4. A.L. Liberman, K.K. Schnabel, T.V. Vasina, B.A. Kasanskii, Kinet. Katal., 2, 446, 1961 ; Dokl. Akad. Nauk. SSSR, 117, 430, 1957.
5. G.A. Mills, H. Heinemann, T.M. Milliken, A.G. Oblad, Ind. Eng. Chem., 45, 134, 1953.
6. J.R. Anderson, B.G. Baker, Nature (London),187, 937, 1960 ; Proc. Roy. Soc. London, A 271, 402, 1963.
7. Y. Barron, D. Cornet, G. Maire, F.G. Gault, J. Catalysis, 2 , 152, 1963.
8. Y. Barron, G. Maire, J.M. Muller, F.G. Gault, J. Catalysis, 5, 428, 1966
9. G. Maire, G. Plouidy, J.C. Prudhomme, F.G. Gault, J. of Catalysis, 4 , 556, 1965

10. J.R. Anderson, N.R. Avery, J. Catalysis, 5, 446, 1966.
11. J.R. Anderson, N.R. Avery, J. Catalysis, 7, 315, 1967.
12. C. Corolleur, S. Corolleur, F.G. Gault, J. Catalysis, 24, 385 1972.
13. C. Corolleur, D. Tomanova, F.G. Gault, J. Catalysis, 24, 385 1972.
14. P. Parayre, V. Amir-Ebrahimi, A. Frennet, F.G. Gault, J. Chem. Soc., Faraday transactions II, in press.
15. F. Garin, F.G. Gault, J. Am. Chem. Soc., 97, 4466, 1975.
16. V. Amir-Ebrahimi, F.G. Gault, J. Chem. Soc., Faraday Transactions II, in press.
17. J. Wei, C.F. Prater, Advances Catal., 13, 203, 1962.
18. D.M. Adams, J. Chatt, R.G. Guy, N. Sheppard, J. Chem. Soc., 738, 1961.
19. F. Garin, Thèse d'Etat, Strasbourg, 1978.
20. W. Kirmse, in "Carbene Chemistry", Academic Press N.Y., p. 457, 1971.
21. C.P. Casey, Organic Chemistry, Volume 33-1, 1976.
22. G. Leclercq, L. Leclercq, R. Maurel, J. Catalysis, 50, 87, 1977.
23. P. Parayre, V. Amir-Ebrahimi, F.G. Gault, J. Chem. Soc., Faraday Transaction II, in press.
24. A. Chambellan, J.M. Dartigues, C. Corolleur, F.G. Gault, N. Jour. de Chimie, Vol. 1, n° 1, 41, 1976.
25. J.W. Byrne, H.V. Blaser, J.A. Osborn, J. Amer. Chem. Soc., 97, 3871, 1975.
26. F. Luck, Thèse de Spécialité, Strasbourg, 1978.
27. J.M. Muller, F.G. Gault, Symposium on mechanisms and kinetics of complex catalytic reactions, paper n° 15, Moscow, 1968.
28. M.A. Mc Kervey, J.J. Rooney, N.G. Samman, J. Catalysis, 30, 330, 1973.
29. J.M. Muller, F.G. Gault, J. Catalysis, 24, 361, 1972.
30. Y. Chauvin, J. L. Herrisson, Makromol. Chem., 141, 161, 1971.
31. G. Maire, Thèse d'Etat, Caen, 1967.
32. F.G. Gault, Thèse d'Etat, Paris, 1958.
33. F.G. Gault, C.R. Acad. Sci., 245, 1620, 1957.
34. J.M. Dartigues, A. Chambellan, F.G. Gault, J. Am. Chem. Soc., 98, 856, 1976.
35. J.M. Dartigues, A. Chambellan, S. Corolleur, F.G. Gault, A. Renouprez, B. Moraweck, P. Bosch-Giral, G. Dalmaï-Imelick, N.J. Chimie, 1979, in Press.
36. F. Weisang, F.G. Gault, J.C.S. Chem. Comm., 519, 1979.
37. G. Maire, F.G. Gault, Bull. Soc. Chim. (France) n° 3, 894, 1967.
38. V. Amir-Ebrahimi, Thèse d'Etat, Strasbourg, 1978.
39. F.G. Gault, J.J. Rooney, C. Kemball, J. Catalysis, 1, 255, 1962.
40. M. Hajek, S. Corolleur, C. Corolleur, G. Maire, A. O'Cinneide, F.G. Gault, J. de Chimie Physique, 71, 1329, 1974.
41. F.E. Shephard, J.J. Rooney, J. Catalysis, 3, 129, 1964.
42. J.K.A. Clarke, J.J. Rooney, Adv. in Catalysis, Vol. 25, p.125, 1976.

43. C. Corolleur, J.M. Muller, F.G. Gault, 20th Meet. Soc. Fr. Chim. Phys.,paper n° 21, sept. 1970.
44. J.J. Rooney, A. Stewart, Catalysis, Vol. 1, p. 277, The Chemical Society.
45. B.H. Davis, J. Catalysis, 15, 363, 1969 ; 29, 398, 1973.
46. F.M. Dautzenberg, J.C. Platteuw, J. Catalysis, 19, 41, 1970.
47. G.R. Lester, J. Catalysis, 13, 187, 1969.
48. E.F.G. Herrington, E.K. Rideal, Proc. Roy. Soc., A 184, 434 and 447, 1945.
49. V. Amir-Ebrahimi, A. Choplin, P. Parayre, F.G. Gault, J. Am. Chem. Soc., 1979, in press.

CATALYTIC REFORMING
THE REACTION NETWORK

H.S. van der Baan

Laboratory for Chemical Technology,
Eindhoven University of Technology, The Netherlands

Just as shown for catalytic cracking a reaction network can be presented for catalytic reforming, that reduces the actual complexity of the reaction network to a rather simple model. Such a simplified model is shown in figure 1.

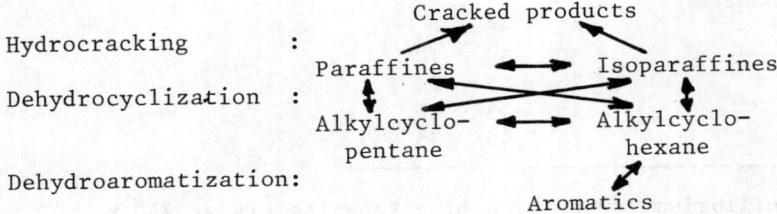

Fig. 1. Simplified reaction network for catalytic reforming.

Reaction	Apprioximate reaction enthalpy kJ/mol
Isomerization	0
Hydrocracking	− 50
Dehydrocyclization	+ 50
Dehydroaromatization (from cyclohexanes)	+ 200

Table 1. Reaction enthalpies for catalytic reforming reactions.

Of these reactions the hydrocracking is the slowest and the dehydroaromatization the fastest.
The reaction enthalpies for a number of these reactions are given in table 1.
These reaction enthalpies are averages. Especially the enthalpies of the dehydrocyclization reaction vary considerable, being only 30 kJ/mol for the conversion of n C_9 into C_3-cyclohexane, but 70 kJ/mol for the conversion of i C_6 into methylcyclopentane. As the same is valid for the Gibbs free energies of these reactions it is clear that the equilibrium composition of hydrocarbon/hydrogen mixtures not only depends on the hydrocarbon/hydrogen ratio but also on the composition of the hydrocarbon fraction of that mixture. When discussing equilibrium compositions, one has to exclude hydrocracking reactions because the hydrocracking products are thermodynamically the most favoured ones, and the final equilibrium compositions will be a mixture of methane and hydrogen only. As however the rate of the hydrocracking reaction is low, exclusion of these reactions still gives useful information.
We will also leave out of consideration the numerous isomerization reactions between the paraffines and isoparaffines and assume that the paraffin fractions will in their isomer composition approach equilibrium, of which table two gives an example.

number of branchings	0	1	2	3
carbon number				
C_4	65	35		
C_5	30	60	10	
C_6	25	45	30	
C_7	15	45	40	
C_8	15	45	35	5

Table 2. Equilibrium composition of alkane isomers at 750 K.

From the equilibrium calculations it follows that for temperatures above 600 K and hydrogen pressures between 1 and 3 MPa the cyclo paraffin content is always below 5 percent and often negligible. At 600 K the mixture consists almost exclusively of paraffins whereas at 800 K aromatics are the only components, except for C_6 under 3 MPa hydrogen. There the temperature has to be above 900 K before the C_6 paraffins are converted into benzene, as has been shown by Kugelman (1976).
Radosz and Kramarz (1975) have calculated the equilibrium aromatics content for a number of feed mixtures (different fractions of C_6, C_7, C_8 and C_9) under the assumption that no cycloparaffins would be present. A number of their results have been reproduced in figure 2.

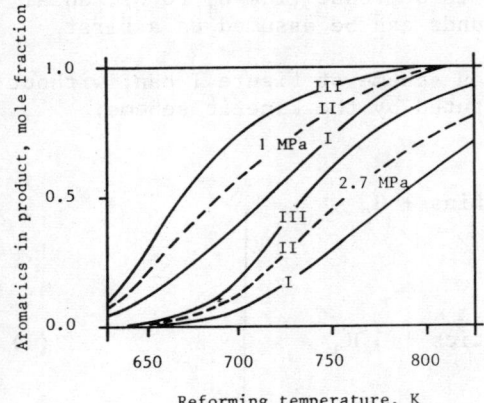

Fig. 2. Effect of feed composition, temperature and pressure on the equilibrium aromatic constant of reformate.

	Feed composition (fraction)			
	C_6	C_7	C_8	C_9
I	.4	.3	.2	.1
II	.25	.25	.25	.25
III	.1	.2	.3	.4

One other remark has to be made as far as the hydrocracking is concerned. Hamsel and Donaldson (1951) have already shown that products of hydrocracking in general do not crack any further within the reaction time available in normal plants. They found for the catalytic reforming of n heptane the following product distribution of the cracked fraction:

	moles/100 moles feed
methane	3.1
ethane	5.1
propane	25.1
butanes	26.7
pentanes	6.1
hexanes	3.0

Table 3. Distribution of light products from the catalytic reforming of n heptane.

The correspondence in yield of C_1 and C_6, of C_2 and C_5 and of C_3 and C_4 is striking and suggests that these are products from the primary hydrocracking reaction that are not broken down further. In commercial operation there is much less difference in the mole fractions of the cracked products i.a. because of the range of components in the feed. Although generally there is a slight tendency for the mole fractions to decrease from C_1 to C_5, equal molar fractions for these compounds can be assumed as a first approximation.

This then means that the reaction scheme of figure 1 can, without much loss of accuracy be substituted by the kinetic scheme:

$$\text{paraffins} \underset{k_{-1}}{\overset{k_1}{\rightleftarrows}} \text{cycloparaffins} + H_2$$

$$\text{cycloparaffins} \underset{k_{-2}}{\overset{k_2}{\rightleftarrows}} \text{aromatics} + 3\,H_2 \qquad (1)$$

$$\text{paraffins} + H_2 \overset{k_3}{\rightarrow} \text{cracked products}$$

As discussed above these rate constants are a function of the number of carbon atoms per molecule and must be determined separately.

In order to describe the conversion in a commercial platformer we can approximate the reactors, generally fixed bed axial flow reactors, by adiabatic pseudo homogeneous plug flow reactors for which we have to develop the massbalances for the various components, the enthalpy (heat) balance and the mechanical energy balance.

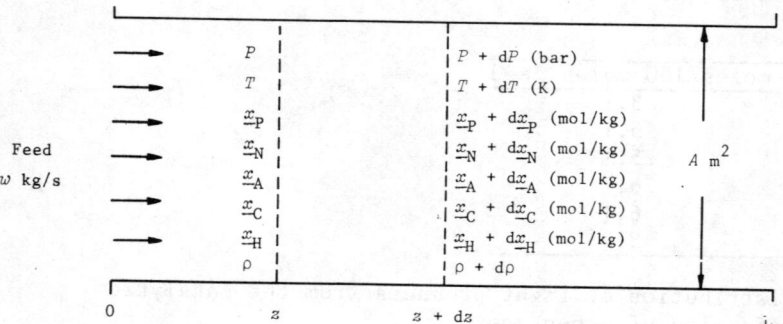

Fig 3. Model of catalytic reforming reactor.

Reaction network in catalytic reforming

We then have as the mass balances for the differential reactor volume of $A \, dz \, m^3$ for the steady state:

for paraffins $\qquad w \, d \, \underline{x}_P = r_P \, A \, dz \, \rho_{cat}$

for cycloparaffins
(naphthenes) $\qquad w \, d \, \underline{x}_N = r_N \, A \, dz \, \rho_{cat}$

for aromatics $\qquad w \, d \, \underline{x}_A = r_A \, A \, dz \, \rho_{cat}$ $\qquad\qquad$ (2)

for cracked prod $\qquad w \, d \, \underline{x}_C = r_C \, A \, dz \, \rho_{cat}$
and
for hydrogen $\qquad w \, d \, \underline{x}_H = r_H \, A \, dz \, \rho_{cat}$

with r_i the rate constant for the formation of i, per kg catalyst and ρ_{cat} the mass of catalyst per m^3 reactor volume.
We further have: (refer to equation (1))

$$r_P = k_{-1} \, C_N \, C_{H_2} - k_1 \, C_P - k_3 \, C_P \, C_{H_2}$$

$$r_N = k_1 \, C_P - k_{-1} \, C_N \, C_{H_2} - k_2 \, C_N + k_{-2} \, C_A \, C_{H_2}^3$$

$$r_A = k_2 \, C_N - k_{-2} \, C_A \, C_{H_2}^3 \qquad\qquad (3)$$

$$r_C = k_3 \, C_P \, C_{H_2}$$

$$r_{H_2} = k_1 \, C_P - k_{-1} \, C_N \, C_{H_2} + 3 \, k_2 \, C_B - k_2 \, C_A \, C_{H_2}^3$$

We eliminate the concentrations with

$$C_i = \underline{x}_i \, \rho \qquad\qquad (4)$$

and (from the ideal gas law)

$$\rho = \frac{P}{\underline{x}_{tot} \, RT} \qquad\qquad (5)$$

The enthalpy balance for the differential reactor volume $A \, dz \, m^3$ reads for the steady state:

$$- w \, \sum_i \Delta H_{r_i} \, d\underline{x}_i = c_p \, dT \qquad\qquad (6)$$

The summation of the right hand side must be done for the three reactions of equation (1) where $d\underline{x}_i$ and ΔH_{r_i} are for the first reaction:

$$d \underline{x}_i = -d \underline{x}_P - d \underline{x}_C \quad \text{and} \quad \Delta H_{r_i} = 50 \text{ kJ/mol}$$

for the second reaction:

$$d \underline{x}_i = d \underline{x}_A \quad \text{and} \quad \Delta H_{r_i} = 200 \text{ kJ/mol}$$

and for the third reaction

$$d \underline{x}_i = d \underline{x}_C \quad \text{and} \quad \Delta H_{r_i} = -50 \text{ kJ/mol}$$

and

$$c_p = \Sigma C_{p_i} \underline{x}_i \tag{7}$$

c_p being the specific heat per kg of reaction mixture and C_{p_i} the molar heat, both at constant pressure.

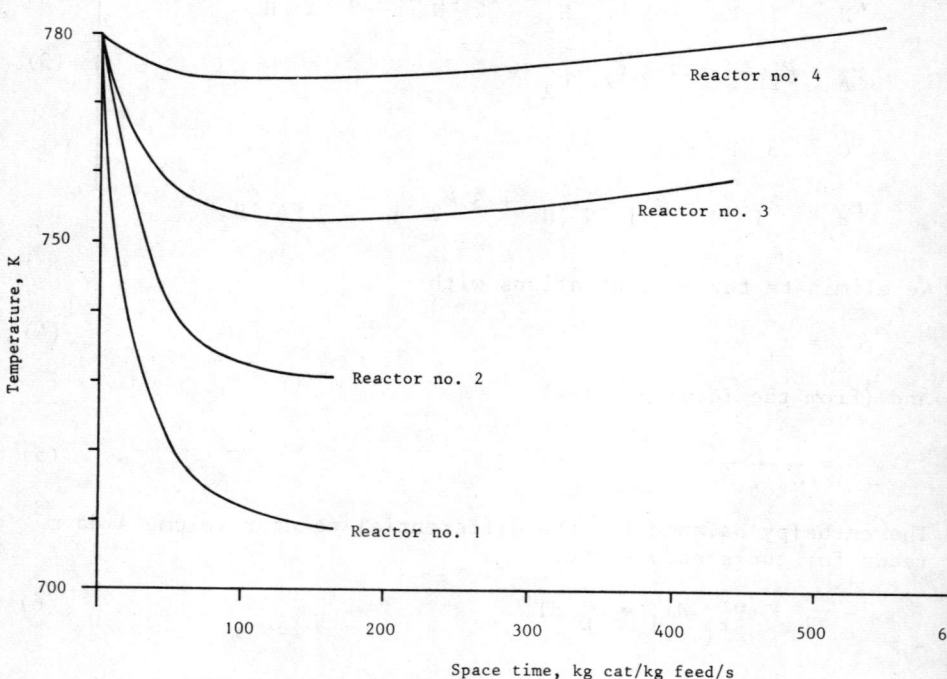

Fig. 4. Calculated temperature profile through a set of four reactors as function of the space time (kg cat/kg feed/s).

Reaction network in catalytic reforming

Finally we have the mechanical energy balance for a packed bed (Ergun, 1952)

$$dP = 150 \frac{(1-\varepsilon)^2}{\varepsilon^3} \frac{\mu v_s}{d_p^2} + 1.75 \frac{1-\varepsilon}{\varepsilon^3} \frac{\mu v_s^2}{d_p} dz \qquad (8)$$

with v_s the superficial flow rate $\frac{m^3}{s\,m^2}$, ε the porosity, μ the absolute viscosity and d_p the particle diameter (m).

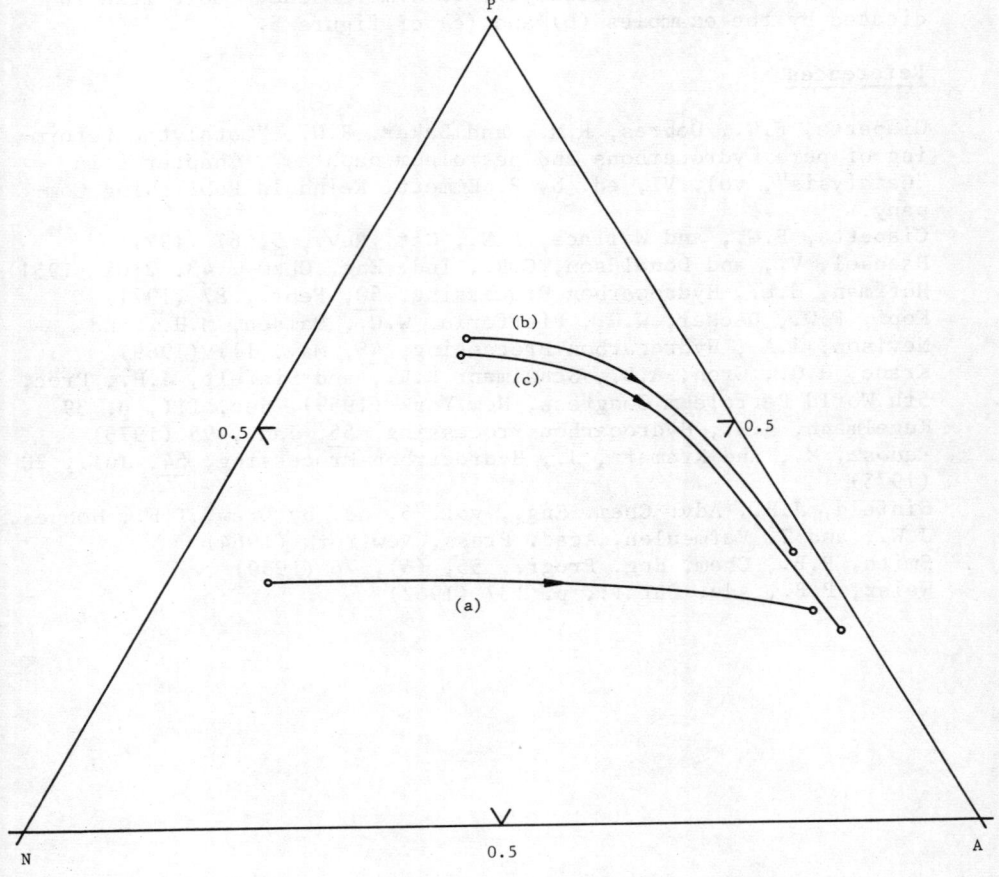

Fig. 5. Reaction paths for catalytic reforming.
    (a) low severity, naphthenic feed. ONF: reformate 95.5
    (b) medium severity, paraffinic feed. ONF: reformate 93.2
    (c) high severity, paraffinic feed. ONF: reformate 99.5

The set of equations obtained in this way can be numerically integrated starting from the given inlet conditions. A numerical example has been given by Smith (1959) for an isobaric case. The temperature profile through the four reactors is given in figure 4.
We see that especially in the third and fourth reactor the exothermic hydrocracking becomes noticeable.
The conversion path is indicated by curve (a) of figure 5, where also are included conversion paths for two more paraffinic feeds. We can see that for the example given by Smith a catalyst has been used with an extremely low conversion rate for paraffins into cycloparaffins. Modern catalytic reformers behave more like indicated by the examples (b) and (c) of figure 5.

References

Ciapetta, F.G., Dobres, R.M., and Baker, R.W., "Catalytic reforming of pure hydrocarbons and petroleum naphtas", Chapter 6 in "Catalysis", vol. VI, ed. by P. Emmett, Reinhold Publishing Company
Ciapetta, F.G., and Wallace, D.N., Cat. Rev., $\underline{5}$, 67 (1971)
Haensel, V., and Donaldson, G.R., Ind. Eng. Chem., $\underline{43}$, 2102 (1951)
Hoffman, H.L., Hydrocarbon Processing, $\underline{50}$, Febr., $\underline{87}$ (1971)
Kopf, F.W., Decker, W.H., Pfefferle, W.C., Dalson, M.H., and Nevison, J.A., Hydrocarbon Processing, $\underline{48}$, May, 111 (1969)
Krane, H.G., Groh, A.B., Schulman, B.L., and Sinfelt, J.H., Proc. 5th World Petroleum Congress, New York (1959), Sec. III, p. 39
Kugelmann, A.M., Hydrocarbon Processing, $\underline{55}$, Jan., 95 (1976)
Radosz, M., and Kramarz, J., Hydrocarbon Processing, $\underline{54}$, Jul., 20 (1975)
Sinfeld, J.H., Adv. Chem. Eng., vol. 5, ed. by Drew, T.B., Hoopes, J.W., and T. Vermeulen, Acad. Press, New York (1964)
Smith, R.B., Chem. Eng. Progr., $\underline{55}$, (6), 76 (1959)
Weisz, P.B., Adv. Catal., p. 137 (1962)

MODERN PROCESSES FOR THE CATALYTIC REFORMING OF HYDROCARBONS

R. Prins

Laboratory for Inorganic Chemistry, Eindhoven University
of Technology, Eindhoven, The Netherlands

SUMMARY. High temperatures and low pressures would be very
desirable process conditions in the catalytic reforming of
naphtha to gasoline, but catalyst deactivation by coke formation
is favoured by the same conditions. Because of that the real
process conditions are a compromise. The introduction of the
very stable bimetallic catalysts at the end of the sixties has
allowed lower pressures to be used. Nevertheless the operating
severity of the process had to go up still. The answer to this
problem has been to regenerate the catalyst more frequently and
in some cases even continuously. A description of the process
scheme of such a continuous catalytic reforming process will be
given.

INTRODUCTION

The design of the process for the catalytic reforming of
hydrocarbons is strongly determined by a combination of
thermodynamic and kinetic factors [1] (cf. Table I). A high
process temperature and a low pressure favour the conversions in
the dehydrogenation and aromatisation reactions, while the
reverse is true for the hydrocracking of paraffines and for the
hydrodealkylation of aromatics. On the other hand temperature
and pressure have almost no influence on isomerisation equilibria.
The reactions of the first group mentioned are highly endothermic,
the reactions of the second group are exothermic, while the
isomerisation reactions are only weakly exothermic. All reactions
mentioned contribute to the upgrading of a naphtha feedstock
to the desired octane number range. The endothermic reactions
dominate the heat balance, though, and heat has to be supplied

Table I Thermodynamic and kinetic data for typical reforming reactions

		P	T	$\Delta H(kJ.mol^{-1})$	log K(500°C)	$\Delta H(kJ.mol^{-1})$
dehydrogenation	$N_6 \rightarrow A$	−	+	−190	6	+190
dehydroisomerisation	$N_5 \rightarrow A$	−	+	−190	5	+190
dehydrocyclisation	$P \rightarrow A$	−	+	−240	5	+240
dehydrogenation	$P \rightarrow O$	−	+	−120	−2	+120
isomerisation	$N_5 \rightarrow N_6$			+10	−1	−10
isomerisation	$P \rightarrow iP$			+5	0	−5
hydrocracking	$P \rightarrow P' + P''$	+	−	+50	4	−50
hydrodealkylation	$A \rightarrow A + P$	+	−	+50	4	−50

N_5, N_6 = 5- and 6-ring naphthenes, A = aromatic, P = paraffin, O = olefin
A + sign in the columns P and T means that an increase in either pressure or temperature increases the conversion of the reaction considered. All values for ΔH and log K are semi-quantitative, a positive sign for ΔH means an endothermic reaction.

Modern processes for the catalytic reforming

to the process. In all existing processes this is done by using several adiabatic reactors in combination with pre- and intermediate heaters. Usually the inlet temperatures of the reactors are more or less equal. Due to the fact that especially some of the endothermic reactions are very fast, the first reactor is always smaller than the second, while the third (and fourth) is even larger still. All this leads to a process scheme as depicted in figure 1.

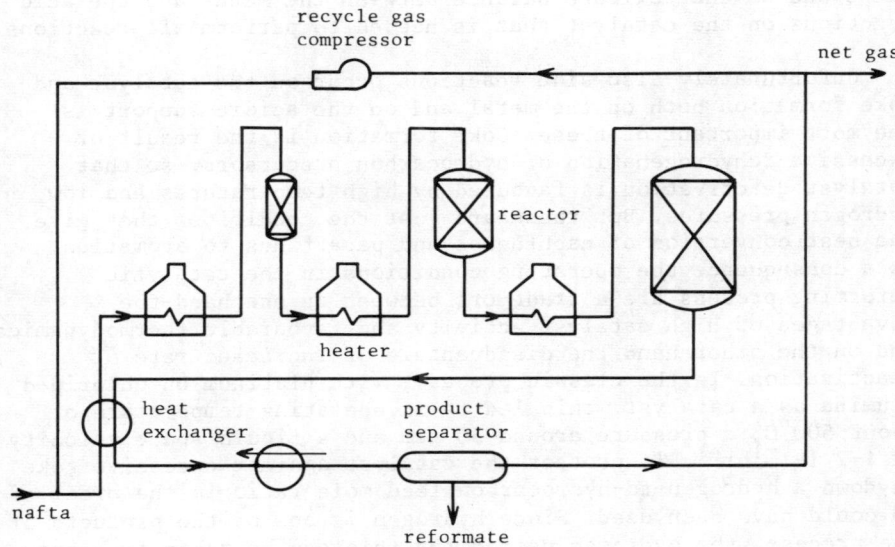

Figure 1   Process flow diagram for a conventional catalytic reforming unit with three reactors

Usually a number of elementary steps are required in reforming reaction chemistry. For instance paraffins are dehydrogenated to olefins, which isomerise and are hydrogenated to isoparaffins. Furthermore ring closure may occur, followed by dehydrogenation to aromatics. The catalyst thus has several functions, such as hydrogenation-dehydrogenation and isomerisation. The first function is performed by a metal like platinum and the second by an acid function like chlorided alumina, together giving a bi-functional catalyst. Cyclisation can be performed by the metal as well as by the acid, but it is usually assumed that under process conditions the metal cyclisation activity is surpressed [2]. Hydrocracking occurs on acid sites mainly, since the hydrogenolysis over the metal is surpressed by presulphiding of the catalyst [3].

As an example of the complexity of the elementary reaction scheme, in figure 2 the possible elementary steps are indicated from n-hexane to 2-methylpentane and to benzene. In this figure the acid catalysed reactions are indicated by a single arrow and the metal catalysed reactions by double arrows. The figure gives an indication of the inter relationship between the elementary steps and of the delicate balance between the metal and the acid functions on the catalyst that is needed to perform all reactions.

Unfortunately also side reactions occur on the catalyst and coke formation both on the metal and on the acidic support is the most important of these. Coke formation is the result of excessive dehydrogenation of hydrocarbon precursors, so that catalyst deactivation is favoured by high temperatures and low hydrogen pressures. But these are just the conditions that give the best conversion of naphthenes and paraffines to aromatics. As a consequence the operating conditions in the catalytic reforming process are a trade-off between on one hand the advantages of high catalyst activity and favourable thermodynamics and on the other hand the disadvantage of increased rate of deactivation. In the classic process, with platinum on chlorided alumina as a catalyst, this led to an operating temperature of about $500^\circ C$, a pressure around 30 atm and a liquid space velocity of 1-2 $1.1^{-1} hr^{-1}$. To protect the catalyst against excessive coke laydown a hydrogen-to-hydrocarbon feed mole ratio in the order of 10 could have been used. Since hydrogen is one of the products of the process, the hydrogen needed for this can be taken from the process itself and is recycled by means of a large gas compressor (cf. figure 1). The compression costs constitute an appreciable part of the total operating costs.

PROCESS DEVELOPMENTS IN THE SIXTIES AND SEVENTIES

In a situation of ample supply of naphtha feedstocks of high quality, the severity of the reforming process can be kept low. As a result the deactivation of the catalyst is slow and so is the temperature increase needed to counteract the loss in product quality. In such a situation, which existed in many plants in the past, the operation could be continued for a long period before either the furnace temperatures became exceedingly high or the product quantity became unacceptably low. When this happened the remedy was to shut the whole process down and to either load a new batch of catalyst or to regenerate the catalyst in the reactors
When the need for more gasoline grew and the availability of

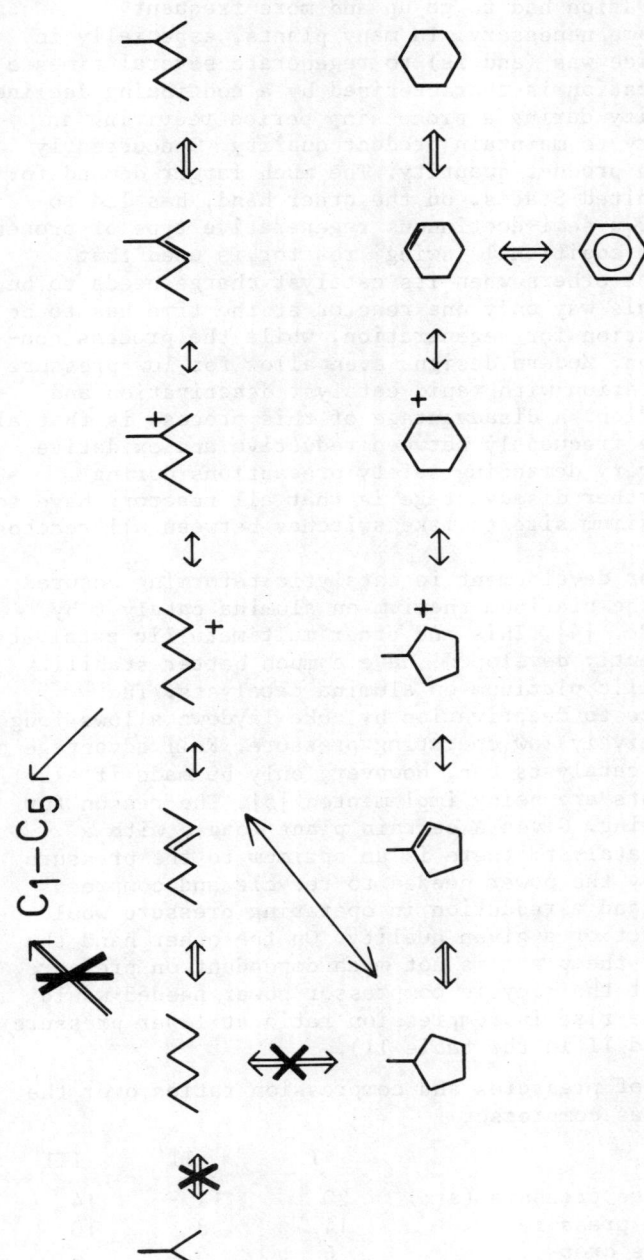

Figure 2 Reaction mechanism for the catalytic reforming of hexane to 2-methylpentane and benzene over a bifunctional catalyst.
Metal catalysed reactions are indicated with a double arrow and acid catalysed reactions with a single arrow.
Presulphiding of the catalyst eliminates hydrogenolysis, but also diminishes the metal catalysed isomerisation and cyclisation.

high quality naphtha feedstocks became limited, the severity of the reforming operation had to go up and more frequent regenerations became necessary. In many plants, especially in Europe, the practice was (and is) to regenerate several times a year. Such an operation is characterised by a continuing decline of catalyst activity during a processing period requiring an increasing severity to maintain product quality. Concurrently there is a loss in product quantity. The much larger demand for gasoline in the United States, on the other hand, has led to the development of a semi-continuous regenerative type of process. In this process an additional 'swing' reactor is used that replaces one of the others when its catalyst charge needs to be regenerated. In this way only one reactor at the time has to be taken out of operation for regeneration, while the process continues in operation. Modern designs even allow for low-pressure high-severity operation with rapid catalyst deactivation and frequent regeneration. A disadvantage of this process is that all reactors alternate frequently between reductive and oxidative conditions, with very demanding safety precautions during switch-over. A further disadvantage is that all reactors have to be of the same maximum size to make switches between all reactors possible.

In 1968 a major development in catalytic reforming occured with the invention of the platinum-rhenium on alumina catalyst by Chevron Research Co. [4]. This and other multimetallic catalysts that were subsequently developed, have a much better stability than the monometallic platinum on alumina catalysts. The enhanced resistance to deactivation by coke laydown allows long runs even at relatively low operating pressure. Full advantage of the multimetallic catalysts can, however, only be made if also process improvements are being implemented [5]. The reason for this is the following. Given a certain plant loaded with a certain batch of catalyst, there is an optimum to the pressure that is dictated by the power needed to recycle and compress hydrogen. On one hand a reduction in operating pressure would provide more product of a given quality. On the other hand the pressure drop over the plant is not much dependent on pressure, and because of that the recycle compressor power needed would increase due to the rise in compression ratio at lower pressure (compare case I and II in the table II).

Table II Examples of pressures and compression ratios over the recycle gas compressor

	I	II	III
compressor discharge pressure (atm)	20	14	14
compressor suction pressure	14	8	10
pressure drop	6	6	4
compression ratio	1.43	1.75	1.40

Modern processes for the catalytic reforming 395

The higher hydrogen-to-hydrocarbon recycle ratio needed to maintain constant run length would even necessitate a further increase in compression power. The compression power being an important cost item, there is an economic lower limit of operation pressure for each catalyst-plant combination, below which the compression power cost rises sharply [5]. When using more stable catalysts a lower pressure can be used while using the same or even a somewhat smaller recycle ratio. Full advantage is only reached when also the compression ratio can be brought down (case III in the table).

In the last decade much effort has therefore been spent on improving the pressure drop over the plant. Changes in the design of the heat exchangers, the furnaces and pipes have resulted in a lower pressure drop [5, 6]. Also a decrease in the liquid space velocity reduced the recycle gas rate needed to maintain constant run length. The decrease in capital and operating costs of a smaller recycle compressor with reduced power requirement more than offsets the increased capital costs for reactors and catalyst [5].

THE MOVING BED CONTINUOUS PROCESS

At the end of the sixties Universal Oil Products Co. started development studies on the concept of really continuous catalytic reforming. The development of such a process arose from the anticipated need to produce unleaded gasoline, which requires reforming to higher octane numbers, and the desire to produce large quantities of high-octane reformate and of high-purity hydrogen on a continuous basis [6]. All these goals could be met by the design of a low-pressure process with continuous regeneration of the catalyst. In this process small quantities of catalyst are continuously withdrawn from an operating reactor, transported to a regeneration unit, regenerated and returned to the reactor system. In 1971 the first plant of this kind was built. It consists of stacked reactors, allowing catalyst transport through the reactors via gravity flow. The catalyst is withdrawn from the bottom reactor, comparable to the third or fourth reactor in a conventional design, and regenerated catalyst is returned to the top reactor, equivalent to the first reactor in figure 1.

A process flow diagram for a continuous catalytic reforming unit with three stacked reactors [7] is given in figure 3. Comparison of this scheme with that of a conventional reforming plant (fig. 1) clearly shows the different reactor configuration and the added regenerator. Furthermore a two-separator system is used instead of a single system. The additional separator is installed at the discharge side of the compressor to take advantage of the more favourable gas-liquid equilibrium at higher pressure [6]. In this way a better recovery of product is achieved and the

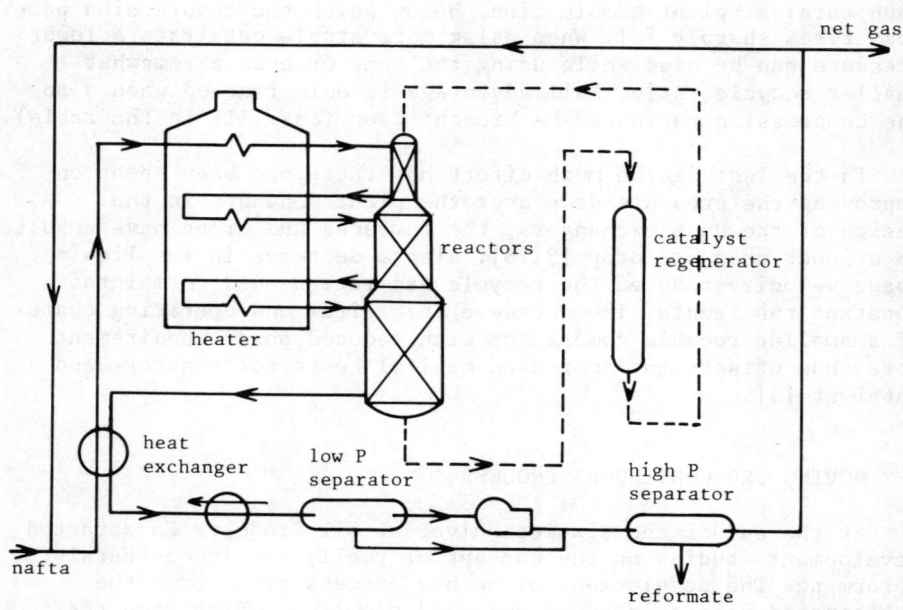

Figure 3 Process flow diagram for a continuous catalytic reforming unit with three reactors operating at low pressure.

hydrogen produced has a higher purity. Also less hydrocarbons are recycled. In the future low-pressure plants may well be equipped with refrigerators for further improvements of these items.

In figure 4 a diagram is given of the catalyst flow through two stacked reactors and through the regenerator and of the catalyst flow between the reactors and the regenerator [6, 8]. Catalyst transport through the reactors and the regenerator is by gravity flow. The spherical alumina particles that are already in use in catalytic reforming are very suited for this. Transport of catalyst from the last reactor to the top of the regenerator and from the bottom of the regenerator to the first reactor is by the gas-lift method. Nitrogen is used for transport to the regenerator, while hydrogen blows the catalyst particles back to the first reactor. Since the atmosphere in the reactors is reducing and that in the regenerator is oxidative, a sluicing system is needed for moving the particles from one atmosphere to the other. For this purpose lock hoppers are used that can handle small batches of catalyst at the time and that can be purged. Actually the catalyst flow is not really continuous but batchwise! The catalyst flow rate for the entire system is controlled in the lock hopper at the outlet of the regenerator and all catalyst transfers are fixed by level-control requirements.

Two principal requirements for the reactor design are low-pressure drop and ease of moving the catalyst through the reactor. The radial flow reactor that is being used in conventional reforming units can meet these requirements very well.
This reactor, with its high length-to-diameter ratio, gives a small pressure drop due to the fact that the reactants flow radially through the catalyst bed and makes a uniform catalyst flow possible with very little catalyst attrition [6]. A schematic drawing of the top reactor of the reactor section is given in figure 5 [8]. The catalyst particles are confined in an annular moving bed formed by spaced cylindrical screens. They descend by gravity through the bed and are transfered to the next reactor by means of transfer conduits. The reactor system thus has a common catalyst bed that moves as a column of particles from top to bottom of the reactor section. The gaseous reactants flow out-to-in radially through the annular catalyst bed and continue downwardly through the cylindrical space surrounded by the inner catalyst screen. Since the reaction is endothermic, the reactor effluent is taken to the heater for reheating and brought back to the entrance of the second reactor for further processing. A suitable pressure drop over the catalyst transfer conduits ensures that only a minimum amount of reactants bypass the heater and pass directly to the next reactor together with the catalyst flow.

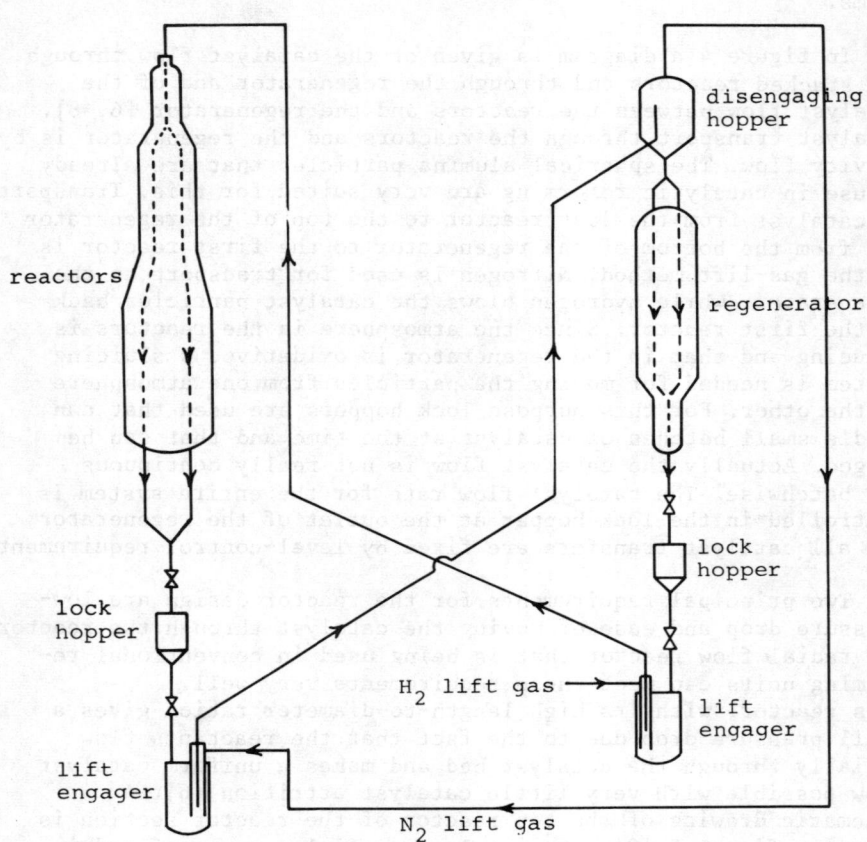

Figure 4 Diagram of the catalyst flow.

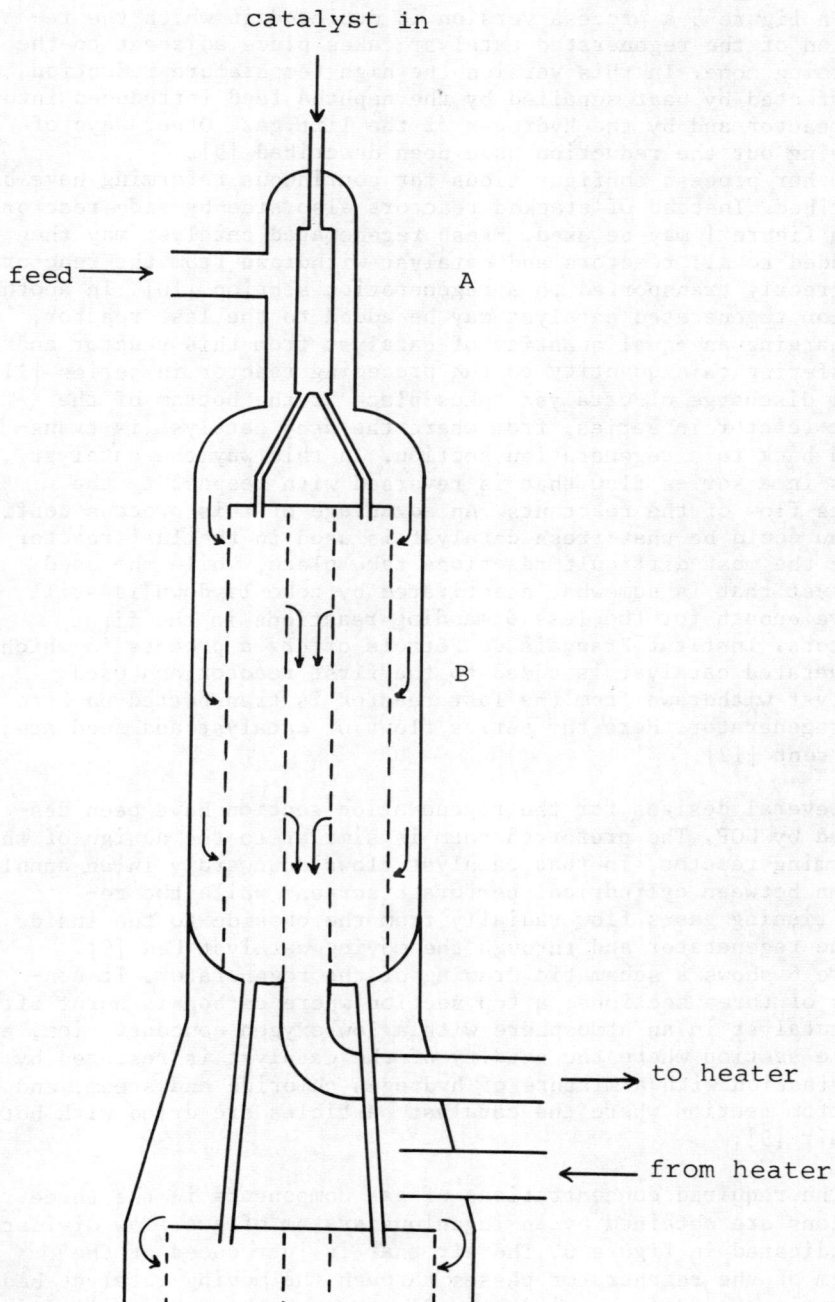

Figure 5 Detail of the reduction (A) and top reactor section (B) of a continuous catalytic reforming unit.

In figure 5 a process version is depicted in which the reduction of the regenerated catalyst takes place adjacent to the reforming zone. In this version the high temperature reduction is effected by heat supplied by the naphtha feed introduced into the reactor and by the hydrogen of the lift gas. Other ways of carrying out the reduction have been described [9].

Other process configurations for continuous reforming have been described. Instead of stacked reactors also side-by-side reactors as in figure 1 may be used. Fresh regenerated catalyst may then be added to all reactors and catalyst withdrawn from the reactors is directly transported to a regeneration section [10]. In another version regenerated catalyst may be added to the last reactor, discharging an equal quantity of catalyst from this reactor and transfering this quantity to the preceding reactor in series [11]. Final discharge of catalyst takes place at the bottom of the first reactor in series, from where the used catalyst is transfered back to a regeneration section. In this way the catalyst flows in a series flow that is reversed with respect to the series flow of the reactants. An advantage of this process configuration could be that fresh catalyst is used in the last reactor where the most difficult reactions take place, while the used catalyst that is somewhat deactivated by coke laydown is still active enough for the less demanding reactions in the first reactors. Institut Français du Pétrole offers a process in which regenerated catalyst is added to the first reactor and used catalyst withdrawn from the last reactor is transported back to the regenerator. Here the series flows of catalyst and feed are cocurrent [12].

Several designs for the regeneration section have been described by UOP. The preferred form is similar to the design of the reforming reactor, in that catalyst flows downwardly in an annular column between cylindrical perforate screens while the reconditioning gases flow radially from the outside to the inside of the regenerator and through the moving catalyst bed [9]. Figure 6 shows a schematic drawing of the regenerator. It consists of three sections: a top section where carbon is burnt off the catalyst in an atmosphere with a low oxygen concentration, a middle section where the acidity of the catalyst is restored by chlorination with a mixture of hydrogen chloride and steam, and a bottom section where the catalyst particles are dried with hot dry air [9].

The required concentrations of gas components in the three sections are obtained by an ingenious system of gas flow dividers as indicated in figure 6. The air that is introduced at the bottom of the regenerator passes through the moving catalyst bed in section C and is carried upwardly through the inner cylindrical space of the chlorination section B. Together with hydrogen chloride and steam flowing through the catalyst bed in section B

Modern processes for the catalytic reforming

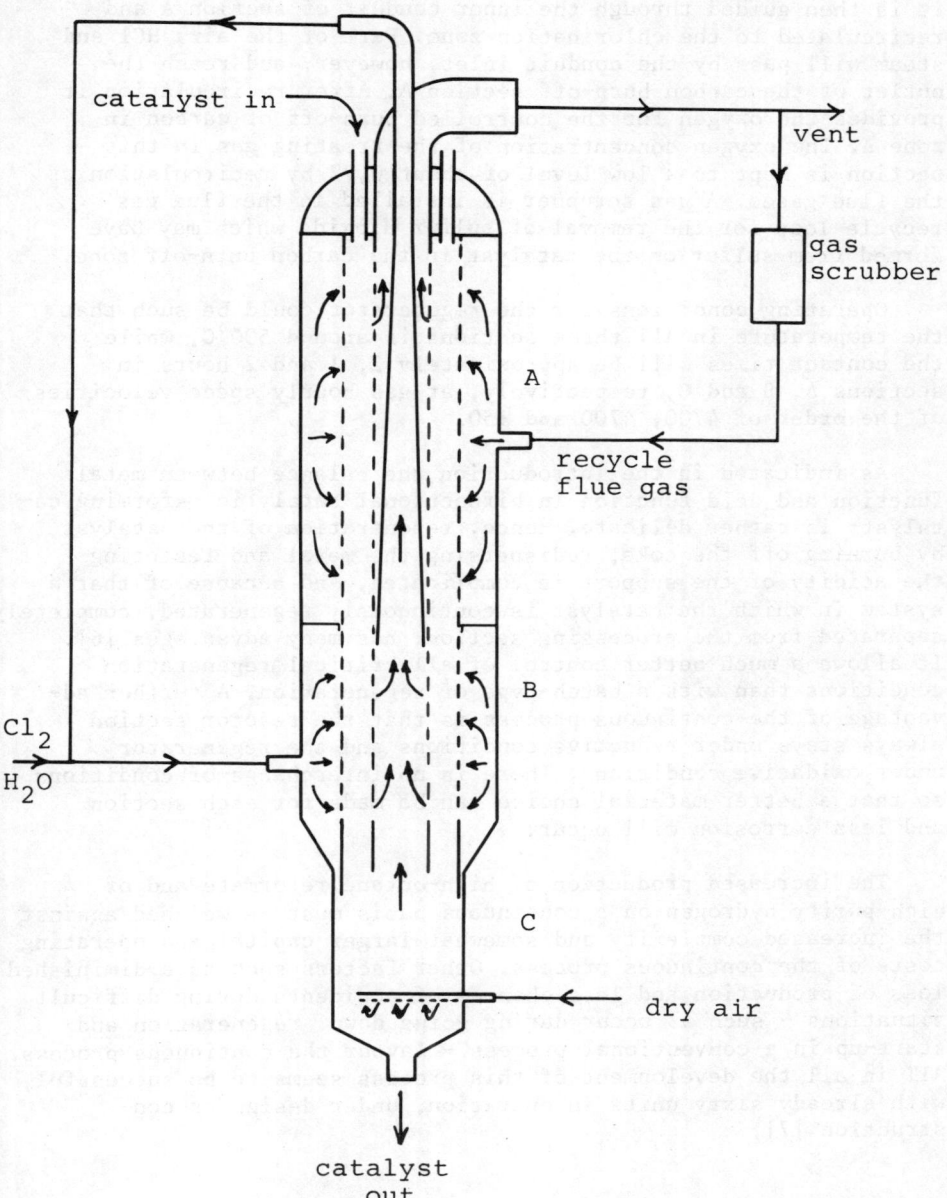

Figure 6 Regeneration unit of a continuous catalytic reformer with carbon burn-off zone A, chlorination zone B and drying zone C.

it is then guided through the inner conduit of section A and recirculated to the chlorination zone. Part of the air, HCl and steam will pass by the conduit inlet, however, and reach the outlet of the carbon burn-off section A. After recirculation it provides the oxygen for the controlled burn-off of carbon in zone A. The oxygen concentration of the treating gas in this section is kept to a low level of about 0,7% by recirculation of the flue gases. A gas scrubber is installed in the flue gas recycle loop for the removal of sulfur dioxide which may have formed from sulfur on the catalyst in the carbon burn-off zone.

Operating conditions for the regenerator could be such that the temperature in all three sections is around $500^0 C$, while the contact times will be approximately 2, 1 and 2 hours in sections A, B and C, respectively, at gas hourly space velocities of the order of 4700, 4700 and 150.

As indicated in the introduction the balance between metal function and acid function in bifunctional catalytic reforming catalysts is rather delicate. Hence, regeneration of the catalyst by burning off the coke, redispersing the metal and restoring the acidity of the support is complicated. And because of that a system in which the catalyst is continuously regenerated, completely separated from the processing section, has many advantages [6]. It allows a much better control of all critical regeneration conditions than with a batch-type of regeneration. A further advantage of the continuous process is that the reactor section always stays under reductive conditions and the regenerator under oxidative conditions. There is no interchange of conditions so that a better material choice can be made for each section and less corrosion will occur.

The increased production of high-octane reformate and of high-purity hydrogen on a continuous basis must be weighed against the increased complexity and somewhat larger capital and operating costs of the continuous process. Other factors such as a diminished loss of production and less chances of accidents during difficult situations - such as occur during going down, regeneration and start-up in a conventional process - favour the continuous process. All in all the development of this process seems to be successful, with already sixty units in operation, under design or construction [7].

FUTURE TRENDS

With the increasing pressure on a better use of raw materials it is clear that the incentive of a better process efficiency will continue to exist in catalytic reforming. Improvements in the

liquid yield of high-octane reformate will be important, as will
be the processing of more difficult feedstocks in situations of
uncertain supply. Therefore the emphasis in catalytic reforming
will stay on going to higher severity, to still lower pressures
and recycle ratios. It may be that these goals can be met by
adaptations to the continuous process but also other designs may
be contemplated, such as the one developed by the French oil
company ELF [13]. In this process isothermal instead of adiabatic
operation is used by placing tubular reactors directly in a
furnace. A lower energy consumption is claimed for this process.
Further advantages could be larger reformate yields at lower
mean temperatures and higher space velocities than obtained in
a conventional reforming unit.

REFERENCES

1. F.G. Ciapetta, R.M. Dobres, and R.W. Baker, in 'Catalysis' VI
   P.H. Emmett (ed), Reinhold, New York, 1958, p. 495.
2. M.J. Sterba and V. Haensel, Ind. Eng. Chem., Prod. Res. Dev.
   15, 2 (1976).
   A.J. Silvestri, P.A. Naro, and R.L. Smith, J. Catal. 14, 386
   (1969).
3. P.G. Menon and J. Prasad, in 'Proc. Sixth Int. Conf. on
   Catalysis', The Chemical Society, London, 1976, p. 1061.
4. H.E. Kluksdahl, U.S. Patent 3, 415, 737 (1968).
5. T.R. Hughes, R.L. Jacobson, K.R. Gibson, L.G. Schornack,
   and J.R. McCabe, Hydrocarbon Processing 55 (5), 75 (1976).
6. E.A. Sutton, A.R. Greenwood, and F.H. Adams, Oil and Gas
   Journal 70, 52 (1972).
7. Hydrocarbon Processing 57 (9), 163 (1978).
8. A.R. Greenwood and K.D. Vesely, U.S. Patent 3, 647, 680 (1972).
9. A.R. Greenwood and K.D. Vesely, U.S. Patent 3, 652, 231 (1972).
10. D.B. Carson, U.S. Patent 3, 470, 090 (1969).
11. A.R. Greenwood and K.D. Vesely, U.S. Patent 3, 725, 248 (1973).
12. Hydrocarbon Processing 57 (9), 161 (1978).
13. M.H. Fromager and J.J. Patouillet, Hydrocarbon Processing
    58 (4), 171 (1979).

Part IV

HOMOGENEOUS CATALYSIS

REACTION MECHANISMS IN HOMOGENEOUS CATALYSIS

R. Prins

Laboratory for Inorganic Chemistry,
Eindhoven University of Technology, The Netherlands

INTRODUCTION

Many chemical products are made in the petrochemical industry by the addition of oxygen or oxygen containing molecules to unsaturated hydrocarbons. For instance ethylene oxide is synthesized from oxygen and ethylene, acetic acid from carbonmonoxide and methanol, aldehydes are made from carbonmonoxide, hydrogen and olefins, while alcohols are obtained by a subsequent hydrogenation. In some of these cases catalytic processes are involved that operate in a homogeneous phase, such as in the hydroformylation process, the Wacker process and in the carbonylation of methanol. In the past decade much knowledge has been gathered on the chemistry involved in homogeneous catalysis and in some areas the level of understanding has even further advanced than in comparable areas in heterogeneous catalysis. Before discussing specific reactions, however, we will start with a review of the elementary reaction steps, such as substitution, insertion and oxidative addition, that take place in nearly all homogeneous reactions.

BASIC REACTIONS

Ligand substitution

Ligand substitution of a ligand by one that takes part in the chemical reaction is a step that occurs in many organotransition metal reactions. Mainly two reaction mechanisms are observed for ligand substitution, an associative and a dissociative

mechanism [1]. Four coordinate square-planar complexes react by
the associative mechanism, a five coordinate complex being the
transition state. As far as they have been studied, reactions
of five coordinate trigonal-bipyramidal complexes are independent
of the concentration of the entering ligand and therefore the
reactions must be of the dissociative type. Six coordinate
octahedral complexes also react by the dissociative mechanism.
An example of ligand substitution on a hexa-coordinated manganese
complex will be encountered in the following section.

Insertion

Insertion reactions are reactions of the type

$$M-L + X \longrightarrow M-X-L$$

in which an organic group X is so-to-speak 'inserted' between the
metal atom and a ligand of a metalorganic compound. One could
also say that the metal containing compound is added to the
organic substrate and one speaks of 1,1 addition when both M and
L are linked to the same atom in the final product and of 1,2
addition if they are linked to neighbouring atoms:

$$M-L + X = Y \longrightarrow M-X-Y-L$$

Addition reactions - and also the reverse reactions, the
elimination reactions - play an important role in the homogeneous
hydrogenation and polymerisation of olefins, in the formation of
metal-carbene complexes, in the rearrangement reactions of metal
carbonyls, and in many other reactions. In the literature of the
last two decades numerous examples of addition reactions can be
found with all kinds of groups X, Y and L and with many
transition-metal atoms. Some examples are given in Table I.
Of the reactions presented in Table I we will discuss in some
detail the 1,1 addition to CO and the 1,2 addition to olefins.

The 1,1 addition of an organometallic compound to CO, or
as this reaction is often called the insertion of CO into a
metal-carbon bond, has been extensively studied. From a mechanistic
point of view especially methyl pentacarbonyl manganese has been
well studied. Addition of strong donor ligands to this compound
induces a reaction that can be called a CO insertion between the
manganese atom and the methylgroup or the 1,1 addition of
methylmanganese to CO:

$$H_3C-Mn(CO)_5 + L \longrightarrow H_3C-\overset{O}{\overset{\|}{C}}-Mn(CO)_4L$$

(L = CO, amine, phosphine)

metal-ligand bond	organic substrate	product
M-C	1,1 CO, CNR, CR_2, $SnCl_2$ 1,2 C_2H_4, C_2H_2, O_2, CO_2	M(COR), M-C(R)=NR' M-C_2H_4-R, M-C(O)-OR
M-H	1,2 C_2H_4	M-C_2H_5
M-NR_2	1,2 CO_2 1,3 CS_2	M-O-C(O)-NR_2 M(S_2C-NR_2)
M-Cl	1,1 CR_2, $SnCl_2$	
M-M	1,1 $SnCl_2$ 1,2 SO_2	M-$SnCl_2$-M M-O-S(O)-M

Table 1 Examples of 1,1 and 1,2 additions of organometallic compounds to organic substrates.

The reaction kinetics indicate that a two-step mechanism is operative. In first instance the methyl group migrates to one of the cis CO ligands and forms an acetyl group, followed by addition of L to the vacated site

Assuming a steady state approximation for the acetyl intermediate and neglecting the back reaction k_{-2} we obtain

$$\frac{d(P)}{dt} = \frac{k_1 k_2}{k_{-1} + k_2(L)} \cdot (R) \cdot (L)$$

For small concentrations of L one indeed finds a first order dependence in both (R) and (L), while at high concentrations of L the reaction is first order in (R) and zero order in (L).

A very important question in the CO insertion reaction is the question: how does the first step take place? Does the CO insert into the Mn-CH_3 bond or does the methyl group migrate to the CO ligand. The latter has been shown to be the case in very clever labelling experiments by Noack and Calderazzo [2].

First they studied the CO insertion and de-insertion by means of infrared spectroscopy.

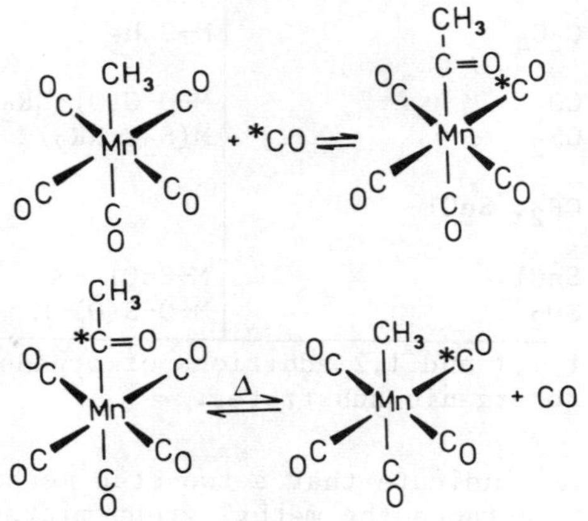

Since the products formed are exclusively the ones indicated in the figure, CO insertion and de-insertion must be intramolecular and regiospecific to the cis coordination site. These experiments in itself do not decide on the two mechanisms, however. An analysis of the decarbonylation products of acetylmanganese pentacarbonyl, with one ^{13}CO group, now allows a distinction to be made between carbonyl insertion and methyl migration. According to the schemes given below the two pathways would give different product distributions. Experimentally a cis-to-trans $CH_3Mn(CO)_4{}^{13}CO$ ratio of 2:1 was found, proving that methyl migration is the correct mechanism.

CO insertion

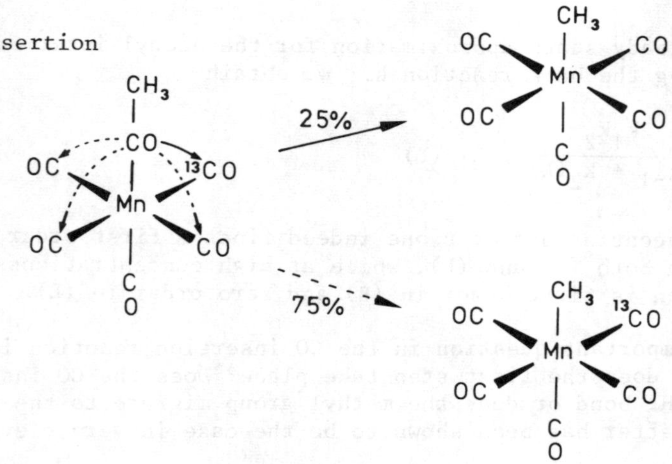

CH$_3$ migration

Furthermore carbonylation of cis-CH$_3$Mn(CO)$_4$^{13}CO with natural CO gave acetylmanganese pentacarbonyl with cis, trans and ^{13}COCH$_3$ products in the ratio 2:1:1, as expected for the methyl migration route. Carbonyl insertion would have given cis and ^{13}COCH$_3$ products in the ratio 3:1 and no trans isomer.

Another elegant proof of alkyl migration is found in the decarbonylation of the following iron compound

The iron acyl compound is chiral, with the iron atom as the center of chirality. Upon decarbonylation the chirality of the complex is inverted. This can only be explained by assuming that the labelled CO group has left the complex and that the alkyl group has migrated to the vacated site [3].

Of the 1,2 addition reactions the addition of a metalhydride to an olefin, or in other words the insertion of an olefin into a metal-hydrogen bond, is studied most. This reaction plays an important role in the hydrogenation of olefins, in the formation of aldehydes from olefins via the hydroformylation reaction, and in the polymerisation of olefins via the Ziegler-Natta reaction. An example of olefin insertion into a metal-hydrogen bond is the reaction of olefins with trans-PtHCl(PEt$_3$)$_2$. At 95^{0}C and 80 atm of ethylene there is a 25% conversion to the σ-bonded ethyl group, while heating in vacuo induces the reverse reaction [4].

$$\text{Cl-Pt(PEt}_3\text{)}_2\text{-H} + C_2H_4 \rightleftharpoons \text{Cl-Pt(PEt}_3\text{)}_2\text{-}C_2H_5$$

Heating the deuterated complex trans-Pt(CD$_2$CH$_3$)Cl(PEt$_3$)$_2$ gave a platinum product with both Pt-H and Pt-D bonds. This demonstrates that the addition or elimination of the olefin proceeds through a symmetrical transition state, the hydrido ethylene platinum complex: The equilibrium between metalhydride and olefin on one hand and the metal-alkyl complex on the other hand lies more to the olefin side for the larger olefins than for ethylene. For instance, with octene-1 no platinumoctyl complex is formed. That olefin addition and elimination does occur, however, was proved by the conversion of trans-PtDCl(PEt$_3$)$_2$ into trans-PtHCl(PEt$_3$)$_2$ when heated with octene-1 at 130^{0}C for 48 hours.

Although it is usually assumed that the 1,2 addition of M-H to C$_2$H$_4$ goes via comples formation between the reactants:

$$L_nMH + H_2C=CH_2 \rightleftharpoons \left[L_nM\cdots\overset{H}{\underset{CH_2}{\|}}_{CH_2} \right] \rightleftharpoons L_nM\cdots\overset{H\cdots CH_2}{\underset{CH_2}{\|}} \rightarrow L_nM\text{-}CH_2CH_3$$

only very few hydrido-ethylene-metal complexes have been isolated. Recently an hydrido-ethylene-rhodium complex has been prepared which in nitromethane solution proved to be in equilibrium with the σ-ethyl complex [5].

$$[(C_5H_5)RhH(C_2H_4)PMe_3]^+ \rightleftarrows [(C_5H_5)Rh(C_2H_5)PMe_3]^+$$

At $-20°C$ the proton-NMR spectrum shows sharp signals for the C_2H_4 and H protons, the former being splitted into two multiplets and the latter into a doublet of doublets by the interaction with the rhodium and phosphorus nuclear spins $\frac{1}{2}$. At room temperature and above the hydrido and ethylene protons appear as broad bands indicating that exchange between these groups occurs. This is confirmed by an experiment in which D_2O is added to the solution at -20 C. The ^{1}H-NMR signals of the hydrido and ethylene protons disappear and analysis of the solution demonstrates that $[(C_5H_5)RhD(C_2D_4)PMe_3]^+$ has been formed. The equilibrium between the hydrido-ethylene-rhodium and ethyl-rhodium complexes is shifted to the right by the addition of ethylene

$$[(C_5H_5)RhH(C_2H_4)PMe_3]^+ + C_2H_4 \rightleftarrows [(C_5H_5)Rh(C_2H_5)(C_2H_4)PMe_3]^+$$

Apparently the ethylene is bonded to the vacant site that is created by the transformation of $H-Rh-C_2H_4$ into $Rh-C_2H_5$.

Hydrido-olefin-metal complexes are rarely observed but also alkyl-metal complexes are hard to isolate. For a long time this has been ascribed to a weak transition-metal-carbon σ-bond, but now it is generally agreed that the reason is kinetic rather than thermodynamic. Alkyl transition-metal complexes are very susceptible to the so-called β-H-elimination reaction, the reverse of the olefin insertion reaction given above. Stable transition-metal alkyls can be prepared but they must either have no β hydrogen atoms, or the coordination around the metal atom must be complete so that the β hydrogen atom cannot be added to the metal atom. For alkyl-metal complexes with vacant coordination sites the β-H-elimination usually is rapid, but bulky ligands may slow it down.

Dimerisation, oligomerisation and Ziegler-Natta polymerisation of olefins have long been assumed to involve insertion of the olefin into the metal-carbon bond of an intermediate metal-alkyl [6]. Green, Rooney and coworkers have pointed out, however, that there are no unambiguous examples where a well characterised alkyl-olefin-metal complex has been observed to react to the expected insertion product [7]. This has led them to suggest an alternative mechanism for apparent insertion reactions which involves α-elimination of hydrogen atoms to form a transient carbene complex and carbon-carbon formation via reaction between the carbene and the π-bonded olefin, analogous to the reaction in metathesis. The polymerisation of propene is then described by the following scheme:

On the other hand, Evitt and Bergman have recently clearly demonstrated that the formation of propene and methane in the reaction of ethylene with η^5-cyclopentadienyl(triphenylphosphine)-dimethylcobalt is due to insertion rather than α-elimination [8]:

$$\text{CpCo(PPh}_3)(\text{CH}_3)_2 + 2\ \text{C}_2\text{H}_4 \longrightarrow \text{CpCo(PPh}_3)(\text{C}_2\text{H}_4) + \text{CH}_4 + \text{C}_3\text{H}_6$$

When the perdeutero cobalt complex was treated with C_2H_4 mass spectral analysis showed CD_3H and propene-d_3 to be formed, while in the reaction of C_2D_4 with the perdeutero cobalt complex CD_4 was formed. Apparently, the hydrogen atom that is added to the methyl group to form methane is derived from the ethylene and not from other hydrogen sources in the system, and a complete methyl group is transferred to the ethylene in the formation of propene. Since in the insertion pathway three of the methane hydrogens derive from the initial methyl group and the fourth from ethylene, whereas in the α-elimination pathway the fourth hydrogen comes from the second methyl group and a methylene is transferred to ethylene, the experimental results demonstrate that insertion rather than α-elimination is responsible for the methylation of ethylene by the cobalt complex.

Although this single example of course does not prove that the α-elimination pathway has to be excluded for all apparent insertion reactions, it certainly has brought back confidence in the 'old' insertion pathway.

The olefin insertion reaction has attracted also the attention of theoreticians and in a recent article Thorn and Hoffmann give a lucid explanation of the factors that make the transition from a π-bonded olefin into a σ-bonded alkyl group possible [9]. They argue that the approach of a hydrido ion to an ethylene molecule, in the absence of a transition metal atom, leads to a repulsive interaction between the H ion and the approaching C atom because of the fact that the highest occupied molecular orbital is an antibonding combination between the ethylene π and the hydrido 1s orbitals

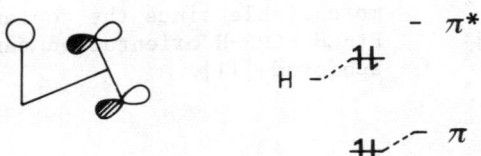

Adding a transition metal atom to the scene leads to a diminished repulsive interaction between the hydrido ion and the approaching C atom. The hydrido-π olefine molecular orbital now forms bonding and antibonding combinations with the $d_{x^2-y^2}$ orbital. At the same time this d orbital will combine with the ethylene π* orbital. This results in some mixing of the π and π* orbitals and in a weakening of the repulsion between the approaching H and C atoms. Concurrently the orbital develops into a bonding Pt-C orbital

The interaction of the $d_{x^2-y^2}$ orbital with both π and π^* orbitals, which is allowed by the low symmetry in the transition state, in effect leads to a rehybridisation of the π and π^* orbitals back into two 2p orbitals localised on the two carbon atoms. The energy increase that would go with the rehybridisation is largely surpressed by the interaction with the d orbitals of the metal atom

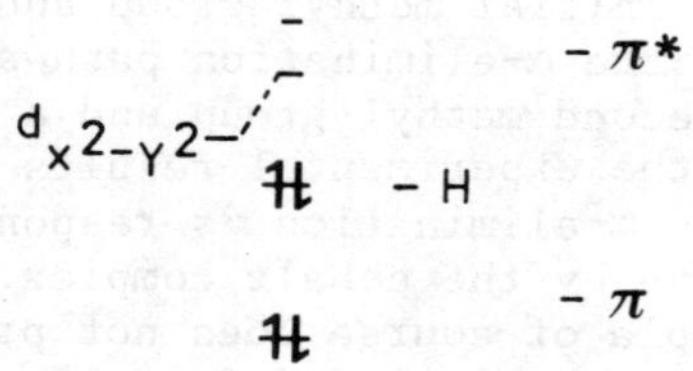

$$d_{x^2-y^2} \begin{matrix} - \pi^* \\ \uparrow\downarrow \quad - H \\ \uparrow\downarrow \quad - \pi \end{matrix}$$

The antibonding combination of $d_{x^2-y^2}$, π and π^* is highly unfavourable. As a consequence this orbital must be empty for the total energy of the system to remain low. This explains why the olefin insertion proceeds especially smoothly in square planar d^8 complexes. It also means that the reverse reaction, the β-H-elimination, can only proceed when the resulting H atom and π-bonded olefin are able to form part of a square planar arrangement around the metal atom. Dissociation of a phosphine ligand from $Pt(C_4H_9)_2(PR_3)_2$ thus is a prerequisite for β-elimination and added phosphine can surpress it entirely [10]. The related macrocyclic compound is much more stable since the coplanar $Pt-CH_2-CHR-H$ orientation cannot be achieved [11].

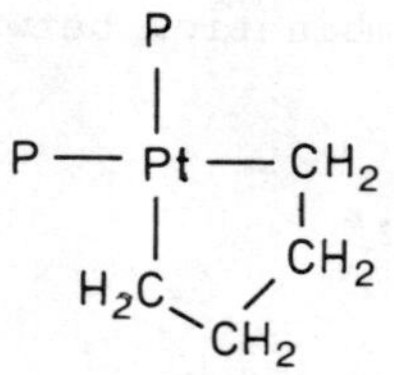

```
        P
        |
  P — Pt — CH₂
        |    |
   H₂C   CH₂
      \CH₂/
```

Oxidative addition

Oxidative addition reactions are a class of reactions of transition-metal complexes in which ligands are added to the metal atom and in which the formal charge of the central metal increases. These reactions can also be described as 1,1 addition to the metal atom

$$XY + M-L \rightleftharpoons X-\underset{\underset{L}{|}}{M}-Y$$

The following reactions constitute examples of oxidative additions:

In the second reaction the oxidative addition of methyliodide is followed by methyl migration (CO insertion). With regard to the stereochemistry of oxidative addition, in all cases studied up to now hydrogen has been found to add in a cis fashion, while the addition of alkylhalides has been reported to be both cis and trans, but most often trans.

The oxidative addition reactions can be understood by looking at the effective number of valence electrons (NVE) around the metal atom. Usually the d^8 complexes of Ru(0), Rh(I), Ir(I), Pd(II) and Pt(II) are square planar complexes with 16 valence electrons or they are trigonal bipyramidal 18 electron complexes. Species such as H_2, X_2, HX, RX, RCOX, RSnX can add to these compounds. Addition to the square planar, 16 electron, d^8 complexes leads directly to octahedral, 18 electron, d^6 complexes whereas addition to the five coordinate, 18 electron, d^8 complexes leads to octahedral complexes only after dissociation of one of the original ligands.

The influence of the metal atom in oxidative addition reactions can be rationalised by taking into account its nucleophilicity. A high electron density on the metal atom favours oxidative addition and the largest metals of the third row, in the lowest oxidation state, therefore show the greatest tendency to undergo oxidative addition [12]. By a similar reasoning it is understandable that strongly donating ligands like carbonyls and phosphines promote the oxidative addition.

The influence of metal basicity is evident when comparing two metals like iridium and rhodium. With $IrCl(CO)(PPh_3)_2$ the addition of CH_3I is complete and irreversible, while for the rhodium analog it is quite reversible. The reverse reaction is called reductive elimination and together with oxidative addition it plays an important role in the homogeneous catalysis of hydroformylation and of the carbonylation of methanol to acetic acid.

REACTION MECHANISMS

Having discussed the basic reactions, we will in this section clarify part of the chemistry involved in homogeneous catalysis and in particular we will discuss the hydrogenation of olefins, the carbonylation reaction (as used in the hydroformylation and methanol carbonylation processes), and the oxidation of ethylene to acetaldehyde.

Olefin hydrogenation

Many metal complexes are able to activate hydrogen, some of the best known are $RuCl_6^{3-}$, $Co(CN)_5^{3-}$, and $RhCl(PPh_3)_3$. Especially the latter complex, the so-called Wilkinson's complex, is very effective in homogeneous hydrogenation. Although some details are still to be clarified, it is generally believed that the activity of this catalyst in the hydrogenation of olefins is explained by the following scheme [13].

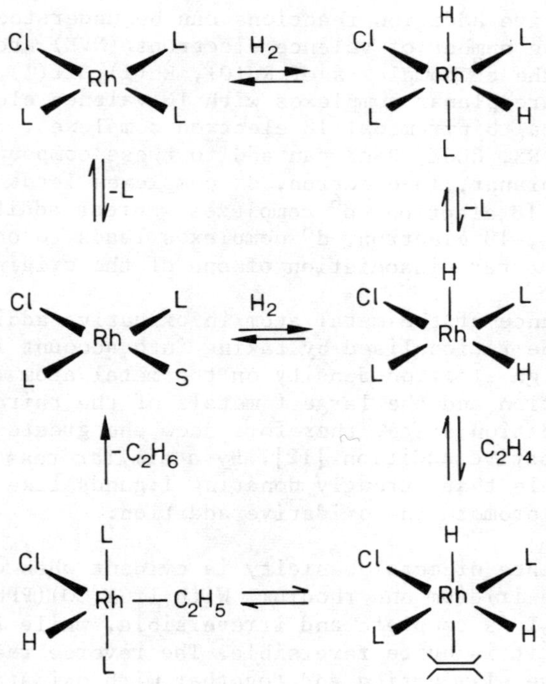

The first step is oxidative addition of H_2 to the rhodium complex followed by ligand dissociation, or vice versa. Then complex formation with the olefin takes place and insertion of the olefin into the Rh-H bond. Reductive elimination of the alkane regenerates the catalytically active species. Note that whenever an oxidative addition occurs in a catalytic cycle a reductive elimination has to follow in order to close the catalytic cycle.

Carbonylation

A reaction that is of great industrial importance is the hydroformylation reaction of α olefins to aldehydes. The aldehydes are further hydrogenated to alcohols which can be used as such as solvents or can be further processed to plasticisers and detergents. Whereas cobalt carbonyl is still the most widely used catalyst new processes use a rhodium catalyst such as $RhCl(CO)(PPh_3)_2$. Rhodium catalysts are very active and, even more important, can be used at much lower pressures. They also give a higher ratio of linear-to-branched aldehydes.

The reaction mechanism is complex and, depending on the concentration of phosphine, involves an associative mechanism in which the olefin adds to a five coordinate rhodium complex or a dissociative mechanism in which the olefin adds to a four coordinate complex [14]. The reaction sequence mainly consists of
1) addition of the olefin followed by olefin insertion into the Rh-H bond
2) migration of the resulting alkyl group to a CO ligand (CO insertion)
3) oxidative addition of H_2 followed by reductive elimination of the aldehyde.

At high triphenylphosphine concentrations the lower catalytic cycle, the associative route, is prefered. The olefin is added to a five coordinated rhodium complex and the rather large steric hindrance induces a high linear-to-branched ratio of the aldehyde product. At too high phosphine concentrations the catalyst activity decreases due to the formation of the inactive $RhH(CO)(PPh_3)_3$ species. At low phosphine concentrations the dissociative mechanism is dominant. Because of the fact that the olefin is now added to a four coordinated species the aldehyde product will have a lower linear-to-branched ratio. Too high pressures of CO lead to the formation of $Rh(COR')(CO)_2(PPh_3)_2$ which is inable to undergo the oxidative addition of H_2. Since the oxidative addition is assumed to be the rate determining step high CO pressures will inhibit the reaction. In agreement with this infrared experiments under high pressure have demonstrated the presence of metal acyl complexes, indicating that CO insertion is relatively fast and oxidative addition of

H_2 is slow. For internal olefins no metal acyl complex is observed but only the metal hydrido complex. Apparently the coordination or the insertion of internal olefins is slow.

The better activity of the rhodium catalyst compared to the cobalt catalyst undoubtedly arises from the larger tendency of rhodium for oxidative addition. Also iridium compounds are active catalysts and $HIr(CO)_3PR_3$ with R = isopropyl has been used by Whyman in a very instructive breakdown of the hydroformylation catalytic cycle in its individual steps [15]. By means of infrared spectroscopy he observed that when treating the iridium compound with 14 atm ethylene at 50°C after 30 minutes conversion into $(C_2H_5)Ir(CO)_3PR_3$ had occurred

$$HIr(CO)_3PR_3 + C_2H_4 \longrightarrow (C_2H_5)Ir(CO)_3PR_3$$

On venting the excess of ethylene from the solution and replacing it with 14 atm CO at 50°C new infrared bands appeared which demonstrated that CO insertion had occurred

$$(C_2H_5)Ir(CO)_3PR_3 + CO \longrightarrow (C_2H_5CO)Ir(CO)_3PR_3$$

Replacing the excess CO by 14 atm H_2 led to the disappearance of the acyl infrared bands, the appearance of a propionaldehyde band and the reappearance of the bands of the original iridium complex

$$(C_2H_5CO)Ir(CO)_3PR_3 + H_2 \longrightarrow HIr(CO)_3PR_3 + C_2H_5CHO$$

The sequence of reactions could be repeated many times by pressurising with the respective gases, the intensity of the aldehyde absorption band increasing at the end of each cycle. This clearly proofs that the above reaction steps form part of the hydroformylation sequence and that the catalyst in this case is a mononuclear iridium carbonylphosphine complex.

The chemistry of the carbonylation of methanol to acetic acid

$$CH_3OH + CO \longrightarrow CH_3C\overset{\displaystyle O}{\underset{\displaystyle OH}{}}$$

is related to hydroformylation chemistry. Also in this case rhodium is a better catalyst than cobalt, giving a higher selectivity at much lower pressure. The mechanism goes via oxidative addition of methyl iodide, insertion of CO into the Rh-methyl bond and reductive elimination of acetyliodide [16]. The acetyl iodide reacts further with methanol to acetic acid and methyl iodide, that reenters the catalytic cycle.
Also here infrared spectroscopy proved helpful in demonstrating the occurrence of the separate reaction steps. When adding excess methyl iodide to a solution of $[Rh(CO)_2I_2]^-$ at room temperature, Forster [17] observed the infrared spectrum of $[(CH_3CO)Rh(CO)I_3]^-$. This anion could even be isolated in crystalline form, in the form of a dimer with a Rh-I bridge, when large organic cations were added as counterions. The reversibility of the CO insertion was demonstrated by heating of this crystalline material under vacuum. On bubbling CO through a solution of the rhodium acetyl complex very rapid formation of a new species is observed, followed by slow decomposition to the original $[Rh(CO)_2I_2]^-$ complex.

Kinetic data have shown that the methanol carbonylation reaction is first order in both rhodium and methyl iodide and zero order in CO, suggesting that the addition of methyl iodide is rate determining. Forster has confirmed this by observing that under in situ conditions only the infrared bands of $[Rh(CO)_2I_2]^-$ were present [17].

$$\left[\begin{array}{c}I \quad CO \\ Rh \\ I \quad CO\end{array}\right]^{-} \xrightarrow{+ CH_3I} \left[\begin{array}{c}CH_3 \\ I \mid CO \\ Rh \\ I \mid CO \\ I\end{array}\right]^{-}$$

$$\uparrow -CH_3COI \qquad \downarrow$$

$$\left[\begin{array}{c}O \\ C \quad CH_3 \\ I \mid \mid C=O \\ Rh \\ I \mid CO \\ I\end{array}\right]^{-} \xleftarrow{+ CO} \left[\begin{array}{c}CH_3 \\ I \mid C=O \\ Rh \\ I \mid CO \\ I\end{array}\right]^{-}$$

$$CH_3COI + CH_3OH \longrightarrow CH_3COOH + CH_3I$$

The Wacker process

The oxidation of ethylene to acetaldehyde with platinum group metals has been known since 1894. Bubbling ethylene through a solution of K_2PtCl_4 first the familiar Zeise's salt is formed $K[PtCl_3(C_2H_4)]$, but heating the solution results in the formation of acetaldehyde and metallic platinum. It lasted until the end of the fifties before Smidt c.s. discovered that the metal formed during the oxidation reaction can be reoxidised in situ by cupric chloride [18]. When combining this with the fast reoxidation of cuprous chloride by oxygen, the reaction sequence is turned into a catalytic cycle. This catalytic reaction is now known as the Wacker process, a very successful industrial process for producing acetaldehyde

$$\begin{aligned}
PdCl_2 + C_2H_4 + H_2O &\longrightarrow CH_3CHO + Pd^0 + 2\ HCl \\
Pd^0 + 2\ CuCl_2 &\longrightarrow PdCl_2 + 2\ CuCl \\
\underline{2\ CuCl + 2\ HCl + \tfrac{1}{2}O_2} &\longrightarrow \underline{2\ CuCl_2 + H_2O} \\
C_2H_4 + \tfrac{1}{2}O_2 &\longrightarrow CH_3CHO
\end{aligned}$$

Nickel and platinum show some activity in this reaction too, but they are inferior to palladium. Most probably nickel binds ethylene too weakly and platinum binds ethylene too strongly. This is another illustration of the rule that some bonding between catalyst and substrate is required for catalysis, but weak bonding is not good enough and too strong bonding is not good either.

The mechanism of the Wacker reaction has attracted a great deal of attention. Kinetic studies indicate that the rate is described approximately by

$$\frac{d[H_3CCHO]}{dt} = \frac{k[C_2H_4][PdCl_4^{2-}]}{[H^+][Cl^-]^2}$$

This rate equation can be explained by assuming that there are three pre-equilibria before the rate determining step:

$$[PdCl_4]^{2-} + CH_2H_4 \rightleftharpoons [(C_2H_4)PdCl_3]^- + Cl^-$$

$$[(C_2H_4)PdCl_3]^- + H_2O \rightleftharpoons [(C_2H_4)PdCl_2(OH_2)] + Cl^-$$

$$[(C_2H_4)PdCl_2(OH_2)] \rightleftharpoons [(C_2H_4)PdCl_2(OH)]^- + H^+$$

The rate determining step is the transformation of the hydroxy ethylene palladium complex into the β-hydroxy-ethyl palladium complex

$$[(C_2H_4)PdCl_2(OH)]^- + H_2O \longrightarrow [(HO-CH_2-CH_2)PdCl_2(H_2O)]^-.$$

The steric course of this rearrangement has long been assumed to go via cis migration of the hydroxyl group to the coordinated olefin (or to put it wrongly, insertion of the olefin into the Pd-OH bond). Recent studies have cast doubt on this, however, and have suggested that the rearrangement involves nucleophilic trans attack of water from outside the coordination sphere of the palladium [19, 20].

Cis and trans additions to ethylene follow different steric courses and by using dideutero-ethylene of a specific configuration it should be possible to distinguish between these two possibilities. Thus it has been found that when cis-1,2-dideutero ethylene is used as a reactant in the Wacker reaction a hydroxylethyl palladium complex is formed that can be trapped with CO into the trans-2,3-dideutero propiolactone [20]. This result implies that the hydroxylation of ethylene has taken place from outside the coordination sphere of palladium.

It may well be that both intra and inter molecular addition are possible and that depending on the reaction conditions one or the other route dominates. In that respect it has to be noted that the insertion step in the Wacker reaction is different from the examples given in hydroformylation and carbonylation chemistry in that it is the only case in which a strongly nucleophilic solvent is used. This may be the reason that in the Wacker reaction trans attack by a solvent molecule can compete with cis attack by a nucleophilic ligand inside the coordination sphere of the metal.

Another point that is as yet unsolved is the rearrangement of the β-hydroxy ethyl group into acetaldehyde. It may well be that it proceeds via a reaction sequence of β-H-elimination, migration of the H atom to the other carbon atom, followed by β-H-elimination from the hydroxy group. In agreement with this sequence the oxidation of C_2D_4 gives exclusively CD_3CDO.

REFERENCES

1. R.F. Heck, 'Organotransition metal chemistry', Acad. Press, New York, 1974.
2. K. Noack and F. Calderazzo, J. Organometal. Chem. 10, 101 (1967).
3. A. Davison and N. Martinez, J. Organometal. Chem. 74, C17 (1974).
4. J. Chatt, R.S. Coffey, A. Gough and D.T. Thompson, J. Chem. Soc. A, 190 (1968).
5. H. Werner and R. Feser, Angew. Chem. 91, 171 (1979).
6. P. Cossee, J. Catal. 3, 80 (1964).
7. K.J. Ivin, J.J. Rooney, C.D. Stewart, M.L.H. Green and R. Mahtab, J.C.S. Chem. Comm. 604 (1978).
8. E.R. Evitt and R.G. Bergman, J. Am. Chem. Soc. 101, 3973 (1979).
9. D.L. Thorn and R. Hoffmann, J. Am. Chem. Soc. 100, 2079 (1978).
10. G.M. Whitesides, J.F. Gaasch and E.R. Stedronsky, J. Am. Chem. Soc. 94, 5258 (1972).
11. J.X. McDermott, J.F. White and G.M. Whitesides, J. Am. Chem. Soc. 98, 6521 (1976).
12. R.S. Nyholm and K. Vrieze, J. Chem. Soc. 5337 (1965).
13. J. Halpern, in 'Organotransition Metal Chemistry' (ed. Y. Ishii and M. Tsutsui), Plenum Press, New York, 1975, p. 109.
14. R.L. Pruett, Adv. Organometal. Chem. 17, 1 (1979).
15. R. Whyman, J. Organometal. Chem. 94, 303 (1975).
16. J.F. Roth, J.H. Craddock, A. Hershman and F.E. Paulik, Chem. Techn. 600 (1971).
17. D. Forster, J. Am. Chem. Soc. 98, 846 (1976); Adv. Organometal. Chem. 17, 225 (1979).
18. J. Smidt, W. Hafner, R. Jira, J. Sedlmeier, R. Sieber, R. Rüttinger and H. Kojer, Angew. Chem. 71, 176 (1959); Angew. Chem. Int. Ed. 1, 80 (1962).
19. J.E. Bäckvall, B. Åkermark and S.O. Ljunggren, J.C.S. Chem. Comm. 264 (1977).
20. J.K. Stille and R. Divakarumi, J. Organometal. Chem. 169, 239 (1979).

CATALYSIS BY METAL CLUSTERS

B. C. Gates

Center for Catalytic Science and Technology
Department of Chemical Engineering
University of Delaware
Newark, Delaware 19711 U.S.A.

1. CATALYSIS BY METAL CLUSTERS IN SOLUTION

Our consideration of homogeneous catalysis by metal complexes in "Chemistry of Catalytic Processes" [1] has been focused on mononuclear complexes, those containing a single metal atom. We now turn to polynuclear metal complexes (called metal clusters) exemplified by $Co_4(CO)_{12}$, which was encountered in our discussion of the hydroformylation (Oxo) process [1]. These compounds have frameworks of three or more metal atoms or ions held together by metal-metal bonds. Hundreds of such compounds are known, many having well-defined structures determined by x-ray crystallography and NMR, infrared, and Raman spectroscopies. The stable compounds incorporate familiar ligands, commonly carbonyls, hydrides, phosphines, and chlorides.

One simple, well-investigated metal cluster is $Ru_3(CO)_{12}$, which has the following structure:

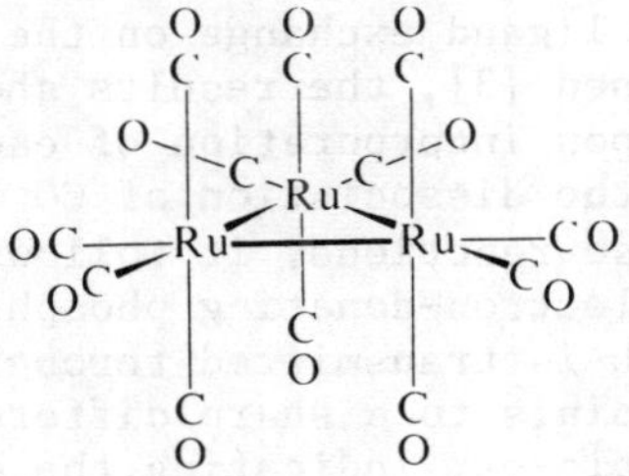

The metal atoms define a triangle, and each metal atom bears four CO ligands, each bonded terminally, i.e., as M-CO. $Os_3(CO)_{12}$ has the same structure.

There are many tetranuclear structures also, e.g., $Ir_4(CO)_{12}$:

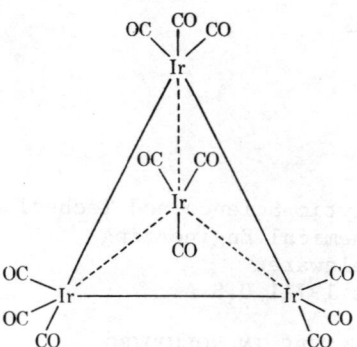

Here, the metal atoms define a tetrahedron, and, again, the CO ligands are terminally bonded. The structure of $Co_4(CO)_{12}$ is slightly different [ref. 1, p. 125]; the metal atoms again define a tetrahedron, but there are bridging as well as terminal CO ligands.

When $Ir_4(CO)_{12}$ in solution with benzene is brought in contact with triphenylphosphine, ligand substitution occurs, giving mono-, di-, and tri-substituted clusters shown in Fig. 1. The incorporation of the first phosphine ligand causes three of the CO ligands to adopt a bridging configuration. The phosphines occupy positions on separate Ir atoms, and being bulky, exert much more steric hindrance than CO, inhibiting further reaction at the metal center.

The kinetics of the ligand exchange on the tetrairidium cluster has been determined [3], the results showing a dramatic increase in reactivity upon incorporation of each successive phosphine group. Since the dissociation of CO was evidently the slow step in each of these reactions, it follows that there is a strong effect of the electron-donating phosphine ligands (labilizing the CO) which is transmitted through the metal-metal bond [3]. This result points to a sharp difference between clusters and mononuclear complexes, indicating the opportunities for influencing the reactivity (and catalytic activity) of a cluster.

Catalysis by metal clusters

There are many other cluster compounds known, including platinum clusters which are oligomers (Fig. 2) and a rhodium cluster (Fig. 3) in which the metal atoms have a hexagonal close packing typical of many crystalline solids. It is clear that the cluster structures provide a transition between mononuclear metal compounds and solid metals. We anticipate that the clusters will

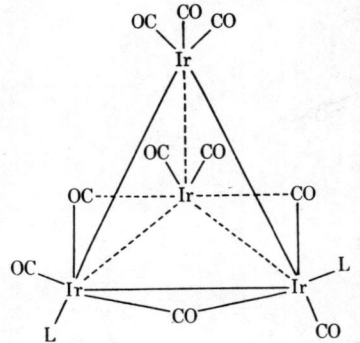
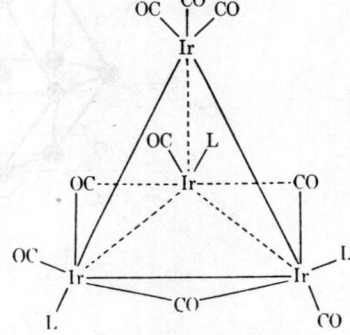

Fig. 1 Structures of $Ir_4(CO)_{11}PPh_3$, $Ir_4(CO)_{10}(PPh_3)_2$, and $Ir_4(CO)_9(PPh_3)_3$ [2,3], L = PPh_3.

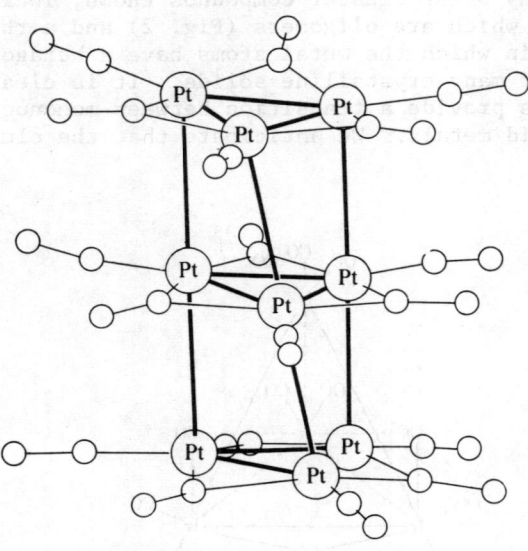

Fig. 2 Structure of $[Pt_3(CO)_3(\mu_2-CO)_3]_3^{2-}$ anion [4].

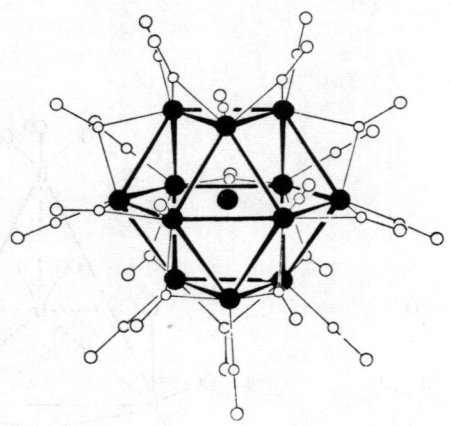

Fig. 3 Structure of $[Rh_{13}(CO)_{25}H_{5-n}]^{n-}$ anions, where n = 2 or 3 [5].

Catalysis by metal clusters

also have catalytic properties intermediate between those of mononuclear complexes and metal surfaces.

Unfortunately, this anticipation cannot be judged critically yet, since the subject of catalysis by metal clusters is new and just now entering a phase of vigorous development [6,7].

Catalysis by metal clusters is difficult to study because the clusters often fragment in solution to give mononuclear complexes (which may be responsible for the catalysis), and they also often aggregate to give small colloidal metal particles or crystallites (which may also be responsible for the catalysis). Most of the reported examples of metal-cluster catalysis must be viewed with skepticism, since the catalyst characterization has been incomplete.

There is one example of metal-cluster catalysis for which intermediates have been isolated and identified and a catalytic cycle firmly established; the catalyst is a triosmium cluster, and the reaction is olefin isomerization. The cycle is shown in Fig. 4. The CO ligands on the cluster are omitted to simplify the depiction and emphasize the interactions of the hydride and hydrocarbon ligands. The triosmium cluster is initially coordinately unsaturated, having as Os-Os double bond. The olefin coordinates there to begin the catalytic cycle.

There is one reaction of potential technological importance for which metal clusters have been implicated; it is the direct conversion of CO and H_2 into ethylene glycol at pressures of hundreds of atmospheres [10]:

$$2\ CO + 3\ H_2 \longrightarrow \underset{CH_2-CH_2}{\overset{OH\ \ OH}{|\ \ \ \ |}} \tag{1}$$

The reaction takes place in the presence of soluble rhodium compounds, and infrared spectra indicate that they include the cluster $Rh_{12}(CO)_{30}^{2-}$, which consists of a framework of two Rh_6 octahedra bonded together at two Rh atoms with bridging CO ligands. The suggestion that the clusters are catalytically active remains to be tested.

We can generalize about the unique opportunities offered by metal clusters in catalysis:

(1) The clusters allow multiple bonding of a single reactant molecule, offering opportunities for conversions not possible with mononuclear species. Some of the opportunites for bonding are illustrated in Fig. 5.

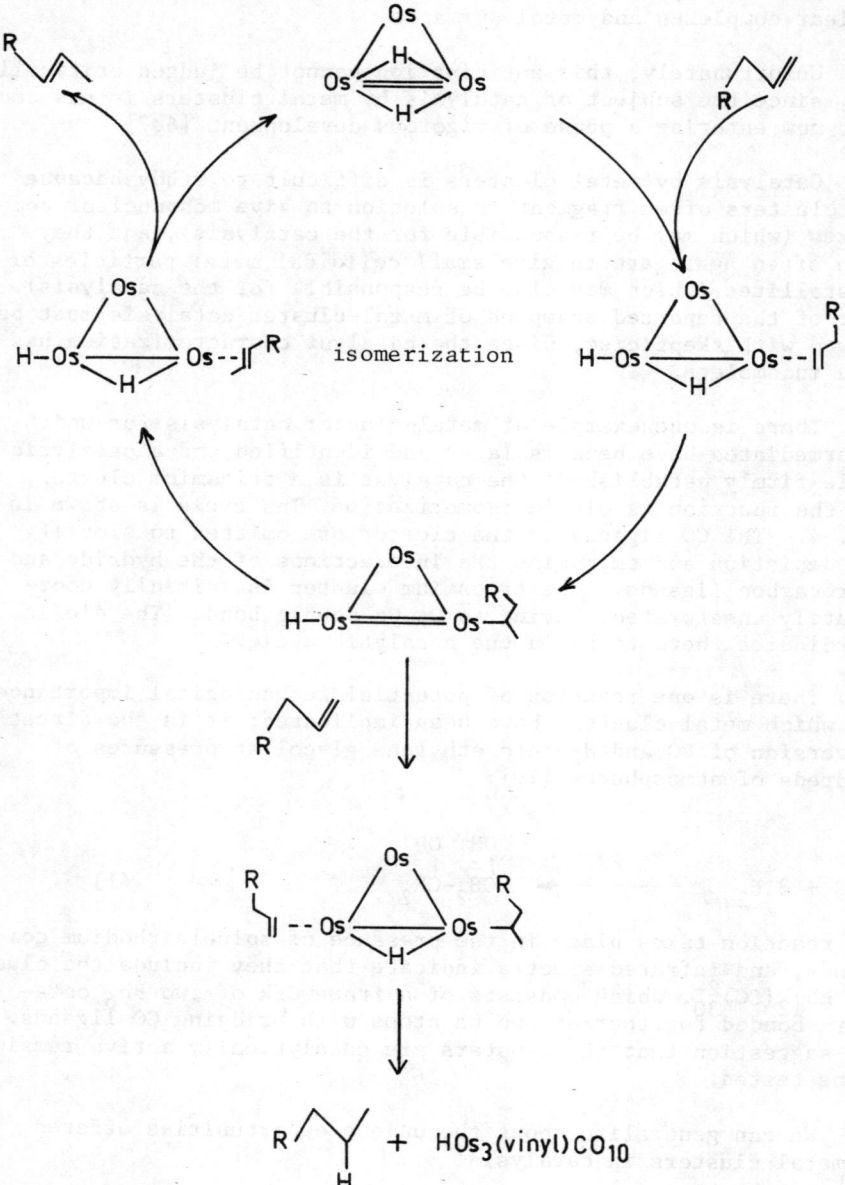

Fig. 4. Catalytic cycle for olefin isomerization catalyzed by triosmium clusters [8,9]. The carbonyl ligands on Os are omitted for simplicity.

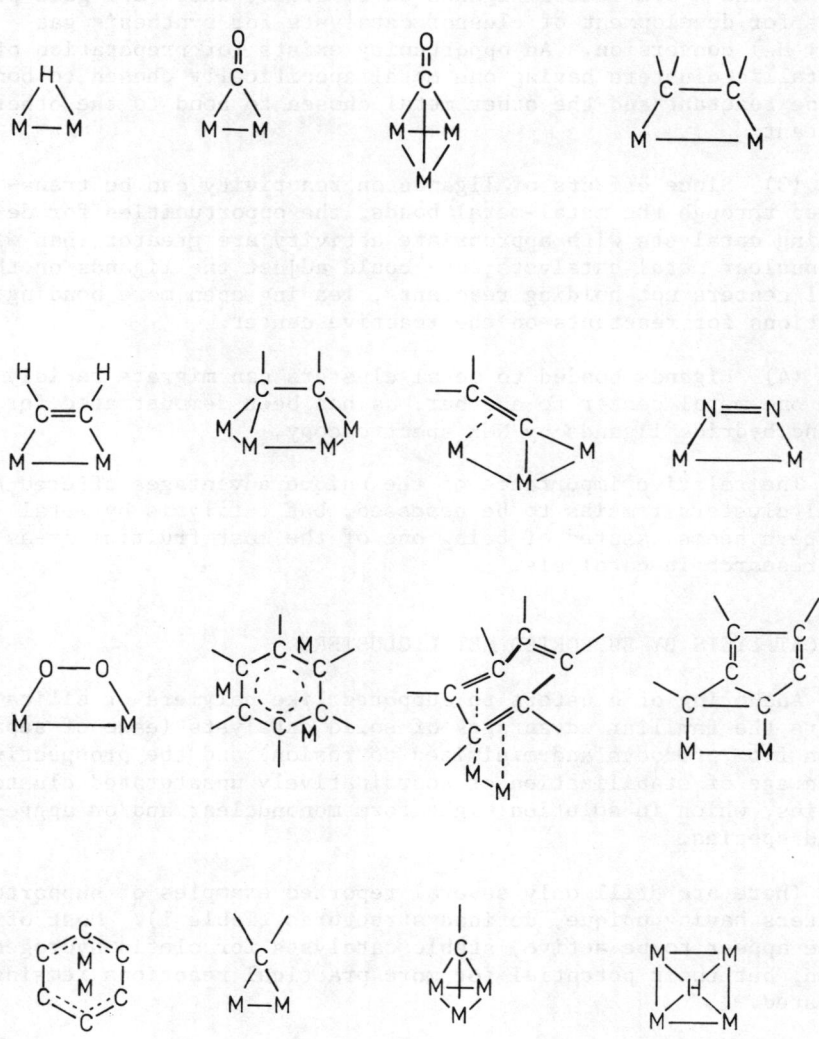

Fig. 5 Some metal-ligand combinations in metal clusters [6].

(2) Having neighboring metal centers, the clusters allow bonding of different reactants in neighboring positions. Since both CO and H are common ligands in clusters, there are good prospects for development of cluster catalysts for synthesis gas (CO + H_2) conversion. An opportunity exists for preparation of bimetallic clusters having one metal specifically chosen to bond to one reactant and the other metal chosen to bond to the other reactant.

(3) Since effects of ligands on reactivity can be transmitted through the metal-metal bonds, the opportunities for designing catalysts with appropriate activity are greater than with mononuclear metal catalysts; one could adjust the ligands on the metal centers not holding reactants, leaving open more bonding positions for reactants on the reactive center.

(4) Ligands bonded to metal clusters can migrate rapidly from one metal center to another, as has been demonstrated for CO and hydride ligands by NMR spectroscopy.

The relative importance of the unique advantages offered by metal clusters remains to be assessed, but catalysis by metal clusters seems assured of being one of the most fruitful areas for research in catalysis.

2. CATALYSIS BY SUPPORTED METAL CLUSTERS

Anchoring of clusters to supports like polymers or silica offers the familiar advantages of solid catalysts (ease of separation from products and minimized corrosion) and the prospective advantage of stabilization of coordinatively unsaturated cluster species, which in solution might form mononuclear and/or aggregated species.

There are still only several reported examples of supported clusters having unique, defined structures (Table 1). Most of these appear to be active, stable catalysts for olefin hydrogenation, but their potential for more practical reactions remains untested.

Table 1. Supported Metal-Cluster Catalysts

Catalyst	Reaction	Reference
$Ir_4(CO)_{11}Ph_2P-\circled{P}$	ethylene hydrogenation	11
$Fe_2Pt(CO)_8(Ph_2P-\circled{P})_2$	ethylene hydrogenation	12
$RuPt_2(CO)_5(Ph_2P-\circled{P})_3$	ethylene hydrogenation	12
$HAuOs_3(CO)_{10}(Ph_2P-\circled{P})$	ethylene hydrogenation	12
$HAuOs_3(CO)_{10}(Ph_2P-SIL)$	butene isomerization	12

Notes: $\circled{P}$ is poly(styrene-divinylbenzene).
SIL is silica.

REFERENCES

1. B. C. Gates, J. R. Katzer, and G. C. A. Schuit, Chemistry of Catalytic Processes, McGraw-Hill, New York, 1979.
2. G. F. Stuntz and J. R. Shapley, J. Amer. Chem. Soc., 99, 607 (1977).
3. K. J. Karel and J. R. Norton, J. Amer. Chem. Soc., 96, 6812 (1974).
4. J. C. Calabrese, L. F. Dahl, P. Chini, G. Longoni, and S. Martinengo, J. Amer. Chem. Soc., 96, 2614 (1974).
5. V. G. Albano, A. Ceriotti, P. Chini, S. Martinengo, and M. Anker, J. Chem. Soc. Chem. Commun., 859 (1975).
6. Muetterties, E. L., Bull. Soc. Chim. Belg., 84, 959 (1975).
7. A. K. Smith and J. M. Basset, J. Mol. Catal., 2, 229 (1977).
8. A. J. Deeming and S. Hasso, J. Organometal. Chem., 114, 313 (1976).
9. J. B. Keister and J.R. Shapley, J. Amer. Chem. Soc., 98, 1056 (1976).
10. R. L. Pruett, Ann. NY Acad. Sci., 295, 329 (1977).
11. J. J. Rafalko, J. Lieto, B. C. Gates, and G. L. Schrader, Jr., J. Chem. Soc. Chem. Commun., 540 (1978).
12. R. Pierantozzi, K. J. McQuade, B. C. Gates, M. Wolf, H. Knözinger, and W. Ruhmann, J. Amer. Chem. Soc., in press.

POLYMER-SUPPORTED CATALYSTS

B. C. Gates

Center for Catalytic Science and Technology
Department of Chemical Engineering
University of Delaware
Newark, Delaware 19711 U.S.A.

1. INTRODUCTION

Although most solid catalysts are inorganic, typically metals, metal oxides, or metal sulfides, we consider organic polymers here because they provide a good transition topic: The most important characteristic of the polymers is that groups bonded to them may be nearly uniform in character and function catalytically much as they do in solution; we can, therefore, build straightforwardly on our knowledge from Chap. 2 of "Chemistry of Catalytic Processes" [1]. The physical properties of polymers can be varied widely, being influenced by the average molecular weight, by the chemical nature of the monomer or the combination of monomers, and by the conditions of polymerization, which affect the arrangement of the polymer molecules and their interactions with each other. An especially simple kind of solid polymer is a gel, a phase in which the long strands of polymer molecules are intertwined randomly.

A gel of linear polystyrene is much like a hydrocarbon liquid. The molecules are in constant random motion; they accommodate high concentrations of molecules having similar chemical character, e.g., benzene, alkylbenzenes, and many other hydrocarbons, but they lack an affinity for water and other polar molecules. Small molecules like benzene can diffuse through the gel rapidly, encountering almost as little resistance to passage between the chains as they would diffusing through a solution of long hydrocarbon molecules. The interactions among the polymer molecules are similar to the interactions among hydrocarbon molecules in solution, and not surprisingly the polymers may dissolve in certain solvents.

Polymers can also be constructed to form solids with physical properties that make them suitable as building materials. To illustrate the range of physical properties, we consider now a copolymer of styrene and a closely related monomer, divinylbenzene:

$$n \underset{\text{styrene}}{\overset{HC=CH_2}{\underset{}{\bigcirc}}} + m \underset{\text{divinylbenzene}}{\overset{HC=CH_2}{\underset{HC=CH_2}{\bigcirc}}} \longrightarrow \tag{1}$$

$$\underset{\text{poly(styrene-divinylbenzene)}}{---HC-CH_2-CH-CH_2-CH-CH_2-CH-CH_2---}$$

We refer to the divinylbenzene as a crosslinking agent and to the monomer unit which ties two chains together as a crosslink. The polymer molecules have a chemical character almost identical to that of the linear polystyrene, but they now are more highly structured three-dimensional molecules.

The divinylbenzene-to-styrene ratio in the monomer mixture has a strong influence on the physical properties of the resulting polymer. If this ratio is only 0.01, the polymer can be produced as solid gellular particles. The molecules are so loosely packed that a particle can swell to about five times its original (shrunken) volume if immersed in a good swelling agent like benzene.

If the divinylbenzene-to-styrene ratio is increased to 0.1, the resultant polymer beads are still flexible and swellable, but the maximum increase in volume is only about 30%. More and more added crosslinks hold the polymer network together more and more tightly. In the limit as the ratio approaches values exceeding 1 or more, the polymer becomes a highly structured solid, nearly impermeable even to moleclues having a strong chemical affinity for it.

Polymers, including especially crosslinked polystyrene, are available commercially with a wide variety of crosslinking contents and physical properties. The physical properties can be varied easily, and tailormade polymers can be designed to hold catalytic groups. The variations in physical properties of the polymers are associated with wide variations in the catalyst performance, a topic discussed in the following pages.

2. ATTACHMENT OF CATALYTIC GROUPS TO POLYMERS

The familiar acidic, basic, and transition-metal-complex groups and metal clusters can be bonded to solid polymers. By bonding the catalytic groups to a support, we can isolate them in their own separate phase. If a stream containing reactants flows over the beads, the familiar catalytic reactions can take place, and the reaction products can flow on through the bed of particles. In this kind of application, typical of industrial catalysis, the practical problems of corrosion by catalysts in solution and of separation of soluble catalysts from reaction products are alleviated. It is primarily for these reasons that most industrial catalysts are solids.

Familiar organic synthesis routes show us how to incorporate catalytic groups into a solid by chemically bonding them to the solid. For example, sulfonic acid groups can be incorporated by direct sulfonation of crosslinked polystyrene:

$$\cdots\text{-CH-CH}_2\cdots\ \text{(C}_6\text{H}_5\text{)} + H_2SO_4 \rightarrow H_2O + \cdots\text{-CH-CH}_2\cdots\ \text{(C}_6\text{H}_4\text{-SO}_3\text{H)} \qquad (2)$$

Amine groups can be incorporated via reaction of $-CH_2Cl$ groups to give, for example,

$$\cdots\text{-CH-CH}_2\text{-}\cdots\ \text{(C}_6\text{H}_4\text{-CH}_2\text{-NH}_2\text{)}$$

Polymers containing redox groups can be made by incorporation of monomer units like

(benzoquinone structure)

Polymers containing groups similar to the familiar ligands used in transition-metal-complex catalysis can also be prepared by familiar routes of organic synthesis. For example,

```
···-CH-CH₂-·-·
     |
     ⌬
     |
     PPh₂
```

is an often-used polymer support, since, when brought in contact with metal complexes like $Rh(PPh)_3Cl$, it undergoes phosphine-phosphine ligand exchange and forms metal-containing polymers like

```
····--CH-CH₂-····
      |
      ⌬
      |
      PPh₂
      |
 Ph₃P-Rh-Cl
      |
      PPh₂
      |
      ⌬
      |
   ----CH-CH₂-···
```

These polymers have chemical and catalytic properties similar to those of their soluble analogs, as one would expect. Many similar metal-containing polymers have been prepared, some of them listed in Fig. 1. In the following paragraphs we demonstrate these similarities and go on to emphasize what is less obvious--the differences in performance distinguishing catalytic groups in solutions and the same groups bonded to a polymer support.

3. CATALYSIS IN POLYMER GELS

Polymer gels, being closely similar to solutions, are among the simplest solid catalysts. Many familiar reactions take place in gels incorporating the appropriate catalytic groups much as they take place in solutions. We consider as a first example catalysis by $-SO_3H$ groups, which are responsible for solution catalysis by p-toluenesulfonic acid. The acid groups in the poly(styrene-divinyl-benzene) gel are so strongly polar that the gel, when suspended in water, becomes strongly hydrated, having perhaps 5 or more water molecules associated with each acid group; the interactions involve hydrogen bonds and resemble solvation. The acid groups in the hydrated gel are dissociated, and specific acid catalysis takes place readily, often with reaction rates being nearly the same as

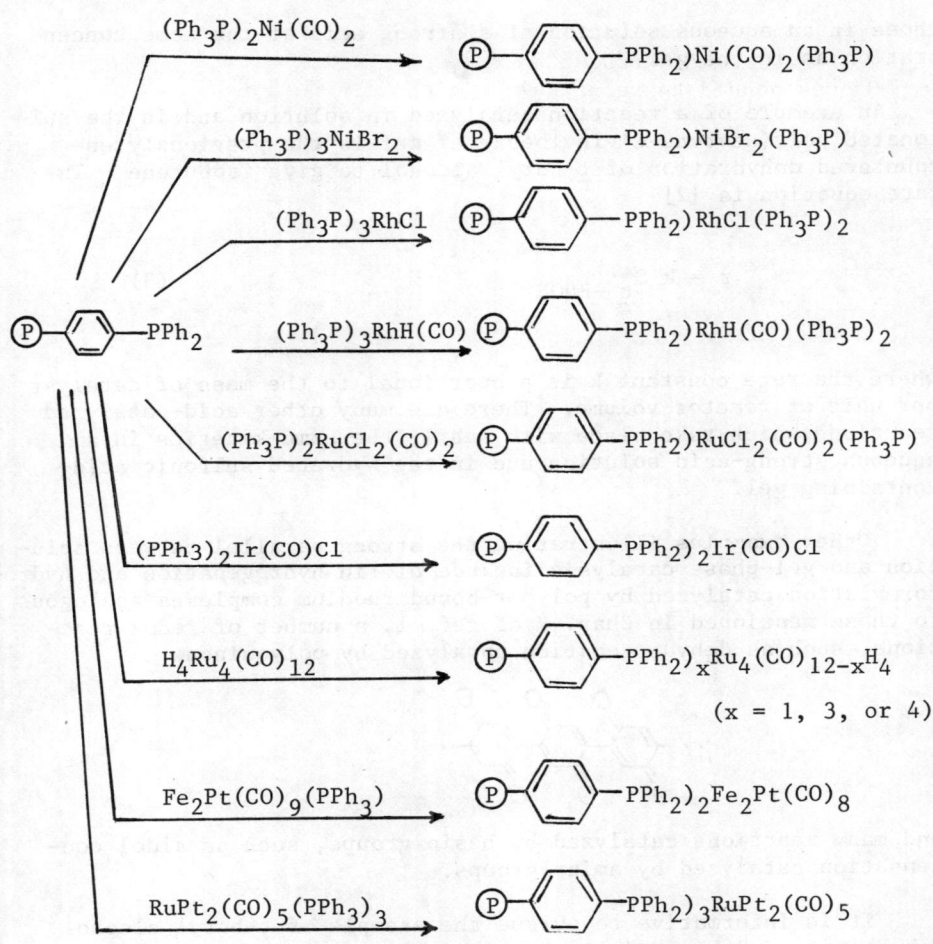

Fig. 1. Metal-containing polymers prepared by ligand exchange of metal complexes and clusters with poly(styrene-divinylbenzene) incorporating phosphine groups. Ⓟ represents the polymer support.

those in an aqueous solution of a strong acid at the same concentration as in the gel.

An example of a reaction catalyzed in solution and in the sulfonated poly(styrene-divinylbenzene) gel is the previously encountered dehydration of t-butyl alcohol to give isobutene. The rate equation is [2]

$$r = k\, C_{\underline{t}\text{-BuOH}} \qquad (3)$$

where the rate constant k is proportional to the mass of catalyst per unit of reactor volume. There are many other acid-catalyzed reactions which take place with nearly the same kinetics in an aqueous strong-acid solution and in the hydrated sulfonic acid-containing gel.

Other examples illustrating the strong parallel between solution and gel-phase catalysis include olefin hydrogenation and hydroformylation catalyzed by polymer-bound rhodium complexes analogous to those mentioned in Chap. 2 of ref. 1, a number of redox reactions, such as dehydrogenation catalyzed by polyquinone

and many reactions catalyzed by basic groups, such as aldol condensation catalyzed by amine groups.

It is informative to pursue the example of t-butyl alcohol dehydration further, since there are more similarities between gel and solution catalysis--as well as some fundamental differences. We consider the situation arising as the liquid medium surrounding particles of the gel containing $-SO_3H$ groups is made to contain smaller and smaller concentrations of water in solution with the reactant alcohol. Reaction rate data (insert, Fig. 2) show that as water concentration is reduced, the rate of the catalytic reaction increases in proportion to the alcohol concentration, consistent with Eq. (3). This pattern, however, holds only over a limited range of water concentrations, as shown in Fig. 2. Complications set in as the water concentration is reduced to low values; in this range, the rate increases much more than linearly with increasing alcohol concentration, rising very sharply as the last traces of water are removed. The data show that water in low concentrations depresses the rate of the catalytic reaction much more than the paraffin methylcyclohexane at the same concentration (Fig. 2).

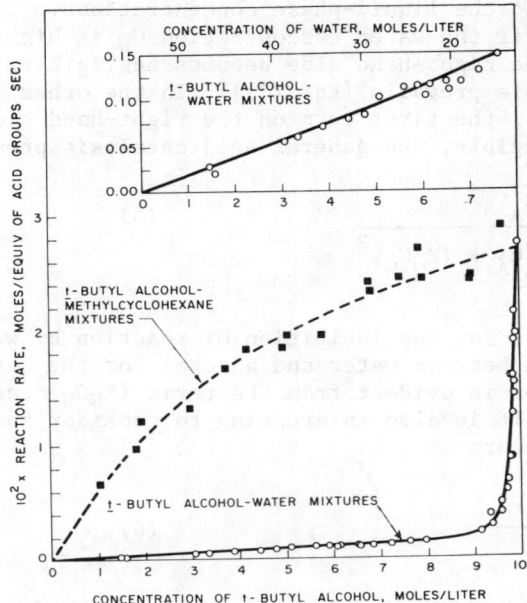

Fig. 2. Kinetics of the dehydration of t-butyl alcohol at $80\pm 2°C$ catalyzed by particles of sulfonated poly-(styrene-divinylbenzene) [2].

It is clear that water present in low concentrations is a reaction inhibitor, although water present in high concentrations, as we have seen, simply dilutes the reactant.

The chemistry of the catalysis is different when only little water is present in the polymer. Without the water, the $-SO_3H$ groups are not dissociated, and general acid catalysis prevails. This example shows a transition from general to specific acid catalysis as water is added to the catalyst. It is evident that the $-SO_3H$ groups are more active catalysts than the hydrated protons. In the case of general acid catalysis, water is a reaction inhibitor, since it competes favorably with the alcohol for the $-SO_3H$ groups, which are the proton donors and sites of catalysis in the polymer.

The rate equation describing the data of Fig. 2 is

$$r = k\, C_A + \frac{k' K_A C_A}{1 + K_A C_A + (K_W C_W)^2} \qquad (4)$$

where the C's refer to the liquid-phase concentrations. There are two limiting cases: if the water concentration C_W is high, then the second term on the right-hand side becomes negligible, and specific acid catalysis prevails [Eq. (3)]. In the other limit, as C_W approaches zero, the first term on the right-hand side of Eq. (4) becomes negligible, and general acid catalysis prevails:

$$r = \frac{k' K_A C_A}{1 + K_A C_A + (K_W C_W)^2} \qquad (5)$$

This equation accounts for the inhibition of reaction by water and shows the competition between water and alcohol for the catalytic sites; the competition is evident from the terms $(K_W C_W)^2$ and $K_A C_A$ in the denominator. It is also interesting to consider the special case for which C_W is zero:

$$r = \frac{k' K_A C_A}{1 + K_A C_A} \qquad (6)$$

The rate increases linearly with increasing C_A at low values of C_A, then the rate vs. C_A curve bends over to a horizontal line, the shape of the curve being that of the Langmuir isotherm. This example illustrates saturation kinetics: at low alcohol concentrations in paraffin, adding more alcohol to the reactant solution proportionately increases the rate of reaction, presumably by proportionately increasing the concentration of alcohol in the catalyst phase, i.e., associated with the catalytic sites. At the highest alcohol concentrations in the liquid phase, further increases hardly increase the rate since the catalytic sites are almost saturated with the reactant alcohol.

In summary, the first special case (large C_W) corresponds to solutionlike catalysis, with the specific acid catalysis described by an equation [Eq. (3)] typical of specific acid catalysis in solution. In the other limiting case (small C_W), catalysis by the individual $-SO_3H$ groups is described by saturation kinetics and a competition between reactant and a product for the catalytic sites. Kinetics of the latter form is typical for heterogeneous catalysis.

At this point it is helpful to consider more carefully where Eq. (5) came from. We consider the simple special case for which C_W is zero and consider the following sequence of steps:

$$H_3C-\underset{CH_3}{\overset{CH_3}{C}}-OH \text{ (solution)} \longrightarrow H_3C-\underset{CH_3}{\overset{CH_3}{C}}-OH \text{ (polymer)} \qquad (7)$$

(This is a transfer of alcohol between phases.)

$$\begin{array}{c}\text{H}_3\text{C}\\|\\\text{H}_3\text{C-C-OH (polymer)} + -\text{SO}_3\text{H (polymer)} \rightarrow\\|\\\text{H}_3\text{C}\end{array} \quad \begin{array}{c}\text{H}_3\text{C} \quad \text{H-O-}\overset{\overset{\text{O}}{\|}}{\text{S}}\text{-}\\|\quad\quad\quad\|\\\text{H}_3\text{C-C-O-H}\quad\quad\text{O}\\|\\\text{H}_3\text{C (polymer)}\end{array} \quad (8)$$

$$\begin{array}{c}\text{H}_3\text{C} \quad \text{H-O-}\overset{\overset{\text{O}}{\|}}{\text{S}}\text{-}\\|\quad\quad\quad\|\\\text{H}_3\text{C-C-O-H}\quad\quad\text{O} \text{ (polymer)} \xrightarrow{\text{slow}}\\|\\\text{H}_3\text{C}\end{array} \begin{array}{c}\text{CH}_2\\\|\\\text{H}_3\text{C-C} + \text{H}_2\text{O}\\|\\\text{CH}_3\end{array} \quad (9)$$

$$+ -\text{SO}_3\text{H (polymer)}$$

We consider the last step to be rate determining.

The first two (fast) steps constitute an equilibration of the alcohol between phases. If the equilibrium is described by the Langmuir isotherm, we have

$$\theta_A = \frac{K_A C_A}{1 + K_A C_A}, \quad (10)$$

where θ_A is the fraction of $-\text{SO}_3\text{H}$ groups bonded to the alcohol. Now, the rate determining step is (9), and therefore

$$r = k'\theta_A$$

$$\text{or} \quad r = \frac{k' K_A C_A}{1 + K_A C_A}, \quad (11)$$

which has the form of Eq. (6). If we now consider the competition between alcohol (A) and water (W) for the acid groups and assume again that the equilibrium is described by the Langmuir isotherm, we have

$$\theta_A = \frac{K_A C_A}{1 + K_A C_A + K_W C_W} \quad (12)$$

This assumption leads to the rate equation

$$r = \frac{k' K_A C_A}{1 + K_A C_A + K_W C_W} \quad (13)$$

this result is similar to the observed rate equation [Eq. 4] for the special case of small C_W, but it is not quite the same, which suggests that the picture presented here is an oversimplification.

Reactions catalyzed by the sulfonic acid resin include many besides alcohol dehydration which are familiar to us, including olefin isomerization and oligomerization, esterification, alkylation of aromatic compounds, and so on. Some have found industrial application, including propylene hydration, phenol alkylation, and bisphenol A synthesis from phenol and acetone. Another is the conversion of methanol and isobutene into methyl-t-butyl ether, a gasoline blending component. For a number of these reactions, the dependence of reaction rate on the concentration of catalytically active acid groups in the polymer has been determined; these catalysts are especially convenient to study in this way, since the catalytic groups ($-SO_3H$ groups) can be replaced in controlled degrees by simple ion exchange, for example by contacting the catalyst with aqueous NaOH to neutralize a fraction of the acid groups and form $-SO_3Na$ groups in their place.

Often it has been observed that the rate of reaction is strongly dependent on the concentration of $-SO_3H$ groups. The data of Fig. 3 demonstrate the effect for the dehydration reaction of t-butyl alcohol. We see that the $-SO_3M$ groups (where M is Na, K, or Rb) all have the same effect--they evidently dilute the catalytically active acid groups. The data show that the reaction is roughly first order in acid-group concentration at low acid-group concentrations and roughly fourth order at high concentrations. The results suggest that the catalytic sites are single $-SO_3H$ groups when the population of groups in the polymer is sparse, and that the catalytic sites are clusters of $-SO_3H$ groups when their population is dense. Infrared spectroscopy shows that the acid groups in close proximity to each other form a hydrogen-bonded network, and a plausible explanation for the high order of reaction in acid-group concentration is suggested by the structures of Fig. 4. According to this model, the reaction involves concerted proton transfers, and both reactant alcohol and inhibitor water can be hydrogen bonded in the network. The high orders of reactions in acid-group concentration have also been observed for general acid catalysis in nonpolar solvents. There is a striking analogy between catalysis in the flexible hydrocarbon polymer matrix (with its strongly polar catalytic groups) and catalysis in a concentrated solution of a strong acid like p-toluenesulfonic acid ($H_3C-\langle\bigcirc\rangle-SO_3H$) in solution with a hydrocarbon like n-hexane.

There is, however, an important distinction between the two kinds of catalysis. The acid groups in solution are unrestricted, having a strong tendency to dimerize and otherwise bond with themselves or other polar molecules. But the groups anchored to the polymer matrix are restricted in their motions and cannot bond so effectively with each other; consequently, they may be more available for bonding with polar reactants like alcohols or with weakly polar reactants like olefins. Groups bonded to relatively rigid polymers (e.g., those having high degrees of crosslinking) may be efficiently held apart and accessible to reactants, whereas groups in more flexible (more solutionlike) polymer gels may be inefficient catalysts because they accommodate reactants poorly.

We now recognize some of the possibilities for catalyst design related to the physical properties of the support, not just the chemical properties of the catalytic groups. We also see the importance of catalytic groups working in combination. Examples follow to illustrate the point that most solid catalysts are characterized by combinations of catalytic groups.

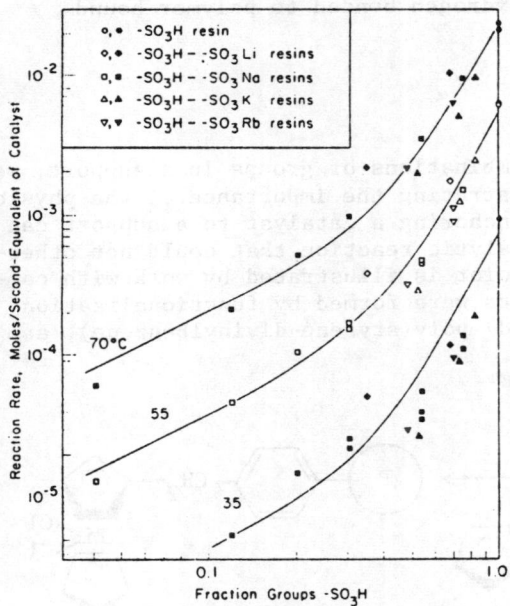

Fig. 3. Dependence of t-butyl alcohol dehydration rate on the composition of sulfonated poly(styrene-divinylbenzene) catalyst [3].

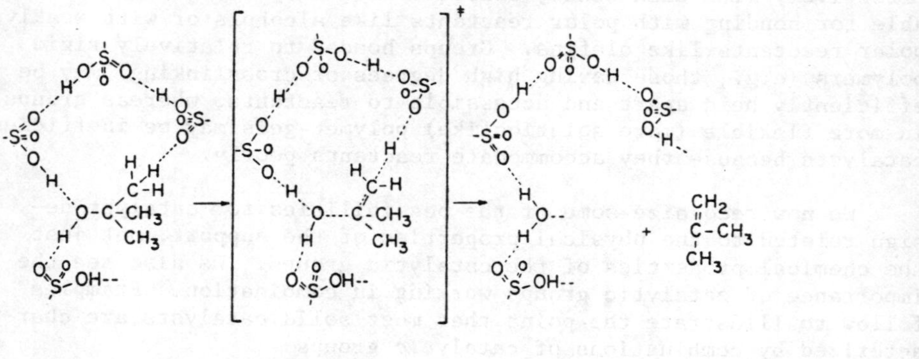

Fig. 4. Suggested mechanism of t-butyl alcohol dehydration involving reactant hydrogen bonded to polymer-bound -SO$_3$H groups [3].

Before discussing the combinations of groups in a support, we consider another example illustrating the importance of the physical properties of the support. Anchoring a catalyst to a support can sometimes make possible a catalytic reaction that could not otherwise even take place. This point is illustrated by work with complexes titanium [4]. Catalysts were formed by functionalization of a rigid (highly crosslinked) poly(styrene-divinylbenzene), as follows:

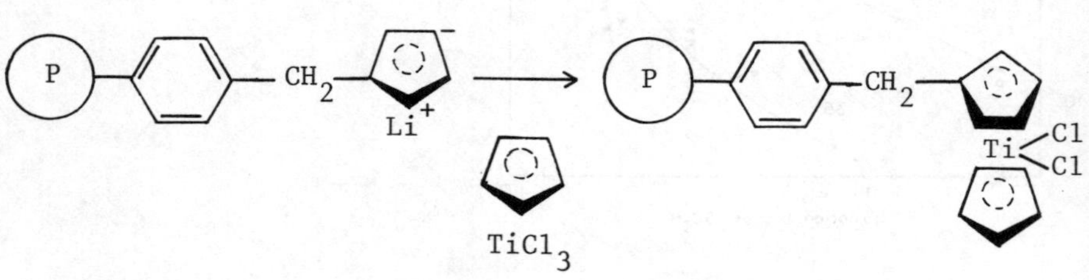

where Ⓟ represents the polymer backbone. Upon reduction with butyllithium, catalyst was formed, presumably having the following structure:

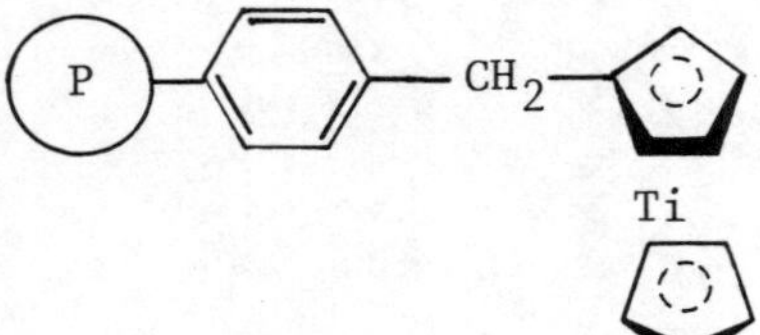

Polymer supported catalysts

The sandwich complex of Ti had the coordinative unsaturation to allow bonding of reactant hydrogen and olefin (or acetylene), and it was catalytically active for hydrogenation. The molecular analog of this species

(titanocene, ) is ineffective as a catalyst in solution,

since Ti-Ti bonds form and it oligomerizes or polymerizes, forming species lacking the necessary coordinative unsaturation. This is an example of self inhibition of a catalyst and is suggestive of the phenomena involved with cobalt hydroformylation catalysts in solution. In the case of hydroformylation, CO ligands bonded to the Co prevent self inhibition by metal-metal bond formation; in the present example, rigid polymeric ligands bonded to the Ti prevent self inhibition by metal-metal bond formation. The physical properties of the polymer support are critical; if the support is too flexible and solutionlike, the catalyst cannot function; if it is rigid and holds the titanocene groups in their own territories, they cannot nullify each other's catalytic activity.

4. BIFUNCTIONAL AND MULTIFUNCTIONAL CATALYSIS

So far we have considered the simple case of a single kind of catalytic group attached to a solid. There are many opportunities for carrying out intricate catalytic reactions and reaction sequences by incorporating just the right combinations of groups in the polymers. One of the simplest and most striking examples of bifunctional catalysis by a solid is provided by the results of Affrossman and Murray [5], who used poly(styrene-divinylbenzene) supports containing $-SO_3H$ groups and $-SO_3Ag$ or $-SO_3K$ groups. The catalytic reactions were

$$H_2O + CH_3-C\underset{OC_3H_7}{\overset{O}{\diagup}} \longrightarrow CH_3COOH + C_3H_7OH \qquad (15)$$

$$H_2O + CH_3-C\underset{\underset{\text{allyl acetate}}{OCH_2CH=CH_2}}{\overset{O}{\diagup}} \longrightarrow CH_3COOH + H_2C=CHCH_2OH \qquad (16)$$

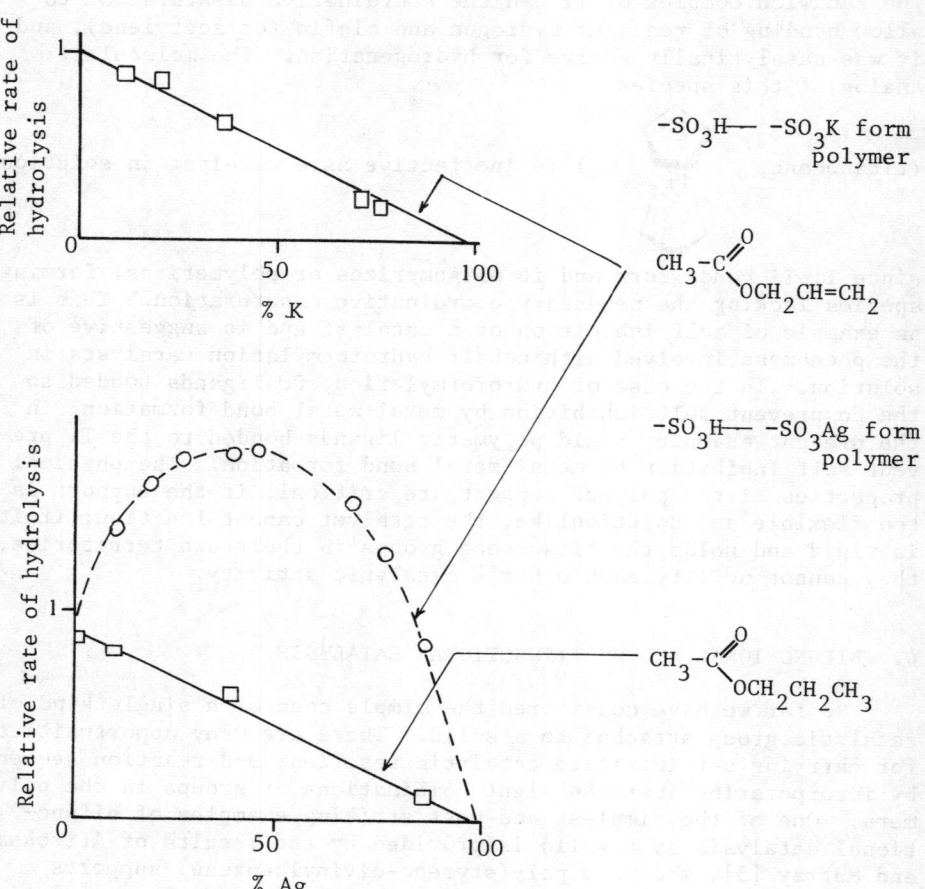

Fig. 5. Kinetics of hydrolysis of n-propyl acetate and allyl acetate at 27.5°C catalyzed by sulfonic acid resin modified by incorporation of -SO$_3$K and -SO$_3$Ag groups. The results show the effectiveness of the -SO$_3$Ag groups in concentrating the allyl ester in the catalyst phase [5].

The reactant solutions contained suspended beads of the polymer. The dependence of reaction rate on -SO$_3$H group concentration is shown for each of the esters in Fig. 5. When the acid groups were diluted with -SO$_3$K groups, the rate was linearly dependent on -SO$_3$H group concentration. The catalytic sites were simply (hydrated) -SO$_3$H groups. But when -SO$_3$Ag groups rather than -SO$_3$K groups were present with the -SO$_3$H groups, the catalytic activity was strikingly higher for allyl acetate but not for propyl acetate

(Fig. 5). It is clear that the $-SO_3Ag$ groups are simply inactive ("diluent") groups for the one reactant but exert a strong influence on the catalysis for the other. The explanation for the role of the $-SO_3Ag$ groups is their ability to coordinate olefins:

$$\underset{Ag^+}{\overset{\displaystyle \rangle C = C \langle}{\vdots}}$$

When the $-SO_3Ag$ groups are present in the polymer, the concentration of reactant allyl acetate molecules in the polymer is greater than when $-SO_3K$ groups are present instead. The silver-containing groups tend to concentrate reactant molecules in the phase containing the catalytically active groups ($-SO_3H$ groups)--they affect the distribution of reactants between phases, i.e., they influence the phase equilibrium and not the reaction mechanism. This example shows how the chemistry of a solid catalyst can be designed by incorporation of groups playing an indirect role in the catalysis.

A further example of the opportunities for catalyst design by selection of combinations of groups in polymer matrices is provided by the aldox process. In the industrial aldox process involving solution catalysis, one could use a cobalt or a rhodium complex for hydroformylation, a solution of KOH for aldol condensation, and a metal like Ni for hydrogenation. As we have already mentioned, a polymer incorporating rhodium-complex groups is active for olefin hydroformylation, and a polymer incorporating basic amine groups is active for aldol condensation. A polymer incorporating both kinds of groups (Fig. 6) has not surprisingly been found to catalyze the sequence of reactions shown in Fig. 7; the reactions are hydroformylation (catalyzed by the Rh complex), aldol condensation (catalyzed by the amine groups), and hydrogenation of the olefinic double bond (catalyzed by the Rh complex). The Rh complex is not an active enough hydrogenation catalyst to reduce the -CHO group to a $-CH_2OH$ group, although it might be if it were coordinated to amine instead of phosphine ligands.

The aldox catalyst is an example of a multifunctional catalyst designed for a sequence of reactions. One of its remarkable characteristics is its high activity in comparison with a combination of monofunctional catalysts, i.e., a fraction of catalyst particles incorporating just Rh-complex groups, with the remainder of particles containing just amine groups. The multifunctional catalyst gave higher rates of reaction under comparable conditions, presumably because the presence of the polar amine groups in the polymer with the Rh-complex groups produced a higher concentration of reactants in the polymer phase, near the catalytic groups. This effect of the amine groups is comparable to the effect of the $-SO_3Ag$

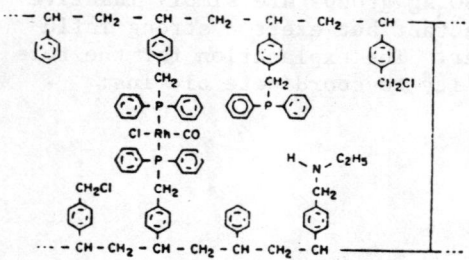

Fig. 6. A representative structural element of the multifunctional aldox catalyst, [6].

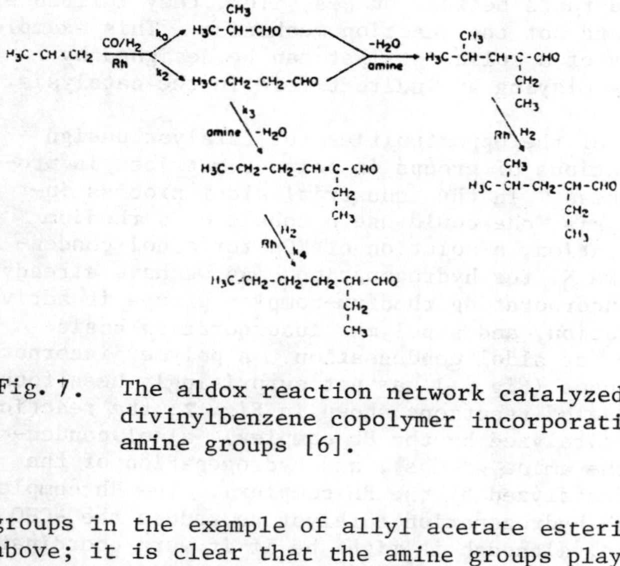

Fig. 7. The aldox reaction network catalyzed by a styrene-divinylbenzene copolymer incorporating Rh-complex and amine groups [6].

groups in the example of allyl acetate esterification discussed above; it is clear that the amine groups play both a direct and an indirect role in the catalysis.

Another kind of bifunctional catalysis is exemplified by recent work [7] with polymers designed to catalyze the Wacker reaction. We recognize from the discussion in ref. [1] that (1) the only metal suitable for the Wacker oxidation of ethylene is palladium and that (2) a second catalyst function is needed to reoxidize the Pd that is formed in the ethylene oxidation step (this is the role played by Cu^{2+} in solution).

A catalyst design meeting these criteria includes palladium sulfonate groups and quinone groups which undergo the quinone-hydroquinone redox cycle:

Polymer supported catalysts

$$\left[\begin{array}{c} \underset{SO_3^-}{} \underset{Pd^{2+}}{} \underset{^-O_3S}{} \end{array} \right]$$

Still another example follows from the discussion [1] of methanol carbonylation. The catalyst, a Rh(I) complex, requires a cocatalyst, which in commercial practice is methyl iodide. A bifunctional solid catalyst might be prepared if groups analogous to the Rh(I) complex and to methyl iodide could be attached to a support. Since methyl iodide itself cannot be supported, groups with similar properties ("pseudohalides") have been supported [8]:

Cl, Cl, Cl, Cl, Cl — SCH_3 a pseudohalide

The polymer functions as a bifunctional catalyst without the need for any added cocatalyst. The roles of the two functions are suggested in the oxidative addition step of Fig. 8.

The bisphenol A synthesis reaction (giving bisphenol A and water from phenol and acetone) provides another example illustrating a combination of functions in a polymer catalyst. The kinetics of the reaction between phenol and acetone in the presence of sulfonated poly(styrene-divinylbenzene) has been measured and represented with the following equation [9]:

$$r = \frac{k K_A C_A K_P^2 C_P^2}{(1 + K_A C_A + K_W C_W)(1 + K_P C_P + K_H C_H)^2} \quad (17)$$

where A represents acetone; P, phenol; W, water; and H, the solvent, n-heptane. The role of heptane indicated by Eq. (17) is that of an inhibitor, not that of a simple diluent (which would fail to appear in the equation). Evidently, the heptane does not play the role of a competitive inhibitor, vying with water and acetone for the polar catalytic sites, the $-SO_3H$ groups. Rather, it appears to compete with phenol for a different kind of site.

We can understand the kinetics better by recognizing how to write Eq. (17) in a form representing the rate in terms of the polymer-phase concentrations of reactants. We follow the development

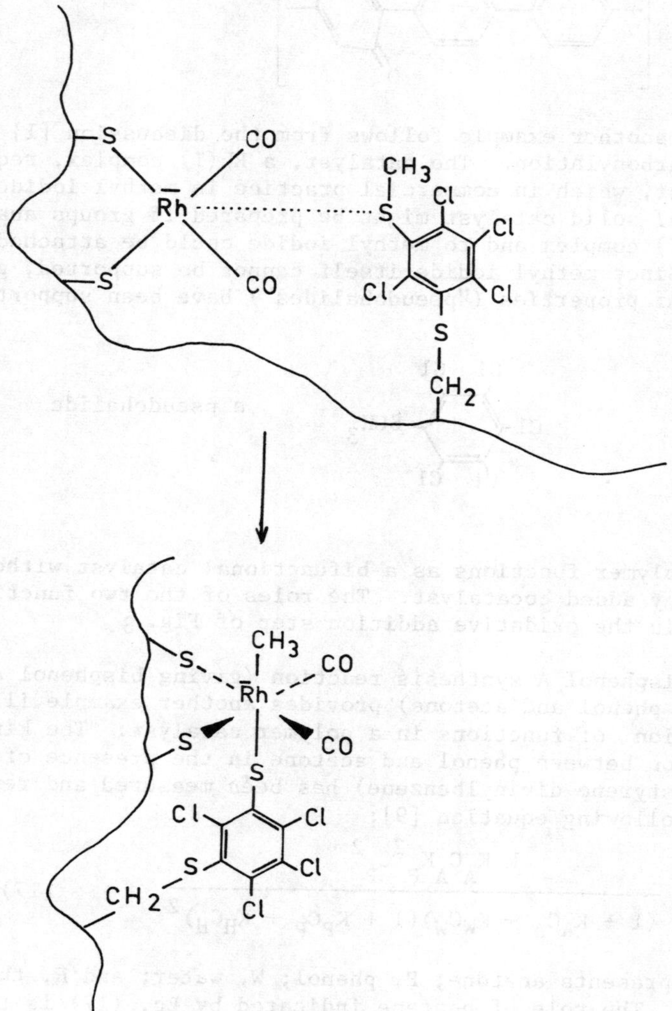

Fig. 8. Suggested oxidative additon step occurring in a bifunctional polymer catalyst for methanol carbonylation [8].

illustrated by the simpler example of alcohol dehydration in the presence of the same catalyst. If the acetone and water compete for the $-SO_3H$ groups and if the Langmuir model describes the adsorption at equilibrium, then

$$\theta_A = \frac{K_A C_A}{1 + K_A C_A + K_W C_W} \tag{18}$$

In principle, we expect any polar molecule to compete with acetone and water for the acid groups, but the data suggest [9] that the term $K_P C_P$ is negligible in comparison with $[1 + K_A C_A + K_W C_W]$.

If the phenol is predominantly bonded to a different kind of site in competition with n-heptane, we may expect that

$$\theta'_P = \frac{K'_P C_P}{1 + K'_P C_P + K'_P C_H} \tag{19}$$

(where the prime denotes a separate polymer phase).

Accepting these suggestions, we recognize that Eq. (17) can be represented as

$$r = k\theta_A^2 \theta'_P \tag{20}$$

The question now arises, what are the sites where phenol and n-heptane bond and the more polar reactants do not? An answer is provided by the hydrophobic or apolar regions of the polymer gel; it is clear that the paraffin has an affinity for the backbone and aromatic rings of the polymer, and the aromatic ring of phenol has an affinity for the same groups. There is an analogy here to the hydrophobic bonding encountered in catalysis by enzymes. We represent the polymer as a bifunctional catalyst, concentrating the reactant acetone at one kind of site and concentrating the phenol at another kind of site. The reaction is facilitated since these two kinds of sites are intimately mixed with each other.

We can now represent the catalytic process as follows:

$$A_{Abs} + -SO_3H \xrightleftharpoons{K_1} A_{Ads} \tag{21}$$

$$A_{Ads} + P_{Abs} \xrightleftharpoons{K_2} I_{Ads} \tag{22}$$

$$I_{Ads} + P_{Abs} \xrightarrow{k_3} H_2O_{Ads} + BPA_{Abs} \qquad (23)$$

If step (23) is rate determining, we have the kinetics of Eqs. (17) and (20).

5. POROUS POLYMERS AND THE IMPORTANCE OF SURFACES

We recognize from the above example that the role of the polymer catalyst in offering sites for bonding one reactant in high concentration and in activating it for reaction is not sufficient--the other reactant must also be accommodated in the catalyst so that it can find access to the first reactant and undergo conversion. We can illustrate this point with another example of an acid-catalyzed reaction, and the example leads us to the recognition of another important opportunity in the design of catalyst properties.

The reaction is the alkylation of benzene with propylene to give isopropylbenzene. When a solution of the two reactants is brought in contact with particles of the sulfonated poly(styrene-divinylbenzene) gel, no catalytic reaction occurs [10]. The difficulty is that the hydrocarbon reactants have such low affinities for the acid groups of the polymer that these groups are tightly bonded with each other through hydrogen bonds like the following [11]:

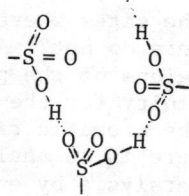

Evidently the hydrophobic interactions are too weak to overcome these polar interactions, and the polymer remains unswollen--the network is collapsed and almost free of reactants.

What is needed to make an acid group effective catalytically is a means for concentrating the reactants in its immediate vicinity. One way of doing this might be to add a good swelling solvent to the reactant solution, but a good polar solvent like water has so much greater an affinity for the $-SO_3H$ groups than the reactants do that it suppresses the catalytic reaction almost completely. The practical alternative is to modify the physical properties of the polymer.

An active catalyst is formed when the polymer has a large fraction of its $-SO_3H$ groups exposed on a surface rather than buried in a sometimes nearly impenetrable mass of hydrocarbon. The surface groups can function effectively because the reactants contact them directly--the groups are present at an interface. Of course a small fraction of the $-SO_3H$ groups are accessible on the peripheral surface of the gel-form polymer as well, but the modified polymer may have 10^3 or 10^4 times as many groups on a surface, having a surface area of perhaps 900 m^2/g.

We refer to the modified polymer as porous or macroporous. Examined under an electron microscope, it appears as a jumble of interconnected microspheres having a typical dimension of 50 or 100 Å. This space can be filled with reactant molecules, which readily find access to the $-SO_3H$ groups on the microsphere surfaces.

The so-called "macroporous" polymers find industrial application for reactions like phenol alkylation. The polymers can be made in a wide range of physical properties including the pore dimensions and surface area. The macroporous polymers are prepared by polymerizing the styrene and divinylbenzene monomers in the presence of n-heptane, for example, which is a good solvent for the monomers but not a good swelling agent for the polymer. The polymer forms a separate solid phase (which becomes the microspherules) and the spaces originally filled with solvent containing the reservoir of monomers remain as pores when the polymerization is completed. The surfaces can be functionalized with any of the acid, base, redox, and metal groups mentioned earlier. The macroporous polymers, typically being highly crosslinked, also offer the practical advantage of greater mechanical stability than the gel-form polymers.

REFERENCES

1. B. C. Gates, J. R. Katzer, and G. C. A. Schuit, Chemistry of Catalytic Processes, McGraw-Hill, New York, 1979.
2. B. C. Gates and W. Rodriguez, J. Catal., 31, 27 (1973).
3. B. C. Gates, J. S. Wisnouskas, and H. W. Heath, Jr., J. Catal., 24, 320 (1972).
4. W. D. Bonds, Jr., C. H. Brubaker, Jr., E. S. Chandresekaran, C. Gibbons, R. H. Grubbs, and L. C. Kroll, J. Amer. Chem. Soc., 97, 2128 (1975).
5. S. Affrossman and J. P. Murray, J. Chem. Soc. B, 1966, 1015.
6. R. F. Batchelder, B. C. Gates, and F. P. J. Kuijpers in Proc. 6th Intl. Cong. Catal., The Chemical Society, London, 1977.
7. H. Arai and M. Yashiro, J. Mol. Catal., 3, 427 (1978).
8. K. M. Webber, B. C. Gates, and W. Drenth, J. Mol. Catal., 3, 1 (1977).

9. R. A. Reinicker and B. C. Gates, <u>Amer. Inst. Chem. Eng. J.</u>, <u>20</u>, 933 (1974).
10. R. B. Wesley and B. C. Gates, <u>J. Catal.</u>, <u>34</u>, 288 (1974).
11. G. Zundel, <u>Hydration and Intermolecular Interaction. Infrared Investigations with Polyelectrolyte Membranes</u>, Academic Press, New York, 1969.

Part V

PARTIAL OXIDATION OF HYDROCARBONS AND THE ACRYLONITRILE PROCESS

CATALYTIC OXIDATION, AN INTRODUCTION

G. C. A. Schuit* and B. C. Gates[†]

*Laboratory for Inorganic Chemistry and Catalysis,
Eindhoven University of Technology, Eindhoven,
The Netherlands
   and
*[†]Center for Catalytic Science and Technology,
Department of Chemical Engineering, University of
Delaware, Newark, Delaware 19711, U.S.A.

1. INTRODUCTION

Catalytic oxidation provides perhaps the most intricate and challenging terrain in the whole field of catalysis. One reason for the complexity is that there is almost always a noncatalytic ("thermal") reaction occurring simultaneously with the catalytic reaction; it may be an almost-negligible background reaction, but it sometimes can accelerate and even overwhelm the catalytic reaction. Oxidation reactions of hydrocarbons are strongly exothermic and consequently tend to be auto accelerating, which makes them difficult to control. If there are two competing reactions, one thermal and another catalytic, control can become even more difficult; catalytic reactions can ignite thermal processes, and homogeneous gas-phase reactions can lead to the combustion of products of catalytic reactions. There is at least one group of important catalytic oxidation processes for which the "catalyst" is nothing but a modifier of the thermal process. All these complications make it necessary to take a hard look at the thermal reactions before considering the subject of pure catalytic oxidation.

2. THERMAL (NONCATALYTIC) OXIDATION AND COMBUSTION

The following description of the mechanisms of noncatalytic oxidation is based largely on the recent survey of Benson and Nangia [1]. Combustion proceeds by a sequence of elementary reactions (steps) leading from one intermediate to another, each intermediate being richer in oxygen than the preceding one. To unravel the sequence of steps, one usually starts with a study of slow reactions proceeding at low temperatures. Thermal oxidations take place via chain reactions, with free radicals serving as chain carriers, and in the lower temperature range the products are chiefly hydroperoxides. A simplified representation of the low-temperature ($\leq 100°C$) chain mechanism is as follows:

Initiation: Initiator $\rightleftarrows$ free radicals (R·) (1)

Propagation cycle:
$$R· + O_2 \underset{b}{\overset{a}{\rightleftarrows}} RO_2·$$ (2a)

$$RO_2· + RH \rightleftarrows ROOH + R·$$ (2b)

Termination: $2RO_2· \rightleftarrows ROOOOR$ (3)

ROOOOR can decompose into products, as follows:

$$ROOOOR \rightarrow RCH_2OH + RCHO + O_2$$ (4)

The initiation reaction is not well understood. It is usually supposed that there is a unimolecular decomposition of the organic reactant involving fission of a C-H bond; because of the high activation energies of such reactions (Table 1), the initiation is very slow. As one would expect, there is often a correlation between increasing tendency of reactants to be oxidized and decreasing strength of the C-H bond in the reactant.

Table 1. X-H bond dissociation energies

Structure	Bond Energy (kcal mole^{-1})	Structure	Bond Energy (kcal mole^{-1})
CH_3-H	103	$CH_2=CH-CH_2$-H	85
$\underline{n}$-C_3H_7-H	99	$Ph-CH_2$-H	85
$\underline{i}$-C_3H_7-H	94	RCO-H	86
$\underline{t}$-C_4H_9-H	90	PhO-H	88
$CH_2=CH$-H	105	ROO-H	90
C_6H_5-H	103		

The rate of oxidation of a reactant can be increased by adding compounds (initiators) that easily decompose into radicals. Such compounds are well known, also serving as initiators for free-radical vinyl polymerization reactions. One of the most reactive initiators is azoisobutyronitrile, $(CH_3)_2C-N=N-C(CH_3)_2$, with CN groups on each central carbon,

which decomposes to form N_2 and two $(CH_3)_2\overset{|}{\underset{CN}{C}}\cdot$ radicals with an activation energy of 30 kcal mole^{-1} and a half-life of 1 h at 85°C. Another initiator is t-butylhyponitrite, t-BuO-N=N-O-t-Bu, which reacts similarly with an activation energy of 28 kcal mole^{-1} and a half-life of 1 h at 60°C.

Peroxides and hydroperoxides belong to this class of initiators, and they are also common reaction intermediates in hydrocarbon oxidation; we return later to the consequences of this important complication of the scheme of oxidation reactions.

The first propagation step (2a) is invariably very fast and takes place with almost no activation energy; its rate is around 10^5 l mole^{-1} sec^{-1}. The bond strength of R-O$_2\cdot$ is only about 28 kcal mole^{-1}, which indicates that the bond formation is a reversible process. The second propagation step (2b) is much slower and is strongly dependent on the nature of RH; the activation energy is about 10-20 kcal mole^{-1}, and the rate constant for some common reactants is about 1 l mole^{-1} sec^{-1} at 30°C.

The termination reaction is puzzling, and Benson and Nangia [1] discussed it in some detail, suggesting an acceptable microscopic formulation. Its formulation (3) is unquestionably real and relevant. It is strongly dependent on the nature of the peroxide, being fast for HO$_2\cdot$ (with a rate constant of about 2×10^9 l mole^{-1} sec^{-1}) but quite slow for tertiary radicals t-RO$_2\cdot$ (with a rate constant of about 10^3 l mole^{-1} sec^{-1} and an activation energy of 8 kcal mole^{-1}).

The total enthalpy of reaction of the hydroperoxide formation

$$RH + O_2 \rightarrow ROOH \tag{5}$$

is around -18 to -24 kcal mole^{-1}. This is exothermic enough that we expect secondary reactions to set in as temperatures increase during reaction.

At temperatures around 100°C, the hydroperoxides begin to react with the radicals appearing in the chain of propagation steps. They do this via reactions such as

$$R\cdot + ROOH \rightarrow RO\cdot + ROH \quad (6)$$

and

$$RCH_2O\cdot \rightarrow R\cdot + CH_2O \quad (7)$$

The occurrence of chain transfer steps changes the distribution of the products but does not substantially change the nature of the chain propagation.

At somewhat higher temperatures (~150°C), however, the hydroperoxides begin to decompose unimolecularly, forming new radicals:

$$ROOH \rightarrow RO\cdot + \cdot OH \quad (8)$$

That is, the products begin to act as initiators, which, formally, amounts to a type of chain branching. This was named "degenerate chain branching" by Semenov to account for the delay between formation and decomposition of the product. The strength of the O-O bond determines the rate of decomposition; for most hydroperoxides the activation energy for bond breaking is about 42 kcal mole^{-1}, but for H_2O_2 it is 57 kcal mole^{-1}. Unimolecular decomposition of most hydroperoxides is therefore rather rapid even at 300°C, but for H_2O_2 it begins to be important only at temperatures $\geqslant$500°C.

To illustrate the consequence of the degenerate branching, it is helpful to consider a simple approximate analysis. The rate of the oxidation reaction is the product of the number of chains (n) operating at a given time and the rate of product formation per chain, p. For each chain there is one radical, so that the number of radicals present at a given time is equal to the number of chains running simultaneously. Therefore, the reaction rate is

$$\frac{d^2x}{dt^2} = p\frac{dn}{dt} \quad \text{where } x = \text{concentration of product molecules (hydroperoxides)} \quad (9)$$

The number of chains being initiated at a certain time is determined by the spontaneous rate of formation of radicals from the reactant (or initiator) (n_0) plus the rate of formation of new radicals resulting from decomposition of the hydroperoxides. The latter reaction is first order in the hydroperoxide concentration x (with a rate constant k), so that

$$\frac{dn}{dt} = n_0 + kx \quad (10)$$

$$\frac{d^2x}{dt^2} = pn_0 + pkx$$

$$\equiv pn_0 + f^2 \cdot x \quad (11)$$

Integration gives

$$x = \frac{n_o}{k}(e^{ft} - 1) \qquad (12)$$

If f is very small $\exp(ft) \cong 1 + ft$, and thus the conversion grows linearly with time. At higher temperatures, however, $\exp(ft)$ is much greater than 1 since the decomposition has a considerable activation energy. In the limiting case,

$$x = \frac{n_o}{k}\exp(ft) \qquad (14)$$

The conversion therefore grows exponentially with time, and since $\frac{n_o}{k}$ is typically very small, some considerable time--called the induction period--will be required before the reaction products become observable. Once the product concentration is appreciable, the conversion increases very fast (Fig. 1), and around 300°C most oxidation reactions of hydrocarbons and aldehydes tend to become explosive, or to give rise to flames (sometimes called "cool flames").

Surprisingly, these reactions often slow down again at higher temperatures, illustrating the well-known negative temperature coefficient of thermal oxidation. At temperatures $\geq 250°C$, the reaction mechanism is no longer the same, the chain of propagation steps being represented instead as

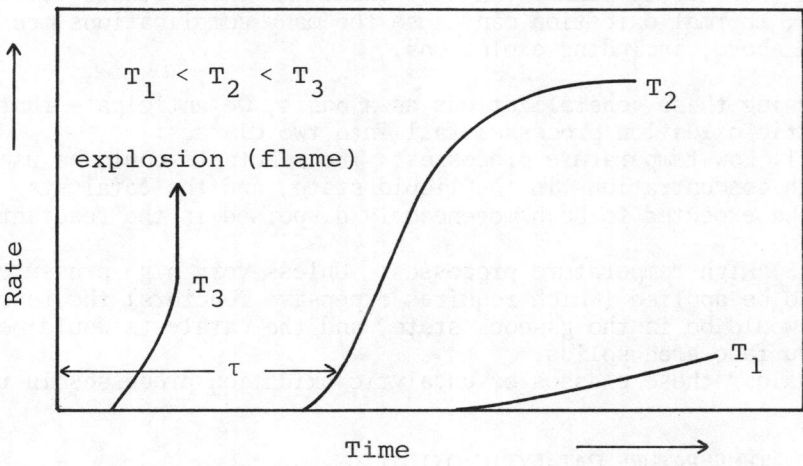

Figure 1. Thermal oxidation of a paraffin in the temperature range 200 to 300°C. τ is the induction period.

$$R\dot{C}H\text{-}CH_3 + O_2 \rightarrow RCH\text{=}CH_2 + \cdot HO_2 \qquad (10a)$$

$$\cdot HO_2 + RCH_2\text{-}CH_3 \rightarrow H_2O_2 + R\dot{C}HCH_3 \qquad (10b)$$

The overall stoichiometry is

$$O_2 + RCH_2CH_3 \rightarrow RCH\text{=}CH_2 + H_2O + 2 \text{ kcal mole}^{-1} \qquad (11)$$

Whereas the low-temperature reaction is decidedly exothermic and therefore potentially autocatalytic, the high-temperature reaction is nearly thermoneutral; it is typically an oxidative dehydrogenation. The hydroperoxide now formed is H_2O_2, which because of its relatively strong O-O bond (57 kcal mole^{-1}), does not decompose readily. At the higher temperature, the rate of branching is decreased, and the rate of termination is presumably increased, since termination by $\cdot HO_2$ recombination is fast. The shift from an organic hydroperoxide to H_2O_2 as the main peroxidic intermediate is ultimately due to the reversibility of reaction (2a); at higher temperatures the superoxide radical ($RO_2\cdot$) does not last long enough to allow a reaction with the hydrocarbon molecule to occur at a significant rate.

From this cursory review of thermal oxidation we learn that there are two temperature ranges where catalytic oxidation might compete successfully with the thermal process: At temperatures $\leq 150°C$ the thermal process is relatively slow, and at temperatures $\geq 350°C$ it is probably not too fast and, more important, it has no strong tendency to lead to an explosion. At intermediate temperatures, thermal oxidation can cause the many complications mentioned above, including explosions.

Using these generalizations as a basis, we anticipate that catalytic oxidation processes fall into two classes:
   (1) Low-temperature processes: The reactants could be used in high concentrations in the liquid state, and the catalysts might be expected to be homogeneously dispersed in the reactant solution.
   (2) High-temperature processes: Unless very high pressures were to be applied (which requires expensive reactors) the reactants would be in the gaseous state, and the catalysts would be high-surface-area solids.
We consider these classes of catalytic oxidation processes in turn.

3. LOW-TEMPERATURE CATALYTIC OXIDATION

There are a number of good reviews of low-temperature catalytic oxidation [2-5], and here we mainly follow the development given by Sheldon and Kochi [2]. There are two principal classes of reaction mechanisms describing the oxidations:

(1) Homolytic processes involve one-electron processes and outer sphere electrons. They are the kinds of reactions we encountered in free-radical oxidation, but now, for the first time, we consider how transition metal ions can play a role. The reactions proceed via a series of one-electron changes, the metals undergoing redox processes such as $Fe^{2+} \rightleftarrows Fe^{3+}$.

(2) Heterolytic processes involve two-electron rather than one-electron changes of metals. They are the kinds of reactions we associate with coordination catalysis, involving reactants bonded in the coordination spheres of transition metals and exemplified by elementary steps like oxidative additions, discussed elsewhere in this volume.

We now consider these two classes of processes in detail.

3.1 Homolytic catalytic processes

To illustrate how transition-metal ions can cooperate in chemical reactions, let us consider Fenton's reagent, a Fe^{2+}-salt used in aqueous solution with H_2O_2 to initiate polymerization reactions by producing radicals:

$$Fe^{2+} + H_2O_2 \rightarrow Fe^{3+} + OH^- + \cdot OH \qquad (12)$$

$$Fe^{3+} + H_2O_2 \rightarrow Fe^{2+} + H^+ + HO_2 \cdot \qquad (13)$$

In a similar type of reaction, organic hydroperoxides can also be involved in the formation of radicals according to the following cycle:

$$ROOH + M^{(n-1)+} \rightarrow RO\cdot + M^{n+} + OH^- \qquad (14)$$

$$ROOH + M^{n+} \rightarrow RO_2\cdot + M^{(n-1)+} + H^+ \qquad (15)$$

The metal M is not consumed here, and its role is therefore catalytic. The result of this catalytic cycle is a branching process which produces two radicals at the expense of one ROOH molecule. The second ROOH molecule is rapidly replenished by the reaction

$$RO_2\cdot + RH \rightarrow ROOH + R\cdot \qquad (16)$$

Kinetically, therefore, this metal-catalyzed process has an effect that is equivalent to the chain branching resulting from unimolecular decomposition in thermal oxidation. The catalytic reaction, even in the presence of only small amounts of the appropriate metal ions, is substantially faster than the thermal process, which indicates that the catalytic reactions (14) and (15) are faster than the thermal branching by unimolecular decomposition reaction (8).

It is important that catalytic branching consists of separate steps, namely (14), in which the hydroperoxide oxidizes the cation, and (15), in which it reduces the cation. Hydroperoxides are usually strong oxidizing agents but only mild reducing agents, and so (14) is presumably faster than (15) under comparable conditions. For those cations that are commonly applied, i.e., Co, Mn, Fe, and Cu, reaction (14) is indeed found to be much faster than the thermal branching, reaction (8). Its activation energy is about 10-12 kcal mole^{-1} (compared with about 40 kcal mole^{-1} for thermal decomposition). (This result applies to reactions in both polar and nonpolar solvents in the presence of very small amounts of catalyst.)

Information about reactions like (15) is sparse, but they are presumably slower than reactions like (14) when the same concentrations of reactants are present for each. For a rough guess as to how the rates of reactions (14) and (15) compare, the redox potentials of the systems $M^{n+} + e^- \rightleftarrows M^{(n-1)+}$ (Table 2) might serve as a guide. The greater this potential, the more the rate of reaction (15) might be expected to approach that of reaction (14).

Although the redox potentials are expected to be only a rough guide to reactivity (and these are not even the correct redox potentials, being applicable to reactions in aqueous solution, whereas the reactions are usually performed in nonpolar hydrocarbon media), it is nevertheless striking that the cations with the higher redox potentials are indeed the ones most often applied as catalysts.

Alternatives to reaction (15) have been proposed for reduction of the cation. Long ago, Haber and Weiss suggested that the catalytic chain for the oxidation of alkyl aromatics consists of an attack of the more highly charged species on the alkyl aromatic:

Table 2. Redox potentials [2]

Reaction	E^o, volt
$Co^{3+} + e^- \rightarrow Co^{2+}$	1.82
$Mn^{3+} + e^- \rightarrow Mn^{2+}$	1.51
$Fe^{3+} + e^- \rightarrow Fe^{2+}$	0.77
$Cu^{2+} + e^- \rightarrow Cu^{+}$	0.15

$$PhCH_3 + Co^{3+} \rightarrow [PhCH_3]^+ \rightarrow PhCH_2\cdot + H^+ + Co^{2+} \qquad (17)$$

where $[PhCH_3]^+$ is a radical cation. After this step, the reaction presumably proceeds via the previously stated route, with Co^{2+} interacting with the hydroperoxide.

Another possibility is given by the extensive formation of peroxides and superoxides by the interaction of low-valent cations with O_2, e.g.,

$$Co^{2+} + O_2 \rightarrow Co^{2+}\cdot O_2 \text{ (or } Co^{3+}\cdot O_2^-) \qquad (18)$$

Provided that the ligand surrounding is appropriate, peroxidation is quite common for these cations and occurs in different forms (with the occurrence of $\sigma\pi$-backbonding, σ-bonding, and monomeric and dimeric species such as CoO_2Co). Although there are undoubtedly examples for which these intermediates are involved, their contribution in the more common processes remains a matter of controversy.

The field of homolytic oxidation proceeding via catalyst-modified reactions encompasses some of the most important commercial oxidation processes, including the following:
  (1) The oxidation of cyclohexane to give cyclohexanol and cyclohexanone;
  (2) the oxidation of p-xylene to give terephthalic acid and its simpler but less important analogue, the oxidation of toluene to give benzoic acid; and
  (3) the oxidation of n-butane to give acetic acid.

These three processes are similar, all being catalyzed by Co- and Mn-salts of organic acids (e.g., acetates and naphthenates). The oxidation of cyclohexane gives products (cyclohexanol and cyclohexanone) which are subsequently oxidized to adipic acid, an intermediate in nylon manufacture. The main product is cyclohexyl hydroperoxide; cyclohexanol and cyclohexanone arise from the Co-catalyzed dissociation of the peroxide. It is generally assumed that the function of the cobalt catalyst is the formation of radicals from the peroxide, and there is considerable doubt about whether attack of Co^{3+} on cyclohexane is important. The hydroperoxide cleavage leads to the radical

$$\underset{CH_2-CH_2}{\overset{CH_2-CH_2}{CH_2}}\!\!\!>\!\!C\!\!<\!\!\overset{H}{\underset{}{O\cdot}}$$

which can produce (1) cyclohexyl alcohol by adding H from cyclohexane or (2) cyclohexanone by cleavage of H·.

The oxidation of n-butane to give acetic acid is believed to involve an attack of Co^{3+} on the hydrocarbon

$$C_4H_{10} + Co^{3+} \rightarrow CH_3\dot{C}HC_2H_5 + Co^{2+} + H^+ \qquad (19)$$

followed by the usual formation of the hydroperoxide. The cleavage of the peroxide seems to be more complicated since it leads also to a cleavage of C-C bonds. Parshall [5] proposed the following mechanism:

$$\underset{C_2H_5}{\overset{CH_3}{>}}CHOOH + Co^{2+} \rightarrow \underset{C_2H_5}{\overset{CH_3}{>}}CHO\cdot + Co^{3+} + OH^- \quad (20)$$

$$\underset{C_2H_5}{\overset{CH_3}{>}}CHO\cdot \rightarrow CH_3C\overset{O}{\underset{H}{<}} + \cdot C_2H_5 \quad (21)$$

(β-bond fission)

and, further (with details left out),

$$\dot{C}_2H_5 \rightarrow C_2H_5\dot{O}_2 \rightarrow C_2H_5OOH \rightarrow C_2H_5\dot{O} \rightarrow C_2H_5OH \text{ etc.} \quad (22)$$

and (via radical intermediates and cationic catalysis),

$$CH_3C\overset{O}{\underset{H}{<}} \rightarrow CH_3C\overset{O}{\underset{OH}{<}} \quad (23)$$

The oxidation reactions of toluene and p-xylene follow the route indicated by reactions (17) and (18), i.e., the Co^{3+} attacks the hydrocarbon. The remainder of the reaction sequence is probably as follows:

$$Co^{2+} + PhCOOH \rightarrow Co^{3+} + OH^- + Ph\dot{C}O \quad (24)$$

$$Ph\dot{C}O + O_2 \rightarrow PhC\overset{O}{\underset{OO\cdot}{<}} \quad (25)$$

$$PhC\overset{O}{\underset{OO\cdot}{<}} + PhCH_3 \rightarrow PhC\overset{O}{\underset{OOH}{<}} + PhCH_2\cdot \quad (26)$$

$$Co^{2+} + PhC\overset{O}{\underset{OOH}{<}} \rightarrow Co^{3+} + OH^- + PhC\overset{O}{\underset{O\cdot}{<}} \quad (27)$$

$$PhC\overset{O}{\underset{O\cdot}{<}} + PhCH_3 \rightarrow PhC\overset{O}{\underset{OH}{<}} + PhCH_2\cdot \quad (28)$$

It has been reported that the reaction does not start unless Co^{3+} is present, which is a strong indication that the first step in the reaction is as given in reaction (17).

A most interesting detail in the oxidation of p-xylene is that it does not proceed further than the oxidation of one methyl group to give $CH_3-C_6H_4-COOH$. However addition of NaBr to the solution (as in the Amoco process) in the presence of Co and Mn cations leads to oxidation of both methyl groups, presumably because of the rate enhancement of the hydrogen addition to $RO_2\cdot$:

$$RO_2\cdot + HBr \rightarrow ROOH + Br\cdot \quad (29)$$

$$Br\cdot + RH \rightarrow HBr + R\cdot \quad (30)$$

The last process we consider in this category is the oxidative dehydrogenation of mercaptans to give disulfides:

$$2RSH + O_2 \rightarrow RSSR + H_2O_2 \quad (31)$$

$$4RSH + O_2 \rightarrow 2RSSR + 2H_2O \quad (32)$$

The process is carried out under the trade name Merox and serves as a sweetening operation. There is no consensus as to whether the reaction is of the one-electron or two-electron type. The cations operating as catalysts are similar to those discussed earlier, i.e., Co, Mn, Fe, and Cu, and this fact seems to support the suggestion of a one-electron process. It is clear that the reaction requires two different catalyst functions; the first step is the loss of a proton to the first function, a base (B):

$$RSH + B \rightleftarrows RS^- + BH^+ \quad (33)$$

In aqueous media the reaction therefore has to be performed in solution of high pH. The second function (the metal, which is involved in electron transfer) then involves the following (provided that the reaction can be assumed to proceed via one-electron steps):

$$RS^- + Cu^{2+} \rightarrow RS\cdot + Cu^+ \quad (34)$$

$$2RS\cdot \rightarrow RSSR \quad (35)$$

$$Cu^{1+} + O_2 \rightarrow Cu^{2+} + O_2^- \quad (36)$$

$$Cu^{1+} + O_2^- \rightarrow Cu^{2+} + O_2^{2-} \quad (37)$$

$$O_2^{2-} + 2H_2O \rightarrow H_2O_2 + 2OH^- \quad (38)$$

So far, all attempts to isolate the superoxy ion (O_2^-) and the radicals by ESR have failed. Neither has one arrived at a satisfactory explanation of the reaction kinetics (half order in oxygen and between first and second order in Cu).

An alternative mechanism would be given by a concerted two-electron process

$$2RS^- + Cu^{2+} + O_2 \rightarrow \left[\begin{array}{c} RS \\ \diagdown \\ \diagup \\ RS \end{array} Cu^{2+} \begin{array}{c} O \\ | \\ O \end{array} \right]^{\ddagger} \rightarrow \begin{array}{c} RS \\ | \\ RS \end{array} + Cu^{2+} + \begin{array}{c} O^- \\ | \\ O^- \end{array} \quad (39)$$

The main objection to this model is that it fails to explain the half-order in oxygen.

The recognition of the functions involved in the solution catalysis (base and metal) has led to the design of solid catalysts for this reaction [6,7]. They are complexes of Co^{3+} with phthalocyanine (CoPc) anchored to polymers incorporating basic groups, $(-CH_2-CHNH_2-)_n$. These bifunctional solid catalysts were found to be 50 times more active per Co atom than CoPc in alkaline solution. The reasons for the high activity are not yet clear, but the important point is that this example provides a simple, easily understood bridge between homogeneous and heterogeneous catalytic oxidation.

3.2 Heterolytic oxidation reactions

The heterolytic processes differ sharply from the homolytic processes, which are so closely tied in with thermal oxidation. The heterolytic reactions are generally far less well understood, but two subclasses are now relatively well defined:

(1) Oxidation reactions catalyzed in aqueous solutions of salts of Pd^{2+} and Cu^{2+} (Wacker type reactions).

(2) Epoxidation reactions of olefins.

We consider these in turn.

The Pd-catalyzed reactions are similar or related to the commercial Wacker process for oxidation of ethylene to give acetaldehyde. The Wacker process proceeds via a sequence of three reactions, and they have been studied separately,

$$Pd^{2+}(Cl^-)_2 + C_2H_4 + H_2O \rightarrow Pd^0 + 2Cl^- + CH_3CHO + 2H^+ \quad (40)$$

$$Pd^0 + 2Cu^{2+} \rightarrow Pd^{2+} + 2Cu^+ \quad (41)$$

Catalytic oxidation

$$4Cu^+ + O_2 + 4H^+ \rightarrow 2H_2O + 4Cu^{2+} \tag{42}$$

Reaction (40) is discussed in some detail elsewhere in this book. It is important to emphasize at this point that the oxygen for the oxidation of Pd^{2+} is provided not by O_2, but instead by H_2O, an oxide; the reaction is typically a concerted process with internal transfer of an electron pair. O_2 is introduced in reaction (42), a well-known reaction, which is, however, mechanistically not well understood. It is expected to be similar to reaction (34) in the mercaptan oxidation. Reaction (41) is the electron transfer step; again we are confronted with the difficulty of having to choose between a concerted two-electron process and a merging of one-electron processes into a pairwise process.

We stress that the Wacker reaction is in many respects a strikingly appropriate model for the high-temperature heterogeneous catalytic reactions, even though it is a homogeneous catalytic reaction. The basis for the comparison is as follows: In both the Wacker process and in surface-catalyzed partial oxidation, one catalyst component takes care of the oxidation of the organic reactant, using oxygen atoms from an oxide. In the high-temperature surface reactions, the oxide is a solid. A second cation takes care of the incorporation of O_2; since in the surface reactions this cation is incorporated in a second solid oxide, the two oxides are typically applied in the form of a binary compound. Electron transfer and transfer of O^{2-} will then occur <u>in</u> this solid.

The epoxidation of olefins is catalyzed by oxygen-containing derivatives of W, Mo, V, Os, Se, and Cr. In the presence of O_2 there is only a slow homolytic reaction as could be expected from the redox-potentials of the metals in their highest valence state (table 3). The reduction of the metal-cation by the hydroperoxides would probably be a slow process. However, in the presence of H_2O_2 or hydroperoxides epoxidation becomes important. Hydroperoxides appear to form peroxides of the catalyst-cation that subsequently react with the olefins to form epoxides (Sheldon [11]).

$$Mo(V) + RO_2H \xrightarrow{fast} Mo(VI) + RO\cdot + OH^-$$

$$Mo(VI) + RO_2H \rightleftarrows [Mo(VI)\cdot RO_2H] \xrightarrow{slow} Mo(V) + RO_2\cdot + H^+$$

$$\xrightarrow{fast} + C_nH_{2n} \longrightarrow \underset{}{>}C\underset{O}{-}C< + ROH + Mo(VI)$$

Mimoun et al [12] propose the following mechanism for the reaction of the inorganic peroxide with the olefin.

Table 3. Redox potentials [2]

Redox couple	E^o, volt
$Mo^{+6} + e^- \rightarrow Mo^{+5}$	~ 0.2
$W^{+6} + e^- \rightarrow W^{+5}$	− 0.03
$V^{+3} + e^- \rightarrow V^{+2}$	− 0.20
$Ti^{+4} + e^- \rightarrow Ti^{+3}$	− 0.37
$Cr^{+3} + e^- \rightarrow Cr^{+2}$	− 0.41

$$\underset{O}{\overset{O}{|}}\underset{O}{\overset{\|}{M}}\underset{O}{\overset{O}{|}} + C=C \longrightarrow \underset{O}{\overset{O}{|}}\underset{\underset{C=C}{O\uparrow}}{\overset{\overset{O}{\|}}{M}}\underset{O}{\overset{O}{|}} \longrightarrow$$

$$\underset{O}{\overset{O}{|}}\underset{\underset{C}{O-C}}{\overset{\overset{O}{\|}}{M}} \longrightarrow \underset{O}{\overset{O}{|}}M=O + C\underset{O}{\overset{}{-}}C$$

4. COMPARISON OF HIGH-TEMPERATURE AND LOW-TEMPERATURE CATALYTIC OXIDATION

We have seen that there are both similarities and some remarkable differences between low- and high-temperature noncatalytic oxidation. We might anticipate parallels with catalytic oxidation, and indeed they exist, as illustrated briefly in the following paragraphs.

There is one high-temperature oxidation process that produces an epoxide: the oxidation of ethylene to give ethylene oxide catalyzed by silver on an alumina support. Kilty et al. [8] proposed a mechanism in which the key intermediate is a surface superoxy species. Since Ag^{2+} has one of the highest known redox potentials (1.98 volt), it is not unexpected that O_2 can form reducing peroxides; once they form, the next step

$$\text{(su)peroxide + olefin} \rightarrow \text{epoxy compound} \qquad (44)$$

is similar to the low-temperature case. The real surprise is that the surface-catalyzed reaction is so selective, but this can be understood from the fact that ethylene is not a particularly aggressive reactant and is more or less restricted to epoxide formation under mild conditions. Moreover, in the commercial process, a subterfuge is used to restrict the number of peroxide sites on the surface by partially covering it with Cl^- ions.

By far the greater number of important high-temperature catalytic oxidation reactions, especially those with olefins as reactants, follow the pattern of bifunctionality illustrated by the Wacker process, with oxygen being delivered to the organic reactant by the oxidic catalyst. The first step in the reaction of the olefin leads to the formation of the allyl radical. One may ask whether this radical is formed in the gas phase or on the catalyst surface. (In the former case the catalytic reaction would be viewed as just a modified thermal reaction.) From the weight of evidence [9,10] it is concluded that the allyl is formed on the surface of the mixed oxide catalyst. Further details of the surface reaction are given by van Hooff in this book.

REFERENCES

1. S. W. Benson and P.S. Nangia, Accts. Chem. Res., 12, 223 (1979).
2. R. A. Sheldon and J. K. Kochi, Oxid. Combust. Rev., 5, 135 (1973).
3. M. M. Taqui Khan and A. E. Martell, Homogeneous Catalysis by Metal Complexes, Academic Press, New York, 1974.
4. J. E. Lyons, Aspects Homog. Catal., 3, 1 (1977).
5. G. W. Parshall, J. Mol. Catal., 4, 243 (1978).
6. L. D. Rollmann, J. Am. Chem. Soc., 97, 2132 (1975).
7. J. Zwart, H. C. van der Weide, N. Broker, C. Rummens, G. C. A. Schuit, and A. L. German, J. Mol. Catal., 3, 151 (1977/78).
8. P. A. Kilty, N. C. Rol, and W. M. H. Sachtler, in Proceedings of the Fifth International Congress on Catalysis, p. 929, North-Holland, Amsterdam, 1973.
9. B. Grzybowska, J. Haber, and J. Janas, J. Catal., 49, 150 (1977).
10. J. D. Burrington and R. K. Grasselli, J. Catal., 59, 79 (1979).
11. R. A. Sheldon, Rec. Trav. Chem., 92, 253 (1973)
12. H. Mimoun, I. Seree de Roch and L. Sajus, Bull. Soc. Chim. France (1969) 1481.

OXIDE CRYSTAL CHEMISTRY AND CATALYSIS

Frank S. Stone

School of Chemistry, University of Bath,
Bath BA2 7AY, England

1. INTRODUCTION

Metallic oxides are important in heterogeneous catalysis both as catalysts per se and as supports. Good examples from the topics already covered in previous lectures are zeolite catalysts for catalytic cracking and alumina-supported platinum for catalytic reforming. It is conventional, and for many purposes very appropriate, to classify oxide catalysts on the basis of general chemical behaviour, such as their acidity, or (in the case of transition metal oxides) their ability to undergo oxidation-reduction cycles, or their propensity for total reduction to metal in the reaction atmosphere, features which can be described in essentially thermodynamic terms. Similar considerations can be applied to oxide supports, for instance in respect of stability towards phase change or crystallization, or solid state reaction with supported material.

There is, however, another kind of classification, one which is based upon crystal structure. This leads to distinctions between oxides on the basis of criteria such as the nature and number of atom neighbours (that is, coordination and ligancy), the presence of layered as opposed to three-dimensional structure, and the occurrence of defects. To appreciate the chemistry and chemical engineering of catalytic processes one needs to be able to grasp the concepts of both classifications. The respective classifications are not mutually exclusive, because they are both underpinned by the theory of chemical bonding, but familiarity with both approaches is the best preparation for dealing with catalysts and understanding catalytic action.

This lecture will be dedicated to the classification of catalysts on the basis of structure. There is a good reason for thinking about catalysts from this standpoint, namely that the processes of heterogeneous catalysis take place on <u>surfaces</u>. There are difficulties in treating surfaces from the thermodynamic point of view, not least because of departure from equilibrium. The adaptation of structure at surfaces, on the other hand, is not fraught with the same uncertainty of extrapolation. Indeed, by adding in the knowledge we have of <u>molecular</u> structure, and particularly the structures of organo-metallic compounds, one has often to make only a small intuitive step to define a model for the adsorbent-adsorbate complex. This affords the best approach so far to the important problem of understanding selectivity in catalysis.

2. OXIDE STRUCTURES

2.1 Close-packing of oxygen ions

The structures of a large number of metal oxides can be understood on the basis of a packing of large spheres (oxygen ions) with small spheres (metal ions) fitting into the remaining spaces.

There are various ways to pack spheres, but the packing most relevant for oxide structures is that in which each large sphere is in contact with 12 other spheres of the same size. This arrangement, known as close-packing, occupies 74% of the total space. There are two main variants, (a) hexagonal close packing (hcp), with a stacking sequence ABABAB ... for the layers, and (b) cubic close packing (ccp), where the sequence is ABCABC ..., thus repeating after three layers instead of two (Fig. 1). Both kinds of close-packing generate two types of hole between the layers, the so-called <u>octahedral holes</u>, bounded by six spheres with centres at the vertices of a regular octahedron, and the <u>tetrahedral holes</u>, bounded by four spheres with centres at the vertices of a regular tetrahedron (Fig. 2). For every $\underline{n}$ spheres in a close-packed array there are $\underline{n}$ octahedral holes and $\underline{2n}$ tetrahedral holes.

Small spheres (metal ions) may now be distributed in these octahedral and tetrahedral holes between the large spheres (oxygen ions), and the various ways in which this can be done afford good models for the structures of many oxides encountered as catalysts. Many sulphide and halide structures also correspond to these models. For example, if all the octahedral holes in a ccp array of O^{2-} ions are occupied by M^{2+} cations, and all the tetrahedral holes are empty, the system is electrostatically balanced with a formula MO: this is the structure of MgO and most other divalent oxides. If the O^{2-} ions are in hcp array and two-thirds of the octahedral holes are occupied by M^{3+} cations, with no

Oxide crystal chemistry and catalysis

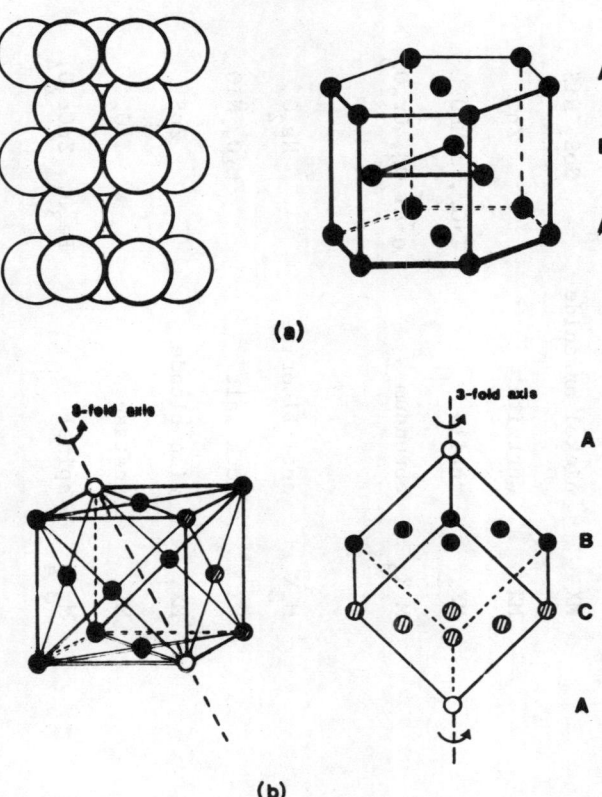

Fig.1. (a) The hexagonal close-packing arrangement. (b) The cubic close-packing arrangement; note that the close-packed layers are stacked normal to the 3-fold axis (body diagonal of cube). [After D.M.Adams, Inorganic Solids, John Wiley & Sons, London, 1974].

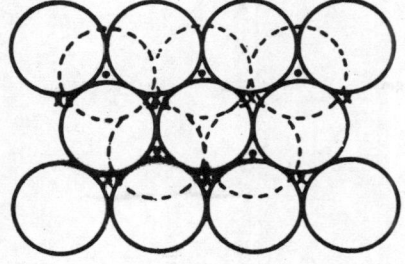

Fig.2. Two layers of close-packed spheres, enclosing octahedral holes (crosses) and tetrahedral holes (black dots).

Table 1

Structures derived from hexagonal close-packing (hcp) and cubic close packing (ccp)

	Fraction of octahedral holes occupied by metal ions	Fraction of tetrahedral holes occupied by metal ions	Formula	Name of structure	Examples
hcp	1	-	MX	nickel arsenide	CoS, NiS
	-	$\frac{1}{2}$	MX	wurtzite	ZnO
	$\frac{1}{2}$	-	MX_2	rutile	TiO_2, $\beta-MnO_2$
	$\frac{2}{3}$	-	M_2X_3	corundum	$\alpha-Al_2O_3$, Cr_2O_3
ccp	-	1	M_2X	anti-fluorite	Na_2O
	1	-	MX	rock-salt	MgO, NiO
	-	$\frac{1}{2}$	MX	zinc blende	ZnS
	$\frac{1}{2}$	-	MX_2	anatase	TiO_2
	$\frac{1}{2}$	$\frac{1}{8}$	M_3X_4	spinel	Fe_3O_4, $ZnCr_2O_4$

Oxide crystal chemistry and catalysis

occupation of tetrahedral holes, there is again charge balance, but with a formula $M_{0.66}O$, i.e. M_2O_3. This is the structure of $\alpha-Al_2O_3$, $\alpha-Cr_2O_3$ and $\alpha-Fe_2O_3$. Table 1 lists these variants and shows a number of other examples of solids of interest as catalysts whose structures can be derived from the symmetry shown by hexagonal close packing and cubic close packing respectively. A simple but useful symbolism for these structures is to indicate by intercalation in ABABAB or ABCABC the type and fraction of occupied hole (o = octahedral, t = tetrahedral). Thus, for example:

hcp   A . B . A . B . A . B   X

      A o B o A o B o A o B   MX   (NiS)

      A $\tfrac{2}{3}$o B $\tfrac{2}{3}$o A $\tfrac{2}{3}$o B $\tfrac{2}{3}$o A $\tfrac{2}{3}$o B $M_{\tfrac{2}{3}}X, M_2X_3$ ($\alpha-Al_2O_3$)

ccp   A . B . C . A . B . C   X

      A t B t C t A t B t C M_2X (Na_2O)

      A o B o C o A o B o C   MX   (MgO)

The structure of spinel is different from the others listed in Table 1 in that <u>both</u> types of hole are partially occupied. Spinel may be described as:

ccp $A(\tfrac{1}{2}o)B(\tfrac{1}{4}o + \tfrac{1}{4}t)C(\tfrac{1}{2}o)A(\tfrac{1}{4}o + \tfrac{1}{4}t)B(\tfrac{1}{2}o)C$ $M'_{tet}[M_2]_{oct}X_4$ ($ZnCr_2O_4$)

To obtain electrostatic charge balance with the full complement of ions in the spinel structure one must either have a mixture of differently charged cations, e.g. $Fe^{2+}Fe_2^{3+}O_4$ for Fe_3O_4, or a mixture of differently charged anions. The usual case is a mixture of differently charged cations. From the fractional occupations given above, and remembering that there are twice as many tetrahedral holes as octahedral holes, the number of cations in tetrahedral holes is seen to be exactly half the number of cations in octahedral holes. However, it does not necessarily follow that the cations in each type of hole are all of the same charge. Whereas in $ZnCr_2O_4$ we do have a <u>normal spinel</u> with tetrahedral Zn^{2+} and octahedral Cr^{3+}, designated as $Zn^{2+}[Cr_2^{3+}]O_4$, in Fe_3O_4 we have an <u>inverted spinel</u>, with all its divalent ions in octahedral sites, i.e. $Fe^{3+}[Fe^{2+}Fe^{3+}]O_4$. Some spinels, e.g. $MgAl_2O_4$, are partially inverted.

In most of the examples in Table 1 the anion array is a somewhat expanded version of true close packing of spheres. Isotropic expansion of the close-packed structure enables electrostatic repulsion between the O^{2-} ions to be reduced. Hence cations which are larger than the size permitted by the geometry of true close

packing frequently occur in octahedral and tetrahedral holes. With
MgO, for instance, the Mg^{2+} cation radius is 0.72Å compared with
0.58Å for the maximum radius of a sphere sited in an octahedral
hole among truly close-packed O^{2-} ions (r = 1.40Å). There are some
cases where this expansion is very large. In zinc oxide, for
example, the anion 'close packing' has been expanded to the point
where the oxygen spheres occupy only about 45% of the total space,
compared with the 74% of true close packing. This enables the quite
large zinc ions (r = 0.7Å) to be accommodated in what are normally
very small tetrahedral holes, holes which allow a maximum radius
of r = 0.32Å in true close packing. These greatly expanded
structures hold together because of a substantial component of
directional (covalent) bonding, e.g. sp^3 hybridization in the
tetrahedral structures.

Although many oxides retain the very high symmetry of the
close-packed arrangement in spite of being expanded structures, it
is not uncommon to find distortion. Thus the octahedral holes, for
instance, in both rutile and anatase forms of TiO_2 no longer have

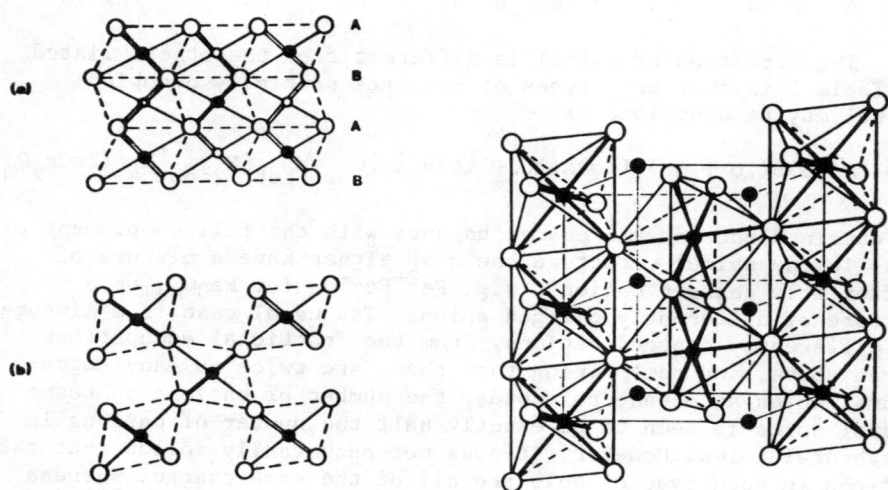

Fig.3. Relationship between the nickel arsenide and rutile
structures. (a) The nickel arsenide structure showing the
AoBoAoB arrangement, with metal atoms as small open and
filled circles; (b) the rutile structure, derived from
NiAs by removing half the metal atoms (those shown as open
circles) and distorting the oxygen layers. The right-hand
drawing shows the rutile structure in elevation. [After
A.F.Wells, Structural Inorganic Chemistry, Oxford
University Press, 1975].

regular octahedral symmetry: four of the oxygen ions around the titanium are closer than the other two (Fig.3). For many purposes, however, there is greater merit in focussing attention on the fact that a structure is near to a regular one than there is in defining its exact symmetry. One need not be too concerned that in rutile each Ti is surrounded by four oxygens at a distance of 1.944Å and two others at 1.988Å. The important thing is to know that in TiO_2 the cations are essentially in octahedral coordination. This is a philosophy which becomes particularly important when one has to deal with complex oxides.

The structure of alumina deserves special mention. It is, after all, the most versatile single oxide in the catalyst armoury, being both a catalyst and a support. α-Al_2O_3, the support used for the silver in an ethylene oxidation catalyst, is a member of the hcp group (Table 1). γ-Al_2O_3, the isomerization catalyst, is a member of the ccp group, but it is not related to α-Al_2O_3 in the simple way that the rutile and anatase forms of TiO_2 are related. γ-Al_2O_3 is more correctly described as

$$\{Al_{2\frac{2}{3}}, \square_{\frac{1}{3}}\} O_4$$

a <u>spinel</u> structure M_3O_4 in which one-ninth of the positions normally occupied by cations are unoccupied. If the alumina has a high surface area and that surface is partially hydroxylated, it is better to represent the solid as having a proportion of its anions OH^- ions and the remainder O^{2-} ions, which will mean a larger proportion of the cation sites is vacated by Al^{3+} ions. The charge-balanced spinel formula is then

$$\{Al_{2\frac{2}{3}}, \square_{(\frac{1}{3}+\frac{3y}{4})}\} O_{(4-y)} (OH)_{2y}$$

or its equivalent

$$\{Al_{(2\frac{2}{3}-\frac{x}{3})}, \square_{(\frac{1}{3}+\frac{x}{3})}\} O_{(4-x)} (OH)_x$$

for a spinel formula containing the conventional total of 4 anions. The dehydroxylation reaction which occurs on rigorous outgassing is:

$$\{Al_{2\frac{2}{3}}, \square_{(\frac{1}{3}+\frac{3y}{4})}\} O_{(4-y)} (OH)_{2y} \rightarrow \{Al_{2\frac{2}{3}}, \square_{\frac{1}{3}}\} O_4 + \frac{y}{2} H_2O$$

in the first case (yielding γ-Al_2O_3 as written above in the conventional spinel formulation with 4 anions), or:

$$\{Al_{(2\frac{2}{3}-\frac{x}{3})}, \square_{(\frac{1}{3}+\frac{x}{3})}\} O_{(4-x)} (OH)_x \rightarrow \{Al_{(2\frac{2}{3}-\frac{x}{3})}, \square_{(\frac{1}{3}-\frac{x}{24})}\} O_{(4-\frac{x}{2})} + \frac{x}{2} H_2O$$

in the second case, again yielding γ-Al$_2$O$_3$ with a spinel structure, but written less conventionally. Some textbooks and reviews represent the hydroxylated γ-Al$_2$O$_3$ as having H$^+$ ions occupying cation sites in the spinel structure, e.g.

$$\{Al_{(2\frac{2}{3}-\frac{x}{3})} H_x \square_{(\frac{1}{3}-\frac{2}{3}x)}\} O_4$$

instead of the formulae using OH, but this is misleading.

The presence of cation vacancies in the spinel structure of γ-Al$_2$O$_3$ raises the question of whether the vacancies are in normally-occupied tetrahedral sites or normally-occupied octahedral sites. Using the convention of writing octahedral sites in square brackets, one may visualise two extremes: <u>either</u> the vacancies are all in the octahedral sites

$$\{Al [Al_{1\frac{2}{3}}, \square_{\frac{1}{3}}]\} O_4$$

or they are all in the tetrahedral sites

$$\{Al_{\frac{2}{3}}, \square_{\frac{1}{3}} [Al_2]\} O_4.$$

Although there is some doubt on the matter, it is very likely that a difference in respect of this occupancy is the feature which distinguishes the two catalytically different aluminas known as γ-Al$_2$O$_3$ and η-Al$_2$O$_3$, respectively. It has been suggested that η-Al$_2$O$_3$ is a form which contains relatively more tetrahedral Al^{3+} than γ-Al$_2$O$_3$. The occupancy characteristics will depend on the type of crystal plane which is most common in the surface of the high-area alumina, and on the degree of hydroxylation of that surface. This will depend in turn on the particular method which is chosen for the preparation and outgassing. Bayerite, Al(OH)$_3$, yields on decomposition at about 250°C, the alumina which the catalyst chemist designates as η-Al$_2$O$_3$, whereas boehmite, AlO(OH), decomposes at higher temperatures ($\sim$ 450°C) to yield the material known characteristically as γ-Al$_2$O$_3$.

2.2 Linked polyhedra

The basis for the description of oxide structures used in the preceding Section was to note that an array of $\underline{n}$ equal-sized close-packed spheres encloses $\underline{n}$ octahedral holes and 2$\underline{n}$ tetrahedral holes, and that together they account for all the space. An equivalent way to account for all the space is to concentrate on the actual octahedra and tetrahedra that encompass the holes, and disregard the spheres. Thus, $\underline{n}$ octahedra and 2$\underline{n}$ tetrahedra properly stacked together exactly fill all space. The only requirement is that the lengths of the sides of the octahedra and

Oxide crystal chemistry and catalysis

Fig.4 Top left - cubic close packing of octahedra and tetrahedra.
Right - exploded view of cubic close packing, revealing the
internal space-filling tetrahedra. Bottom left - exploded
view of hexagonal close packing. [After L.V.Azároff,
Introduction to Solids, McGraw Hill, New York, 1960].

tetrahedra must be the same, and equal to the diameter of the spheres from which one started. When viewed in this way an important distinction emerges between cubic close packing and hexagonal close packing (Fig.4). In ccp, there are no common faces between tetrahedra and no common faces between octahedra. In hcp, on the other hand, common faces occur between tetrahedra and there are also faces shared between octahedra. It is this fact which rules out total occupancy of tetrahedral holes by cations in hcp. There is no AtBtAtB analogue of the AtBtCtAtBtC structure of antifluorite (Table 1) because repulsion of the adjacent cations at the close distance across a common face is too prohibitive. For an equivalent reason, the octahedral structure AoBoAoB is only found with markedly covalent solids (e.g. sulphides) and not with strongly ionic solids (e.g. oxides).

This alternative description of space-filling enables the structures in Table 1 to be seen as consisting of MX_6 or MX_4 tetrahedra which are linked together in different ways. Polyhedra (octahedra or tetrahedra) which are unoccupied by M ions are ignored. Thus, MgO and other oxides with the rock-salt (ccp) structure can be described as having MO_6 octahedra in which all 12 edges are shared with other MO_6 octahedra. Each oxygen is therefore common to six MO_6 octahedra, and hence the formula is $MO_{6/6}$, i.e. MO. In the antifluorite (ccp) structure, the MO_4 tetrahedra share all six edges with other MO_4 tetrahedra: each oxygen is common to eight MO_4 tetrahedra, so the formula is $MO_{4/8}$, i.e. M_2O. In α-Al_2O_3 and other oxides with the corundum structure the hexagonal close packing and the two-thirds occupation lead to each MO_6 octahedron being joined to one other across a face, to three others by a common edge, and to nine more by a common corner. Finally, in spinel the MO_6 octahedra are joined by edges to other occupied octahedra and by corners to occupied tetrahedra. The type of joining defines the M-M distances and the MOM and OMO angles.

Fig.5a shows a fragment of MgO structure exemplifying the edge-sharing of MO_6 octahedra in a ccp structure. Fig.5b shows a pair of face-shared MO_6 octahedra, such as occurs in the hcp structure of α-Al_2O_3. Fig.5c illustrates an element of the spinel structure, showing edge-sharing of MO_6 octahedra but corner-sharing of MO_6 octahedra and MO_4 tetrahedra. Fig.3 shows that in rutile the MO_6 octahedra share edges in the c direction, producing vertical chains (see the elevation drawing), and that these chains fit together by corner-sharing of the MO_6 octahedra (see the plan view of Fig.3b).

The linking of occupied polyhedra is thus another way to describe the structures of close-packed oxides. However, this type of description has the particular merit that one may discuss in a rather straightforward way not only close-packed oxides but also

Oxide crystal chemistry and catalysis

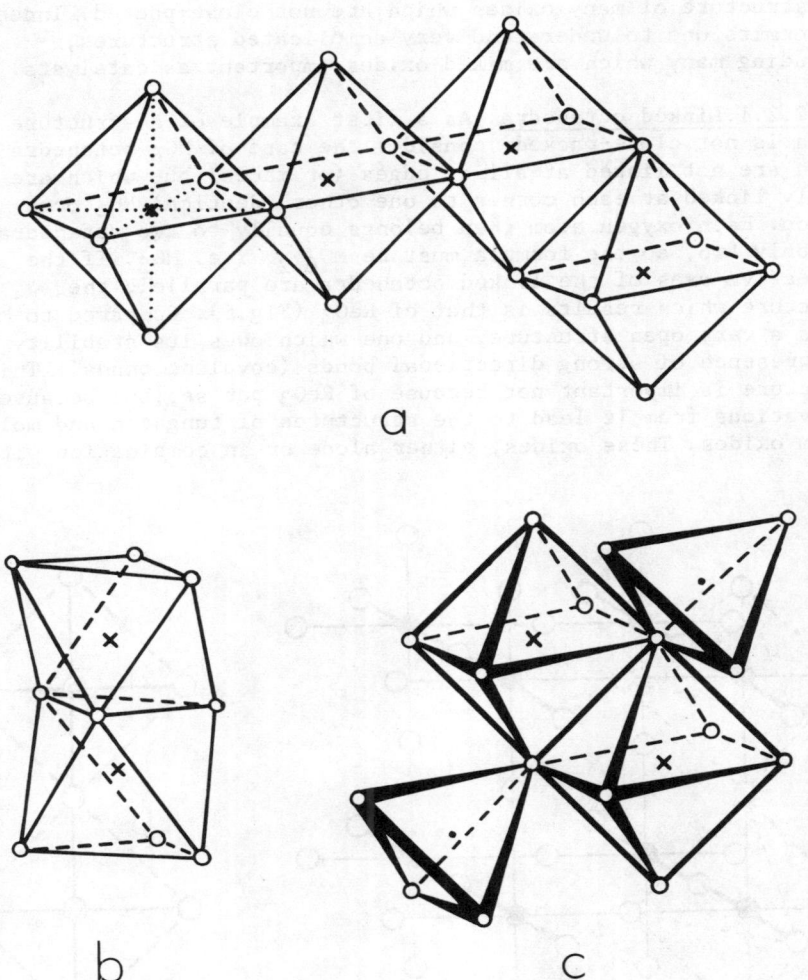

Fig.5 The linking of cation-occupied polyhedra in (a) MgO, (b) corundum and (c) spinel. In MgO, the cations (×) are octahedrally coordinated and all MO_6 octahedra are edge-shared; four such octahedra are shown. In corundum cations (×) are octahedrally coordinated and each MO_6 octahedron shares a face with one other octahedron; the resulting rather short distance between the cations in these pairs causes some distortion of the octahedra (not shown in the drawing). In spinel, two-thirds of the cations are octahedrally coordinated (×) and one-third are tetrahedrally coordinated (•); MO_6 octahedra share edges with other MO_6 octahedra and corners with MO_4 tetrahedra.

the structure of many oxides which are not close-packed. Indeed, it permits one to understand very complicated structures, including many which are mixed oxides important as catalysts.

2.2.1. <u>Linked octahedra</u>. As a first example of a structure which is not close-packed, consider the case of MO_6 octahedra which are not linked at all at edges (or faces) but which are merely linked at each corner to one other identical MO_6 octahedron. Each oxygen atom then belongs equally to two octahedra, and only two, so the formula must be $MO_{6/2}$, i.e. MO_3. If the respective axes of the linked octahedra are parallel, the structure which results is that of ReO_3 (Fig.6). Compared to MgO it is a very open structure, and one which owes its stability to the presence of strong directional bonds (covalent bonds). This structure is important not because of ReO_3 per se, but because derivations from it lead to the structures of tungsten and molybdenum oxides. These oxides, either alone or in combination with

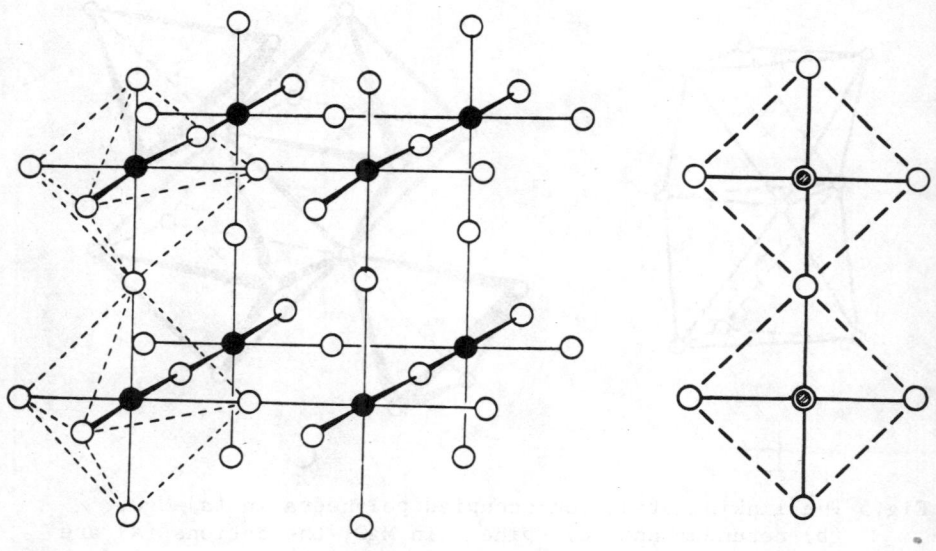

Fig.6 The structure of ReO_3. All cations (solid circles) are octahedrally coordinated to oxygen (open circles) and each MO_6 octahedron is linked at every corner to an identical MO_6 octahedron. If the structure is viewed normal to the front cube face, the two corner-shared octahedra showed in dashed outline in the left-hand diagram appear as shown on the right. The central metal ions (small shaded circle) are now beneath oxygen ions and the octahedra are accordingly represented as crossed squares.

others, frequently feature as catalysts suitable for selective oxidation or oxidative dehydrogenation.

A most intriguing property of WO_6 and MoO_6 octahedra in WO_3 and MoO_3 is their ability to switch from corner-sharing to edge-sharing by a shear process along particular planes. This process, known as crystallographic shear (CS), is illustrated in Fig.7.

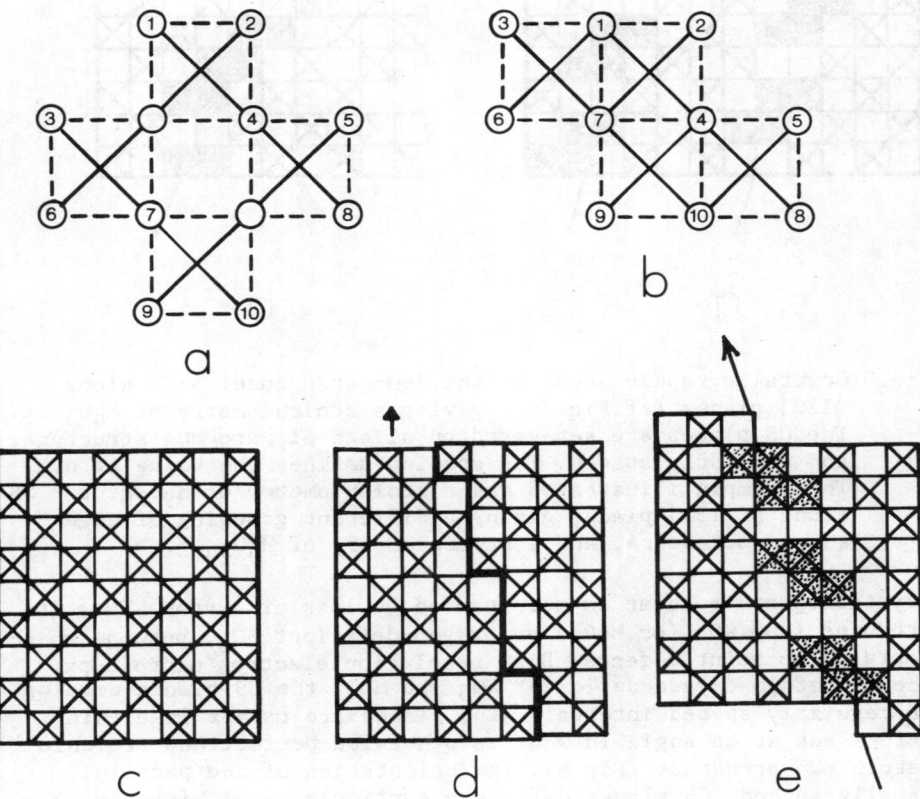

Fig.7 The principle of crystallographic shear (CS). (a) Plan view of four corner-shared MO_6 octahedra (rotate Fig.6 by 45º). Remove two oxygen atoms (unnumbered circles) and transform from corner-sharing to edge-sharing to give (b). This basic process is elaborated in the lower diagrams to show CS along a $\{120\}$ plane. (c) A slab of ReO_3 structure, cf.top left, with MO_6 octahedra indicated as crossed squares. (d) Divide the slab at $\{120\}$ plane. Remove oxygen atoms and move up left half a distance of one octahedral edge (small arrow). (e) Completion of the shear. Large arrow indicates shear plane with groups of edge-shared octahedra, shown stippled.

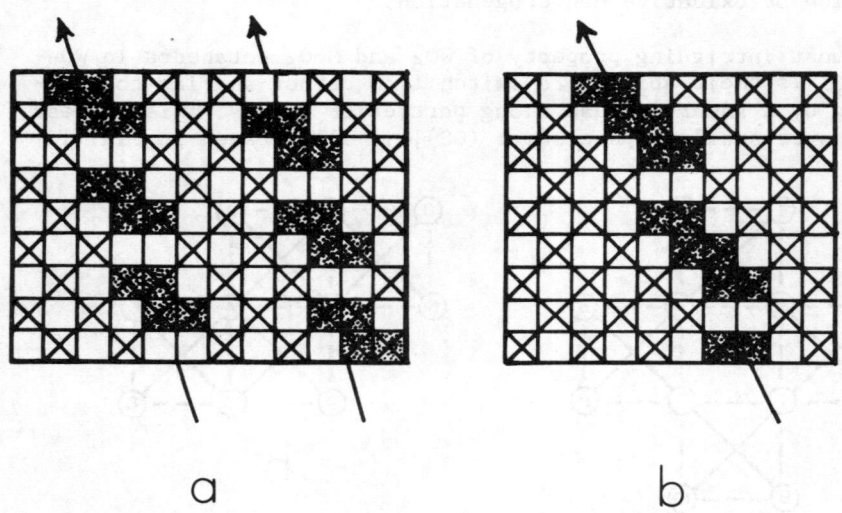

Fig.8 Crystallographic shear in the ReO_3 structure. a-CS along $\{120\}$ planes (cf.Fig.7e), giving a stoichiometry of M_nO_{3n-1}. The CS planes are separated by slices of pure MO_3 structure, and regularly spaced. The spacing defines the value of n. The example illustrated has a stoichiometry of M_9O_{26}. b - CS along a $\{130\}$ plane, giving a different grouping of edge-shared octahedra, and a stoichiometry of M_nO_{3n-2}.

Crystallographic shear is accompanied by loss of oxygen. Thus it produces from WO_3 (or MoO_3) an oxygen-deficient MO_3, but one which contains no point defects. High resolution electron microscopy shows that in oxygen-deficient WO_3 and MoO_3 the CS planes develop at regularly spaced intervals. The planes are rather like thin wafers set at an angle in what is otherwise perfect and stoichiometric MO_3 structure (Fig.8). The orientation of the parallel, equally-spaced, CS planes defines a particular stoichiometry, e.g. M_nO_{3n-1} as in Mo_8O_{23} with CS along $\{120\}$, or M_nO_{3n-2} as in $W_{20}O_{58}$ with CS along $\{130\}$. The width of the slice of MO_3 structure between the CS planes defines the value of n, so there is a homologous series, each with a precise stoichiometry.

The subject can be taken further if the rutile structure is considered. As is evident from Fig.3, rutile (TiO_2) is composed of linked TiO_6 octahedra, joined partly at corners and partly at edges. Just as WO_6 octahedra in WO_3 can switch from corner-sharing to edge-sharing, with loss of oxygen atoms, so TiO_6 octahedra in TiO_2 can switch from edge-sharing to occasional face-sharing. The process again demands release of oxygen atoms from the structure

Oxide crystal chemistry and catalysis

and precisely-defined crystallographic shear planes can be recognised. A homologous series of oxygen-deficient structures Ti_nO_{2n-1} (where n is an integer, $4 < n < 9$) is well known, and another with n lying between 16 and 36 has recently been established by electron microscopy. These oxygen-deficient rutiles are therefore composed of slabs of rutile structure separated by regularly-spaced shear planes at which the TiO_6 octahedra share faces instead of edges. The linking of octahedra at the shear plane is the same as that of face-shared MO_6 octahedra in hcp corundum, which is in fact the structure adopted by Ti_2O_3. The shear structure can thus be described as TiO_2 slabs separated after 2n oxygen planes by a wafer of Ti_2O_3, i.e. $Ti_2O_3 \cdot (n-2)TiO_2$, and hence the formula Ti_nO_{2n-1}.

Crystallographic shear planes will also form in vanadium oxides, so the behaviour is one which is common to tungsten, molybdenum, vanadium and titanium oxides, either with respect to switches from corner-sharing to edge-sharing (e.g. WO_3), or from edge-sharing to face-sharing (e.g. TiO_2). The fact that oxygen exists as O^{2-} ions in the oxides but is released as <u>uncharged</u> oxygen when these switches occur means that excess electrons are produced in the oxides. This additional electron density is accommodated in metal <u>d</u> orbitals, and the shear structures are almost certainly stabilized by some metal-metal bonding involving normally unoccupied <u>d</u> orbitals. The <u>mechanism</u> of shear structure formation is at present obscure. However, the formation must be initiated at the surface, either by high temperature and high vacuum conditions, or by attack of a reducing species. The switch in the linking of the octahedra is then a process for releasing single oxygen atoms in a stereo-specific way. The combination of a reductive attack with localized release of oxygen atoms is what is needed for selective oxidation of adsorbates in a catalytic reaction. Moreover, the act of building a shear plane in a crystallite is equivalent to extracting oxygen in depth without the solid losing its bonding strength or losing its ability to be reconstructed.

A property of many selective oxidation catalysts is an ability to act as a reservoir of oxygen until a steady state for oxygen supply by a route through the lattice can be set up. The occurrence of crystallographic shear structures in the parent oxides of tungsten, molybdenum, vanadium and titanium, and their interpretation as due to switching of the linkages of constituent MO_6 octahedra, is thus an important concept that is strongly suggestive of the mode of action of selective oxidation catalysts. The transformation from corner-shared to edge-shared octahedra is probably more relevant than that from edge-shared to face-shared octahedra in the context of selective oxidation catalysis, since the former process will probably be easier at temperatures of 400-500°C.

Coherent groups of edge-linked MO_6 octahedra of finite size occur in the complex ions of tungsten, molybdenum and vanadium. Typical of these are the ions $Mo_7O_{24}^{6-}$, $Mo_8O_{26}^{4-}$ and $V_{10}O_{28}^{6-}$, which are compact units resembling small blocks of rock-salt structure (Fig.9). There is some distortion of the octahedra, the M-O

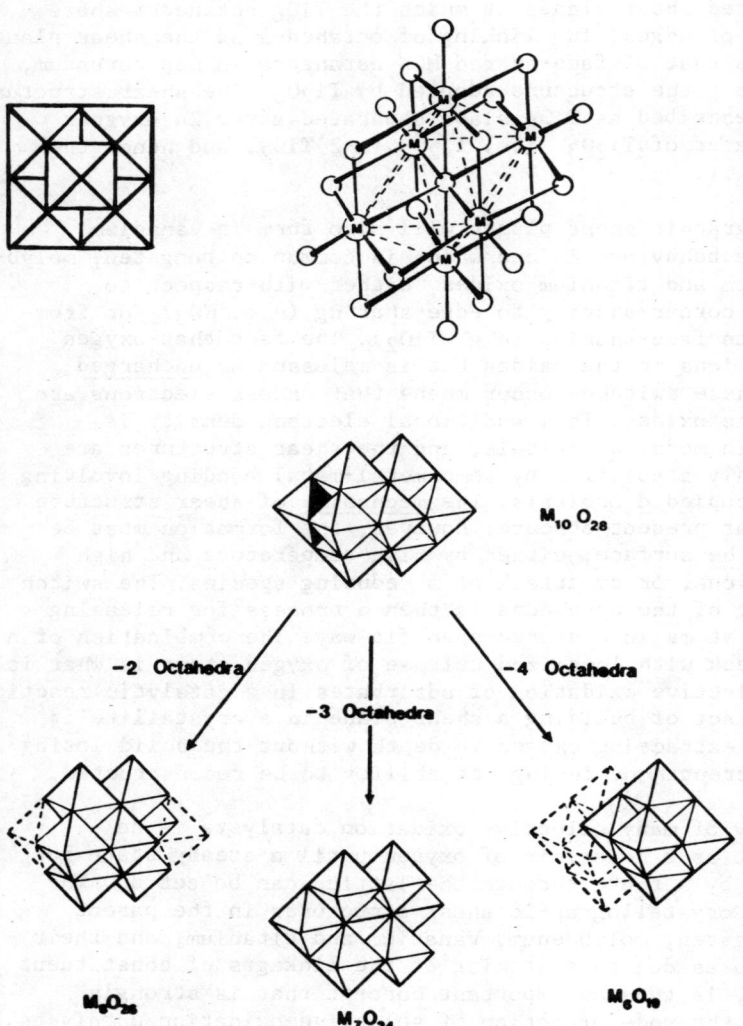

Fig.9 Edge-shared octahedra in polyanions. Top diagrams show the arrangement of six edge-shared MO_6 octahedra in $Nb_6O_{19}^{8-}$. Lower diagrams show schematically how other polyanions are structurally related (After D.L.Kepert, Inorganic Chemistry, 8, 1556 (1969) and D.M.Adams, Inorganic Solids, Wiley, 1974).

distances varying depending on whether the oxygen is on the outside or towards the centre of the group of linked polyhedra. The distortion of MO_6 octahedra when linked other than by corners can sometimes be severe, leading to situations which make the coordination of the metal with oxygen quite difficult to describe. A good example is the structure of V_2O_5. This oxide is variously quoted in the literature as having a coordination of V with O which is square pyramidal, trigonal pyramidal, or distorted octahedral. The V-O distances in fact vary from 1.6Å to 2.8Å for the six oxygens in the distorted octahedron, so there is some ground for scepticism in using the term octahedral. Yet there are some features of V_2O_5 which are best understood by starting with the premise that it is composed of octahedral MO_6 units, and certainly this is true when relationships between molybdenum and vanadium oxides are being considered, e.g. in the formation of oxide solid solutions.

The oxide crystal chemistry of tungsten, molybdenum and vanadium is not, however, limited to MO_6 octahedral units and derivations from linked octahedra. Tetrahedral coordination MO_4 occurs in the so-called ortho salts, e.g. $CaWO_4$, $PbMoO_4$, with the scheelite structure, and structures built from linkages of octahedral MO_6 and tetrahedral MO_4 are also known. $Na_2Mo_2O_7$ and $K_2Mo_3O_{10}$ are examples of structures containing both MoO_4 and MoO_6 polyhedra, as also is $VOMoO_4$.

2.2.2. Linked tetrahedra. MO_4 tetrahedra never link across faces because of the prohibitive repulsion between the metal ions, and even linking across edges, although found in the antifluorite structure, is rare. Linking of MO_4 tetrahedra through corners, on the other hand, is an extremely important structural principle. The stoichiometry of a structure in which every corner is shared in a tetrahedral MO_4 system is necessarily MO_2. There are many ways in which this can be done in a regular fashion, and the various forms of crystalline silica (quartz, tridymite, cristobalite), all of which have three-dimensional arrays of linked SiO_4 tetrahedra, are the classic examples.

From the point of view of catalysts, however, it is intergrowth of silica with alumina that is important. Mixed structures containing corner-linked AlO_4 and SiO_4 tetrahedra are very common. Since Al^{3+} has one less positive charge than Si^{4+}, external species (e.g. M^{x+} cations, H^+ or alkyl ammonium ions) must be present to provide charge balance. The analogue of silica, with tetrahedra joined at every corner, will accordingly have the formula $(Al_nSi_{1-n})O_2, M^{x+}_{n/x}$. Amorphous silica-alumina catalysts will contain elements of such structure, but only ordered over short distances. Of much greater structural interest are crystalline zeolites, especially Zeolite A, faujasite (in the form of its synthetic variants Zeolite X and Zeolite Y), mordenite and

Zeolite ZSM-5. These zeolites are very important as shape-selective catalysts or adsorbents in hydrocarbon processing, so their structures deserve attention.

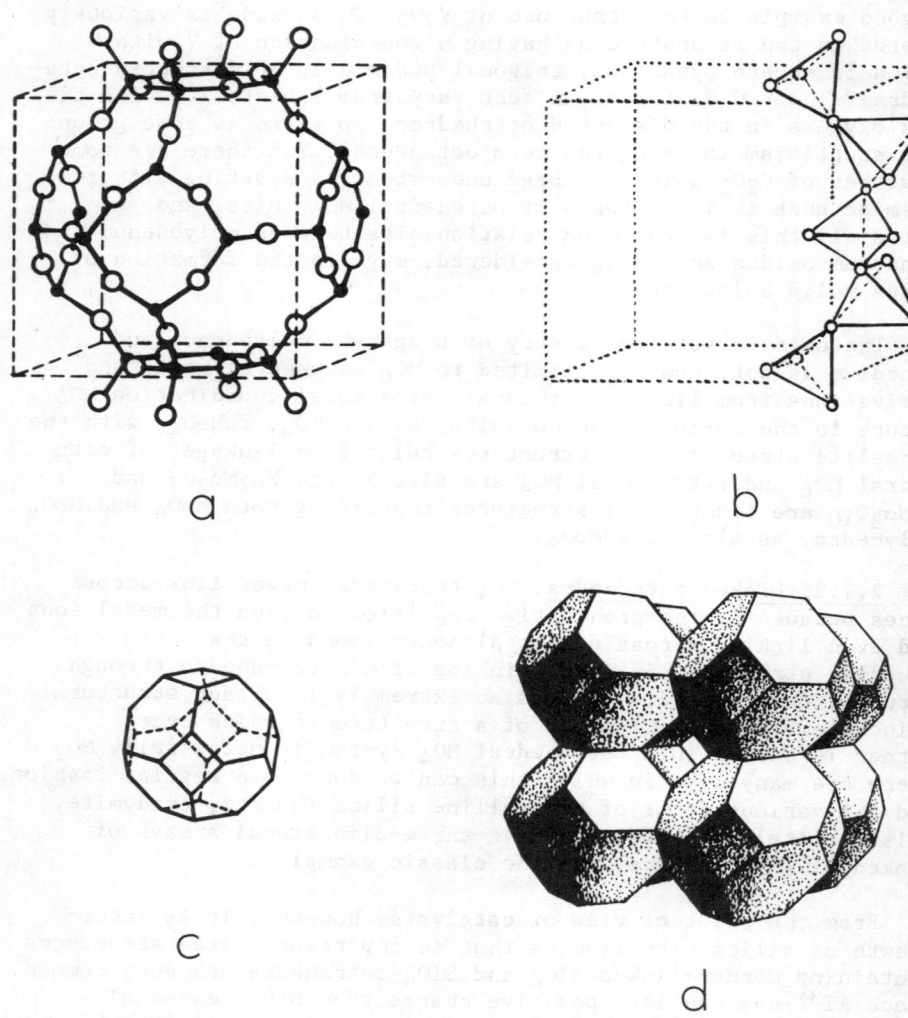

Fig. 10 The structure of sodalite. a-the cubic unit cell, showing oxygen atoms (open circles) and M atoms (small solid circles); b-front right of unit cell, redrawn to show the corner-linking of the MO_4 tetrahedra; c-unit cell, redrawn to emphasize the truncated octahedron (M atoms at vertices); d-the stacking of the truncated octahedra (sodalite units) so as to fill all space, as in sodalite.

Oxide crystal chemistry and catalysis

Zeolite A and the faujasite zeolites (X,Y) have structures which are based on sodalite units, the truncated octahedra which are the unit cells of the zeolite mineral, sodalite. The tetrahedra in these zeolites are linked in a rigid framework of well-defined symmetry. Fig.10 shows the structure of sodalite. MO_4 tetrahedra, alternately SiO_4 and AlO_4 are corner-linked so as to form regular $M_{24}O_{36}$ truncated octahedra in which the vertices are M atoms and where each side is close to the M-O-M path joining adjacent Si(Al) atoms. These truncated octahedra are space-filling polyhedra, although one should remember that the polyhedra themselves are rather hollow. If they are stacked in the space-filling configuration (fusing together at common hexagonal faces and common square faces), the resulting structure is that of sodalite, which is not of catalytic interest. If, however, the truncated octahedra are linked through prisms on either the square faces or the hexagonal faces, a very open structure of interconnecting cavities is developed into which gas molecules can penetrate. These are the framework structures of Zeolite A and of faujasite (Zeolite X,Y) respectively (Fig.11). Almost half the volume of the crystal is accessible pore volume.

The formula of Zeolite A is close to $(Al_{0.5}Si_{0.5})O_2,Na^+_{0.5}$, with AlO_4 and SiO_4 alternating. In Zeolite Y, a typical unit cell formula is $(Al_{56}Si_{136})O_2,Na^+_{56}$, though the numbers here do not define any particular topology for the distribution of aluminium and silicon. The cavities ('supercages') in the two structures are similar in diameter (~ 12Å), but the important difference is that

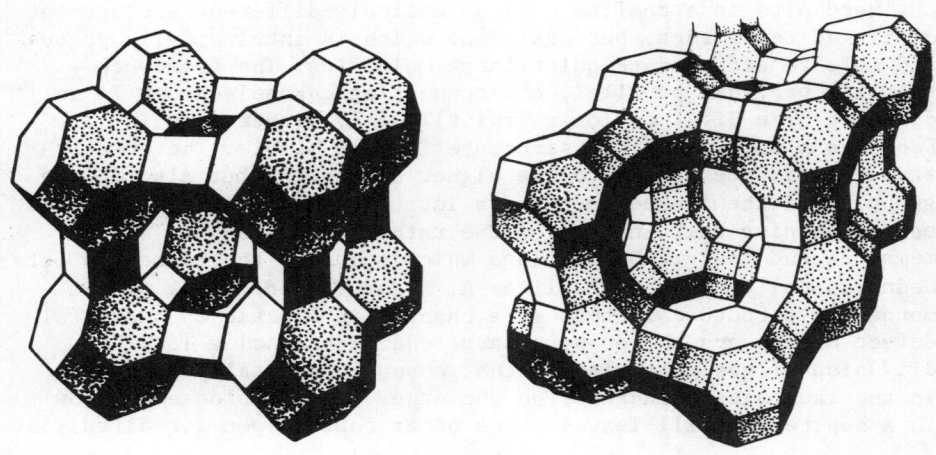

Fig.11 The structure of Zeolite A (left) and of faujasite (Zeolite X and Zeolite Y) (right).

the windows connecting one supercage with another are very much
smaller in Zeolite A (4.2Å) than in faujasite (7.5Å). This
property makes Zeolite A useful as a selective adsorbent, since
ethylene, for instance, can penetrate into the supercages but
branched hydrocarbons are excluded. The solid is a molecular
sieve. However, its potential as a catalyst is severely
restricted because large molecules cannot enter the pores.
Zeolites X and Y, on the other hand, can admit molecules as bulky
as cyclohexane or even naphthalene, and the large internal surface
area of the interconnecting pores now enables a great variety of
catalytic reactions to occur. Zeolites X and Y have the same
framework (Fig.11), but differ in their Si:Al ratio. For Zeolite
X it is near unity, but is in the range of 2-3 for Zeolite Y;
thus Zeolite Y, with its lower Al content, has fewer charge-
balancing cations per unit cell than Zeolite X. The charge-
balancing cations occupy sites in the prisms or close to the faces
of the truncated octahedra, either in the sodalite units or in the
supercages. Because they are not strongly bonded, they can be
readily exchanged one for another, and much of the versatility of
zeolites as catalysts stems from this. Sodium forms, for instance,
are poor catalysts, but forms in which Na^+ ions have been
exchanged for Ca^{2+}, La^{3+} or H^+ ions are often very active
catalysts. The aluminosilicate framework remains essentially
intact during cation exchange. It is a difficult matter to specify
exactly the location of the cations in a cation-exchanged zeolite,
and there is the added complication that the cation distribution
is apt to be affected by the presence of sorbed gases, especially
when they are molecules with polarisable bonds.

Mordenite is a zeolite with an entirely different arrangement
of linked tetrahedra, but again one which is intrinsically porous
and able to accommodate quite large molecules. The framework
(Fig.12) provides parallel, non-connecting channels about 7Å in
diameter. The Si:Al ratio is typically 5:1, higher than in
Zeolites X and Y, and the structure is more stable. The greater
stability is due partly to the higher Si content but also to the
grouping of the linked tetrahedra into five-membered rings as the
basic building unit instead of the rather more strained six-
membered and four-membered rings which comprise the truncated octahedral sodalite unit of Zeolites A, X and Y. A drawback of the
mordenite structure is that if a channel is blocked by a crystal
defect during growth, or by a large charge-balancing ion,
diffusion of gas molecules in that channel is totally blocked:
in the faujasite structure, on the other hand, a blocked window
in a supercage still leaves three other routes open for diffusion.

Faujasite and mordenite are known as minerals, but they would
not be of interest as catalysts if it were not for the fact that
they can be synthesized. Recently a new zeolite prepared by
synthesis has begun to assume prominence, the ZSM-5 zeolite of the

Mobil Corporation. The structural make-up is shown in Fig.13. The characteristic feature is a chain formed by fusing five-membered rings (Fig.13b) of corner-shared tetrahedra. These chains, which run in the z-direction, are cross-linked in both the x-direction (Fig.13c) and in the y-direction (Fig.14a). The result is to produce vertical channels intersected by zig-zag horizontal

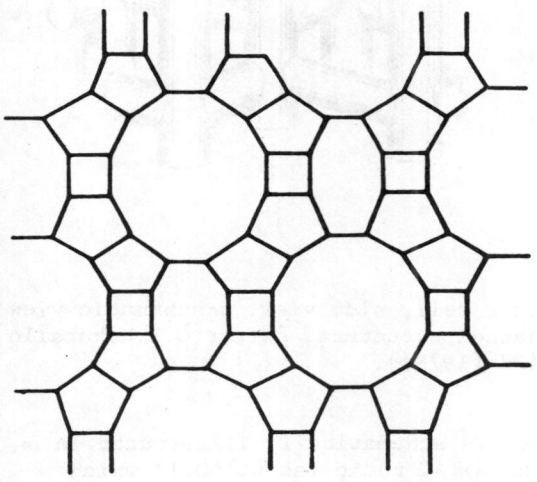

Fig.12 The framework of mordenite, perpendicular to the channels.

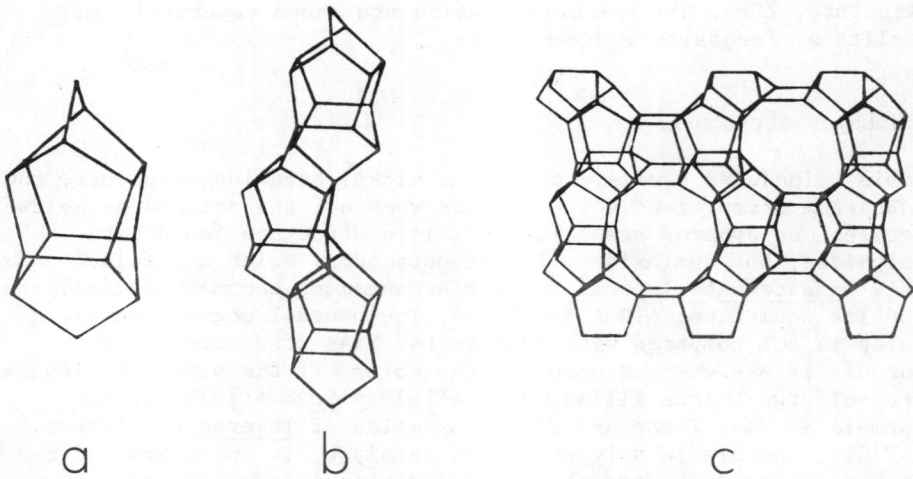

Fig.13 Zeolite ZSM-5. a-the basic unit; b-the chain; c-the unit cell, front view. [After G.T.Kokotailo, S.L.Lawton, D.H.Olson and W.M.Meier, Nature, 272, 437 (1978)].

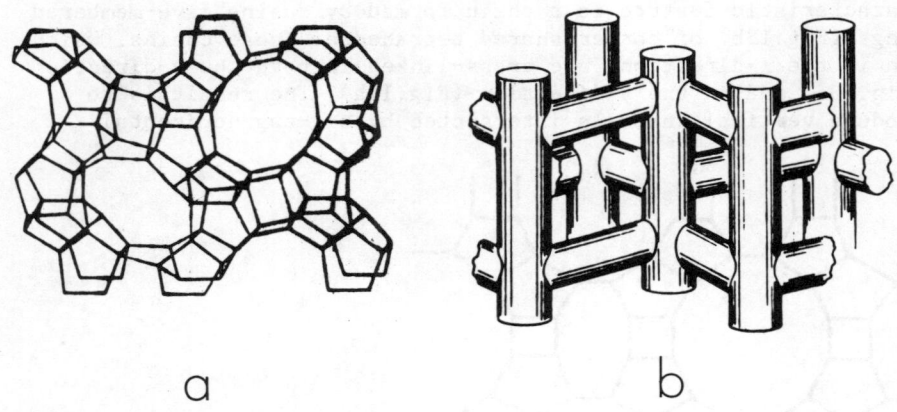

Fig.14 Zeolite ZSM-5. a-the unit cell, side view. b-schematic view of the intersecting channel structure. [After G.T.Kokotailo et al., Nature, 272, 437 (1978)].

channels. The channel structure is schematically illustrated in Fig.14b. The zeolite has a high Si:Al ratio (about 40:1) which together with the predominance of strong 5-membered rings confers high thermal stability. The intersecting nature of the pores ensures against adventitious blocking, and their diameters are small enough to provide shape selectivity. Thus in a single structure, ZSM-5 has the merits which are shown separately in Zeolite A, faujasite and mordenite.

2.3 Layer structures

Table 1 includes a reference to the nickel arsenide structure, the AoBoAoBoA arrangement which results when all the octahedral holes between hcp spheres are filled. It is a structure found with sulphides, but not oxides. If the octahedral holes are filled only in <u>alternate</u> layers, the hcp arrangement becomes AoB.AoB.AoB .. (or its equivalent A.BoA.BoA.B ..). The overall occupation of holes is 50% compared with 100% in the NiAs structure, so the formula is MX_2 where M occupies the holes. If the alternate layers are only two-thirds filled, viz. $A(\frac{2}{3}\underline{o})B.A(\frac{2}{3}\underline{o})B.A(\frac{2}{3}\underline{o})B$ .., the formula is MX_3. These are simple examples of <u>layered</u> structures. α-$TiCl_3$, the olefin polymerisation catalyst, is an example of the MX_3 structure just described. There are analogous layered structures with the X atoms in ccp array, e.g. A$\underline{o}$B.C$\underline{o}$A.B$\underline{o}$C .., the structure of $NiCl_2$.

As with the close-packed structures, layer structures occur as expanded variants. However, due to the absence of metallic species between every other layer, the bonding between those 'empty' layers is only of the weak, van der Waals type. The distances between the adjacent X layers are then apt to be greater between the 'empty' layers than between 'occupied' layers. For structures in which the X atoms are easily polarizable sulphur atoms, the additional variant appears of AoB.AoB.AoB ... being replaced by ApA.BpB.ApA ..., where the X atoms enclosing occupied holes are now in the same orientation, thereby converting the sixfold coordination around the M atom from an octahedron (o) to a trigonal prism (p). This MX_2 variant, shown in Fig.15, is the structure of MoS_2 and WS_2. The bonding is quite covalent (which is why a description 'atoms' is used here instead of 'ions'), and this has led to the thought that it might be possible to intercalate atoms of a foreign element between the A and B layers which have their octahedral holes unoccupied. A speculative model invoking this type of intercalation has been proposed for finely-divided nickel tungsten sulphide (Ni_xWS_2), which is a catalyst for hydrodesulphurization. However, the model is by no means universally accepted, and indeed there are serious problems in speculating on sulphide structures because the interplay of covalent bonding, ionic bonding and van der Waals bonding makes for a very facile shift between one kind of structure and another. For example, the sulphur-poor phase Co_9S_8 has ccp S atoms with Co atoms mostly in tetrahedral holes, but at the 1:1 stoichiometry (CoS) there is hcp stacking with Co atoms in octahedral holes.

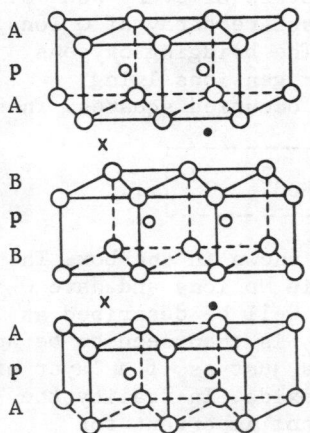

Fig.15 The layer structure of MoS_2 and WS_2. Large open circles denote S atoms, small open circles Mo or W atoms in trigonal prism sites; crosses signify octahedral holes and small solid circles tetrahedral holes.

With rather more sulphur at Co_3S_4 stoichiometry, the S atoms are back again in ccp (spinel structure) with one-third of the cobalt tetrahedrally coordinated, whilst in CoS_2 the S atoms are covalently paired as S-S units and all the Co atoms are octahedrally surrounded. With the oxides of cobalt, on the other hand, there is strong ionic character and only two compounds (CoO and Co_3O_4), both of which have ccp O^{2-} ions.

Layer structures also include those in which a two-dimensional sheet of one structure is present between layers of another structure, but where the bonding of the respective layers to one another is strong. An example is the crystallographic shear structures of Ti_nO_{2n-1} with corundum-type sheets in slabs of rutile. Layered structures also occur when a mixed oxide is formed containing two metallic elements M and M' whose coordination behaviour in their parent oxides is markedly different. The element M forms a sheet of stoichiometry MO_n by linking its polyhedra at edges or corners (or both): the element M' likewise forms a sheet composed of linked polyhedra of its own kind, giving a stoichiometry $M'O_m$. The layers are then fused together by bridging with oxygen atoms. The principle can be illustrated with reference to bismuth molybdate Bi_2MoO_6 (koechlinite), a solid of considerable interest as a selective oxidation catalyst.

Bi_2MoO_6 (Fig.16) can be regarded as consisting of sheets of Bi_2O_2 stoichiometry separated by sheets with MoO_2 stoichiometry, and with oxygens bridging between them. The Bi_2O_2 sheets can be thought of as O ions arranged in a square array with Bi ions alternately just above or just below the centres of every square. The MoO_2 sheet can also be thought of as a square array of O ions, but only __half__ the squares contain Mo ions. The bridging oxygens which link the sheets above and below are oxygen ions lying directly above and below the Mo ions in the occupied squares. Thus the layer sequence is

$$\ldots Bi_2O_2^{2+}, O^{2-}, MoO_2^{2+}, O^{2-}, \boxed{Bi_2O_2^{2+}, O^{2-}, MoO^{2+}, O^{2-}} \ldots$$

giving the overall stoichiometry of Bi_2MoO_6 shown in the box. The O ion squares in the MoO_2 sheet which contain Mo ions and have O ions directly above and below could equally well be described as MoO_6 octahedra, and the square array of MoO_2 is then seen to be an array of corner-linked MoO_6 octahedra. It is just as if a layer of ReO_3 structure separated sheets of bismuth oxide. In reality the sheets are somewhat distorted, but this is not important for understanding the basis of the structure. Note the similarity between the MoO_2 layer and the configuration in Fig.7c.

Other bismuth molybdates are known, namely $Bi_2Mo_2O_9$ and $Bi_2Mo_3O_{12}$. One might reasonably have hoped Nature to provide

Oxide crystal chemistry and catalysis

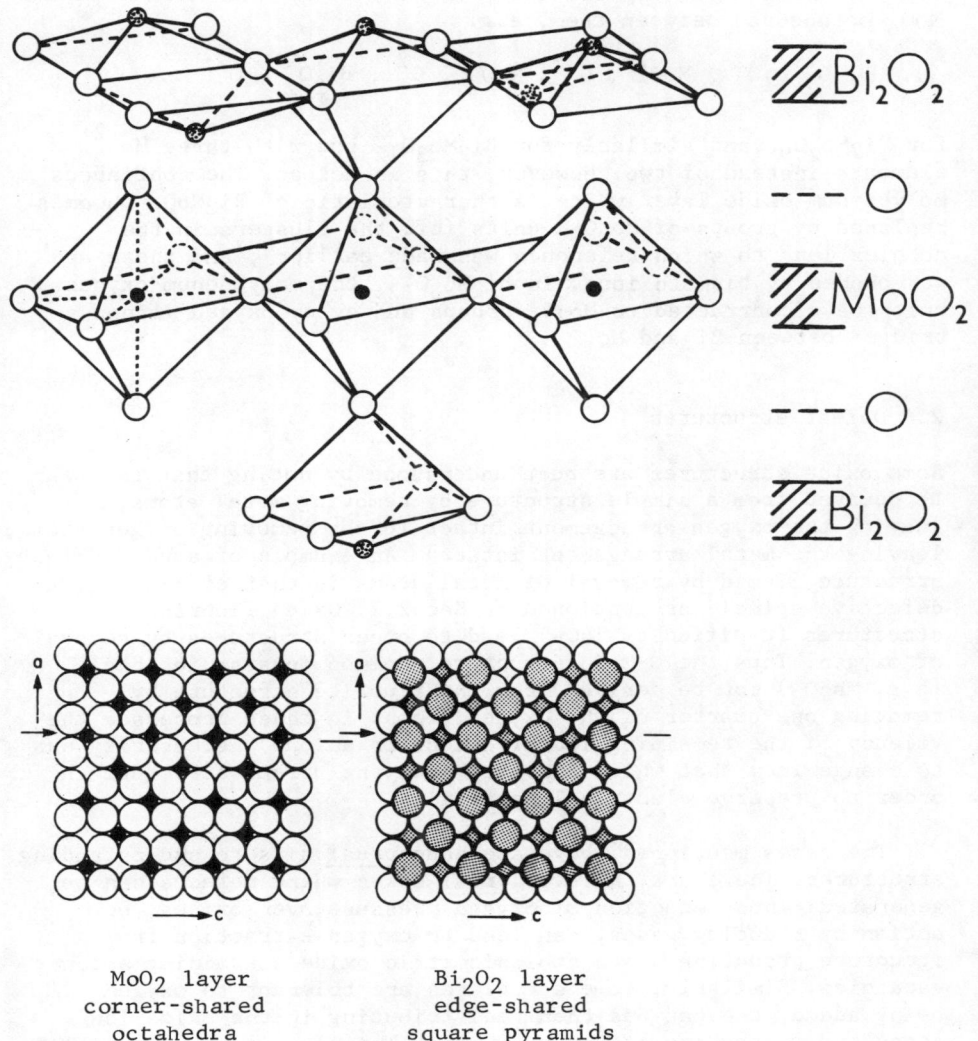

Fig.16 The layer structure of bismuth molybdate, Bi_2MoO_6 (koechlinite). Top diagram-in elevation; bottom diagrams-in plan. Open circles-oxygen; shaded circles-bismuth; solid circles-molybdenum. Mo ions are octahedrally coordinated. [Adapted from B.C.Gates, J.R.Katzer and G.C.A.Schuit, Chemistry of Catalytic Processes, McGraw Hill Inc., New York, 1979. See also A.F.van den Elzen and G.D.Rieck, Acta Cryst., B29, 2436 (1973)].

structures for them which consisted of the same Bi_2O_2 sheets as before but with thicker slabs of ReO_3 structure (the corner-linked MoO_6 octahedra) between them, e.g.

$$\ldots Bi_2O_2^{2+}, O^{2-}, MoO_2^{2+}, O^{2-}, MoO_2^{2+}, O^{2-}, Bi_2O_2^{2+}, O^{2-} \ldots$$

for $Bi_2Mo_2O_9$, and similarly for $Bi_2Mo_3O_{12}$ but with three MoO_2^{2+} elements instead of two. However, this is not so. The continuous molybdenum oxide layer which is characteristic of Bi_2MoO_6 becomes replaced by groups of Mo_4O_{16} units (cf. the clusters in the complex ions to which reference was made earlier), and these are surrounded by bismuth ions. In $Bi_2Mo_3O_{12}$, the molybdenum-oxygen units have contracted to Mo_2O_8 groups and every oxygen is now bridged between Bi and Mo.

2.4 Defect structures

Some oxide structures are best understood by noting that they can be derived from a simple structure by removing metal atoms, leaving the oxygen arrangement intact (or by removing oxygen atoms, leaving the metal arrangement intact). An example of such a defect structure formed by removal of metal atoms is that of γ-Al_2O_3, a defective spinel, as mentioned in Sec.2.1. Oxide fluorite structures (typified by ThO_2) lead to other structures by removal of oxygen. Thus the C-M_2O_3 structure, common in sesquioxides (e.g. Mn_2O_3) can be derived from the fluorite structure by removing one-quarter of the oxygen atoms. In these processes the valency of the remaining ions (cations or anions) necessarily has to change from that which characterizes the parent structure in order to preserve electrical neutrality.

The cases mentioned above are <u>conceptual</u> aids to understanding structures. There are, however, <u>real</u> cases where defects can be generated. Thus reduction of oxygen pressure over oxides, or action by reducing gases, can lead to oxygen extraction from a structure producing a non-stoichiometric oxide containing anion vacancies. Similarly, some structures are tolerant to oxygen being added, the cations then redistributing in the oxide ion array and generating cation vacancies. The classic example is TiO with the rock-salt structure: this oxide will tolerate deviations from stoichiometry between the limits of $TiO_{0.7}$ (predomination of anion vacancies) and $Ti_{0.8}O$ (predomination of cation vacancies). Bismuth molybdate structures are tolerant to the removal of oxygen, the charge balance being preserved by reduction of the Mo^{6+} ions. The production of non-stoichiometry with oxygen deficit accommodated by cation vacancies (e.g. $TiO_{0.7}$ from TiO) is to be distinguished from the case where oxygen deficit is accommodated by crystallographic shear (e.g. Ti_nO_{2n-1}

derived from rutile TiO_2). In some cases, however, there is room for doubt as to which mechanism operates.

An important variant on this theme is the generation of vacancies by replacing one ion in a structure by another of different but stable valency. The most appropriate example is substitution in the scheelite ($CaWO_4$) structure, since these solids can be used as catalysts for selective oxidation and ammoxidation of olefins. In the scheelite structure, which can be written in general as AMO_4, the cation M is tetrahedrally coordinated to oxygen (Fig.17). The MO_4 tetrahedra, which are all separated from one another, are arranged around the A cation in such a way that the A cation is eight-coordinated to oxygen (an AO_8 polyhedron), with each of these oxygens coming from a different MO_4 tetrahedron. Thus AO_8 and MO_4 polyhedra are corner-

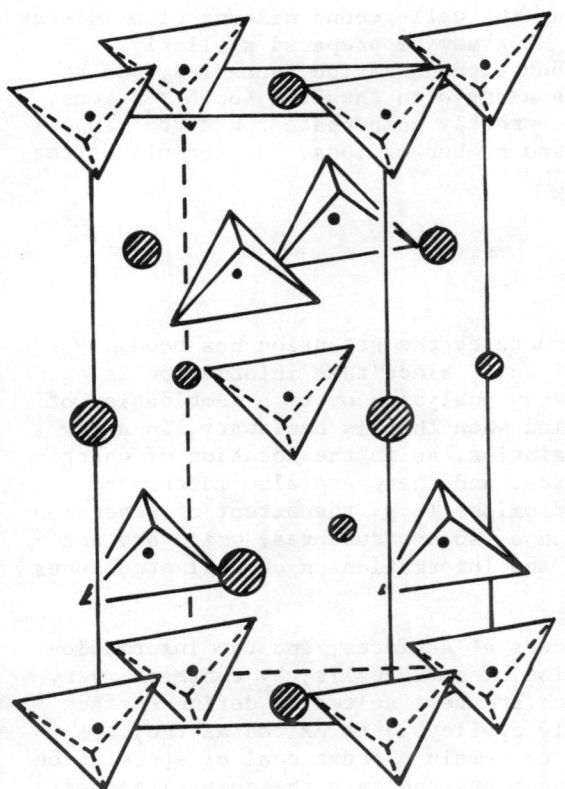

Fig.17 The scheelite structure, AMO_4. Oxygen atoms lie at the corners of MO_4 tetrahedra. A atoms are shown shaded.

shared. The structure, which is a relatively open one, readily tolerates substitution by foreign cations, and the ions do not have to be of the same valency as in the parent oxide. Thus although many scheelites are like $CaWO_4$ with the 8-coordinate cation a divalent ion and the 4-coordinate cation a hexavalent ion ($A^{2+}M^{6+}O_4$), others have trivalent and pentavalent cations ($A^{3+}M^{5+}O_4$). Among the ions which may be incorporated in the scheelite structures are bismuth (as Bi^{3+}) and molybdenum (as Mo^{6+} or Mo^{5+}). By choosing an example of the $A^{2+}M^{6+}O_4$ type and then replacing three A^{2+} ions by two A^{3+} ions, a cation vacancy is generated at the A cation site. In this way the concentration of cation vacancies in the structure is controlled by the substitution of the appropriate amount of A^{3+} ions.

A specific example is the scheelite $PbMoO_4$, which can be converted to $Pb^{2+}_{1-3x}Bi^{3+}_{2x}\square_x Mo^{6+}O_4$ by incorporation of some Bi_2O_3 into the mixture of PbO and MoO_3 when the scheelite is being prepared (normally by firing the well-ground mixture of oxides at about 700°C). $Pb^{2+}_{1-3x}La^{3+}_{2x}\square_x MoO_4$ may be prepared similarly. Corresponding scheelites, but without cation vacancies, can be prepared by adding Na^+ ions along with the Bi^{3+} (or La^{3+}) ions, since $Na^+ + Bi^{3+}$ together correctly compensate for 2 Pb^{2+} ions both in respect of charge and number of ions. The formula of the solid is then $Pb^{2+}_{1-2x}Na^+_x Bi^{3+}_x Mo^{6+}O_4$.

3. CONCLUSION

In this review of oxide structures the attention has been concentrated on the crystal bulk, since this information is definitively provided by X-ray analysis, or by a combination of X-ray and neutron diffraction when that is necessary. In a few instances there are uncertainties, as in the location of charge-balancing cations in zeolites, and there are also matters of detail which arise in mixed oxides (e.g. the extent of inversion in spinels, or the formation of superstructures) which are not always easily defined, but the information on crystal structures is generally unequivocal.

Catalysis, however, occurs at surfaces, and the information available on the conformation of surfaces is, by contrast, very equivocal. The methods which are best suited to define surface structure are not as readily applicable to oxides as they are to metals, and there is bound to remain a great deal of speculation concerning the atomic arrangements and even the composition of oxide surfaces. Zeolites are an exception, since here the surface, being predominantly internal, is defined by the bulk crystal structure. In general, however, the surface of an outgassed oxide must be a surface containing truncated polyhedra,

Oxide crystal chemistry and catalysis

and chemisorption is a means to restore the coordination. In these circumstances, the acquisition of a corpus of knowledge about the arrangements preferred at the local level by atoms in oxides, such as have been described in this review, is the only sure basis on which to develop models for the arrangements at oxide surfaces, and in turn the arrangements of chemisorbed species concerned in the catalytic act.

4. BIBLIOGRAPHY

General texts with discussions of oxide crystal chemistry suitable for supplementing this lecture are:

1. A.F.Wells, Structural Inorganic Chemistry, 4th Edition, Oxford University Press, 1975.
2. H.Krebs, Fundamentals of Inorganic Crystal Chemistry, McGraw Hill Pub.Co.Ltd., 1968.
3. D.M.Adams, Inorganic Solids, John Wiley, 1974.
4. B.C.Gates, J.R.Katzer and G.C.A.Schuit, Chemistry of Catalytic Processes, McGraw Hill Inc., 1979.

INDUSTRIAL CATALYTIC PARTIAL-OXIDATION PROCESSES

J.H.C. van Hooff

Laboratory for Inorganic Chemistry, Eindhoven University of Technology, Eindhoven, The Netherlands

INTRODUCTION

Catalytic partial-oxidation reactions are of considerable importance in the chemical industry. Roughly speaking the industrial partial-oxidation processes can be divided into two groups,
i. reactions in the gasphase proceeding at high temperatures over solid catalysts,
ii. reactions in solution catalyzed by soluble inorganic complexes, and thus confined to lower temperatures.

We shall in the following only discuss the first group and of all processes in this group we will ultimately focuss the attention on the acrylonitrile process.

Survey of the field of the high temperature partial-oxidation process

1. One of the oldest processes in this group is the oxidation of naphtalene to phtalic-anhydride.

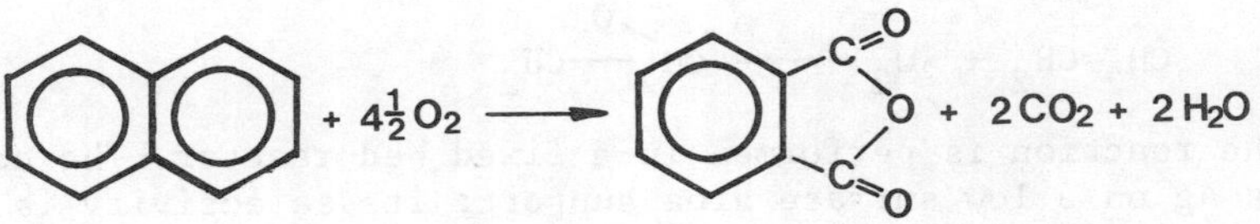

$$\text{naphtalene} + 4\tfrac{1}{2} O_2 \longrightarrow \text{phtalic anhydride} + 2 CO_2 + 2 H_2O$$

The catalyst is V_2O_5, supported on some low surface area, highly porous support. Because of the large amounts of heat evolved, the transfer of heat away from the reactor is of vital importance; for this reason the reactor is either a multitubular reactor or else a fluid bed reactor.
An interesting detail is that the catalyst contains K_2SO_4 that apparently causes the V_2O_5 to be in the liquid state in the pores of the support

2. A similar process is the conversion of o-xylene to phtalic-anhydride.

3. The oxidation of benzene to maleic-anhydride.

$$C_6H_6 + 4\tfrac{1}{2} O_2 \longrightarrow \text{maleic anhydride} + 2CO_2 + 2H_2O$$

Reactors applied are of the multitube type and the catalyst is a mixture of V_2O_5 and MoO_3 on a support.

4. The oxidation of methanol to formaldehyde. This is a typical oxidative dehydrogenation. In the past the reaction

$$CH_3OH + \tfrac{1}{2}O_2 \longrightarrow HCHO + H_2O$$

was performed on a copper metal catalyst but the more modern versions apply a $Fe_2(MoO_4)_3$ catalyst.
Since the heat evolved is not so great as for the former three processes, a simple fixed bed reactor can be used.

5. The oxidation of ethene to ethene oxide.

$$CH_2{=}CH_2 + \tfrac{1}{2}O_2 \longrightarrow CH_2\underset{\displaystyle O}{\overline{}}CH_2$$

The reaction is performed in a fixed bed reactor. The catalyst is Ag on a low surface area support; its selectivity is considerably enhanced by adding small amounts of chlorine containing compounds to the feed.

6. The oxidation of propene to acrolein and the ammoxidation of propene + NH_3 to acrylonitrile.

Partial oxidation processes

$$CH_2=CH-CH_3 + O_2 \longrightarrow CH_2=CH-CHO + H_2O$$
$$CH_2=CH-CH_3 + NH_3 + 1\tfrac{1}{2}O_2 \longrightarrow CH_2=CH-CN + 3H_2O$$

Both reactions are operated on a technical scale, the catalysts for both the oxidation and ammoxidation being quite similar (see later). The reactor for the ammoxidation is almost always a fluid-bed reactor; the oxidation of propene however is often performed in a fixed bed reactor.
Acrolein is rarely end product but usually oxidized in a second reactor to acrylic acid. The catalyst for the latter reaction consists of a mixture of transition metal molybdates.

It is noteworthy that the catalysts for these reactions are all oxides (even the Ag catalyst is in an oxidic state during operation) More often than not the catalyst consists of a mixture of oxides, or binary compounds of oxides. In most cases one of the oxides is a transition metal oxide.
Each reaction seems to need a special catalyst and a good catalyst for one process may be unsatisfactory for another. For instance, the V_2O_5 catalyst used for the selective oxidation of aromatics causes excessive combustion of propene, while the catalysts used for the latter process are not very active for the naphtalene oxidation. As to reactor technology it is not surprising that with the large amounts of heat to be transferred, reactors are usually of the fluid bed or multitube type.

CATALYSTS AND REACTORS

This lecture is almost entirely concerned with the catalysts for the partial oxidation reactions. Prof. Froment on the other hand will devote the greater part of his lecture to the design of the reactors for these processes and in particular to the problem of an adequate removal of the heat developed during the reaction.
One might ask whether an ideal reactor designed for optimal heat transfer would not be sufficient to ensure that all partial-oxidation reactions could be performed using only one type of catalyst. Experience has shown that this is probable not feasible; we do not really know why.
We can however offer some plausible reasons. One of the more attractive suggestions is that although all are oxidation reactions, their mechanisms may be different. For instance, breaking an aromatic C-C bond in the oxidation of naphtalene may need a basically different reaction path, with different types of intermediates as the epoxidation of ethene. Indeed this is not so strange an assumption since V_2O_5 and Ag are widely different types of catalysts.
However, suppose that we might succeed in classifying the various oxidation reactions into types that have related mechanisms, we

still remain confronted with a choice as to the best catalyst
from a family such as for instance that of the molybdate catalysts.
Now, partial oxidation means that the introduction of only a few
oxygens should be fast while a further introduction should be slow.
'Fast' and 'slow' might be translated into 'low activation energy'
and 'high activation energy'. Let us now try to see whether we
can fit such a concept into reactor theory: we shall do this in
a superficial manner since Prof. Froment shall give a far more
complete discussion later on.

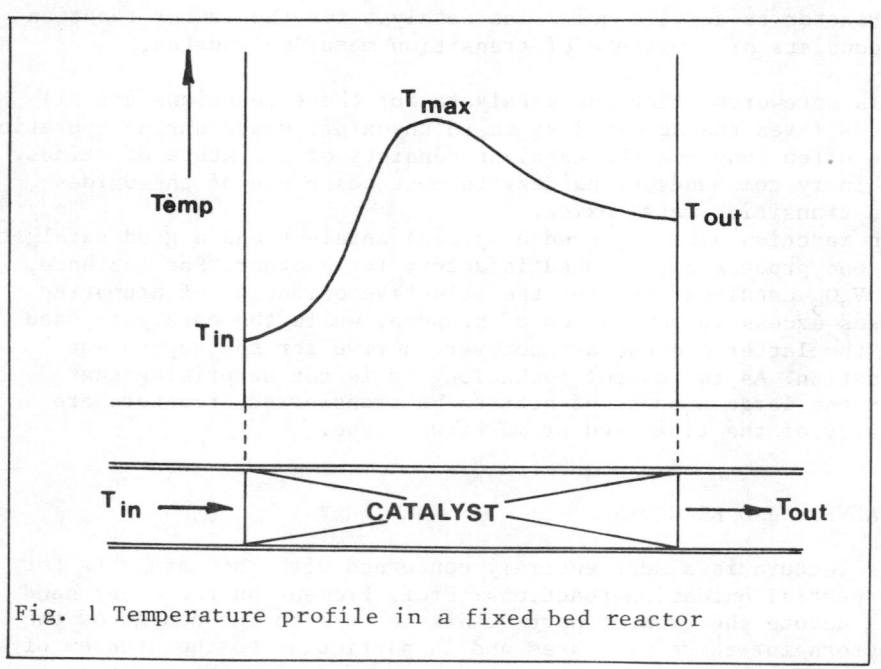

Fig. 1 Temperature profile in a fixed bed reactor

Consider first a single tube containing the catalyst (see fig. 1)
The feed enters the reactor at the left side. Because of the
exothermicity of the reaction heat is evolved and the temperature
increases: simultaneously heat will flow out of the reactor in a
direction perpendicular to the direction of the gasflow. There
will now arise a competition between heat production and heat
removal: the faster the latter the lower the temperature rise.
Since heat production diminishes as the feed becomes exhausted,
the axial temperature profile will be as shown in the figure.
It will show a maximum somewhere along the axis and this maximum
may be high and narrow or low and broad. (If it is high we call
it a 'hot spot'). It is difficult a priori to predict the height
of the peak the more so since we assumed above that there are two
reactions, one that starts easily but has a low activation energy

Partial oxidation processes

and another that has a high activation energy. (Remember that we would like only the first to become operative.)
Some qualitative information can be obtained by considering what might happen when the reaction is performed in a fluidized bed. The temperature in the fluidized bed reactor can be assumed constant over the entire length. We can however simulate the phenomena that lead to a 'hot spot' in the fixed bed by increasing the inlet temperature; we shall suppose that this temperature is equal to that of the cooling coils.

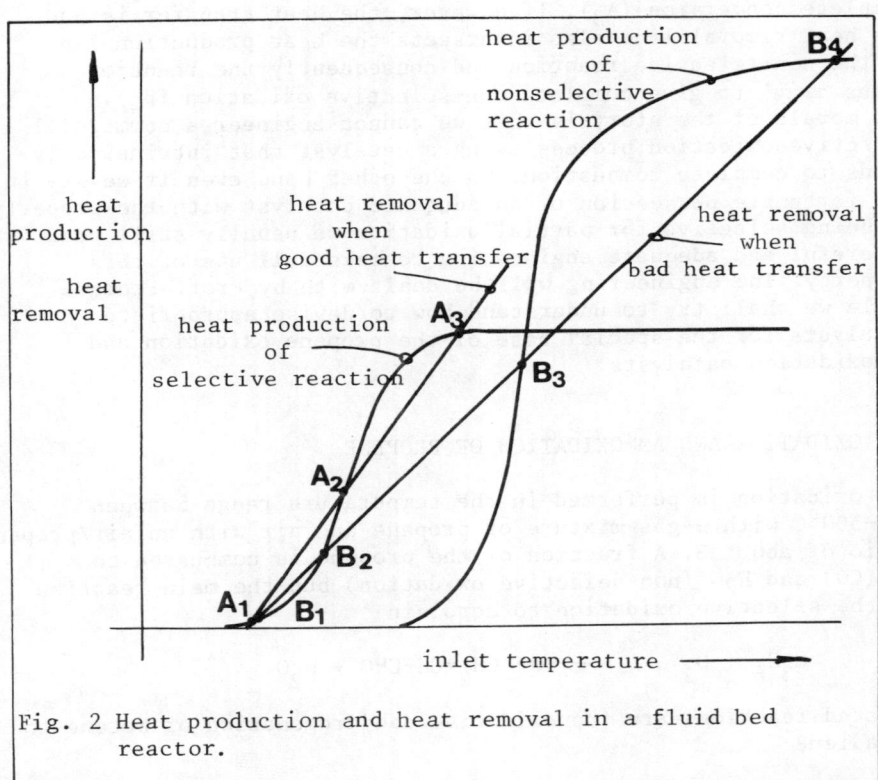

Fig. 2 Heat production and heat removal in a fluid bed reactor.

Fig. 2 gives a picture of the production of heat because of the reaction and of the heat removal because of heat transfer from reactor contents to cooling coils, both as a function of the inlet temperature. The rate of the reaction increases exponentially with this temperature (Bolzmann-Arrhenius factor) and so does the heat produced. However, there is an upper limit to the rate given by the total amount of feed introduced per unit of time; the curve is therefore sigmoidal. The heat removal is supposed to

be linear in ΔT, the difference in reactor temperature and cooling temperature, resulting in straight lines in the figure.
The slope of these lines depends on the extent of the heat transfer. (The better the heat transfer the steeper the line.) The intersections between the two lines indicate the states where there is equilibrium between the heat production and the heat removal. Normally there are three of these intersections (A_1, A_2 and A_3) of which two (A_1 and A_3) correspond with stable states and one (A_2) with an unstable situation. Consequently there is either a reaction with low conversion (A_1) or another with almost complete conversion (A_3). If however, the heat transfer is bad the heat removal line also intersects the heat production line of the non-selective reaction and consequently the reaction 'runs away' to give complete non-selective oxidation (B_4).
The morale of the story is that we cannot engineer a commercial selective oxidation process using a catalyst that intrinsically tends to complete combustion. On the other hand even if we are in the fortunate possession of an adequate catalyst with the property of being selective for partial oxidation we usually still need a careful and adequate engineering to make full use of this property. The engineering will be dealt with by Prof. Froment. while we shall try to understand how to devise appropriate catalysts for the special case of the propene oxidation and ammoxidation catalysts.

THE OXIDATION AND AMMOXIDATION OF PROPENE

The oxidation is performed in the temperature range between 400-500°C with a gas mixture of propene and air with an air/propene ratio of about 8. A fraction of the propene is combusted to $CO_2(CO)$ and H_2O (non-selective oxidation) but the main reaction is the selective oxidation to acrolein

$$C_3H_6 + O_2 \longrightarrow CH_2=CH-CHO + H_2O$$

Related reactions are the oxidative dehydrogenation of butene to butadiene

$$C_4H_8 + \tfrac{1}{2}O_2 \longrightarrow CH_2=CH-CH=CH_2 + H_2O$$

and the ammoxidation of propene

$$C_3H_6 + NH_3 + 1\tfrac{1}{2}O_2 \longrightarrow CH_2=CH-CN + 3H_2O$$

which is also basically a dehydrogenation.
Without a catalyst and depending on the temperature there is either hardly any reaction or a non-selective and 'deep' oxidation. The task of the catalyst therefore is not only to increase the rate but at the same time to limit the depth of oxidation i.e. to

Partial oxidation processes

make the reaction more selective.

Metal Oxide	$T_{50\%}$ °C	Selectivity %
MnO_2	~ 250	< 10
Fe_2O_3	~ 400	~ 50
SnO_2	~ 500	~ 20
Bi_2O_3/MoO_3	~ 375	> 90

Table 1 Activity and selectivity of some metal oxides for the oxidation of butene-1.

Table 1 shows how activities and selectivities for the butene-butadiene reaction are changing with the oxide used as a catalyst. The activity of the catalyst is given in terms of the temperature where 50% of the butene is oxidized within a certain residence time. The selectivity is expressed as the percentage of butene oxidized to butadiene. MnO_2 apparently is very active; 50% of the butene is already oxidized at 250°C. However, the selectivity is very low (~5%). A less active oxide Fe_2O_3 needs a temperature of around 400°C to obtain 50% conversion but is more selective (butadiene selectivity of about 50%). However, if a truly selective catalyst such as the combination of Bi_2O_3 and MoO_3 (or Bi-molybdate) is tried the activity is about comparable to that of Fe_2O_3 but the selectivity is far superior. What then is the reason for these differences?

REDOX-MECHANISM (Mars-Van Krevelen mechanism)

Mars and Van Krevelen concluded from experiments on the oxidation of naphtalene over V_2O_5 that catalytic oxidation reactions most often take place in two steps viz.
1. a reaction between the oxide and the hydrocarbon in which the hydrocarbon is oxidized and the oxide reduced, followed by
2. an oxidation of the reduced oxide to re-establish the initial state.

In a condensed notation (HC = hydrocarbon)

$$HC + cat. \, O^{2-} \longrightarrow HCO + cat. \, [2e] \tag{1}$$

$$cat. \, [2e] + \tfrac{1}{2}O_2 \longrightarrow cat. \, O^{2-} \tag{2}$$

The two electrons [2e] might be donated to one or more cations such as for instance:

$$2Fe^{3+} + 2e \longrightarrow 2Fe^{2+}$$
$$Mo^{6+} + 2e \longrightarrow Mo^{4+}$$

It is sometimes assumed that they are actually placed in the conduction band of the solid.

Intuition tells us that oxide catalysts with a weak metal-oxygen bond will easily donate oxygen to the hydrocarbon which means that donation already occurs at relatively low temperature; such catalysts therefore are probably active. It also means however, that donation may occur in numbers of oxygens that are greater than we require for our purpose: the catalyst is then also non-selective. If the catalyst follows an orthodox Mars-Van Krevelen pattern it should not be of great importance whether oxygen gas is present or not provided oxygen depletion of the catalyst does not interfere. Typical examples of such catalysts are Ag_2O or MnO_2.

On the other hand, oxidic catalysts with strong metal-oxygen bonds are not readily reduced and need high temperatures to do so: they are not active and because of the high temperatures also non-selective. Examples should be for instance TiO_2 or SnO_2.

Catalysts that are both reasonably active but also selective represent compromises: as Sachtler and De Boer postulated they are oxides with intermediate metal-oxygen bond strengths. Such oxides are Fe_2O_3 ($\to$ FeO) and MoO_3 ($\to MoO_2$). To show this we define the bond strengths thermochemically as being equal to the reaction enthalpy

$$MO_n + H_2 \rightleftharpoons MO_{n-1} + H_2O \qquad \Delta H_r^o \qquad (3)$$

If we now plot $T_{50\%}$ for the oxidation of butene-1 to butadiene as a function of ΔH_r^o we find fig. 3a showing a nice fit. To illustrate the selectivity pattern we choose an intermediate temperature and plot the % conversion to butadiene as a function of ΔH_r^o (fig. 3b). At low ΔH_r^o we should expect considerable but predominantly non-selective conversion. At high values conversion should be low and because of the high temperatures needed for conversion also quite non-selective. At intermediate values we expect to find a maximal selective conversion and this is indeed what happens. The maximum is situated for this particular reaction approximately at ΔH_r^o is 0-10 kcal/mole. i.e. The metal-oxygen bond $MO_n \to MO_{n-1} + \frac{1}{2}O_2$ should be around 50-60 kcal/mole^{-1}. However, even here conversions are not particularily selective. If on the other hand we apply Bi_2MoO_6 a binary oxide we find a similar activity ($T_{50\%}$) but a far superior selectivity. Characterization of metal-oxygen bond strengths as given in eq. (1) when applied to Bi_2MoO_6 leads to ambiguities: which bond strength do we choose Mo-O-Mo, Mo-O-Bi or Bi-O-Bi? Let us as a rough assumption consider

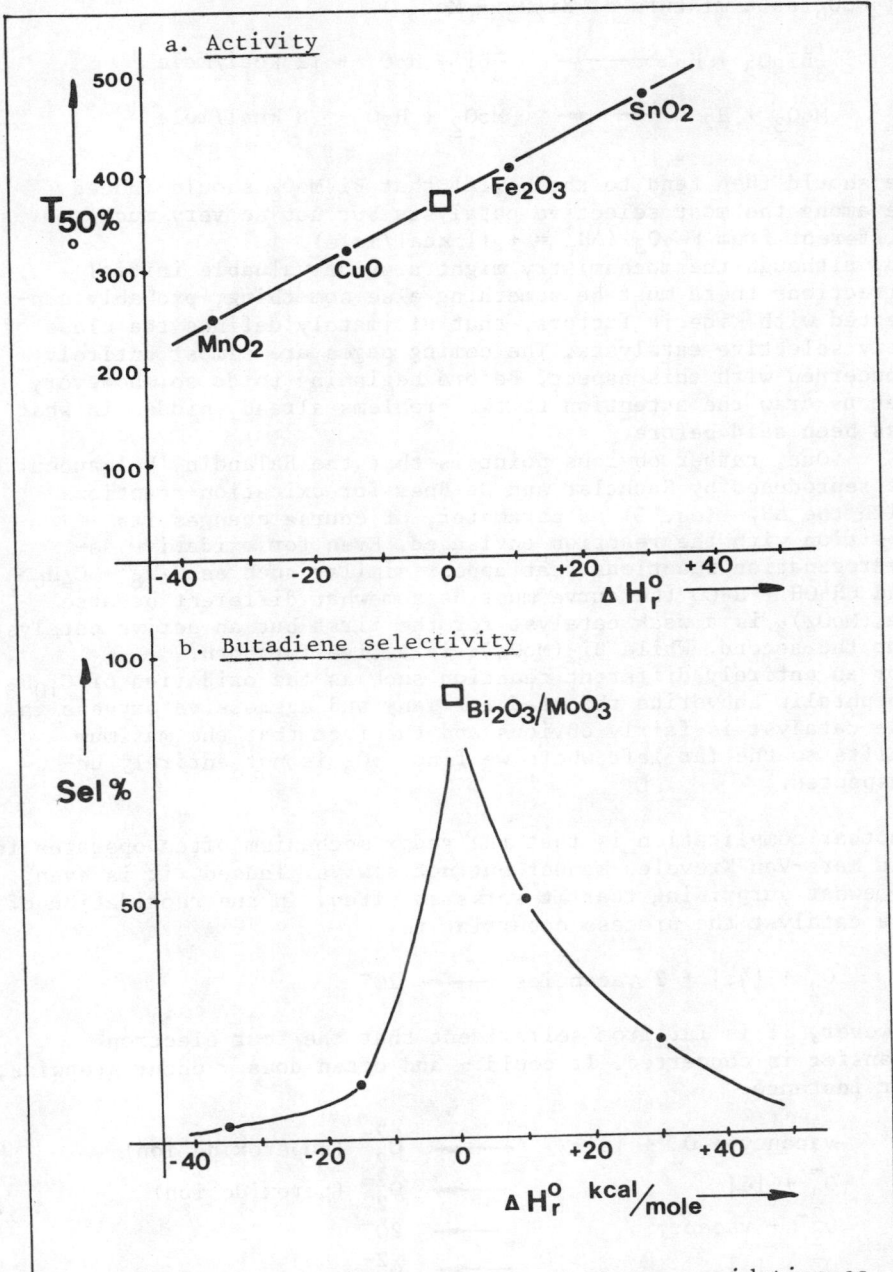

Fig. 3 Activity and Selectivity for the butene oxidation as function of the metal-oxygen bond strength of the catalyst.

Bi_2MoO_6 as a mixture of Bi_2O_3 + MoO_3

$$\frac{1}{3}Bi_2O_3 + H_2 \longrightarrow \frac{2}{3}Bi + H_2O + 12 \text{ kcal/mole}^{-1}$$

$$MoO_3 + H_2 \longrightarrow MoO_2 + H_2O - 8 \text{ kcal/mole}^{-1}$$

We should then tend to the belief that Bi_2MoO_6 should indeed be among the most selective catalysts but not so very much different from Fe_2O_3 (ΔH_r^0 = + 11 kcal/mole).
So, although thermochemistry might provide valuable initial directions there must be something else something, probably connected with kinetic factors, that ultimately defines the class of very selective catalysts. The coming pages are almost entirely concerned with this aspect. Before beginning to do so, however, let us draw the attention to two problems already hidden in what has been said before.

One, rather obvious point is that the Balandin 'Volcanocurve' as reproduced by Sachtler and De Boer for oxidation reactions with the ΔH_r^0 (eq. 3) as parameter, of course changes its position with the reaction envisaged. Even for oxidative dehydrogenation reactions that appear similar such as $C_4H_8 \to C_4H_6$ and $CH_3OH \to H_2CO$ the curve must be somewhat different because $Fe_2(MoO_4)_3$ is a weak catalyst for the first but an active catalyst for the second. While $Bi_2(MoO_4)_3$ is active for both.
For an entirely different reaction such as the oxidation of $C_{10}H_8$ to phtalic anhydride the need for many and aggressive oxygens in the catalyst is fairly obvious and the fact that the maximum shifts to the far left where we find V_2O_5 is not entirely unsuspected.

Another complication is that the redox mechanism often operates in the Mars-Van Krevelen manner but not always. Indeed, it is even somewhat surprising that it works so often. In the reoxidation of the catalyst the process occurring is

$$O_2 + [4e] + 2 \text{ vacancies} \longrightarrow 2O^{2-}$$

However, it is far from selfevident that the four electron transfer is concerted. It could - and often does - occur stepwise, for instance

$$\text{vacancy} + O_2 + [e] \longrightarrow O_2^- \text{ (superoxide ion)}$$
$$O_2^- + [e] \longrightarrow O_2^{2-} \text{ (peroxide ion)}$$
$$O_2^{2-} + \text{vacancy} \longrightarrow 2O^-$$
$$O^- + [e] \longrightarrow O^{2-}$$

The intermediates in this chain, the superoxide ion, the peroxide and O^- are very reactive species that have a tendency at the higher

Partial oxidation processes

temperatures to interact with the hydrocarbon in a non-selective manner. This means that the oxide in the presence of O_2 behaves quite differently than in its absence. Compounds such as TiO_2 and SnO_2 are not very active in the absence of oxygen but become quite virulent in a non-selective manner when there is oxygen present. Now, two of the intermediates viz. O_2^- and O^- can be detected by electron spin resonance; and these have indeed been found as surface species for the two oxides (among others) but not on the typically highly selective ammoxidation catalysts.

REACTION MECHANISM FOR THE CATALYTIC OXIDATION OF PROPENE TO ACROLEIN

The hydrogen atoms in an olefin at carbon positions next to a double bond may be supposed a priori to be relatively easily dissociated because their removal leads to allylic structures (carbocation, radical or carbanion) that are resonance stabilized Consequently, if an olefin becomes involved in a substitution reaction such as oxidation the evident focus of the attack is on the 'allyl hydrogens'. Already very early in the study of the catalytic oxidation reactions it was postulated by Sachtler that the first step in the network of catalytic oxidation reactions consisted in the formation of an allylic structure. Kinetical evidence and measurements of isotope effects by Adams and Jennings showed that it was even the rate determining reaction. A most important detail in the mechanism was revealed by Sachtler's ^{14}C labelling experiments (fig. 4) which showed that not only should there be an allyl but the way of bonding of the allyl to the surface should leave the terminal carbon atoms equivalent since they have an equal chance of collecting an oxygen.

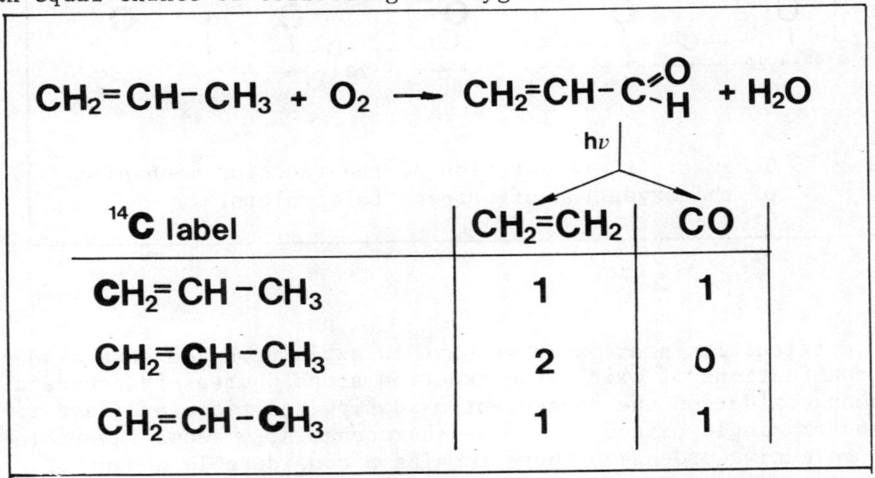

$$CH_2=CH-CH_3 + O_2 \longrightarrow CH_2=CH-C\overset{O}{\underset{H}{\diagup}} + H_2O$$

^{14}C label	$CH_2=CH_2$	CO
CH_2 $=CH-CH_3$	1	1
$CH_2=$**CH**$-CH_3$	2	0
$CH_2=CH-$**CH_3**	1	1

Fig. 4 Schematic representation of Sachtler's ^{14}C-labelling experiments.

This can be realized dissociating a proton from the CH_3 group, attaching the proton to an oxygen on the surface and binding the residual carbanion via acceptor π-bonding to a transition metal cation next to a surface anion vacancy. The π-bonding takes care of the equivalence of the terminal C-atoms while the interaction between the p-orbitals of the allyl and the d-orbitals of the transition metal cation causes the allyl to become bonded to the surface. Because of the σπ-double bond there will occur some transfer of negative charge (electrons) to the cation equivalent to an incipient reduction. Simultaneously oxygen anions of the surface (fig. 5) will move towards the terminal C-atoms because it is at these atoms that the radical character is strongest. The initial steps in the heterogeneous catalytic oxidation are therefore reasonably well understood.

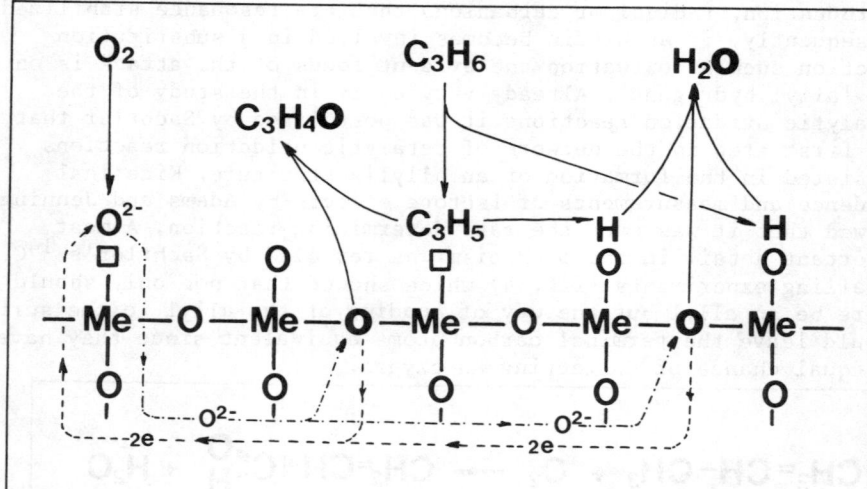

Fig. 5 Schematic representation of the reaction mechanism of the oxydation of propene to acrolein.

Our difficulties start when we have to explain why on some oxides or combinations of oxides the reaction stops whereas in non-catalyti gaseous oxidation the subsequent oxidation reactions are faster. Even for single oxides where the thermochemistry seems favorable for selective processes there remains a considerable amount of more extensive oxidation. One of the difficulties in explaining these observations is that we lack the necessary information as

Partial oxidation processes

to the nature of the subsequent steps in the more extensive oxidation.

There is however a simple explanation for the extraordinary selectivity. Let us suppose that the oxygen ions of the oxidic catalyst can be classified in types, one of which contains only relatively small numbers of oxygens that however are very reactive. Their high activity combined with their small concentration cause the catalyst to be moderately active. However, in the absence of gaseous oxygen the selectivity should be very high since there are simply not enough oxygens in the environment of the allyl to allow a more extensive oxidation. Such a model was first proposed by Callahan and Grasselli and later on worked out in more detail by Grasselli and Suresh to explain the properties of their USb_3O_{10} catalyst. They could show that reduction of the catalyst by propene occurred in two different manners. A small amount of oxygen (A-oxygen) was rapidly removed and produced acrolein. By far the larger amount (B-oxygen) reacted only slowly but the products were CO and CO_2. Another proof for the existence of small quantities of very active oxygen was given by Matsuura. The oxidation of butene is inhibited by its product butadiene and the activation of propene inhibited by acrolein. Oxidation catalysts adsorb butadiene and acrolein in an activated but strong adsorption (A-type adsorption). This adsorption can be eliminated by a small prereduction of the catalyst and is reestablished by an ensuring oxidation. The adsorption site therefore is an O^{2-}-ion and this ion must be the active oxygen otherwise adsorption would not inhibit the propene or butene oxidation. The surface concentration of the active oxygens is indeed surprisingly small: Matsuura found one active oxygen per 150 $Å^2$ for $Bi_2Mo_2O_9$ but one per 1800 $Å^2$ for USb_3O_{10}. While Grasselli considers the active oxygens as still belonging to the normal catalyst structure, Matsuura postulated that they are surface defects. He showed that Bi_2MoO_6 is not active but can be activated by adding small amounts of excess MoO_3. Simultaneously herewith the A-type of adsorption (absent on pure Bi_2MoO_6) grows linearly with the activity.

Defects as the origin of the catalytic activity for the selective oxidation of olefins were first proposed by Sleight. He doped the Scheelite $PbMoO_4$ with small amounts of Bi to give $Pb_{1-3x}Bi_{2x}\square_x$ (MoO_4) where $\square$ is a cationvacancy: the activity of doped catalysts grows linearly with x provided x remains small. A very interesting aspect of the doping is that variations such as $Pb_{1-2x}Bi_xNa_x$ (MoO_4) or $Pb_{1-3x}La_{2x}\square_x$ (MoO_4) remain inactive. This appears to show that an active molybdate catalyst needs Bi^{3+}-cations but also cation vacancies. Matsuura's active sites in Bi_2MoO_6 follow this principle because he assumes that the excess MoO_3 is incorporated in the Bi_2O_2-layers by a replacement of $2Bi^{3+}$-cations by a Mo^{6+} cation and a cation vacancy.

It is often assumed that the defects remain predominantly situated near or on the surface although this is not necessarily always true. (The doped Scheelites as prepared by Sleight have the defects uniformly dispersed over bulk and surface.)
However, for catalyst systems such as $SnO_2 + Sb_2O_4$ (Distillers), $FeSbO_4 + Sb_2O_4$ (Yoshino et al.) and even $USbO_5 + Sb_2O_4$ (Grasselli and Suresh) there are reasons to assume that the outer layers contain an excess of Sb. One of Grasselli and Suresh's catalysts was prepared by impregnating $USbO_5$ (active but not selective) with Sb to effect a surface enrichment of Sb. Indeed the catalyst became selective, presumably because of a formation of USb_3O_{10} (selective on the surface. However, it soon lost its selectivity because the excess Sb diffused into the solid.
SnO_2 is a reasonably active but non-selective catalyst (see before) but small amounts of Sb suffice to make it selective. Similarly, $FeSbO_4$ is active but non-selective; adding Sb_2O_4 causes it to become selective.
Together, all these observations appear to confirm that the 'isolated oxygen' is a relevant and valuable concept in explaining the selectivity of more prominent catalysts for olefin oxidation It is however not entirely consistent with some other commonly accepted facts. One is the observation that products such as butadiene, acrolein and acrylonitril are adsorbed rather strongly and therefore remain attached for considerable time to the surface. Moreover, it is also generally observed that the reoxidation of the reduced catalysts is fast so that depleted sites are rapidly replenished. Experiments by Keulks and collaborators may perhaps be of relevance in this connection. They ran mixtures of C_3H_6 and ^{18}O-enriched oxygen over Bi-molybdate catalysts containing only ^{16}O. Following the ideas mentioned above viz. rapid replacement of oxygen earlier removed from the surface by oxygen molecules from the gasphase one would expect the ^{18}O to become observable in the products acrolein and CO_2 almost immediately. However, for a catalyst such as Bi_2MoO_6 almost the entire bulk of the catalyst has to be enchanged before the ratio's of gasphase ^{18}O and product ^{18}O have become equalized. Keulks' interpretation of this behaviour is that oxygen has to diffuse for a considerable length through the bulk from the site where it is incorporated to where the structural oxygen is going to be removed by the reduction. An appropriate model might then be

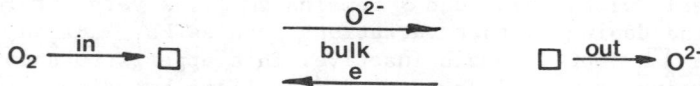

so that the entrance of O_2 is at a different site than the exit. Indeed, maintaining a spatial distance between the two sites has been occassionally proposed as an appropriate model to explain selectivity. It has also been postulated that entrance and exit are connected with the pair of cations as present in almost all selective species. Sleight for instance assumes the oxygen to become removed from two MoO_4-ions next to a cation vacancy while O_2 is entering via Bi. O^{2-} and electrons are travelling through the bulk from one cation to the other in opposite directions. Haber prefers the allyl to be formed on Bi and the oxygen to come in at Mo. Matsuura beleives that formation of the allyl occurs on the Mo but that this migrate to Bi. The actual directions of migration of electrons and oxygen anions is therefore not clear but there is good reason to accept the concept of the two cations being separately operative in the two basic tasks of the catalyst viz. donation of O to the hydrocarbon and accepting oxygen from O_2 molecules: the catalyst is therefore truly bi-functional.

THE ACRYLONITRILE PROCESS

H.S. van der Baan

Laboratory for Chemical Technology,
Eindhoven University of Technology, The Netherlands

1. The reaction model

A major event in the history of oxidation catalysis was the discovery of bismuth molybdate (Idol, 1959) (Bi_2O_3 + MoO_3) as a selective catalyst for the partial oxidation of propene and also, in a one step operation, for the ammoxidation of propene. Other catalysts are oxides of uranium and antimony or of iron and antimony. The latter catalysts may be promoted by potassium or phosphor. The latest development are the bismuth molybdate multicomponent catalysts that contain also one or more of the elements cobalt, iron, nickel, potassium and phosphor. The catalysts that have been studied most are the bismuth molybdates. These are active within the composition range Bi_2MoO_6 to $Bi_2Mo_3O_{12}$.

Under normal reaction conditions i.e. at temperatures above 675 K the catalysts behave according to the Mars-van Krevelen model: the oxygen from the catalyst is used in the oxidation or ammoxidation reaction, after which the reduced catalyst is reoxidized. This follows conclusively from the work of Keulks (1970), Wragg et al. (1971) and others, who have shown that in reactions with ^{16}O in the catalyst and $^{18}O_2$ in the gasphase it takes some time before ^{18}O is found in the reaction products, acrolein and water. The stoichiometric reaction equation for the catalytic ammoxidation of propene:

$$C_3H_6 + NH_3 + 1\tfrac{1}{2} O_2 \rightarrow C_3H_3N + 3 H_2O \tag{1}$$

shows that the propene molecule must loose three hydrogen atoms, and that the dehydrogenated intermediates must be bonded to the catalyst in such a way that they are protected against direct oxidation by gaseous oxygen.

A model of the whole reaction sequence is given by Lankhuijzen (1979):

$$C_3H_6 \underset{}{\overset{1}{\rightleftarrows}} (C_3H_6)_a \overset{2}{\rightarrow} (C_3H_5)_a \overset{3}{\rightarrow} (C_3H_4)_a \underset{}{\overset{4}{\rightleftarrows}} (C_3H_4O)_a \underset{}{\overset{5}{\rightleftarrows}} C_3H_4O$$

$$6 \downarrow \quad \swarrow 9$$

$$(C_3H_4NH)_a$$

$$7 \downarrow$$

$$(C_3H_3N)_a$$

$$8 \updownarrow$$

$$C_3H_3N$$

Fig. 1. Reaction scheme for the formation of acrolein and acrylonitrile.

Reaction step no.	Matsuura	Haber	Otsubo	Peacock	Sleight	Lankhuijzen
2 {1st H^+ abstraction / + C_3H_5 formation	Mo	Bi	Bi	Mo	MoO_4	Mo
3 {2nd H^+ abstraction / C_3H_4 formation on	Bi	Mo	Bi	Mo	MoO_4	Mo
4 {formation of C_3H_4O / with O from	Bi	Mo	Bi	Mo	MoO_4	Mo

Table 1. Location of active ensembles in the oxidation of propene; Matsuura and Schuit (1970)(1971), Haber (1973), Otsubo et al. (1975), Peacock et al. (1969), Sleight (in press), Lankhuijzen (1979)

The acrylonitrile process

Much attention has been devoted to the question on what active surface ensembles the reaction steps of figure 1 take place. A number of the results are given in table 1.

As far as the oxygen is concerned most authors expect it to enter the catalyst at an ensemble connected with Bi.
Of the four main reactions that take place:

$$C_3H_6 + NH_3 + 1\tfrac{1}{2} O_2 \rightarrow C_3H_3N + 3 H_2O \tag{1}$$

$$C_3H_6 + O_2 \rightarrow C_3H_4O + H_2O \tag{2}$$

$$C_3H_4O + NH_3 + \tfrac{1}{2} O_2 \rightarrow C_3H_3N + 2 H_2O \tag{3}$$

$$2 NH_3 + 1\tfrac{1}{2} O_2 \rightarrow N_2 + 3 H_2O \tag{4}$$

the reaction orders are shown in table 2.

Reaction	order with respect to			
	C_3H_6	NH_3	C_3H_4O	O_2
(1)	1-0*	0	-	0
(2)	1-0*	-	-	0
(3)	-	0**	1-0	0
(4)	0**	1	-	0

Table 2. Reaction orders in the ammoxidation of propylene.
  * Depending on temperature and reactant concentration (Lankhuijzen 1979).
  ** actually: slightly negative.

The zero order in oxygen indicates complete coverage of the re-oxidation ensembles, i.e. the oxygen supply to the reaction sites is not influenced by variations in the gas phase oxygen concentration.
The zero order in ammonia for the acrylonitrile formation (both from propene and from acrolein) indicates that the reacting nitrogen surface species have completely covered the sites available to them. As the nitrogen formation is first order in ammonia there also must exist a nitrogen containing species on other sites, where the degree of coverage is rather low. Lankhuijzen assumes that the latter sites are the same as those used by the intermediates derived from propene thus causing the slightly negative order in ammonia for the acrylonitrile formation.

Insert instead:

The competition between ammonia and propene for the same sites is also found in the somewhat negative order exerted by propene on the nitrogen formation, as shown in table 2. That on the other hand, also ammonia and oxygen have common features follows from Lankhuijzen's observation that in pulse experiments where a propene-helium mixture is pulsed over the catalyst the conversion per pulse is increased in exactly the same way by addition of oxygen as by addition of the same quantity of ammonia.
In this connection we recall the observation of Sancier et al who found that in pulse experiments the production of O^{16} acrolein was enhanced by the addition of O_2^{18}.
This all suggests that propene and "first order" ammonia absorb on one type of site, related to Mo and that oxygen and "zero order" ammonia are connected with Bi.

The acrylonitrile concentration has a small but noticeable negative influence on the reaction rate, i.e. on the propene conversion.
The main byproducts of the reaction are H_2O, CO_2/CO, CH_3CHO, N_2, CH_3CN and HCN. The latter two have commercial value and are generally recovered from the product stream.

2. The acrylonitrile process

For the choice of the most appropriate reactor the following data are of interest:
-Most of the reactants and products are flammable and have explosive properties over wide ranges when mixed with air, as shown in table 3.

	Ignition temperature K	Explosive range % by vol. in air
C_3H_6	770	2 -11
NH_3	924	15 -28
C_3H_4O	551	2.8-31
C_3H_3N	754	3.1-17

Table 3. Ignition temperature and explosive range for the main reaction components of the acrylonitrile synthesis (Sax, 1975).

-The reaction produces about 20 MJ/kg acrylonitrile.
-The catalyst is stable and can be used for years without much

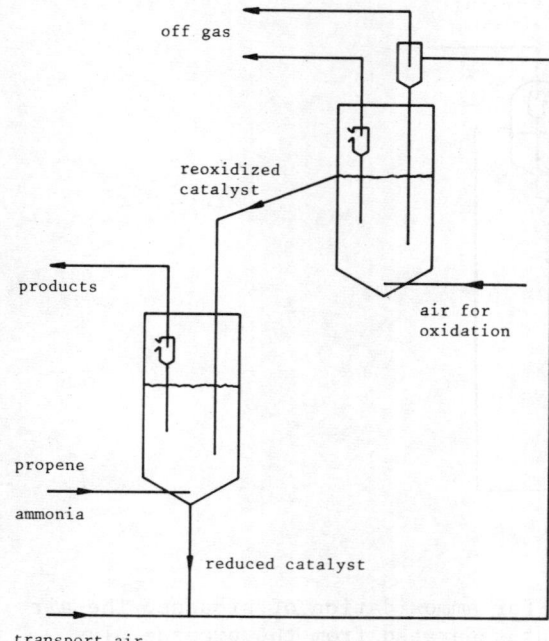

Fig. 2. Two reactor design for ammoxidation of propene. Complete separation of air and propene plus ammonia.

loss of activity, if the temperature is kept within reasonable limits, say below 800 K.
-Acrylonitrile is not completely stable under reaction conditions.

The high heat of reaction requires a reactor with an efficient cooling system, especially as the combination of the inflammable reaction mixture and hot spots would lead to dangerous situations. Although initially fixed bed tubular reactors of the heat exchanger type were used, fluid bed processes developed at an early stage. As fluid bed reactors are very efficient in equalizing temperature differences, the possibility of hot spots is eliminated in such reactors, provided that they are properly designed i.e. that no quantities of stagnant catalyst can build up in corners or on internal beams.
A fluid bed requires a catalyst that is mechanically stronger than the pure oxides. Admixture of a SiO_2 carrier, however, decreases the propene selectivity somewhat.

Initially these fluid beds were also designed with the object of separating the oxygen containing stream from the combustible feed components. This is e.g. shown in figures 2 and 3. The system of figure 2 has a separate reoxidation reactor, while in the

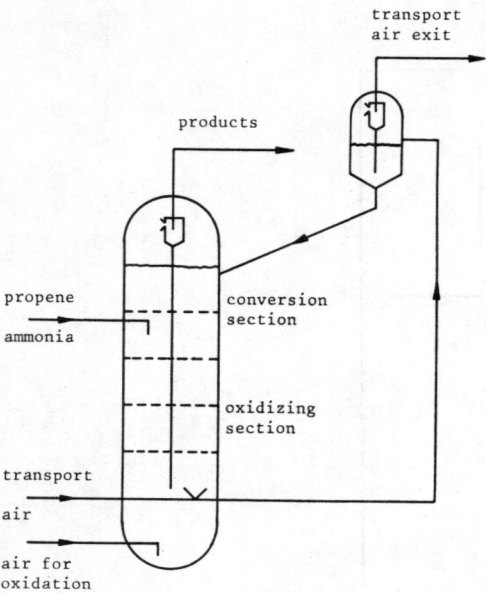

Fig. 3. One reactor design for ammoxidation of propene. The air is to a large extent separated from the propene plus ammonia (Calahan, 1969).

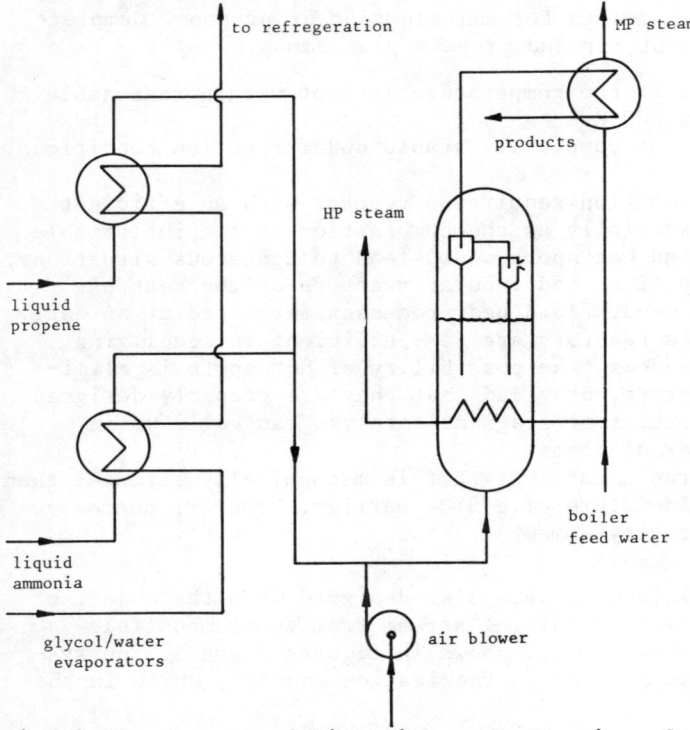

Fig. 4. One reactor design without separation of air and propene plus ammonia (Pujado et al., 1977).

reactor of figure 3 the reoxidation takes place in the lower
part of the fluid bed, so that air with a much reduced oxygen
content is contacted with the propene-ammonia mixture higher up
in the bed. Also steam has been added to the feed to reduce the
explosiveness of the mixture of reactants.
In the modern reactors however air, propene and ammonia enter the
reactor together as shown in figure 4, in accordance with the
experience gained in other oxidation reactors, e.g. the fluidized
bed oxidation of O-xylene to phtalic anhydride.
To withdraw the considerable heat of reaction, the reactors are
all equiped with internal coolers, often arranged in the form of
horizontally spiralized tubes, located at regular intervals along
the height of the fluid bed.
This form of the coolers has the added advantage that the back
mixing in the dense phase of the fluid bed is somewhat reduced.
This has a beneficiary effect on the acrylonitrile yield as the
further conversion of acrylonitrile is somewhat reduced.

3. A mathematical model of the conversion in the fluid bed.

3.1. Of the models available, such as those of van Deemter (1961),
Kunii and Levenspiel (1965), Partridge and Rowe (1969) or of
Fryer and Potter (1972) we will use the model of Partridge and
Rowe of 1969, because it is simple and contains two assumptions
that reflect properties of a fluid bed with a number of horizontal cooling grids, viz:
1. The gas flow is plug flow upward through the dense phase.
   Because the grids suppress backmixing the backmixing models
   of van Deemter and Fryer and Potter will be less appropriate;
2. The volume fraction of bubbles is constant over the bed height.
   The bubbles are fully segregated and do not coalesce. The
   grids are acting as redistribution trays, on which on all
   trays bubbles are formed with the same diameter. The mean
   bubble diameter can be set at 1.5 to 2 times the distance between the elements of the grid, say $d_b \sim 0.06$-0.1 m. This
   leads to a linear bubble rise velocity of 0.5-0.7 m/s (see
   equation R (4)*).
Further assumptions of the Partridge and Rowe model are:
3. The (empty) bubbles are surrounded by a "cloud" and a "wake"
   moving with the bubble (see R figures 12 and 6 respectively).
   The whole ensemble is called the cloud. The outer part is
   called the dense phase-cloud phase overlap or shortly the
   overlap. Thus:

$$\text{cloud} = \text{bubble} + \text{overlap} \qquad (5)$$

---

* This notation indicates equation (4) of the chapter of P.N.
  Rowe on basic fluidization.

4. The gas in the whole cloud is ideally mixed;
5. There is only resistance to mass transport between the dense phase and the overlap;
6. The volume fraction of the bubble phase δ (volume of bubbles per unit volume of reactor) is constant over the bed height;
7. The linear gas velocity in the dense phase is equal to the minimum fluidization velocity;
8. The reactions take place in the dense phase and in the overlap phase. In these two phases the catalyst fraction is the same.

3.2. The mathematical expressions for the Partridge and Rowe fluid bed model.

We will indicate linear velocities (m/s) by v and superficial velocities (m^3/m^2 s) by v_s.

3.2.1. The ratio of cloud volume to bubble volume.

We define α:

$$\alpha = \frac{v_b}{v_{mf}} \qquad (6)$$

in which v_b = linear velocity of the bubbles and v_{mf} = linear velocity at min. fluidization or

$$\alpha = \frac{v_{s,b} \ mf}{v_{s,mf}} \qquad (7)$$

From equation 5 we have

$$V_c = V_b + V_o \qquad (8)$$

Partridge and Rowe derived:

$$V_o/V_b = 1.17 \, (\alpha-1) \qquad (9)$$

or with (8)

$$V_c = V_b \frac{\alpha + 0.17}{\alpha - 1} \qquad (10)$$

3.2.2. The gas transport via the dense phase and the cloud phase.

For the superficial feed velocity $v_{s,f}$ we can write

$$v_{s,f} = v_{s,d} + v_{s,c} \qquad (11)$$

Further

$$v_{s,d} = v_{mf}(1-\delta) \qquad (12)$$

Thus

$$v_{s,c} = v_{s,f} - v_{mf}(1-\delta) \qquad (13)$$

3.2.3. The overall mass transport coefficient between dense phase and overlap k_o follows from

$$Sh = \frac{k_o d_c}{D} \qquad (14)$$

The Sherwood number follows from the correlation:

$$Sh = 2 + 0.69\, Re^{\frac{1}{2}}\, Sc^{1/3} \qquad (15)$$

with

$$Re = \frac{\rho\, v_r\, d_c}{\eta} \qquad (16)$$

and

$$Sc = \frac{\eta}{\rho D} \qquad (17)$$

which is valid for $30 < Re < 2000$.
For the relative bubble velocity v_r Partridge and Rowe have used

$$v_r = v_b - v_{mf} \qquad (18)$$

v_b and v_{mf} can be found with figure R 9 and equation R (2) respectively.
From equation (10) we have

$$d_c = d_b \sqrt[3]{\frac{\alpha + 0.17}{\alpha - 1}} \qquad (19)$$

This allows the computation of k_o. The effective transport coefficient is $\varepsilon\, k_o$, because a fraction $(1-\varepsilon)$ of the exchange surface is occupied by catalyst.

3.2.4. The reaction rate r can be expressed as

$$r = k\, C_{propene} = k\, \rho\, \underline{x}_p \quad \text{(mol/s kg cat)} \qquad (20)$$

This gives for the rate in the dense phase:

$$r_d = r\,\rho_c\,(1 - \varepsilon)\ \text{mol/s m}^3\ \text{dense phase} \qquad (21)$$

where ρ_c = the catalyst particle density and in the cloud phase:

$$r_c = r\,\rho_c\,(1 - \varepsilon)\,\frac{V_b}{V_c}\ (\text{mol/s m}^3\ \text{cloud phase}) \qquad (22)$$

3.2.5. The differential equations.

As we can assume that the process is isothermal, and as we neglect the pressure drop over the bed no further data are required. The reactor can then be modelled according to figure 5. and from this figure we can derive for the dense phase:

$$v_{s,d}\,\frac{d\,\underline{x}_d}{d\,z} + \frac{\varepsilon\,k_o\,\pi\,d_c^{\,2}\,\delta}{V_c\,(1-\delta)}\,(\underline{x}_d - \underline{x}c) - r_d = 0 \qquad (23)$$

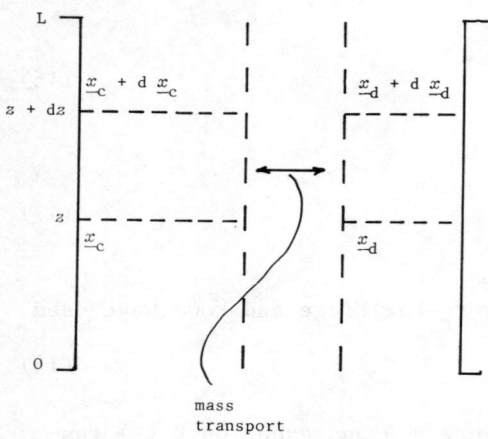

Fig. 5. Model of a fluid bed reactor for the ammoxidation of propene. $\underline{x}$ Represents the specific propene concentration (mol/kg).

and for the cloud phase:

$$v_{s,c} = \frac{d\,\underline{x}_c}{d\,z} + \frac{\varepsilon\,k_o\,\pi\,d_c^{\,2}}{V_c}\,(\underline{x}_c - \underline{x}_d) - r_c = 0 \qquad (24)$$

which can be solved by straight numerical integration from the initial values $x_{c,o} = x_{d,o} = x_o$, the propene inlet concentration. With a constant selectivity of 68 percent the acrylonitrile concentration is everywhere:

$$\underline{x}_A = 0.68 \, (\underline{x}_o - \underline{x}) \tag{25}$$

and for the outlet concentration of acrylonitrile we have finally

$$\underline{x}_{A,e} = \frac{v_{s,d} \, \underline{x}_{A,d,e} + v_{s,c} \, \underline{x}_{A,c,e}}{v_{sf}}$$

Literature

1. Calahan, J.L., Nilberger, E.C., USP 3, 472, 892 (1969)
2. Haber, J., Z. Chem. 13, 241 (1973)
3. Idol, J.D. (Standard Oil Co.) U.S. Pat. 2 904 580 (1959)
4. Keulks, G.W., J. Catal. 19, 281 (1970)
5. Kunii, D., Levenspiel, O., Ind. Eng. Chem. Fund. 7, 446 (1968)
6. Lankhuijzen, S.P., The acrolein and acrylonitrile synthesis over a bismuth molybdate catalyst, Thesis, Eindhoven 1979
7. Levenspiel, O., 'Chemical Reaction Engineering' 2nd ed., John Wiley and Sons Inc., New York (1972)
8. Matsuura, I., Schuit, G.C.A., J. Catal. 20, 19 (1971) and J. Catal. 25, 314 (1972)
9. Murray, J.D., J. Fluid. Mech. 21, 465 (1965) and 22, 57 (1965)
10. Otsubo, T., Miura, H., Morikawa, Y., Shirasaki, T., J. Catal. 36, 240 (1975)
11. Partridge, B.A., Ronn, D.N., Trans. Inst. Chem. Engrs. 44, 335-348 (1969)
12. Peacock, J.M., Sharp, M.J., Parker, A.J., Ashmore, Ph., Hockey, J.A., J. Catal. 15, 379; 398 (1969)
13. Pujado, P.R., Vora, B.V., Krueding, A.P., Hydrocarbon Processing , (5), 169 (1977)
14. Sancier, K.M., Wentreck, P.R., Wise, H., J. Catal. 39, 141 (1975)
15. Sax, N.I., 'Dangerous properties of Industrial materials", 4th ed. v. Nortrand Reinhold CY, New York (1975)
16. Schuit, G.C.A. in B.C. Gates, J.R. Katzer and G.C.A. Schuit: 'Chemistry of Catalytic Processes'; McGraw Hill, New York, chapter 4, p. 325 (1979)
17. Sleight, A.W., in J.J. Benton and R.L. Garten (eds.), 'Advanced materials in Catalysis', Acad. Press, New York (in press)
18. van Deemter, J.J., Chem. Eng. Sci. 13, 143 (1961)
19. Wragg, R.D., Ashmore, P.G., Hockey, J.A., J. Catal. 22, 49 (1971)

HOT SPOTS AND RUNAWAY IN FIXED BED TUBULAR REACTORS

Prof.dr.ir. G.F. Froment

Laboratorium voor Petrochemische Techniek
Rijksuniversiteit, Gent, Belgium.

Exothermic reactions carried out in multitubular catalytic reactors with heat exchange generally lead to temperature profiles exhibiting a maximum, called hot spot. The maximum temperature has to be limited to avoid losses in selectivity, catalyst deterioration and excessive sensitivity or instability leading to runaways of the reactor.

I. PARAMETRIC SENSITIVITY AND RUNAWAY ASSOCIATED WITH IDEAL PSEUDO-HOMOGENEOUS ONE-DIMENSIONAL MODELS FOR TUBULAR REACTORS

Consider first a situation that may be described by the model equations (1)-(3) of Froment's contribution on "Fixed Bed Catalytic Reactors" in the present Proceedings - in words : a one dimensional model for a reactor with a single fluid phase, assuming plug flow, no radial gradients in the bed, no interface and no intraparticle gradients. For pseudo first order kinetics suggested by o.xylene oxidation and constant pressure throughout the reactor Van Welsenaere & Froment [1] simulated the temperature profiles shown in Fig. 1.

Hot spots like those shown in Fig. 1 may develop because of high reaction rates, in particular when associated with high activation energies or because of limited heat exchange. From a certain range of inlet conditions and wall temperatures onwards slight variations in partial pressure or temperature cause an important increase in the hot spot temperature, leading eventually to runaway. For the situation considered here

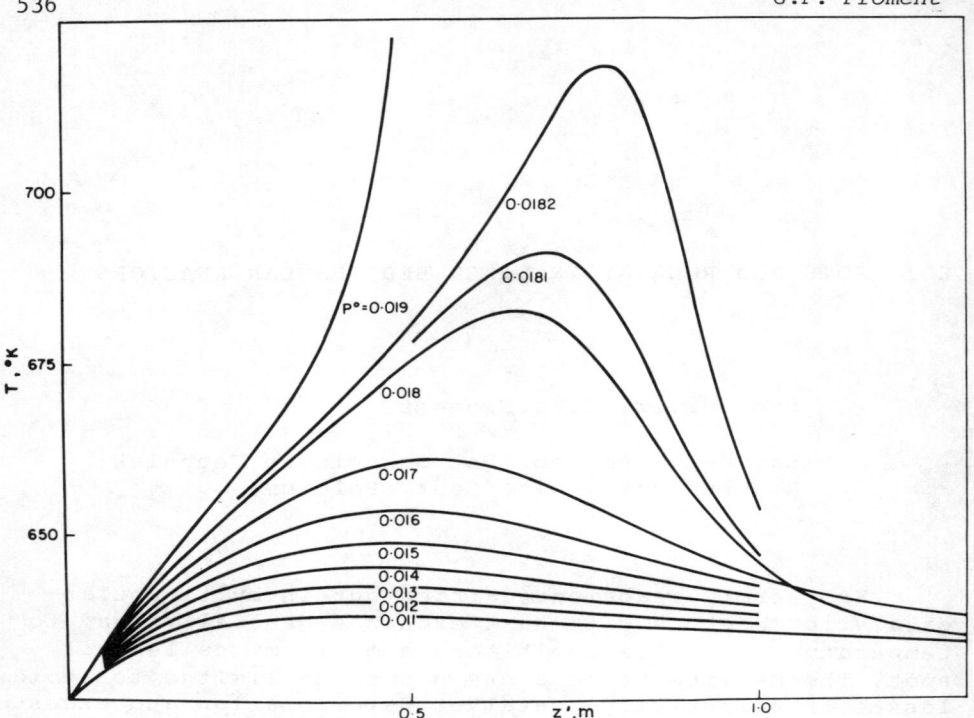

Fig.1. Temperature profiles in a tubular reactor for a strongly exothermic reaction. From [1].

this would be an example of runaway caused by "parametric sensitivity" : when the perturbation is removed the original profiles are recovered. Parametric sensitivity is not to be confused with "instability" in the restricted sense. Instability implies that the removal of the perturbation does not necessarily restore the original profiles. This phenomenon will be dealt with in later sections of this paper.

It is convenient to have criteria for predicting runaway conditions which could be used to analyze such situations prior to any computer simulation of the reactor. There are two approaches to such criteria. The first is to limit the peak temperature on the basis of "external" elements, like catalyst deterioration, safety or materials strength. The second approach is directly based on the properties of the reaction system, as expressed by the model equations and leads to what could be called "intrinsic criteria". Evidently, the second approach cannot ignore the specific constraints mentioned above. Van Welsenaere & Froment [1] derived such an analytical intrinsic criterion for pseudo first order reactions and for constant p_t in the reactor.

Hot spots and runaway in fixed bed tubular reactors

Notice that they used partial pressures instead of concentrations. The analysis starts from the relation between the partial pressure and the temperature in the maximum of the temperature profile in the reactor. Setting $dT/dz = 0$ in the energy equation leads to the relation :

$$p_m = \frac{T_m - T_w}{\frac{B}{C} A_o \exp(-\frac{E}{RT_m})} \qquad (1)$$

For a given reaction the only parameter in this equation is the wall temperature, T_w, which is assumed to equal the inlet temperature, T_o. For each temperature profile a set of values (T_m, p_m) is obtained. The locus of (T_m, p_m) in a phase-plane (T,p) is a curve called "maxima-curve", corresponding to equation (1). This curve too has a maximum, obtained by setting $dp_m/dT_m = 0$. The temperature corresponding to that maximum, T_M, is given by :

$$T_M = \frac{1}{2} (\frac{E}{R} - \sqrt{(\frac{E}{R})^2 - 4 \frac{E}{R} T_w}) \qquad (2)$$

The maxima-curve also exhibits a minimum, but for temperatures far beyond those of practical importance.

The first intrinsic criterion of Van Welsenaere & Froment is based upon conclusions derived from Fig. 1. Translated into the (T,p)-phase plane : trajectories (curves representing the relation between p and T in the reactor) corresponding to sensitive operating conditions intersect the maxima curve beyond the maximum (Fig.2). Consequently, the critical trajectory is that

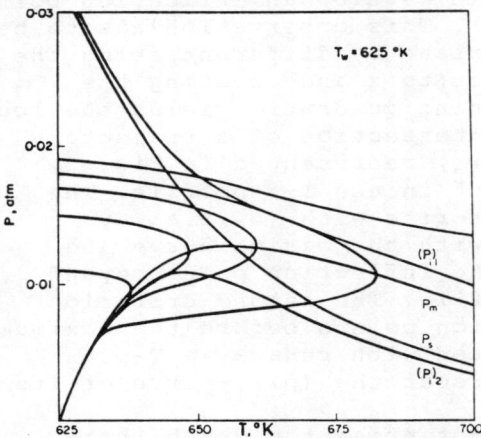

Fig.2. (T,p) plane, showing trajectories, maxima-curve, loci of inflection points and the "simplified" curve, p_s.
From [1].

curve which intersects the maxima-curve in its maximum.
Obtaining the trajectory itself requires numerical integration, but the maximum of the maxima-curve is obtained analytically, as evidenced by (2). What remains to be done is to determine the corresponding critical feed conditions i.e. for given $T_o=T_w$, the critical inlet partial pressure of the reactant. To avoid numerical integration a simple extrapolation procedure is required. The most obvious one is an adiabatic extrapolation. Since at the reactor inlet, i.e. for $T_o=T_w$, trajectories always have a slope smaller than that of the adiabatic trajectory this backward extrapolation will lead to a lower and safe limit for the critical reactant partial pressure in the feed. It is of interest also to define an upper limit, above which runaway will certainly occur. This is arrived at by observing the tangents at the trajectories in a vertical through T_M. These intersect the ordinate through $T_o=T_w$ at values which decline first as trajectories closer to the critical are considered, but suddenly rise because of the strong curvature of trajectories very close to the critical : the tangent at the critical trajectory taken in T_M is parallel to the ordinate. The minimum value of the intersection on the ordinate is also easily obtained and is the upper limit for the reactant partial pressure in the feed. The arithmetic average of upper and lower limits is remarkably close to the true critical inlet concentration, obtained by numerical back integration, from the maximum of the maxima curve onwards.

A second intrinsic criterion was based upon the observation that the temperature profiles in the reactor close to the runaway profile develop an inflection point before the maximum (Fig. 1). This observation has to be translated into the (T-p)-plane by differentiating the energy equation with respect to z and equating the results to zero. The resulting quadratic yields the loci of inflection points. An intersection of a trajectory with the loci $(p_i)_1$ or $(p_i)_2$, represented in Fig. 2, leads to an inflection point in the T-z-profile. The intersection of the trajectories with $(p_i)_2$ always occurs after intersection with the maxima curve and therefore corresponds to the inflection point beyond the maximum in the T-z profile. The second criterion states : "To avoid inflection points before the maximum of the T-z profile associated with runaway a T-p trajectory should not intersect the $(p_i)_1$-curve or its approximation p_s".

Van Welsenaere & Froment present a graph that permits predicting the point of tangency between the

critical trajectory and $(p_i)_1$ or p_s [1]. Again two ways
of extrapolation from this critical point onwards lead
to upper and lower limits for the reactant partial
pressure in the feed. The band width is even somewhat
smaller than that based on the first criterion.

In Fig. 3 the predictions of Van Welsenaere and
Froment [1] are compared with those of Barkelew [2] and
other investigators [3,4]. It may be worthwhile adding
that Barkelew's curve resulted from 750 numerical integrations and that the Arrhenius temperature dependency
of the rate coefficient was approximated by a more
convenient expression.

Agnew and Potter [7] derived sets of diagrams
similar to that of Barkelew for cases whereby radial
temperature and concentration gradients have to be
accounted for (Model A.III of Froment's classification
[8,9]). These inevitably necessitated numerical integration.

II. EFFECT OF AXIAL MIXING. MULTIPLE STEADY STATES AND TRUE INSTABILITY

It has been argued that flow conditions in fixed
bed reactors deviate from plug flow and that the presence of packing creates a certain degree of axial
mixing. This mixing in longitudinal direction causes a
heat and mass flux that has to be superposed upon that
by plug flow and which is expressed in terms of
Fourrier's and Fick's laws. The model equations are
(Model A.II of Froment's classification [8,9]) :

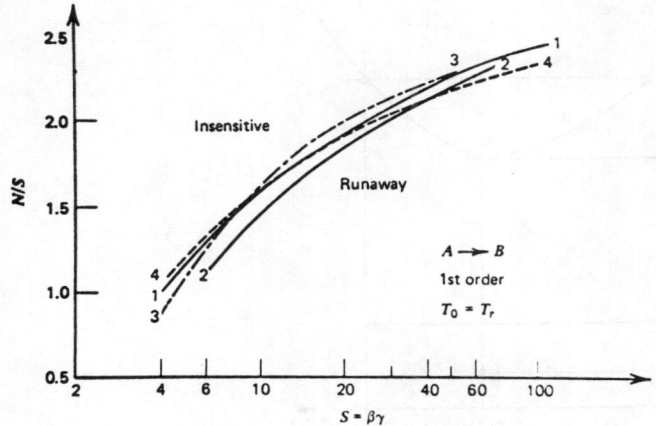

Fig.3. Runaway diagram. From [5].
Curve 1:Barkelew [2]; curve 2:Dente&Collina [3];
curve 3:Hlavacek et al.[4]; curve 4:Van Welsenaere&Froment [1].

$$\varepsilon D_{ea} \frac{d^2 c}{dz^2} - u_s \frac{dc}{dz} - r_A \rho_B = 0 \qquad (3)$$

$$\lambda_{ea} \frac{d^2 T}{dz^2} - u_s \rho_g c_p \frac{dT}{dz} + (-\Delta H) r_A \rho_B - 4 \frac{U}{d_t}(T-T_w) = 0 \qquad (4)$$

with boundary conditions : $u_s(c_o - c) = -\varepsilon D_{ea} \frac{dc}{dz}$

for $z=0$

$$u_s \rho_g c_p (T_o - T) = -\lambda_{ea} \frac{dT}{dz}$$

$$\frac{dc}{dz} = \frac{dT}{dz} = 0 \qquad \text{for } z=Z$$

For simplicity consider the adiabatic version of (4) and let $\lambda_{ea} = \varepsilon D_{ea}$:

$$\frac{d^2 T}{dz'^2} - Pe_a \frac{dT}{dz'} + f(T) = 0 \qquad (5)$$

where $z' = \frac{z}{Z}$; $Pe_a = \frac{u_i Z}{D_{ea}}$ and $f(T) = \frac{k_o Z^2}{\varepsilon D_{ea}} \rho_B (T_{ad} - T) \exp \frac{E}{RT_o}(1 - \frac{T_o}{T})$

(6)

Equation (5) may lead to a monotonic relation between $T(Z)$ and T_o or to a relation as shown in Fig.4.
With a relation like the one shown in Fig. 4 three steady state profiles are possible for a certain range

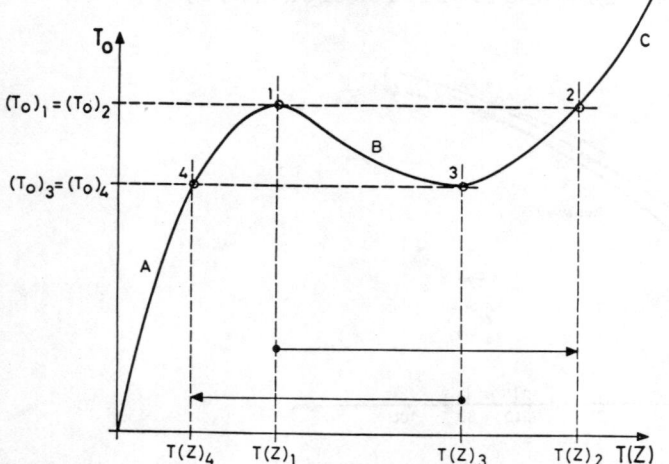

Fig.4. Relation between T_o and $T(Z)$ in an adiabatic fixed bed reactor with axial mixing leading to the possibility of three steady state profiles.

of inlet conditions, $(T_o)_3 < T_o < (T_o)_1$. Of the three values of $T(Z)$ corresponding to a T_o inside that range only the outer two are stable and can be observed without external action. A different type of runaway is associated with this multiplicity of steady states. Indeed, as T_o is increased along branch A the exit temperature $T(Z)$ increases progressively up to $T(Z)_1$. For a value slightly exceeding $(T_o)_1$ the exit temperature $T(Z)$ jumps to $T(Z)_2$. Notice the discontinuity in $T(Z)$: whereas with parametric sensitivity any exit temperature would be possible in the present case the range $T(Z)_1 < T(Z) < T(Z)_2$ is excluded for the chosen c_o and flow rate. This is clearly an instability in the restricted sense. Indeed, if the inlet temperature is lowered from $(T_o)_2$ onwards $T(Z)$ will decrease progressively along branch C untill a value $T(Z)_3$ is reached. If T_o is reduced further $T(Z)$ will jump downwards to $T(Z)_4$ which is different from $T(Z)_3$. Typical for an unstable case the temperature history exhibits a hysteresis.

When instabilities sensu strictu are possible runaway criteria first have to detect the regions of multiplicity. Several criteria for multiplicity of steady states in adiabatic tubular reactors with axial mixing have been proposed [10,11,12]. Briefly, multiplicity would only be possible for fast, strongly exothermic reactions with a very high activation energy carried out in reactors with considerable axial mixing. In industrial fixed bed reactors with a single fluid phase the effect of axial mixing on the reactor performance is not noticeable, unless the reactor is very shallow [6,8]. No industrial reactor is so shallow as to satisfy the length criterium, however. Finally, it is doubtful whether the hydrodynamics of shallow beds can be represented by the model equations (3)-(4), anyway.

It may therefore be stated that in normal industrial tubular fixed bed reactors axial mixing will not cause runaway by unstable behavior. That does not rule out runaway by parametric sensitivity, of course. Any axial mixing will, however, reduce the sensitivity of the reactor with respect to the plug flow conditions assumed in model A.I of Froment's classification [8,9].

III. EFFECT OF INTERFACIAL GRADIENTS

Another situation that has received considerable attention in the literature is that whereby interfacial gradients occur. This is modeled by the set of equations (6)-(9) of the chapter on "Fixed bed catalytic reactors" of these Proceedings.

Interfacial gradients develop when the rate of reaction is high, the heat effect important and the turbulence in the fluid relatively low. The most likely gradient is a temperature gradient. This situation is generally of no concern in industrial reactors, but it is a permanent problem with bench scale experimentation.

Again, runaway can result from two sources: parametric sensitivity and instability.

Instabilities caused by interfacial gradients were first described by Wicke [13] and by Liu & Amundson [14]. Consider a single reaction and a catalyst particle with uniform temperature T_s and uniform concentration of reactant, C_s. Let the bulk fluid phase around that particle have a temperature T and a concentration c. At steady state the heat generated by the reaction $Q_R = r_A(-\Delta H)\rho_B$ has to equal the heat transferred from the particle to the bulk fluid: $Q_T = h_f a_v (T_s - T)$. In Fig. 5 it can be seen that Q_R has a sigmoid form when plotted vs T_s whereas Q_T leads to a straight line. Intersection of Q_R and Q_T in branch A occurs with low values of T and leads to low values of T_s, whereas intersection on branch C requires higher T and leads to high T_s. For a certain range of T three intersections are possible, the outer two corresponding to stable, the intermediate to unstable operation. Which one of the two stable steady states will be selected by the system depends upon the fluid temperature and the

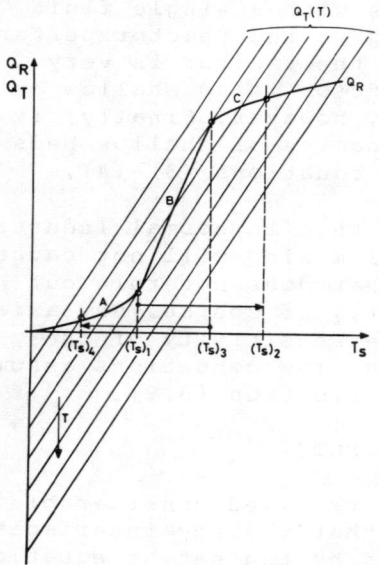

Fig.5. Heat production and heat transfer rates for various fluid and particle temperatures.

previous history. Suppose the system operates in a point on branch A. As soon as the fluid temperature is increased above a value leading to $(T_S)_1$ the solid temperature will jump to a value $(T_S)_2$. For the history described the range $(T_S)_1$, $(T_S)_2$ is not a steady range for T_S. If the particle is initially at $(T_S)_2$ and its temperature is lowered slightly below $(T_S)_2$ the temperature will further drop to $(T_S)_4$ and extinction will occur. In this case $(T_S)_3$, $(T_S)_4$ is not a steady range for T_S. The hysteresis is shown in Fig. 6.

Extending this now from a single particle to a fixed bed reactor leads to the conclusion that the concentration and temperature profiles in a reactor would not only depend upon the feed conditions and the wall temperature, but also upon the temperature profile at the time the feed is initially admitted to the reactor.

Fig. 7 shows temperature profiles in an adiabatic reactor with interfacial gradients, calculated by Liu and Amundson [14]. Fig. 7a relates to a situation in which multiplicity is not possible. Runaway would solely result from parametric sensitivity. Fig. 7b relates to multiple steady states. Notice the much higher feed concentration required to achieve this. In case A the initial bed temperature is lower than 393°C. Up to 0.27 m the particles are in the lower steady state. The temperature then rises very steeply over a few layers to the upper steady state. In case B the initial bed temperature is 560°C. The feed is rapidly heated from 393°C onwards and after a few centimeters the temperature rises to the upper steady state values. A similar although less drastic behavior would be calculated for order higher than one. For the more general

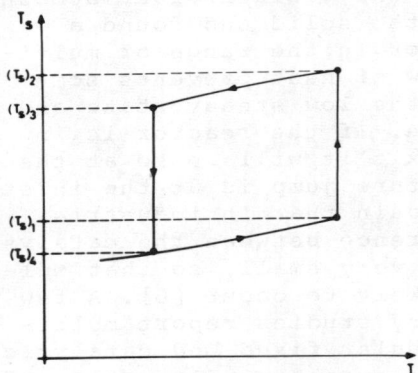

Fig.6. Temperature hysteresis in particle temperature upon variations in fluid temperature.

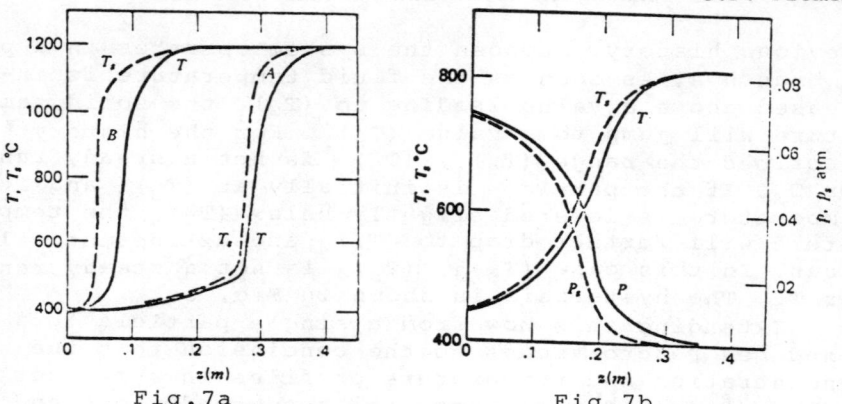

Fig.7a Fig.7b

Temperature profiles in an adiabatic tubular reactor with interfacial gradients. First order reaction. Fig. 7b : unique steady state case ; $p_o=0.007$ atm , $T_o=449°C$; Fig. 7a : non-unique steady state case : $p_o=0.15$ atm ; $T_o=393°C$, initial $T_s=A \leqslant 393°C$, B:560°C. After [14]. From [6,8].

Hougen & Watson type rate equations which account for adsorption of the reacting components the results may differ in some aspects [15].

A priori criteria for testing for the multiplicity were derived by Van den Bosch and Luss [16]. For a first order reaction e.g. $\beta\gamma > 4$. The region in parameter space for which multiplicity can occur decreases with increasing reaction order.

Notice that the model does not provide any axial coupling between the particles : axially heat is only transferred through the fluid. This is a serious simplification when such steep temperature gradients occur over a few layers of particles only. Eigenberger accounted for heat transfer through the solid and found a significantly different behavior in the range of multiplicity [17]. The backward flow of heat prevents some parts of the reactor to be in the low steady state and others in the high steady state. If the reactor is in the high steady state at the exit it will be so at the inlet too, so that the temperature jump is at the inlet.

It should be emphasized again that in industrial reactors the temperature difference between the catalyst and the bulk fluid is normally very small, so that multiple steady states are not likely to occur [6]. A few specifically designed laboratory studies report multiplicity of steady states in tubular fixed bed catalytic reactors. They are listed in an extensive discussion of the problem by Schmitz [18] and mainly deal with CO oxidation on Pt/Al_2O_3. It is not clear, however, whether the multiplicity is caused by interfacial gradients, by intraparticle gradients, by both or by factors related to the phenomena occurring at the catalytic surface. These possibilities will be discussed next.

IV. EFFECT OF INTERFACIAL AND INTRAPARTICLE GRADIENTS

Temperature gradients inside a catalyst particle are very unlikely but there are many examples of reactions carried out on porous catalysts whereby intraparticle concentration gradients occur, leading to effectiveness factors lower than one. This has nothing to do with the flow conditions and is true in industrial reactors as well as in laboratory reactors. Since the activation energy for effective diffusion inside a catalyst particle is low it is clear that the parametric sensitivity will be reduced with respect to the situation of uniform concentration, implied in the analysis of Van Welsenaere & Froment. Applying their criteria to such a situation would evidently lead to very conservative values for the operating conditions.

Intraparticle concentration gradients also decrease the region in parameter space for which multiplicity can occur. Luss recalculated the reactor simulated by Liu and Amundson [viz. Fig. 7] and added internal concentration gradients. Multiplicity and the associated runaway were eliminated [19]. Even if there were also temperature gradients multiplicity would still be unlikely with realistic values of the transport and kinetic parameters, at least for reactions with a positive order. Multiplicity is possible, however, even with isothermal particles for the more general rate equations of the Hougen & Watson type i.e. accounting for the adsorption of the reacting components. Luss has recently reviewed a priori criteria for uniqueness [19].

Rajadhyaksha e.a. [20] and McGreavy & Adderley [21] considered situations in which an interfacial temperature gradient and an intraparticle concentration gradient occur. The reaction was of 1st order and irreversible.

The equations for a reactor in which interfacial and intraparticle gradients are accounted for [Model B.II in Froment's classification] were given already in Froment's contribution on "Fixed Bed Catalytic Reactors" in the present Proceedings [Equations (10)-(15)]. The use of the effectiveness factor η permits writing the equations in a substantially different and simpler form. Expressing the compositions in terms of partial pressures leads to :

$$u_s \frac{dp}{dz} = - \frac{M_m p_t e_B}{\rho_g} \eta r_A \qquad (7)$$

$$u_s \frac{dT}{dz} = \frac{(-\Delta H)\rho_B}{\rho_g c_p} \eta r_A - \frac{4U}{\rho_g c_p d_t}(T-T_w) \qquad (8)$$

$$\frac{k_g}{R'T} a_v (p-p_s^s) = \eta r_A \rho_B \qquad (9)$$

$$h_f a_v (T_s^S - T) = (-\Delta H)\eta r_A \rho_B \qquad (10)$$

with $r_A = A_o \exp(-\frac{E}{RT_s}) p_s$

The pellet is taken to be isothermal so that $T_s^S = T_s$. When there is no external concentration gradient $k_G a_v / R'T \rightarrow \infty$.

Rajadhyasksha e.a. followed the Van Welsenaere & Froment approach and came to the following equations for the maxima curve :

$$p_m = \frac{\frac{4U}{\rho_g c_p d_t}(T_s - T_w)}{\rho_B \eta A_o \exp(-\frac{E}{RT_s}) \left[\frac{(-\Delta H)}{\rho_g c_p} + \frac{4U}{\rho_g c_p d_t} \frac{(-\Delta H)}{h_f a_v}\right]} \qquad (11)$$

$$T_m = T_s - \eta A_o \left[\exp(-\frac{E}{RT_s})\right] \rho_B \frac{(-\Delta H)}{h_f a_v} p_m \qquad (12)$$

Intraparticle gradients shift the maxima curve upwards, while interfacial gradients tend to reduce the critical partial pressure. T_s appears as a parameter in this equation. When there are no intraparticle gradients $\eta=1$ and $(T_s)_M$ is given by the R.H.S. of (2). This value has to be substituted into (11) and (12) to yield the values of p_M and T_M. From these values the extrapolations proposed by Van Welsenaere & Froment lead to upper and lower limits for the inlet partial pressure of the reactant.

In the presence of significant intraparticle gradients $(T_s)_M$ cannot be obtained any longer by analytical differentiation of (11) with respect to T_m and an additional parameter θ has to be introduced : the temperature independent part of the Thiele modulus, which is a constant for the system

$$\Theta = \frac{d_p}{2}\sqrt{\frac{A_o \rho_B R'T}{D_e(1-\varepsilon)}} \qquad (13)$$

The parameter Θ is related to the Thiele modulus by the relation

$$\phi = \Theta\sqrt{\frac{RT}{E}\exp(-\frac{E}{RT})} \qquad (14)$$

The frequency factor A_o is evaluated at E/R. Increasing Θ means larger intraparticle gradients. The results of the numerical differentiation are represented in Fig.8.

The curve for the pseudo homogeneous situation analyzed by Van Welsenaere & Froment is also shown. At very low T_w all the curves coincide with that curve. The larger Θ i.e. the smaller the effective diffusivity D_e, the earlier the Θ-curves bifurcate from it. At high T_w all the curves merge again into an asymptotic curve.

From Fig. 8 or the R.H.S. of (2) Rajadhyaksha e.a. calculate for given kinetics, a wall temperature $T_w=615K$ and no intraparticle gradients a $(T_S)_M$ of 645K and from (11) and (12) : $p_M = 0.043$ atm and $T_M = 637.8K$. The temperature drop over the film is 8K and the critical inlet partial pressure 0.0523 atm. The value obtained by numerical back integration is 0.05 atm, while that calculated on the basis of the pseudo homogeneous model of section I is 0.066 atm. In another example intraparticle gradients are included. For a wall temperature of 615K again and a value of Θ of $1.93.10^6$ Fig. 8 leads

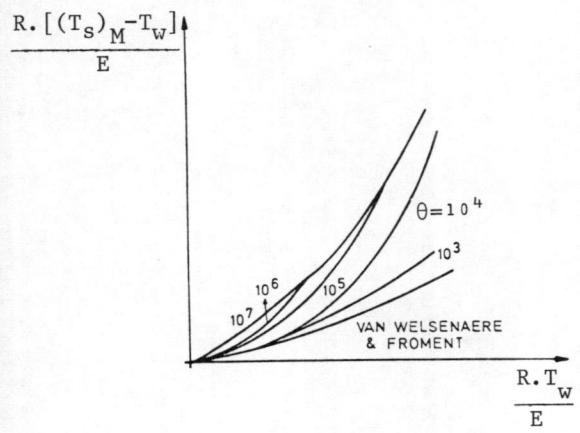

Fig.8. Relation between the solid temperature corresponding to the maximum of the maxima curve and the wall temperature. After [20].

to $(T_s)_M = 675.5 K$. The effectiveness factor is calculated to be 0.2165 at that temperature and $p_M = 0.1042$ atm. The critical partial pressure at the inlet is found to be 0.1064 atm. If the calculations were based upon a pseudo-homogeneous model a strongly conservative value of 0.033 atm would be obtained. These results illustrate the conclusions already derived from the equations (11) and (12) concerning the influence of external and internal gradients on runaway limits imposed because of parametric sensitivity.

For very high values of ηr_A both Δp and ΔT over the external film and internal gradients of partial pressure or concentration and temperature would occur. Then equations (11) and (12) are no longer valid and the curves of Fig. 8 have a different aspect, as shown in Fig. 9.

Clearly, for values of RT_w/E lower than 0.064 and θ lower than 10^5 two values of $(T_s)_M$ correspond to one value of T_w. That means that the maxima-curve $p_m - T_m$ will also have a minimum reasonably close to the maximum. By a reasoning analogous to that leading to Fig. 6 it then follows that for a given T_w and a given range of inlet partial pressures multiple steady states and therefore instabilities are possible. A diagram of McGreavy and Adderley in [21] is eloquent in illustrating ranges in which parametric sensitivity and instability occur with first order reactions. It is shown in Fig. 10. The particle is isothermal but both ΔT and Δp can occur in the external film

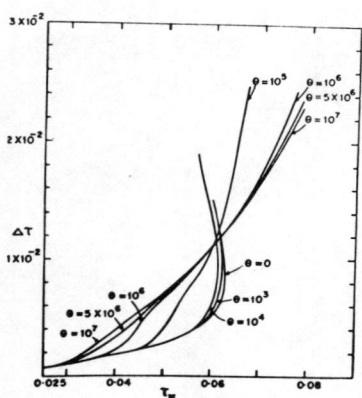

Fig.9. Relation between the solid temperature corresponding to the maximum of the maxima curve and the wall temperature for $A_o \rho_B R'T/k_g a_v = 10^6$.

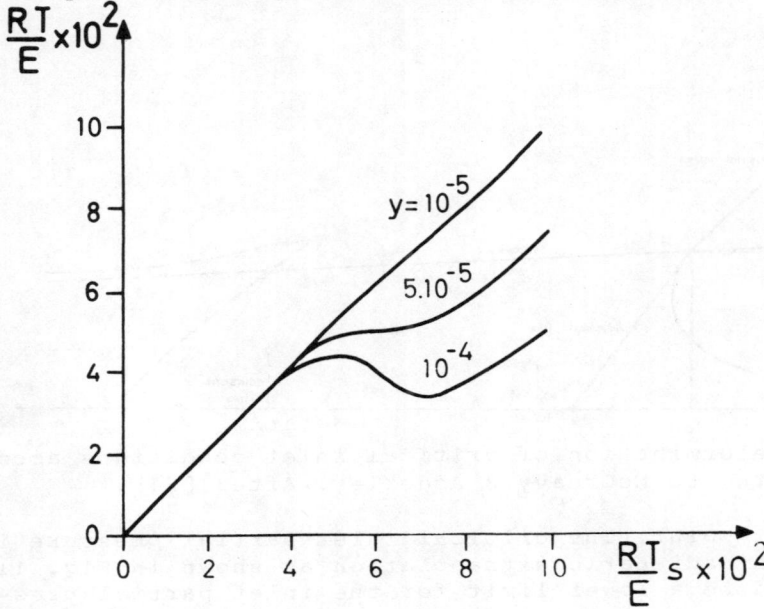

Fig. 10. Relation between fluid and solid temperatures for various values of $Y = \dfrac{(-\Delta H) D_e R}{d_p h_f E} C_{A_o}$

$$\theta = \dfrac{d_p}{2}\sqrt{\dfrac{A_o \rho_B R'T}{D_e(1-\varepsilon)}} = 10^6 \qquad \dfrac{d_p k_g}{D_e} = 500$$

For a given reaction the curve for $Y = 5.10^{-5}$, which has an inflection point, leads to an important increase of T_s for a relatively small change in T. This is runaway caused by parametric sensitivity. Only for much higher inlet concentrations ($Y=10^{-4}$) a hump develops in the curve. This will lead, for a certain range of T-values, to three possible intersections, therefore three possible T_s, the outer two of which are stable. The figure is quite analogous to Fig. 4 and again hysteresis can be observed. The runaway criterion derived by McGreavy & Adderley was based upon the rapid increase of T_s-T near runaway conditions. This enabled a runaway line to be drawn in the Y-T diagram, which is essentially a(T,p)-plane as used by Van Welsenaere & Froment. Any trajectory crossing the runaway line leads to runaway. A trajectory tangent to the runaway line and intersecting the maxima curve in this point of tangency may be considered as a critical trajectory. The point of intersection of the runaway line and the maxima curve is

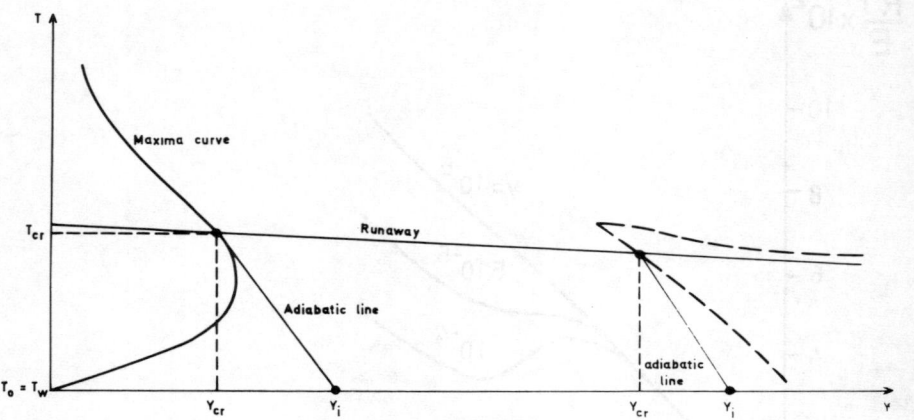

Fig.11. Determination of critical inlet conditions according to McGreavy & Adderley. After [21].

a critical point. The critical inlet partial pressure is obtained by adiabatic extrapolation as shown in Fig. 11. This would be a lower limit for the inlet partial pressure leading to runaway caused by parametric sensitivity. No upper limit was derived.

The criterion seems overly conservative, however, when compared with the maximum criterion applied by Rajadhyaksha e.a. to this heterogeneous situation.

McGreavy & Adderley also address themselves to the question which Y-value will lead to runaway by instability. From Fig. 10 it follows that the T-bounds on the non unique region correspond to $\frac{dT}{dt} = 0$. For first order reaction kinetics these bounds can be obtained analytically for any Y and given Θ and $\frac{d_p k_g}{D_e}$. The intersection of the runaway line with the bound of the non unique region in the Y-T diagram permits determining the critical inlet partial pressure as shown in Fig. 11.

Hegedus e.a. [22] observed multiple steady states in an isothermal tubular reactor for the catalytic oxidation of CO over Pt/Al_2O_3 and convincingly modeled the instability in terms of the interaction between rates of surface reaction and internal mass transfer.

Finally, for extremely fast reactions the global rate is determined by the mass transfer over the external film. The temperature rise is nearly adiabatic. The maxima curve then becomes a straight line and there is no true parametric sensitivity any more. Also there is no corresponding instability. Instability may then arise from axial mixing effects or from an effect not yet discussed : the nature of the catalytic reaction.

V. KINETICALLY INDUCED INSTABILITY

Evidently, this type of instability is not limited to extremely rapid reactions only. It has been proposed by Wicke and co-workers [23] to explain experimental observations in a laboratory tubular reactor for CO-oxidation on a Pt/Al_2O_3 catalyst. A rate equation of the type $r_{CO} = \dfrac{k\, p_{CO}}{1+K p_{CO}}$ is known to be valid for this reaction. The parameters in this equation are such that r_{CO} goes through a maximum as p_{CO} is increased, whereas the rate of adsorption decreases linearly with the adsorbed CO-concentration, therefore with p_{CO}. The situation is represented in Fig. 12.

In addition, the system was also found to oscillate continuously between the upper and lower steady state, under isothermal conditions and without any external forcing. Oscillation requires an additional mechanism on the catalyst surface : Hugo and Jakubith suggested a shift between two forms of chemisorbed CO [24], Eigenberger a slow variation in the surface concentration of some unreactive chemisorbed oxygen [25]. Oscillations in catalytic reactions is a very hot subject nowadays. It has been reviewed by Sheintuch & Schmitz [26] and by Wicke [27].

VI. ADIABATIC REACTORS WITH FEED-EFFLUENT HEAT EXCHANGE

So far the discussion did not consider tubular reactors with external feedback of heat. Yet, reactors of this type are quite common in industry : they are

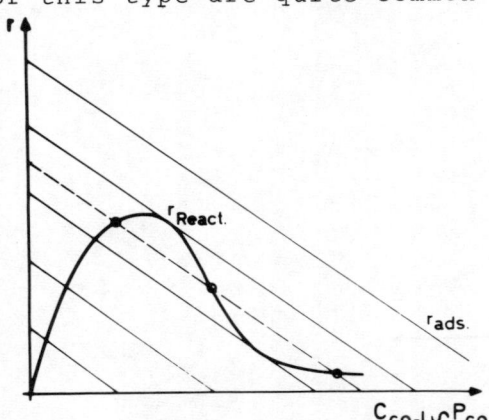

Fig.12. Rate of surface reaction of CO and rate of adsorption as a function of adsorbed CO-concentration or gas phase CO-partial pressure. After [23].

encountered e.g. in phthalic anhydride-, ammonia- and methanol-synthesis. Autothermal processes were dealt with in the paper on "Fixed Bed Catalytic Reactors" in these Proceedings and the model equations are given there.

The feed of an adiabatic reactor may be preheated by the effluent or by continuous heat exchange with the reacting gases. A feed back of heat can of course induce runaway by parametric sensitivity, but also by multiplicity of steady states, even when none of the phenomena mentioned so far : axial mixing, transport effects or catalyst surface effects are operative.

In an adiabatic reactor the following relation exists between the conversion and the temperature variation :

$$\Delta x = \frac{m\, c_p}{F_{A_o}(-\Delta H)} \Delta T = \lambda \Delta T = \lambda [T(Z)-T(0)] \qquad (15)$$

Let the reaction $A \rightleftarrows B$ be reversible and e.g. first order in both directions so that

$$r_A = A_o p_t \left[\exp\left(-\frac{E}{RT}\right) \right] \left[(1-x) - \frac{x}{K} \right]$$

Then, by integrating the continuity equation over the adiabatic bed

$$x(Z)-x(0) = f[T(0)] \qquad (16)$$

If the feed enters the external heat exchanger at T_i and exchanges heat with the effluent it can easily be shown

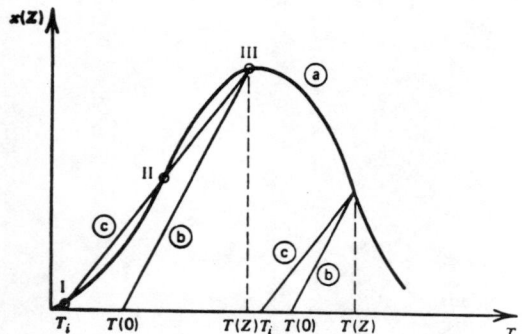

Fig.13. Possible steady-state operating points in a system adiabatic reactor + effluent/feed heat exchanger. After 29 . From [6].

[6] that

$$\Delta x = \frac{\lambda [T(Z) - T_i]}{1 + \frac{U\pi d_t}{mc_p}L} \quad (17)$$

The situation is represented in Fig. 13. Equation (16) yields the bell shaped curve (a), equation (15) the straight line (b). With the addition of an external heat exchanger (17), represented by line (c), has to be satisfied. For the case situated in the right part under the bell shaped curve only one intersection, only one steady state leading to an effluent temperature T(Z) is possible. For a given configuration and flow rate a certain range of feed temperatures T_i will lead to multiplicity of steady states. Of the three steady states only the two external ones will be stable. Therefore, in that range slight modifications of T_i may carry the reactor from the lower steady state and low T(Z) to the high steady state and high T(Z) : runaway is caused by unstable operating conditions. A simulated example for a two-bed adiabatic ammonia synthesis reactor with intermediate quench and external heat exchanger has been given by Shah [28]. Two stable steady states are found as shown in Fig. 14. Notice also that there is no solution when T_i is too low : autothermal operation then becomes impossible. It also turns out that, because of the bell shape of curve (a) the blow-off conditions are close to the optimal operating conditions, as can be deduced from Fig. 13 already.

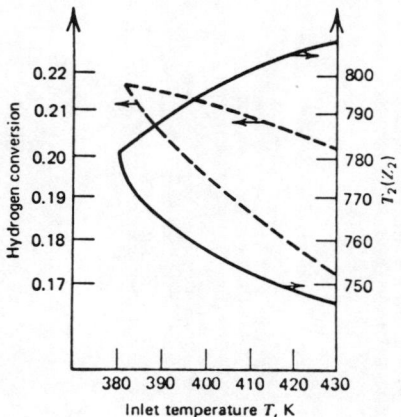

Fig.14. Hydrogen conversion and effluent temperature in a 2 stage adiabatic ammonia synthesis reactor. From [28].

VII. MULTITUBULAR REACTORS WITH INTERNAL FEED-EFFLUENT HEAT EXCHANGE

An analogous situation is encountered with multitubular reactors with heat exchange between feed and reacting gases [29,30]. The simulation of such a reactor has been performed by Baddour e.a. [31] for an ammonia synthesis reactor and by Ampaya & Rinker [32,33] for a multitubular water-gas shift reactor. Fig. 15 is taken from the latter work and illustrates the multiplicity that can be encountered with the autothermal version.

Inoue analytically derived stability criteria for a multitubular autothermal reactor based on the pseudo-homogeneous ideal model A.I and using the Barkelew approach [34]. For stable operation with a first order irreversible reaction $\frac{N}{\beta\gamma}$ has to exceed $\beta\gamma-1$ and for a n-th order reaction, $\beta\gamma-n$. The lower $\beta\gamma$ the narrower the non unique region. When $\beta\gamma<8$ the system is always stable. Equally simple criteria are given for n-th order reversible reactions.

VIII. HOT SPOTS INDUCED BY FLOW MALDISTRIBUTION OR OBSTRUCTIONS.

In practice local hot spots may develop in packed bed reactors because of flow maldistribution caused e.g. by physical obstructions or catalyst fines. Very little has been published on this matter. An excellent analysis of such a situation is presented by Jaffe [35]. It deals

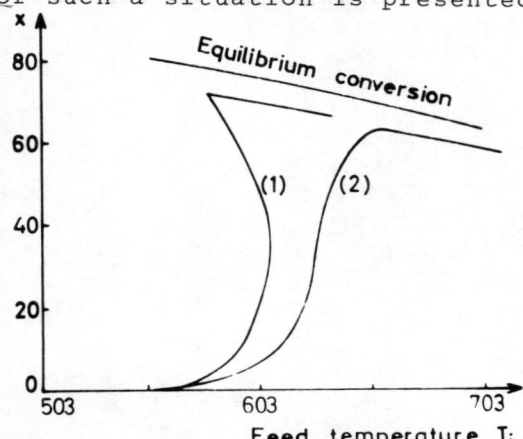

Fig.15. Comparison of the conversion as a function of the feed temperature for a multitubular reactor with internal heat exchanger (1) and an adiabatic bed without feed preheat (2). Parameters chosen from water gas-shift reaction. From [33].

with a commercial hydrogenation reactor with two fluid phases. A local hot spot was detected in one stage of a multi-bed adiabatic reactor, although normally in an adiabatic bed for an exothermic reaction only monotonically increasing temperature profiles are possible. The local thermocouple sheath did not allow measuring the extent of the hot spot in the bed. The simulation led to an estimated T_{max} of 120°C. The radius of the affected region must have been 8,5 cm. Questions that can be answered by such simulations are : under what conditions would the local hot spot propagate to the full size of the bed and cause reactor runaway ? How would the reactor internals have to be modified to avoid regions of low flow velocities leading to such local hot spots ?

IX. CONCLUSION

When designing fixed bed reactors for strongly exothermic reactions with a high activation energy special care has to be taken to provide the means of keeping the hot spot under control. This can be achieved by the choice of the operating conditions or by the judicious lay-out of the heat exchanger. Adequate and accurate models are required to avoid confusion between two possible causes of runaway : parametric sensitivity and multiplicity of steady states. Apparently, in industrial tubular reactors without feedback only runaway by parametric sensitivity has been observed so far. True instabilities caused by non ideal flow or interfacial gradients are not very likely in industrial reactors. Kinetically induced instabilities could arise in a rather narrow range of operating conditions with some special reactions. When internal or external feedback of heat is applied multiplicity of steady states is not uncommon. Transient simulation of start up procedures and perturbations may be advisable in such cases.

REFERENCES

[1] Van Welsenaere R.J. and Froment G.F., Chem.Eng.Sci. 25-1503-1970.
[2] Barkelew C.R., Chem.Eng.Progr., Symp.Ser. 55(25)-38-1955.
[3] Dente M, Collina A., Chim.y Industria, 46-752-1964.
[4] Hlavacek V., Marek M., John T.M., Coll.Czechoslov. Chem.Comm., 34, 3868, 1969.
[5] Froment G.F., Chem.Ing.Techn., 46-374-1974.
[6] Froment G.F. and Bischoff K.B., "Chemical Reactor Analysis and Design" J. Wiley, N.Y., 1979.
[7] Agnew J.B. and Potter O.E., Trans.Instn.Chem.Engrs., 44-T216-1966.
[8] Froment G.F., Adv.Chem.Ser., 109-1-1972.
[9] Froment G.F., Proc.5th Eur./2nd Int.Symp.Chem.React. Engng., Elsevier, Amsterdam 1972.
[10] Luss D. and Amundson N.R., Chem.Eng.Sci., 22-253-1967.
[11] Luss D., Chem.Eng.Sci., 23-1249-1968.
[12] Hlavacek V. and Hofmann H., Chem.Eng.Sci., 25-173-1970 ; 25-187-1970.
[13] Wicke E., Acta Techn.Chim., Acad.Naz.Lincei,Varese, Sept. 1960.
[14] Liu S.L. and Amundson N.R., Ind.Eng.Chem.Fund. 1-200-1962 ; 2-183-1963.
[15] Cardoso M.A. and Luss D., Chem.Eng.Sci., 24-1699-1969.
[16] Van den Bosch B. and Luss D., Chem.Eng.Sci., 32-203-1977.
[17] Eigenberger G., Chem.Eng.Sci., 27-1909-1972 ; 27-1917-1972.
[18] Schmitz R.A., Chemical Reaction Engineering Reviews, Ed. H.M. Hulburt, Adv. in Chem.Ser. 148, Am.Chem. Soc., Washington D.C., 1975.
[19] Luss D., Chemical Reaction Engineering, Proc.4th Int.Symp.Chem.React.Eng., p.487, Dechema, Heidelburg 1976.
[20] Rajadhyaksha, K. Vasudeva and Doraiswamy L.K., Chem.Eng.Sci., 30-1399-1975.
[21] McGreavy C. and Adderley C., Chemical Reaction Engineering II, Adv.Chem.Ser. 133-1974.
[22] Hegedus L., Oh, S.H., Baron K., A.I.Ch.E. Journal, 23-632-1977.
[23] Bensch H., Fieguth P. and Wicke E., Chem.Ing.Techn. 44-445-1972.
[24] Hugo P. and Jakubith M., Chem.Ing.Techn., 44-383-1972.
[25] Eigenberger G., Chem.React.Engng., Proc.4th Int. Symp., p.290, Dechema, Heidelberg 1976.

[26] Sheintuch M. and Schmitz R.A., Cat.Rev.Sci.Eng., 1977.
[27] Wicke E., Chem.Ing.Techn., 46-365-1974.
[28] Shah M.J., Ind.Eng.Chem., 59-No.1-72-1967.
[29] Van Heerden C., Ind.Eng.Chem., 45-1242-1953.
[30] Liljenroth F.G., Chem.Met.Eng., 19-287-1918.
[31] Baddour R.E., Brian P.L.T., Logeais B.A. and Eymery J.P., Chem.Eng.Sci.,20-281-1965.
[32] Ampaya J.P. and Rinker R.G., Ind.Eng.Chem.Proc.Des. Devpt.,16-65-1977.
[33] Ampaya J.P. and Rinker R.G., Chem.Eng.Sci., 32-1327-1977.
[34] Inoue H., J.Chem.Engng. Japan, 11-40-1978.
[35] Jaffe S.B., Ind.Eng.Chem.Proc.Des.&Devpt. 15-410-1976.

NOTATION

A_o frequency factor (n=1 for first order; n=2 for pseudo first order)   kmol/kg cat s $(\frac{N}{m^2})^n$

a_v external particle surface area per unit reactor volume m_p^2/m_r^3

$B = \frac{(-\Delta H)\rho_B p_B^o}{c_p \rho_g}$, a group in (1) $\frac{kg\ cat.atm.m_f^3 K}{kmol\ m_r^3}$

$C = \frac{4U}{c_p d_t \rho_g}$, a group in (1) $m_f^3/m_r^3 s$

c concentration $kmol/m_f^3$

c_{A_o} inlet concentration of reactant A $kmol/m_f^3$

c_p specific heat of fluid   kJ/kg K

D_e effective diffusivity inside particle $m_f^3/m_p \cdot s$

D_{ea} effective diffusivity in axial direction $m_f^3/m_r s$

d_p particle diameter m_p

d_t tube diameter m_r

E activation energy   kJ/kmol

F_{A_o} inlet molar flow rate of reactant A   kmol/s

h_f heat transfer coefficient fluid/particle $kJ/m_p^2 s\ K$

$(-\Delta H)$ heat of reaction   kJ/Kmol

K	equilibrium constant based on conversions	
k	rate coefficient	$kmol/kg\ cat.s.N/m^2$
k_o	value of rate coefficient at inlet temperature	
k_g	mass transfer coefficient	$m_f^3/m_p^2 s$
M_m	average molecular weight of fluid	$kg/kmol$
m	mass flow rate	kg/s
$N = \dfrac{4U}{d_t c_p M_m} \dfrac{1}{\rho_B k p_B^o p_t}$,	a group used in Fig. 3	
p	partial pressure	N/m^2
p_m	partial pressure of reactant corresponding to maximum in temperature profile in reactor	N/m^2
p_t	total pressure	N/m^2
R	gas constant	$kJ/kmol\ K$
R'	gas constant	$m_f^3\ atm/kmol\ K$
r_A	rate of reaction of component A per unit catalyst mass	$kmol/kg\ cat\ s$
T	fluid temperature	K
T_{ad}	adiabatic temperature rise at exit of reactor	K
T_m	fluid temperature in the maximum of the temperature profile	K
T_o	temperature at reactor inlet	K
T_s	solid temperature	K
T_w	wall temperature	K
U	overall heat transfer coefficient bed/coolant	$kJ/m^2 s\ K$
u_i	superstitial fluid velocity	m_r/s
u_s	superficial fluid velocity	$m_f^3/m_r^2 s$
x	conversion	
$Y = \dfrac{(-\Delta H) D_e R C_{A_o}}{d_p h_f E}$,	a dimensionless group in Fig. 10	
Z	total length of reactor	m
z	axial coordinate in reactor	m

GREEK LETTERS

$\beta = \dfrac{(-\Delta H) p_o}{M_m p_t c_p T_o}$ dimensionless adiabatic temperature rise used in Fig. 3

$\gamma = \dfrac{E}{RT}$ dimensionless activation energy used in Fig. 3

ε void fraction of the bed m_f^3/m_r^3

η effectiveness factor for solid catalyst

$\lambda = m\, c_p / F_{A_o}(-\Delta H)$, a group in (15)

λ_{ea} effective axial diffusivity of bed

ρ_B bulk density of bed   kg cat/m_r^3

ρ_g density of fluid kg_f/m_f^3

$\theta = \dfrac{d_p}{2}\sqrt{\dfrac{A_o \rho_B R'T}{D_e(1-\varepsilon)}}$, a group in (13)

ϕ Thiele modulus

Part VI

COAL CONVERSION

CATALYSIS IN COAL GASIFICATION

James R. Katzer

Center for Catalytic Science and Technology
Department of Chemical Engineering
University of Delaware
Newark, Delaware 19711

INTRODUCTION AND THERMODYNAMICS

As early as 1670 it was known that gas could be produced by heating coal in a retort. It was not until about 100 years later that the manufactured gas industry appeared in England where gas produced by coal pyrolysis was used for lighting. In 1855 Bunsen invented the atmospheric gas burner, opening the potential of the gas heating market. This inspired the development of the cyclic carburetted water gas process for gas product in 1875. In this process, hot coke is reacted with steam to produce water gas (H_2 + CO). The gas was enriched by cracking oil in a downstream vessel filled with hot checker-brick. Heat for the process was supplied by intermittently interrupting the flow of steam through the coke and introducing air to burn a portion of the coke. Gas from by-product coke ovens supplemented the supply of carburetted water gas.

Very little development beyond the carburetted water gas process occurred in the United States due to discoveries of vast supplies of natural gas and oil. In Europe, however, coal was the major available fuel and incentive existed to develop improved coal-to-gas processes. Development of the Winkler fluid-bed process began in 1921, and in 1926 of a large-scale air-blown gasifier was built at Leuna. By 1929, there were five Winkler generators at that location producing a total of up to 240 million SCF/day of producer gas from steam and air. In 1960, 16 Winkler plants had been constructed; six of these plants are in Germany, four in Japan, two in Spain, and others in Czechoslovakia, Yugoslavia, Turkey and India. All are oxygen-blown to produce water gas for fuel or synthesis gas for chemical manufacture. The first Lurgi moving-bed gasifier was built in 1936. By 1966 there were 13 Lurgi plants in Europe, England, Australia, Pakistan, and Korea. In the United States no plants were built, and little process development work was done well into the 1960s.

Coal gasification involves typically the molecular species H_2O, H_2, CO, CO_2 and CH_4, and their reactions with solid carbon. There are five molecular species and two elemental material balances, and therefore, three independent stoichiometric equations can be written. The choice is arbitrary except that each species must be represented and no redundancy must occur. Table 1 gives three appropriate equations and the associated thermodynamic information: (a) enthalpy change of reaction $(-\Delta H_R)$ at 1200 K (927°C), (b) temperature above which the equilibrium constant exceeds 1, and (c) at that temperature, the percent increase in K_p per percent increase in temperature, i. e., the temperature sensitivity of the equilibrium constant.

The major gasification reaction of interest is the reaction between H_2O and carbon to produce CO and H_2 (Reaction 1, Table 1). This reaction is highly endothermic requiring large amounts of thermal energy input into the reaction system. The H_2-carbon reaction (Reaction 3, Table 1) is of potentially greater importance since it produces CH_4 directly. However, this reaction is too slow at temperatures where it is favored by equilibrium unless catalyzed (Table 2). At higher temperatures where $C-H_2$ reaction rates are rapid enough, the reaction is equilibrium limited. In fact the $C-H_2O$ reactions and the $C-H_2$ reaction are almost mutually exclusive since high temperature favors the former and low temperature the latter. For most gasification situations methane is a desired product if not the most desired product and process economics are highly favored by methane formation in the gasification step. Much gasification process development has focused on enhancing methane formation.

Thus, the overall gasification process is typically highly endothermic requiring the addition of heat. This can be achieved by adding oxygen and taking advantage of the highly exothermic oxidation reaction,

$$C + \tfrac{1}{2} O_2 \rightarrow CO \qquad \Delta H_R = 26,637 \text{ cal/mole} \qquad (4)$$

At all temperatures and compositions of interest, the thermodynamic tendency of Reaction (4) is to go all the way to the right; K_4 is very large. Thus, equilibrium is approached and oxygen can be considered absent from the products. Because of the high relative rate of this reaction it occurs preferentially in the presence of other gases.

Table 1
Thermodynamic Functions of Gasification Reactions

Reaction	$-\Delta H_R$ at 1200 K, cal/g·mole	Temperature at which $K_p > 1$ [a]	$\dfrac{d\ln K_p}{d\ln T} = \dfrac{+\Delta H_R}{RT}$
(1) $C + H_2O_{(v)} \rightleftarrows CO + H_2$	$-32{,}457$	above 947 K (675°C)	17.2
(2) $CO + H_2O_{(v)} \rightleftarrows CO_2 + H_2$	$+7{,}838$	below 1100 K (827°C)	-3.7
(3) $C + 2H_2 \rightleftarrows CH_4$	$+21{,}854$	below 819 K (546°C)	-12.2

[a] K_p is the equilibrium constant in terms of partial pressures of reactants and products.
(Hottel and Howard, 1971).

Table 2
Approximate Relative Rates of the Gas-Carbon Reactions at 800°C and 0.1 Atmos Pressure

Reaction	Relative Rate
$C-O$	1×10^5
$C-H_2O$	3
$C-CO_2$	1
$C-H_2$	3×10^{-3}

Walker, Shelef and Anderson, 1968.

Equilibrium conversion to CH_4 is favored by high H/O ratios and by increased pressure. The H_2/CO ratios are also enhanced by high H/O ratios. Thus introduction of heat by Reaction (4) is undesirable and is to be minimized. The advantage of increased pressure in addition to higher rates is that it allows substantial conversion to methane at higher temperatures, which is highly desirable because of the low rates at lower temperatures. At sufficiently high temperature, independent of the pressure or H to O ratio the system contains only CO and H_2. The overall reaction tends to be highly endothermic:

$$C + H_2O_v \rightarrow CO + H_2 \qquad \Delta H_R = -32{,}457 \text{ cal/mole.}$$

For temperatures below about $500^\circ C$ the overall reaction tends to be almost thermally neutral

$$C + H_2O \rightarrow \tfrac{1}{2} CH_4 + \tfrac{1}{2} CO_2 \qquad \Delta H_R = -1345 \text{ cal/mole} \qquad (5)$$

This product composition is ideal having a medium heating value (400-800 Btu/ft^3), as is the thermal balance on the system, but unfortunately at this low temperature the rate of reaction is too slow to be of practical interest.

At 955°C where the rate is acceptably high, the equilibrium gas composition for H to O equal 2 and at 68 atm is about 9% CH_4, 9% CO_2, 35% CO, 35% H_2 and 12% H_2O; and the overall reaction is quite highly endothermic. The gas is low in heating value (250 Btu/ft^3) and requires upgrading. These compositions are not necessarily achievable because of kinetic limitations (Table 2); they only represent the potential. These numbers are for graphite, a very stable form of carbon.

Since the water gas shift reaction is essentially at equilibrium under gasification conditions, the hydrogen concentration is a function of the amount of excess steam present.

NON-CATALYTIC GASIFICATION REACTIONS

Gasification reactions include all reactions leading to gaseous products and includes reaction of O_2, CO_2, H_2 and H_2O with coal. The essential first feature involves chemisorption of the gas on the carbon surface. On most carbons only a small fraction (<10%) of the total surface for amorphous carbons and charcoals chemisorb gases. The fraction of coal that is reactive provided gases can get to it and its reactivity is probably very high for coal as it is undergoing pyrolysis but then decreases with conversion to semichar and char. It is expected that the fraction of char surface that is reactive is small. The reaction kinetics and mechanisms of the $C-O_2$, $C-CO_2$,

Catalysis in coal gasification

C-H$_2$O and C-H$_2$ reactions will be considered in that order below.

<u>Carbon-Dioxygen Reaction</u>: Even at low temperatures dioxygen adsorbs dissociatively on coal, charcoal and activated carbon. Most of it desorbs not as O$_2$ but as CO with some CO$_2$. Under conditions where oxygen reaches the carbon surface CO is the main product (Walker et al., 1968).

$$2 C_f + O_{2(g)} \rightarrow 2 C(O) \tag{6}$$

$$C(O) \rightarrow CO_{(g)} \tag{7}$$

where C_f represents an active or free carbon site (a small fraction of the total carbon atoms) and $C(O)$ represents chemisorbed oxygen. Some CO$_2$ also appears to be formed by

$$C_f + O_{2(g)} \rightarrow C(O_2) \tag{8}$$

$$C(O_2) \rightarrow CO_{2(g)} \tag{9}$$

where $C(O_2)$ represents chemisorbed dioxygen. Reactive carbon sites are probably on the edges of aromatic lamellae rather than on the basal planes which are very unreactive.

Coal oxidation occurs by what appears to be a common intermediate for most gasification reactions. This intermediate is a carbon oxide in which the oxygen atom is bound to a carbon atom which is yet attached to other carbon atoms of the structure. This stable intermediate is confirmed by the appearance of carbonyl

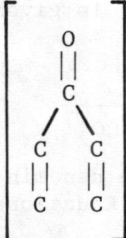

stretching frequencies when dioxygen adsorbs on coal at temperatures below 200°C. At sufficiently high temperatures these surface carbon groups decompose giving CO and generating a fresh active site. This active site (a) can chemisorb another oxygen atom or (b) can thermally aneal out.

At low temperatures the rate of reaction is zero order in dioxygen pressure, the surface is covered with oxide species and the rate-controlling step is the thermal decomposition of these species. At high temperatures the reaction is first order in dioxygen partial pressure suggesting that the surface is only partially covered with oxide species. The resultant rate equation for this type of behavior is

$$r = \frac{kKP_{O_2}}{(1 + KP_{O_2})} \tag{10}$$

Apparent activation energies are 47 to 58 kcal/mole. Under more characteristic oxidation conditions the rate involves diffusional limitations in the char particle or mass transfer to the external surface.

Carbon-Carbon Dioxide Reactions: Carbon dioxide reacts with carbon to give CO

$$CO_{2(g)} + C_f \rightarrow 2CO_{(g)} \tag{11}$$

The reaction occurs via the network (Walker et al., 1968).

$$CO_{2(g)} + C_f \rightarrow C(O) + CO_{(g)} \tag{12}$$

$$C(O) \rightarrow CO_{(g)} \tag{13}$$

Reaction (12) occurs on char at 500 to 600°C with $CO_{(g)}$ being formed but the surface carbon oxide is typically decomposed [Reaction (13)] only at considerably higher temperatures. The surface oxide species formed from $C-CO_2$ reaction is apparently the same as results from interaction of dioxygen with the surface of carbon and the reaction network is surprisingly similar.

The rate of gasification by CO_2 is given by

$$r = \frac{k_1 \, P_{CO_2}}{(1 + K_{CO}P_{CO} + K_{CO_2}P_{CO_2})} \tag{14}$$

The presence of a term for CO in the denominator suggests that Equation (15) must be included with Equations (11)-(13)

$$CO_{(g)} + C_f \rightleftarrows C(CO)$$

This suggests that $CO_{(g)}$ competes for active sites on the carbon surface reducing the number of sites available for CO_2 chemisorption.

Activation energies are typically between 73 and 97 kcal/mole.

Carbon-Steam Reaction: The reaction mechanism for the reaction of water with carbon is (Walker et al., 1968):

$$C + H_2O_{(g)} \rightleftarrows C(H_2O) \tag{16}$$

$$C(H_2O) \rightarrow C(O) + H_2 \tag{17}$$

$$C(O) \rightarrow CO_{(g)} \tag{18}$$

$$C_f + H_2 \rightleftarrows C(H_2) \tag{19}$$

It has been proposed that the chemisorption step involves dissociation of water at the carbon surface into a hydrogen atom and a hydroxyl radical which adsorbs on adjacent carbon sites:

$$2C_f + H_2O_{(g)} \rightarrow C(H) + C(OH) \tag{20}$$

$$C(H) + C(OH) \rightarrow C(O) + C(H_2) \tag{21}$$

$$C(O) \rightarrow CO_{(g)} \tag{22}$$

$$C(H_2) \rightleftarrows H_{2(g)} + C_f \tag{23}$$

The operative mechanism is not clear, but there are intrinsically good reasons to favor the latter. The dissociation of water normally proceeds with the formation of hydroxyl species which are extremely active oxidizing agents. This mechanism will be seen in catalysis of gasification and of the water gas reaction. Again the carbon oxide intermediate is involved. Hydrogen atoms readily diffuse across carbon at typical reaction temperatures (Robell et al., 1964).

The rate equation for the $C-H_2O$ reaction is (Long and Sykes, 1950):

$$r = \frac{K P_{H_2O}}{(1 + K_{H_2}P_{H_2} + K_{H_2O}P_{H_2O})} \tag{24}$$

at temperature and pressure characteristic of gasification.

Activation energies vary from 60 to 80 kcal/mole. The rates at 800°C are about three times those for the $C-CO_2$ reaction (Table 2).

Carbon-Hydrogen Reaction: The carbon-dihydrogen reaction is slowest of the gasification reactions (Table 2) and must involve an entirely different mechanism. It is often referred to as hydrogasification. Hydrogen must attach the sides of aromatic lamellae since the basal planes are unreactive. Successive addition of two hydrogen molecules break the double bond forming methyl groups which are finally hydrogenated off as methane.

The kinetic expression which fits the C-H reaction rate data over the range of one to 30 atmos and 1500 to 1700°F is

$$r_{CH_4} = \frac{k\, P_{H_2}^2}{(1 + K_{H_2} P_{H_2})} \tag{28}$$

Second-order dependence on hydrogen partial pressure in coal char hydrogasification in the slow rate region has been reported. At low hydrogen pressures the rate of hydrogenation to methane has also been reported to be first order in dihydrogen pressure and one-half order in hydrogen atom concentration.

The theoretical order in dihydrogen depends on which reaction step is rate limiting. Second order would be expected if the first reaction step [Reaction (25)] were in equilibrium as might be expected at high hydrogen pressure. The rate limiting step may change with decreasing pressure leading to lower order dependencies on dihydrogen pressure.

The activation energy varies between 10 and 50 kcal/mole with degree of conversion and degree of adsorption of dihydrogen. Table 3 summarizes reported activation energies for the gasification reactions.

<u>Hydrogasification of Coal and Char</u>: In the case of coal and char which has not been pyrolyzed at temperatures below 600°C, simultaneous pyrolysis and hydrogasification reactions occur as the char is heated to higher temperatures. These hydrogasification reactions are slow at low temperatures but are quite exothermic. Little reaction occurs with hydrogen below 600°C.

Pyrolytic of bond breaking reactions produce highly active free radical sites for gasification reactions to preceed at. Thus during the stages of coal or char heating the carbon exhibits high reactivity to reactive gases such as CO_2, H_2O and H_2 leading to high gasification rates as compared to thermally annealed char or after significant char conversion. Observed increases in activation energy with conversion are consistent with this decreasing reactivity of the reacting coal char.

This variation in reactivity suggests that hydrogasification of char should be more appropriately modeled by an expression involving a variation in activation energy. Anthony <u>et al</u>. (1976) have modeled hydrogasification using the approach they applied to coal pyrolysis. They defined the fraction of coal converted to liquids and gases in an inert atmosphere as nonreactive volatiles and that converted in the presence of dihydrogen in excess of the nonreactive volatiles as reactive volatiles. The

Table 3

Activation Energies for Carbon Gasification Reactions

Reaction	Activation energy,[a] kcal/mole
$C + CO_2 \rightarrow 2CO$	86
$C + H_2O \rightarrow CO + H_2$	~80
$C + \tfrac{1}{2}O_2 \rightarrow CO$	50-60
$C + 2H_2 \rightarrow CH_4$	10-50[b]
$C + N_2O \rightarrow CO + N_2$	40-50

[a] Average values reported estimated to be free of diffusional limitations.

[b] For fast-rate carbon (from char) the rate is largely controlled by pyrolysis reactions having an average activation energy of about 56 kcal/mole, most studies on slow carbon give values between 30 and 35 kcal/mole.

mean activation energy was 54.8 kcal/mole with a standard deviation (Gaussian) of σ = 17.2 kcal/mole. This suggests that during the rapid-rate hydrogasification region the rate-limiting step is the thermal decomposition of coal giving volatiles and active sites for reaction with hydrogen. This is consistent with the role of pyrolysis in coal liquefaction by solvent extraction.

The reaction with annealed coal char as with charcoal and graphite probably occurs by the mechanism discussed earlier and is considerably slower.

Reaction of Char with CO_2 and H_2O: The reactivity of char with CO_2 and air decreases with the rank of the parent coal from which it was derived. Interestingly enough the reactivity in air and in CO_2 correlate well. Some of this reactivity difference was due to the presence of diffusional limitations in the char particles. The reactivity correlates well with Ca and with Mg content in the char but not with Na or K which, however, does not vary much. The mineral matter already in coal clearly catalyzes gasification reactions although quantitative understanding of this effect is not yet available. Addition of certain metal salts catalyzes the gasification of char. This is considered further below.

Reaction of low-temperature coal char with water is not as dependent on conversion as with hydrogasification. Rates of gasification with H_2 are about 10 times that with H_2O during the rapid-rate period, but in the slow-rate period the rate of char gasification by H_2O is at least 10-fold higher than that by H_2.

The activation for the steam-char reaction varies from 30 to 60 kcal/mole. Activation energies at 10 atmos range from 30 kcal/mole for lignite to 50 kcal/mole for reactive char. Char has a higher rate per unit surface area than graphite where the total surface area of char and graphite was measured by BET. The higher specific rate for the char is probably due to more sites per unit area because of the smaller size of the lamellae in the char, i.e., more edges, than in the graphite.

CATALYTIC GASIFICATION REACTIONS

Many attempts at the application of catalyst to coal gasification have been made to achieve sufficiently high rates of reaction at reduced temperature so that the more ideal product distribution at these lower temperatures can be achieved. Marson and Cobb (1926) and Von der Hoeven (1945) have written reviews of early work in this area. Walker, Shelef and Anderson (1968) review recent basic studies of catalysis in carbon gasification.

The minerals already present in coal undoubtedly contribute catalytically to coal gasification. However, quantitative understanding of such effects does not exist. If greater understanding did exist, it should be possible to apply appropriate materials to the coal before gasification or possibly choose coals for gasification processes based on inorganic constituents which could lead to improved gasification. In this vein, studies of how catalytic materials enhance gasification are important to improving gasification processes.

Transition Metal Catalysts: The role that catalysts play in the gasification mechanism is not well established and work on model systems is desirable to enhance our understanding. Below we will proceed from more basic work to coal-applied catalysts.

Rates of gasification with H_2 have been found to vary significantly with Pt content and by orders of magnitude with carbon type [Rewick et al. (1974)]. Rates were not normalized to unit carbon or metal surface area. Thus a low rate for graphite (Graphon) was probably mainly due to low surface area.

Catalysis in coal gasification

At higher temperatures the onset of diffusional limitations was observed for H_2 gasification catalyzed by Pt; however, the steam-carbon reaction did not exhibit such behavior for similar rates, and different metals showed different behavior for the H_2-carbon reaction, indicating that other factors may also play a role. Methane was the only product formed for H_2 gasification. The overall activation energy was 55 kcal/mole for all carbons, suggesting that the same rate-limiting step was operative for all carbons. The rate depended on the hydrogen partial pressure to the one-half power, consistent with a process involving the dissociative adsorption of H_2 on the metal. The similarity of the activation energy to the H-H bond dissociation energy suggests that this is the rate-limiting step. The rates of methane formation per unit metal surface area were about fourfold higher on more highly dispersed Pt than for larger Pt crystallites suggesting that intimacy of contact between metal and carbon is an important parameter. The mechanism of catalyzed hydrogasification may be written as:

$$H_2 + 2M \rightleftarrows 2M\text{-}H \qquad (29)$$

$$M\text{-}H + C \rightleftarrows M + C\text{-}H \qquad (30)$$

$$C\text{-}H + M\text{-}H \rightarrow M + C\text{-}H_2 \qquad (31)$$

$$\text{etc.} \quad M\text{-}H + C\text{-}H_3 \rightarrow CH_4 + M \qquad (32)$$

The hydrogen atoms may actually diffuse some distance across the surface before they react. Activated surface diffusion of hydrogen on carbon occurs at much lower temperatures than gasification temperatures (Robell, Ballou and Boudart, 1964).

Table 4 gives the rates of H_2 gasification for several carbons containing different transition metals. Transition metals markedly catalyze methane formation and the catalytic activity of the metals varies greatly. At 496°C Ru is 27 times more active than Pt and Rh is 23 times more active than Pt; Pd, Co and Ni catalyzed no methane formation over and above that observed for the carbon without the metal. This observation is hard to understand.

Table 5 shows that these metals also catalyze the steam-carbon reaction but with much lower effectiveness than they catalyzed the H_2-carbon reaction. There was also less difference between the activities of the metals with steam, and Pt was the most active catalyst. Equal molar quantities of CO and H_2 were found; CH_4 was not a significant product consistent with equilibrium. The activation energy for Pt in the steam-carbon system was about the same as that for the H_2-carbon system. This similarity may mean that the rate is again controlled by bond

Table 4

Net Methane Formation Rates for H_2-Carbon Reaction
Catalyzed by Metals (Rewick et al., 1974).

Net rate* of CH_4 formation (cm^3/min g carbon)

Metal Present/Carbon type		Norit			Sterling FT			Graphon		
Type	Wt%	496°C	552°C	607°C	496°C	552°C	607°C	496°C	552°C	607°C
None	0	0	0.07	5.8	0	0	0.06	0	0.02	0.04
Pt	0.8	--	--	--	0.07	2.0	8.4	0	0.12	1.6
Pt	5	2.7	33	57	0.21	6.6	21	--	--	--
Ru	5	73	10	7	--	--	--	--	--	--
Rh	5	62	68	9	--	--	--	--	--	--
Pd	5	0	0	0	--	--	--	--	--	--
Co	5	0	0	0	--	--	--	--	--	--
Ni	5	0	0	0	--	--	--	--	--	--

* Difference between observed rates of CH_4 formation with and without added catalyst, except for the entry in which no catalyst was added.

Table 5

Net Rates of Steam-Carbon Reaction Catalyzed by Metals
(Rewick et al., 1974).

Net gas formation rate^a (cm^3/min g carbon)

Catalyst	Loading (wt%)	524°C	552°C	580°C	607°C	635°C
None	0	0	0.20	0.22	0.65	1.9
Pt	5	--	--	4.6	18	38
Pt	0.8	0.37	1.1	2.8	7.0	15
Ru	0.8	0.41	0.80	2.1	5.3	11
Ni	0.8	0.10	0.43	1.1	2.2	3.1
Co	0.8	--	--	0.22	0.97	1.6
Fe	0.8	--	--	0	0.09	0.26

^{a}Net rate is the rate in excess of that in the absence of catalyst and is in terms of net CO or H$_2$ formation; 2.4% H$_2$O in He, sterling FT carbon, one atmos total pressure.

dissociation, this time of water. The energies are similar [D (H-H) = 109 kcal/mole as compared with D(HO-H) = 119 kcal/mole].

Otto and Shelef (1975) reported that 300 ppm Ni increased the H$_2$O-char gasification rate 100-fold; its effectiveness decreases with conversion probably because of loss of contact of the Ni with the carbon surface. This suggests that a good catalyst must not sinter, must redisperse itself or be liquid or vapor at reaction conditions to maintain effective contact with the surface.

Ni, Pt, and Ru remain metallic under the H$_2$O-char gasification conditions. Fe and Co are converted to their oxides and are observed to not catalyze the reaction (Table 5), showing that the catalyst must remain in an appropraite active form under reaction conditions.

Although Fe does not catalyze the steam-carbon reaction, it does catalyze the CO_2-carbon reaction. The unextracted charcoal contained 0.21 wt% Fe_2O_3, 0.39 wt% Na_2CO_3 and 1.23 wt% K_2CO_3 which apparently catalyzed the reaction (initial rate 60 mm/min for CO_2-C) since extraction markedly reduced the rate (initial rate 4.0 mm/min for CO_2-C). Amariglio (1960) reported for 110 ppm concentration levels the following rate enhancement of the catalyzed rate of combustion over the uncatalyzed rate at 600°C:

Pb	4.7×10^5
Mn	8.6×10^4
Ag	1.3×10^3
Cu	500
Au	240
Na	230
Co	4
Al	3
Be	1

This shows that very small amounts of catalyst can be effective as does the results of Shelef and Otto for H_2O-carbon reaction.

The role of the transition metal in catalyzing carbon gasification probably involves three factors. First, metal appears to lower the activation energy for reaction. Without catalyst the activation energy for the O_2-carbon reaction is about 60 kcal/mole (Walker, Shelef and Anderson, 1968) from a large number of studies; the activation energy varies from 10 to 59 kcal/mole for a number of catalysts. For the CO_2-carbon reaction the activation energy is about 86 kcal/mole without catalyst from a number of studies and is lowered by from 20 to 40 kcal/mole by added catalyst. For the H_2O-carbon reaction the activation energy is about 80 kcal/mole (Long and Sykes, 1950); catalysts of very different types, Pt and Na, reduce it to about 55 kcal/mole.

Second, metals can alter the rate limiting step in the mechanism leading to a higher rate or change the coverage of the surface by reactive or reacting species. This is particularly true where a catalyst decomposes H_2 or H_2O into reactive species. Activation of the relative nonreactive gas (H_2O, H_2, CO_2) forming highly reactive species that can attack the carbon structure may well be the most important mechanism whereby the rate is increased by catalysts.

Third, metals induce pit formation in the basal plane of the carbon, thereby exposing additional edge planes for reaction. The basal plane is notoriously unreactive; whereas,

Catalysis in coal gasification

the edges of these planes are reactive, and increasing the numbers of such edges should enhance the rate of reaction. The metal particles primarily make contact with the basal planes of carbon which make up the majority of the surface (over 96% for graphite). Studies have shown that there are as many pits after partial oxidation as there were metal particles present initially (Thomas, 1966; Thomas and Walker, 1965).

The mineral matter in char particularly sulfur behaves as a serious poison to transition metal catalyst in gasification just as in coal liquefaction.

The metal may interact with the fused aromatic ring structure of the carbon such as to lower the average C=C bond strength in the immediate region of the metal resulting in easier attack of these bonds by reactive gas species and easier desorption of products formed.

<u>Non-Transition Metal Catalysts</u>: Marson and Cobb (1926) studied the effect of various inorganic materials on the reaction between steam and coke at 1000°C (Table 6). Alumina, silica and fireclay were inert to catalyzing gasification. CaO, Fe_2O_3 and Na_2CO_3 promoted the gasification reaction; Na_2CO_3 promoted the rate of reaction to a much greater extent since no significant decrease in steam conversion was observed for a four-fold increase in the steam feed rate. White and co-workers (Fleer and White, 1936; Fox and White, 1931; Weiss and White, 1934) have shown that Na_2CO_3 catalyzes very effectively the gasification of carbon with air, with CO_2, with steam and with steam-O_2 mixtures. The effectiveness of sodium and potassium carbonates over most other materials in catalyzing coal gasification has been confirmed repeatedly since.

Sodium catalyzes both the water-carbon reaction and the CO_2-carbon reaction. Again the influence of the mineral matter already present in the charcoal is evident. Calcium and aluminum have no catalytic effect.

The mechanism of the catalytic promotion is not well defined. Fox and White (1931) argued that it was due to the reduction of Na_2CO_3 to metallic sodium and CO by the carbon:

$$Na_2CO_3 + 2C \rightarrow 2Na^° + 3CO \tag{33}$$

The transient $Na^°$ then reacts with gaseous species

$$Na^° + CO_2 \rightarrow NaO + CO \tag{34}$$

$$NaO + CO_2 \rightarrow NaCO_3 \tag{35}$$

Table 6
Effect of Catalysts on Activation Energy for Gasification

Reaction	Activation Energy, kcal/mole		Catalyst Type
	Without Catalyst	With Catalyst	
$C + CO_2 \rightarrow 2CO$	86	46–66	Na, Fe, Al, Mn, V
$C + H_2O \rightarrow H_2 + CO$	~80	50–60	Na, K, Fe, Pt, Ru
$C + \frac{1}{2}O \rightarrow CO$	50–64 (61 ave)	10–59	Fe, V, Na, Cu, Mn, etc.
$C + 2H_2 \rightarrow CH_4$	30–35	55 (29–26)	Pt (K)

$$\overset{\circ}{Na} + H_2O \rightarrow NaOH + H\cdot \rightarrow NaO + H_2 \quad (36)$$

$$NaO + C \rightarrow \overset{\circ}{Na} + CO \quad (37)$$

and the cycle is repeated.

Since alkalis promote the water gas shift reaction ($CO + H_2O \rightleftarrows CO_2 + H_2$) which does not involve carbon in aromatic structures it is more probable that the role of the alkali metal ion is to promote the dissociation of the water molecule into ·OH and H· species which then react with carbon. This is consistent with the mechanism for the uncatalyzed steam-carbon reaction.

Inorganic constituents not only catalyze the overall rate of gasification but also affect the selectivity of coal gasification reactions (Table 7). Na, K and Li compounds (Na_2CO_3, NaCl, KCl, Li_2CO_3 and K_2CO_3) catalyze coal gasification the most effectively of all materials tested to date. However, they greatly promote the steam-carbon reaction, Reaction (1), without significantly increasing the methane formation. The cation is important, the anion is of lesser importance. Other materials catalyze the overall gasification less effectively but enhance the methane formation rate more relative to their enhancement of the CO and H_2 formation rates. The conclusions of the Bureau of Mines work (Haynes et al., 1973) is that: (a) a significant increase in the gasification rate and production of desirable products can be effected by a large number of inorganic compounds, (b) a significant increase in CH_4 production is achieved with Raney

Table 7

Catalytic Promotion of Steam-Coke Reaction

Steam Rate liters/hr	% of Steam Decomposed			
	Pure Coke	CaO Coke	Fe_2O_3 Coke	Na_2CO_3 Coke
5	68	84	94	99
10	61	82	91	99
15	56	81	90	98
20	51	78	87	98

[a] Steam passed through a bed of coke mixed with indicated catalyst to the extent of 5 wt%.

(Marson and Cobb, 1926)

---

nickel, Li_2CO_3, Pb_3O_4, Fe_3O_4 and MgO, (c) catalytic effectiveness decreases for temperatures above 750°C and (d) ash containing potassium compounds can be recycled to effectively catalyze gasification.

Table 8 compares the rates of catalyzed steam-carbon gasification observed by Rewick, Wentrcek and Wise (1974) for Pt, Ni and Co with the rates of catalyzed steam-coal gasification observed by Haynes, Gasior and Forney (1973) for several metal oxide catalysts. For cobalt, the rate enhancement at atmospheric pressure is more impressive, but the poisons and higher pressure of the actual cool system give a more realistic evaluation.

The objective of producing mainly methane and/or hydrocarbons directly from coal would appear to be best achieved by catalyzing the rate of gasification of char to CO and H_2 to allow a reduction in the temperature required and then reacts the CO and H_2 to methane and hydrocarbons. The high temperatures prevent formation of higher hydrocarbons, but methane formation can be effectively catalyzed at intermediate temperatures. To achieve such a dual-functional catalyst system is required

Table 8

Comparison of Effect of Catalysts on Steam-Carbon Gasification Rate of Graphite vs. Real Coal Char at 852°C

Additive, 5 wt%	Relative increase in gasification rate $(R_{cat} - R_{carbon})/R_{carbon}$	Reference
Pt	28	(Rewick et al., 1974)[a]
Ni	3.4	
Co	1.5	
Li_2O_3	0.48	
PbO_2, Pb_3O_4	0.35	
BaO	0.33	(Haynes et al., 1973)[b]
Bi_2O_3	0.31	
CoO	0.29	

[a] Carbon = Sterling FT; H_2O = vol%, total pressure = 1 atm (1 atm = 101.3 kPa).

[b] Carbon = bituminous coal (Bruceton, Pa.); H_2O = 6.9 vol%; total pressure = 20 atm.

(Rewick et al., 1974)

(a) to catalyze gasification, e.g., Na_2CO_3 or K_2CO_3, and (b) to catalyze $CO + H_2 \rightarrow CH_4$, e.g., Ni.

Sulfur and other poisoning substances, along with the amount and cost of catalyst required, represent serious problems in applying catalyst to coal gasification processes. Durable catalyst and/or regeneration procedures would represent potentially great advancement in coal gasification technology.

LITERATURE CITED

Amariglio, H., Thése Docteur Science, University of Nancy, 1960.
Anthony, D. B., and Howard J. B., AIChE Journal 22, 625 (1976).
Fleer, A. W., and White, A. H., Ind. Eng. Chem. 28, 1301 (1936).
Fox, A. D., and White, A. H., Ind. Eng. Chem. 23, 259 (1931).
Haynes, W. P., Gasior, S. J., and Forney, A. J., "Catalysis in Coal Gasification at Elevated Pressure," Preprints, Div. of Fuel Chemistry, Amer. Chem. Soc., 18 (2), p. 1-28 (1973).

Hottel, H. C., and Howard, J. B., "New Energy Technology,"
    MIT Press, Cambridge, 1971.
Long, F. J., and Sykes, K. W., J. Chim. Phys. 47, 361 (1950).
Marson, C. B., and Cobb, J. W., Gas J. 175, 882 (1926).
Otto, K., and Shelef, M., paper presented at 4th North American
    Catalysis Society Meeting, Chicago, April, 1977.
Rewick, R. T., Wentrcek, P. R., and Wise, H., Fuel, 53, 274 (1974).
Robell, A. J., Ballou, E. V., and Boudart, M. J., J. Phys.
    Chem. 68, 9 (1964).
Thomas, J. M., in "Chemistry and Physics of Carbon," Vol. 2,
    P. L. Walker, ed., Dekker, New York, p. 1-49, 1966.
Thomas, J. M., and Walker, P. L., Jr., Carbon 2, 434 (1965).
Van der Hoeven, B. J. C., "Producers and Producers Gas," in
    "Chemistry of Coal Utilization," Vol. II, ed. H. H. Lowery,
    p. 1036, Wiley and Sons, New York, 1945.
Walker, P. L., Jr., Shelef, M., and Anderson, R. A., "Catalysis
    of Carbon Gasification," in "Chemistry and Physics of
    Carbon," ed. P. L. Walker, Jr., Vol. 4, p. 287, Marcel
    Dekker, New York, 1968.
Weiss, C. B., and White, A. H., Ind. Eng. Chem. 26, 83 (1934).

MECHANISM OF HYDROCARBON SYNTHESIS OVER FISCHER-TROPSCH CATALYSTS

W.M.H. Sachtler

Koninklijke/Shell-Laboratorium, Amsterdam
(Shell Research B.V.)

ABSTRACT. Experimental data obtained from (1) chain length distribution, (2) reaction kinetics, and (3) isotopic labelling provide clues for the mechanism of hydrocarbon synthesis from carbon monoxide and hydrogen with heterogeneous catalysts such as nickel and ruthenium. Of particular relevance are results of experiments where a catalyst surface is first covered with well-defined quantities of ^{13}C atoms, deposited there by decomposition of ^{13}CO, and then exposed to gaseous $^{12}CO + nH_2$. It is concluded that the hydrogenation of surface carbon C_{ads} on nickel is fast enough to account for methane formation and for the formation of the CH_x groups which are then incorporated into growing alkyl chains to give Fischer-Tropsch hydrocarbons. It is very likely that the same holds for the Fischer-Tropsch catalyst Fe, Co, Ni and Ru, while on the other transition metals CO dissociation seems to be a less important route.

For the former metals some speculations are made with respect to the nature of the CH_x group. The idea is discussed that CH_x is an alkylidene group. Chain growth in Fischer-Tropsch processes would then proceed by formation of a C-C bond between an alkylidene and an alkyl group, both being attached to the same metal atom. This is basically the same type of mechanism as was proposed by Gault et al. for the formation of the new C-C bond in (dehydro)cyclization. In this case both C atoms and the surface metal atom are part of a six-membered ring:

In this view, these processes belong to a larger group of reactions by which new C-C bonds are formed between two ligands attached to the same metal atom; this group also includes olefin metathesis and Ziegler polymerization.

1. INTRODUCTION

The term Fischer-Tropsch catalysis [1] is used in different ways by different authors. Sometimes the totality of catalysed reactions between carbon monoxide and hydrogen is described by this term, but more often the processes which lead to methanol or ethylene glycol as the major products are separated from the Fischer-Tropsch process proper. Most authors regard cobalt, iron and ruthenium as "typical" Fischer-Tropsch catalysts, accepting that under technical conditions significant quantities of oxygen-containing products are obtained with the former two elements. Another possible definition would limit the term to the catalysed production of hydrocarbons from synthesis gas, and in this case it is still possible to include or exclude processes where methane is the only reaction product. In the present paper we shall confine ourselves to hydrocarbon synthesis and assume that the distributions of alkanes e.g., with respect to their chain length, can be described by a simple formalism. Within this scope ruthenium is a typical catalyst for the processes discussed.

The production of hydrocarbon chains from molecules containing one carbon atom only resembles a polymerization process in so far as it is meaningful to distinguish initiation, propagation, and termination steps. However, in a true polymerization process the monomer is present before the catalytic reaction starts. In the Fischer-Tropsch process, however, the propagation "step" is a reaction where the chains are lengthened by one or more CH_2 groups:

$$R_n + m\ CH_2 \longrightarrow R_{n+m} \tag{1}$$

but these CH_2 groups are not offered to the catalyst by the surrounding medium, which consists, at least initially, only of synthesis gas. The actual chain growth by m members is therefore given by the overall reaction:

$$R_n + m\ (CO + 2\ H_2) \longrightarrow R_{n+m} + m\ H_2O \tag{2}$$

which is evidently not an elementary step but the sum of a number of reactions occurring at the catalyst surface. The chemistry of these and of the initiation and termination steps is complicated and is at present the subject of intensive research in various laboratories.

The information which is now available and has helped to reduce the number of possibilities under discussion stems mainly from measured data of:

(1) chain length distribution
(2) reaction kinetics
(3) isotopic labelling
(4) IR spectroscopy

Besides these factual data much use has been made of analogies with other reactions where either CO is "inserted" between a metal and a carbon atom, or where carbon-carbon bonds are formed on catalysts, for instance in (dehydro-)cyclization reactions catalysed by transition metal catalysts. In the present paper a brief and incomplete review is given of the information available.

2. CHAIN-LENGTH DISTRIBUTION

Since R_n and R_{n+m} are supposed to be homologues, differing only in their chain lengths, it follows that the propagation process as described by Equation (1) or (2) can be repeated many times. The easiest way - although not logically necessary - is to assume that in each propagation step the parameter m has the same value. For instance, if $m = 2$ (chain growth by ethylene incorporation) and the smallest chain, created in the initiation step, has an even (odd) number of carbon atoms, all chains will have an even (odd) number of carbon atoms. It is further necessary to assume the occurrence of reactions that separate the chains from the catalyst surface and thus terminate their growth:

$$R_n + ? \longrightarrow P_n + ? \qquad (3)$$

where P_n is the reaction product containing n carbon atoms. The question marks are used to avoid specific assumptions at this stage. If, for instance, R_n is an alkyl group adsorbed to the metal by one terminal carbon atom, termination can be visualized as either addition of a hydrogen atom, leading to a n-alkane molecule, or a β-hydrogen abstraction, yielding an α-olefin.

The rates of propagation and termination would then be

$$r_p = k'_p \theta_{R_n} \qquad (4a)$$

and
$$r_t = k'_t \theta_{R_n} \tag{4a}$$

where the pseudo-rate constants k'_p and k'_t still include the concentration of the partner and θ_{R_n} is the surface concentration of the chain with n carbon atoms. Since the reactivity should not depend on the chain length (with the possible exception of the smallest members) the ratio

$$\frac{k'_p}{k'_p + k'_t} = \alpha \tag{5}$$

should be constant. It follows that the distribution of the product concentrations should be governed by the simple law:

$$c_{P_{n+m}} = \alpha \, c_{P_n} \tag{6}$$

The chance to undergo x insertion steps and thus obtain a product P_x is then equal to α^x. Therefore, a linear relationship holds between x and the logarithm of the concentration of such product. This was derived for polymerizations in general some four decades ago by Schulz [2] and Flory [3].

For the special case m = 1 it is possible to approximate x by i, the number of carbon atoms in the product molecule. The so-called Schulz-Flory plot is then obtained by plotting the molar fraction (or the weight fraction divided by i) logarithmically versus i,

$$\log c_{P_i} = A + i \, (\log \alpha) \tag{7}$$

This relation is found valid in fair approximation for typical Fischer-Tropsch processes, and this lends some justification to the basic assumption m = 1, i.e. chain growth by insertion of one carbon atom (or CH_2 group). This evidence does not exclude, however, that such a main process is accompanied by other reactions where e.g. olefins become incorporated.

A typical Schulz-Flory plot, recently obtained by Rautavuoma [4] for a cobalt catalyst is shown in Fig. 1. The slope of the straight line being $\log \alpha$, it follows from Equation (5) that the ratio of propagation and termination rate constants can be obtained from the product distribution. A small value of α characterizes a situation where termination is a relatively fast process and the product contains much methane. The opposite case of large α is typical for a heavy product

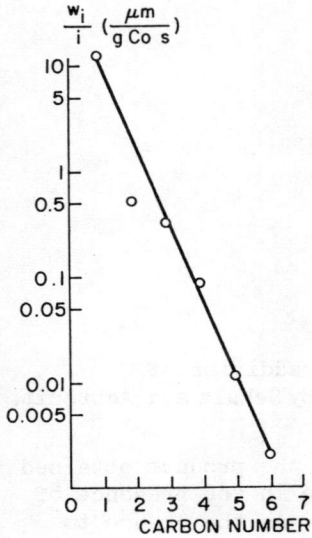

Fig. 1. Schulz-Flory plot (logarithm of the weight of product fraction divided by carbon number, weight of catalyst and contact time, plotted versus carbon number), as obtained by Rautavuoma. Catalyst: Co/Al_2O_3; T = 523 K; flow rate: 3.5 $cm^3 s^{-1} g^{-1}_{oxide}$.

containing much waxy material. The value of α depends on the catalyst, the temperature, the pressure and the H_2/CO ratio. Nickel is typically a methanation catalyst [5,6], while ruthenium is renowned for its ability to produce very high molecular weight wax, chemically identical with polyethylene [7]. For a given catalyst it is clear that a high H_2/CO ratio and a high temperature favour chain termination.

For a technical process intended for the manufacture of liquid hydrocarbons in the boiling range of gasoline, the inevitable co-production of gases and a high molecular weight product is a liability of the Fischer-Tropsch route. Often, the waxy co-products which can subsequently be hydrocracked are considered the minor evil and the process is directed towards a product with high value of α, which means that the termination rate constant is much smaller than the rate constant of chain propagation.

The insertion of olefins in growing chains has been the subject of elegant studies by Pichler and Schulz [8,9], and Schulz and Achtsnit [10], who added α-olefins, labelled in the terminal CH_2 group by the radioisotope C-14, to synthesis gas. These authors proved that ethylene e.g. is readily incorporated

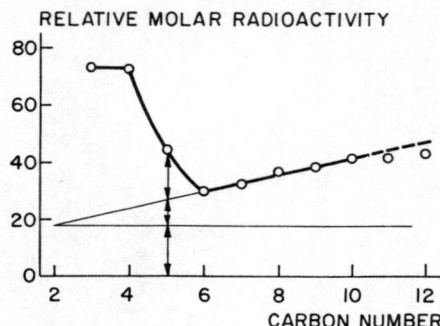

Fig. 2. Molar radioactivity of products upon addition of $^{14}CH_2=CH_2$ to synthesis gas, as found by Schulz and Achtsnit.

in the growing chains. Fig. 2 indicates that the product obtained with cobalt catalysts at atmospheric pressure in the presence of $^{14}CH_2=CH_2$ shows a considerable radioactivity. As the chance to incorporate an ethylene molecule increases linearly with the number of growth steps, the molar radioactivity of the product is found to increase linearly with chain length. As the straight line defined by the larger chains does not pass through the origin, the authors conclude that ethylene can also initiate chain growth, while a high molar radioactivity of C_3 and C_4 appears to show that smaller chains can also be terminated with ethylene.
Dry [11] and Somorjai et al. [12] also reported an increased Fischer-Tropsch activity in the presence of ethylene.

As incorporation of higher olefins in a growing chain would result in branched molecules (with the possible exception of alpha-olefin insertion followed by termination), the predominance of straight-chain molecules [9] seems to show that insertion of higher olefins is not a frequently occurring phenomenon. Indeed, Schulz-Flory plots often show a C_2 point below the line defined by higher products, indicating that the chance of inserting olefins is largest for C_2H_4. However, the work by Pichler and Schulz [8,9] who added ^{14}C-labelled alpha-n-hexadecene to synthesis gas, showed that some incorporation of higher olefins does occur.

An interesting observation in this work was that molecules with fewer than 16 carbon atoms also showed a radioacitivity increasing with chain length. Schulz and Achtsnit [10] ascribe this to an <u>in situ</u> metathesis and we shall discuss the mechanistic implications of this observation below.

3. REACTION KINETICS

The Fischer-Tropsch synthesis is a rather slow catalytic reaction. The rate of growth is often in the order of one CO molecule consumed per surface atom and per minute [6,13]. The reaction order is usually positive in hydrogen, but negative in carbon monoxide [6] as might be expected from the known difference in heats of adsorption of these two gases. Under reaction conditions the accessible and participating sites are more or less saturated with CO and the chance of adsorbing hydrogen on such sites increases with the partial pressure of hydrogen but decreases with that of carbon monoxide. The predominant reaction product being water, the ratio p_{H_2O}/p_{H_2} in the gas will increase with conversion and might reach values where iron surfaces e.g. should be oxidized in equilibrium. For ruthenium catalysts the kinetics have been described by Bond and Turnham [14] by the formula:

$$k = A\ p_{H_2}^x\ p_{CO}^y\ e^{-E/RT}$$

with $x = 1.17$; $y = -0.43$; $E = 20.5$ kcal/mol.

With respect to the question which step is rate determining, we mentioned already that conditions are often chosen such that the rate constant for termination is smaller than that of propagation. The measured value of α provides the ratio of these constants. If we wish to know their individual values, data from non-steady state conditions are required. The two rate constants can then be compared with the overall rate in order to see whether the initiation step is fast or rate determining. This has been attempted by Dautzenberg et al. [13] with ruthenium catalysts, applying pulse conditions. A pulse where synthesis gas was admitted at 210 °C was followed by a pulse where hydrogen terminated all chains on the catalyst surface at 300 °C. The authors found that with periods of the order of 10 or 15 minutes of the (CO + H$_2$) pulse the reaction product differed significantly from that obtained in the steady state. By solving the differential equations for steady state and for pulse conditions they were able to obtain the values for propagation and termination rate constants and, of course, the overall rate. They concluded that initiation is fast, propagation slower and termination slowest under the conditions chosen. In the steady state a large faction of the active surface atoms is covered by growing chains, the remainder mainly by carbon monoxide. These results are, however, not fully supported by the infra-red data of Dalla Betta & Shelef [15], Ekerdt and Bell [16] and Tamaru [17] who conclude that the majority of surface atoms is covered with CO, while a minority carries growing hydrocarbon chains. For Co/Al$_2$O$_3$

catalysts Rautavuoma [4] concludes, likewise, that only a small fraction of the surface Co atoms are carrying growing chains under steady state conditions.

4. REACTION MECHANISM

4.1. Initiation and formation of methane

The mechanism for methane formation on Ni e.g. as discussed in the literature fall essentially into two groups, which assume either that C-H bonds are formed before the C=O bond is completely broken, or that a complete rupture of the C=O bond precedes the formation of the C-H bonds. Possible reaction sequences in the former case include [6,18,19,20,21]:

$$CO_{ads} + 2\, H_{ads} \rightarrow H\diagdown_{\underset{\|}{C}}\diagup OH \rightarrow CH_4 + H_2O \qquad (9a)$$

or

$$CO_{ads} + H_{ads} \rightarrow H\diagdown_{\underset{|}{C}}\diagup O \rightarrow CH_4 + H_2O \qquad (9b)$$

In the second case [22] formation of a "surface carbide" C_{ads} is essential:

$$CO_{ads} \rightarrow C_{ads} + O_{ads} \qquad (10a)$$

The C_{ads} species is then hydrogenated to methane:

$$C_{ads} + 4\, H_{ads} \rightarrow CH_4 \qquad (10b)$$

while the adsorbed oxygen can either react with CO or with hydrogen, yielding CO_2 or H_2O. The formation of CO_2 as a true primary product is very unlikely if CO_{ads} is hydrogenated prior to the rupture of the carbon-oxygen bond.

Until recently most authors seemed to prefer the view that the first group of mechanisms [5,6], involving some aldehydic or enolic surface intermediate, is the more probable. This preference was based on the observation that the hydrogenation of bulk metal carbides to methane and the Boudouard reaction

$$2\, CO \rightarrow C + CO_2 \qquad (11a)$$

are known to be rapid only at much higher temperatures, which suggested that both the dissociation of CO and the hydrogenation or carbon are too slow to be intermediate steps in the methanation reaction. However, these arguments do not rule out the possibility that the formation of the "surface carbide" C_{ads} and its subsequent hydrogenation to CH_4 are fast enough to be the preferred pathway for the formation of methane on the surface of a nickel catalyst. Indeed, in 1948, Kummer et al. [22] had shown that labelled surface carbides can yield labelled methane under the conditions of the Fischer-Tropsch synthesis. The problem was recently reinspected by Wentrcek, Wood and Wise [23] in Stanford, by Araki and Ponec [24] in Leiden, and by Rabo, Risch and Rabo [25] at Union Carbide. The Stanford group observed that on an alumina-supported Ni catalyst the reaction

$$2\ CO \longrightarrow C_{ads} + CO_2 \tag{11b}$$

occurs at a detectable rate at temperatures as low as 350 K. In pulse experiments at 553 K CO was rapidly converted to $C_{ads} + CO_2$. Subsequent pulses of hydrogen at the same temperature rapidly converted the C_{ads} quantitatively to CH_4. The Stanford workers conclude that the dissociative chemisorption of CO on nickel provides an energetically possible mechanism for methanation and they stress that the surface carbon species, unlike Ni_3C, is reactive towards hydrogenation.

Araki and Ponec [24] also found that disproportionation of CO to C_{ads} and CO_2 can be rapid. On nickel films exposed to low pressures of $CO + H_2$ the first product observed in the gas phase was actually CO_2, while CH_4 formation appeared to be preceded by an induction period. Since dissociation of CO is thus demonstrated to occur under the conditions of CO hydrogenation on these films, the pertinent question is: which adsorbed species is faster converted to methane, C_{ads} or CO_{ads}? To answer this question, the authors covered their films with labelled $^{13}C_{ads}$ obtained by pretreatment with ^{13}CO and exposed this film to a gas mixture of 5 H_2 + ^{12}CO at 77 Pa and 523 K. As shown in Fig. 3 it was found that initially much more $^{13}CH_4$ than $^{12}CH_4$ was formed, indicating that under these conditions methane formation via dissociation of CO seems to be the preferred pathway. The same conclusion was drawn by Rabo et al. [25], who found that C_{ads} on Ni is readily hydrogenated to CH_4 even at room temperature where $CO + H_2$ would not be converted to methane. For Co and Ru films it was observed by J.W.A. Sachtler et al. [26] that predeposited ^{13}C is hydrogenated to $^{13}CH_4$, but this carbon also undergoes a deactivation.

Araki and Ponec [24] also found that the rate of methane formation on copper-nickel alloy films is much lower than on nickel films, the turnover number strongly decreasing with increasing copper concentration. As the rate of hydrogenation of surface carbon on these alloy films is not much lower than on

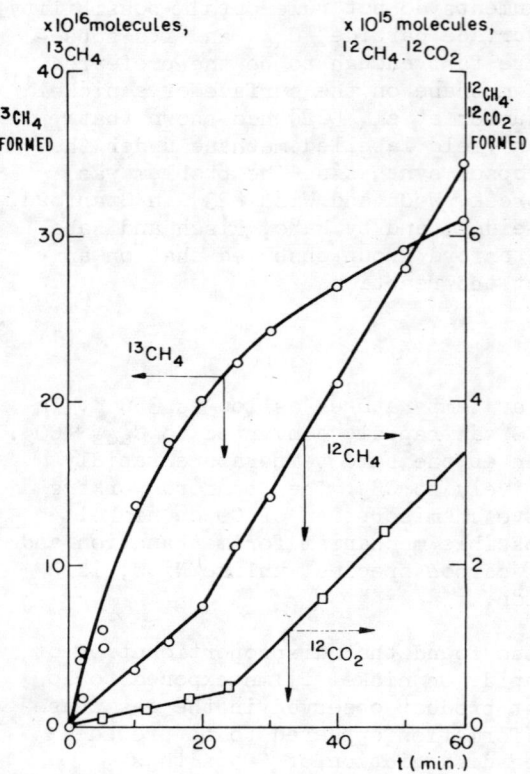

Fig. 3. Reaction products obtained by Araki and Ponec upon exposing a Ni film covered with $^{13}C_{ads}$ to $^{12}CO + 5\ H_2$ at 523 K.

nickel, it appears that alloying strongly lowers the rate of the carbon-oxygen bond fission, in agreement with IR data [27]. The selectivity and activation energy on the alloy films are similar to those on nickel, suggesting that the same type of sites is active, but the number of these sites is much smalller on the alloy surfaces. Similar results were found by Bond et al. [14] for Ru metal and RuCu alloys. The behaviour is typical for the "ensemble effect" of alloy catalysis, i.e. a reaction which requires large ensembles of active atoms is orders of magnitude slower on a surface where the active atoms are diluted with inactive atoms, because the concentration of the required ensembles decreases with a high power of the surface concentration of the active component [28].

4.2. Propagation and formation of n-alkanes

The mechanisms under discussion can be grouped in three categories with respect to the nature of the monocarbon species C_1, forming the new C-C bond:

(a) C_1 is an adsorbed CO molecule, i.e. there is a <u>carbon-oxygen bond, but no carbon-hydrogen bond</u>. In this case the step where the new C-C is formed is a CO insertion, as proposed, for instance, by Pichler und Schulz [8]:

$$R-M + M = CO \longrightarrow \underset{M}{\underset{|}{R\diagdown C}}=O \qquad (12a)$$

followed by

$$\underset{M}{\underset{|}{R\diagdown C}}=O + 2H \longrightarrow \underset{M}{\underset{|}{R\diagdown HC}}-OH \xrightarrow{2H} \underset{M}{\underset{|}{R\diagdown CH_2}} + H_2O \qquad (12b)$$

(b) C_1 contains both <u>carbon-oxygen and carbon-hydrogen bonds</u>. It is, for instance, an enol formed by a reaction of CO with hydrogen, as was proposed by Anderson et al. [29]:

$$\underset{M}{\underset{\|}{R\diagdown C \diagup OH}} + \underset{M}{\underset{\|}{H\diagdown C \diagup OH}} \xrightarrow[-H_2O]{+2H} \underset{M}{\underset{\|}{R\diagdown CH_2 - C \diagup OH}} \qquad (13)$$

(c) C_1 has <u>no carbon-oxygen bond</u>. In this category it can be imaged that CO is either dissociatively adsorbed, as was suggested by Jones et al. [31] and Joyner [32] or that its reaction with adsorbed hydrogen results in the formation of an oxygen-free complex. For instance, this complex might be an alkylidene (conventionally called: carbene) as previously proposed by us [33], and react (in one or more steps) with an alkyl group attached to the same metal atom:

$$H_2C \underset{M}{\diagdown \diagup} CH_2-R \longrightarrow H_2C - CH_2-R \atop \diagdown M \qquad (14)$$

Incorporation of CH_2 had already been proposed by Eidus [30] who assumed, however, that initiation involved an oxygen containing complex. The classification used here has the advantage of being amenable to experimental verification. The experiment should decide whether oxygen-free carbon on the catalyst surface can be incorporated in growing chains at a rate consistent with the known kinetics of the Fischer-Tropsch process.

This question has recently been tackled by Biloen et al. [33]. A supported metal catalyst was first covered with carbon in submonolayer quantities by means of the Boudouard reaction (Eq. (11b)), utilizing ^{13}C-labelled CO. After removing the ^{13}CO and $^{13}CO_2$ the catalyst was exposed to synthesis gas ($^{12}CO + 3\ H_2$) of 1 bar at 423 K. It should be noted that this pressure is three orders of magnitude higher than in the experiments of Araki and Ponec [24], the CO/H_2 ratio is higher and the temperature lower. These factors favour a higher value of α in the reaction product, which is necessary for obtaining sufficient amounts of higher hydrocarbons to analyse them for their content of ^{13}C atoms. The experiments were carried out under conditions where the number of exposed metal atoms and the number of molecules in the gas phase were comparable (10^{20}). While nickel was most intensely studied, cobalt and ruthenium appear to follow the same pattern. Three relevant results from this work have to be mentioned:

(1) The paraffin products show the Schulz-Flory distribution but with an undershoot of C_2.

(2) The specific rate of the Boudouard reaction (Eq. (11)) in the absence of hydrogen is significantly lower than the rate of CO hydrogenation.

(3) A considerable fraction of the hydrocarbons contains <u>several</u> ^{13}C atoms (see Table I. Note that there are propane and butane molecules with three ^{13}C atoms!).

The first of these results supports the view that these data are relevant for "real" Fischer-Tropsch synthesis of hydrocarbons with these catalysts. The second result shows that conversion of CO to the incorporable surface species CH_x does not necessarily proceed via a "naked" species C_{ads}; however, C_{ads} prepared in a separate experiment will easily be hydrogenated to CH_x. The reaction courses for chain growth under normal Fischer-Tropsch conditions and under the conditions where a surface covered with active carbon is offered to the gas can, schematically, be written as follows:

$$\begin{array}{c} CO \\ \searrow^{+H} \\ CH_x \xrightarrow{+\ chain} \text{incorporation} \\ \nearrow_{+H} \\ C_{ads} \end{array} \qquad (15)$$

The third result, which in the present interest we consider most important, shows that oxygen-free species are incorporated in the growing chains at a high rate. It suggests that in Fischer-Tropsch

TABLE I

ISOTOPIC COMPOSITION OF METHANE AND HIGHER HYDROCARBONS
4 %w Ni/SiO_2; T = 170 °C; P = 1 bar; H_2/CO molar ratio = 3

Series	Batch	θ_{13} [a]	CH_4		C_2H_6			C_3H_8				C_4H_{10}					
			0	1	0	1	2	0	1	2	3	0	1	2	3	4	[b]
11	1	0.62	44	56	45	46	9	16	30	37	17						
	2	0.50	42	58	48	39	13	32	34	27	7						
	4	0.32	76	24	61	29	10	61	26	13	-						
	5	0.24	82	18	75	25	-	63	26	11	-						[c]
5	1	0.36	38	62	49	30	21	31	27	28	14						
	2	0.27	54	46	51	30	19	55	26	14	6						
20	2	0.20	75	25	60	23	17	61	25	14	-	38	28	21	13	-	
	3	0.13	86	14	78	22	-	73	21	6	-	48	26	24	-	-	

a) $\theta_{13} = \dfrac{\text{number of }^{13}\text{C atoms deposited}}{\text{number of exposed metal (from hydrogen chemisorption data)}}$

b) number of ^{13}C atoms per molecule.

c) %m.

synthesis carbon-oxygen bond rupture precedes carbon-carbon bond formation. It further suggests that incorporation of a surface carbene as shown in Eq. (14) is a realistic possibility to visualize the propagation step in Fischer-Tropsch synthesis. It cannot be ruled out, at present, that in the C_1 species to be incorporated, the composition is CH_x with $x \neq 2$, but the carbene with $x = 2$ appears a reasonable proposition for a number of reasons.

It was mentioned above that Schulz and Achtsnit [10] explain the formation of radioactive molecules with fewer than 16 carbon atoms upon adding $^{14}CH_2=CH-C_{14}H_{29}$ to synthesis gas by assuming that a metathesis reaction is occurring on the catalyst surface. It is known, however, that metathesis is a chain reaction, as proposed by Chauvin [34], involving a metal carbene and an olefin, and the reaction requires only one transition-metal atom, for instance tungsten:

$$\begin{array}{c} CH_2 \\ \| \\ W \end{array} + \begin{array}{c} CH_2 \\ \| \\ CH_3 \end{array} \longrightarrow \begin{array}{c} CH_2 = CH_2 \\ + \\ W = CH_3 \end{array} \qquad (16)$$

The formation of carbon-carbon bonds in cyclization and isomerization of paraffins on the surface of transition metal catalysts has been very carefully studied by Gault and co-workers

[35,36], who found that the cyclization of n-pentane and its isomerization via the so-called cyclic mechanism involve a metallocyclohexene (I), which can be transformed via a metallocyclohexene (II) into a monoadsorbed cyclopentane (III):

$$M + \begin{array}{c} H_2C\overset{CH_2}{\diagup}CH_2 \\ | \quad\quad | \\ H_3C \quad\quad CH_3 \end{array} \xrightarrow{-3\,H} \begin{array}{c} H_2C\overset{CH_2}{\diagup}CH_2 \\ | \quad\quad | \\ HC \quad\quad CH_2 \\ \diagdown\!\!\diagup \\ M \end{array} \xrightarrow{-H}$$

(I)

$$\begin{array}{c} H_2C\overset{CH_2}{\diagup}CH_2 \\ | \quad\quad | \\ HC \quad\quad CH \\ \diagdown\!\!\diagup \\ M \end{array} \xrightarrow{+H} \begin{array}{c} H_2C\overset{CH_2}{\diagup}CH_2 \\ | \quad\quad | \\ HC \!-\!-\! CH_2 \\ \diagdown \\ M \end{array} \quad\quad (17)$$

(II)   (III)

In this scheme, intermediate (I) is identical with what is proposed in Equation (14) for Fischer-Tropsch, except that in that case the carbene and alkyl groups are not interconnected by a second bridge. The schemes of Equations (14), (16) and (17) and the Cossee scheme [37] for Ziegler polymerization all have in common that new C-C bonds are formed between groups attached to the same metal atom, at least one of the bonds being unsaturated (see Table II).

The main result of this work, proving that the species to be inserted in the propagation step can be an oxygen-free CH_x group, is in good agreement with the results reported by Poutsma et al. [38], who found that metals which are unable to dissociate CO, such as Pd, yield methane at a low rate, but at elevated pressures the main product of CO hydrogenation is methanol. It follows that metals which easily convert CO to the surface species CH_x will readily produce hydrocarbons. Those metals, however, which do not easily break the carbon-oxygen bond to hydrogenate CO to methanol under conditions where this is thermodynamically possible. Ni, Co and Ru are typical representatives of the former and Pd, Pt and Ir of the latter group of catalysts, which also includes the oxides commercially used as catalysts for methanol production.

5. ACKNOWLEDGEMENT

Numerous elucidating discussions with P. Biloen, V. Ponec and J.N. Helle are gratefully acknowledged.

TABLE II

CATALYSED C-C BOND FORMATIONS INVOLVING ONE METAL ATOM

Reaction	Proposed mechanism
1) Ziegler-Natta (Cossee)	$\begin{array}{c} R \\ | \\ H_2C \\ | \\ M \end{array} \overset{\pi}{\leftarrow} \begin{array}{c} CH_2 \\ \| \\ CH_2 \end{array} \longrightarrow \begin{array}{c} R \\ | \\ H_2C-CH_2 \\ | \quad\;\; | \\ M-CH_2 \end{array}$
2) Metathesis (Chauvin)	$\begin{array}{cc} H_2C & CH_2 \\ \| & \| \\ M & CHR \end{array} \longrightarrow \begin{array}{cc} H_2C=CH_2 \\ M = CHR \end{array}$
3) Dehydrocyclization (Muller & Gault)	(six-membered ring with M) $\longrightarrow$ (five-membered ring with M)
4) Fischer-Tropsch propagation (Sachtler)	$H_2C \diagdown_M \diagup CH_2-R \longrightarrow H_2C-CH_2-R \diagdown_M \diagup$

REFERENCES

1. F. Fischer and H. Tropsch, Chem. Ber., 59, 830 (1926); Brennstoffchemie, 7, 97 (1926).
2. G.V. Schulz, Z. Phys. Chem., B 43, 25 (1939).
3. P.J. Flory, Principles of Polymerisation Chemistry, Oxford Univ. Press, 1953.
4. A.O.I. Rautavuoma, The Hydrogenation of Carbon Monoxide on Cobalt Catalysts, Thesis, Eindhoven, Sept. 1979.
5. G.A. Mills and F.W. Steffgen, Catal. Rev. Sci. Eng., 8, 159 (1973).
6. M.A. Vannice, Catal. Rev. Sci. Eng., 14, 153 (1976).
7. H. Pichler, Advances in Catalysis, 4, 271 (1952).
8. H. Pichler, H. Schulz and F. Hojabri, Brennstoffchemie, 45, 215 (1964).
9. H. Pichler and H. Schulz, Chem. Ing. Technik, 42, 1162 (1970).

10. H. Schulz and H.D. Achtsnit, Proc. VIth Iberoam. Congr. Catalysis, Lisboa, 1976.
11. M.E. Dry, Ind. Eng. Chem. Prod. Res. Dev., 15, 282 (1976).
12. D.J. Dwyer and G.A. Somorjai, J. Catal., 56, 249 (1979).
13. F.M. Dautzenberg, J.N. Helle, R.A. van Santen and H. Verbeek, J. Catal., 50, 18 (1977).
14. G.C. Bond and B.D. Turnham, J. Catal., 45, 128 (1976).
15. R.A. Dalla-Betta and M. Shelef, J. Catal., 48, 111 (1977).
16. J.G. Ekerdt and A.T. Bell, J. Catal., 58, 170 (1979).
17. K. Tamaru, private communication, 1979.
18. M.A. Vannice, J. Catal., 37, 462 (1975).
19. G.M. Kozub, M.T. Rusov and V.M. Vlasenko, Kinet. Katal., 6, 244 (1969;
20. J.L. Bousquet, P. Gravelle and S.J. Teichner, Bull. Soc. Chim. France, 1969, 3693.
21. T. van Herwijnen, H. van Doesburg and W.A. de Jong, J. Catal., 28, 391 (1973).
22. J.T. Kummer, T.W. de Witt and P.H. Emmett, J. Amer. Chem. Soc., 70, 3632 (1948).
23. P.R. Wentrcek, B.J. Wood and H. Wise, J. Catal., 43, 363 (1976).
24. M. Araki and V. Ponec, J. Catal., 44, 439 (1976).
25. J.A. Rabo, A.P. Risch and M.L. Poutsma, J. Catal., 53, 295.
26. J.W.A. Sachtler, J.M. Kool and V. Ponec, J. Catal., 56, 284 (1979).
27. W.L. van Dijk, J.A. Groenwegen and V. Ponec, J. Catal., 45, 277 (1976).
28. W.M.H. Sachtler and R.A. van Santen, Adv. Catalysis, 26, 69 (1977).
29. H.H. Storch, H. Golumbic and R.B. Anderson, The Fischer-Tropsch and Related Syntheses, J. Wiley, New York (1951).
30. Ya.T. Eidus, Uspekhi Khim., 20, 54 (1951); Russ. Chem. Rev., 36, 338 (1967).
31. A. Jones and B.D. McNicol, J. Catal., 47, 384 (1977).
32. R.W. Joyner, J. Catal., 50, 176 (1977).
33. P. Biloen, J.N. Helle and W.M.H. Sachtler, J. Catal., 58, 95 (1979).
34. J.L. Herrisson and Y. Chauvin, Makromolekulare Chemie, 141, 161 (1971); see also: N. Calderon, E.A. Ofstead and W.A. Judy, Angewandte Chemie, 88, 433 (1976).
35. F. Garin and F.G. Gault, J. Amer. Chem. Soc., 97, 4466 (1975).
36. J.M. Muller and F.G. Gault, J. Catal., 24, 361 (1972).
37. P. Cossee, J. Catal., 3, 80 (1964).
38. M.L. Poutsma, L.F. Elek, P. Ibarbia, A. Risch en J.A. Rabo, J. Catal., 52, 157 (1978).

THE CONVERSION OF METHANOL TO HYDROCARBONS USING A NEW TYPE OF ZEOLITE AS CATALYST. (MOBIL PROCESS)

J.H.C. van Hooff

Laboratory for Inorganic Chemistry
Eindhoven University of Technology
Eindhoven, Netherlands

INTRODUCTION

Mobil R&D Corporation recently announced a new advance in the technology for hydrocarbon production from synthesis gas. They introduced a process for the conversion of methanol to hydrocarbons using a new shape-selective catalyst. (1)
The product distribution is a function of process-variables and catalyst composition and can be altered with respect to its aliphatic and aromatic content.
More significantly, because of the unique structural characteristics of the catalyst, no hydrocarbons beyond C_{11} are produced. Also catalyst activity maintenance is significantly greater than that which has been observed for other similar catalytic materials at the operating conditions of the process.
These characteristics have enabled Mobil scientists to develop a process that yields aliphatic and aromatic hydrocarbons which are predominantly in the gasoline boiling range. (2)
As can be seen from table 1. the gasoline so produced has an unleaded Research-Octane-Number (RON) of about 95 and therefore is superior in both quality and yield to that produced by the SASOL process.
The chemistry and catalysis involved in the Mobil technology has a number of novel aspects which have implications for the synthesis of hydrocarbons from synthesis gas. In this section the science and technology concerning the methanol-to-hydrocarbons process are discussed in some detail. This includes a description of the process, the nature of the catalyst, and some mechanistic considerations.

Table 1. Comparison of the productdistributions of the Mobil and SASOL processes

Product wt %	Fischer-Tropsch process(3) fixed-bed	fluid-bed	Mobil process(4)
Lightgases (C_1+C_2)	11	23	2
L.P.G. (C_3+C_4)	11	29	22
Gasoline (C_5^+)	25	34	76
Fuel Oil (C_{11}^+)	51	5	traces
Oxygenates	2	9	traces
RON of gasoline	75	75	95

DESCRIPTION OF THE PROCESS

The preferred configuration for the Mobil methanol-to-gasoline process consists of a two stage fixed-bed reactor system. (5) (6) (See figure 1)
The crude methanol (contains 17 wt% H_2O) is first passed through the dehydration reactor, filled with a conventional dehydration catalyst, such as γ-Al_2O_3, to form an equilibrium mixture of methanol, dimethylether and water. The effluent is diluted with recycle gas and passed through the conversion reactor filled with a shapeselective zeolite catalyst such as that described as H-ZSM-5 (7).
The recycle gas provides a heat sink to pick up the heat of reaction and limit the temperature rise.
The reactor effluent is condensed, water and liquid hydrocarbon phases are separated, and gas is recycled.
Typical process conditions and yields are shown in table 2.

Table 2. Process conditions for Mobil methanol-to-gasoline process.

Temperature, inlet °C	360
outlet °C	415
Pressure atm.	20
Recycle ratio mole/mole	9
Space velocity WHSV	2
Yields wt% of charge	
methanol + ether	0
hydrocarbons	36.2
water	63.2
CO, CO_2	0.3
coke	0.2

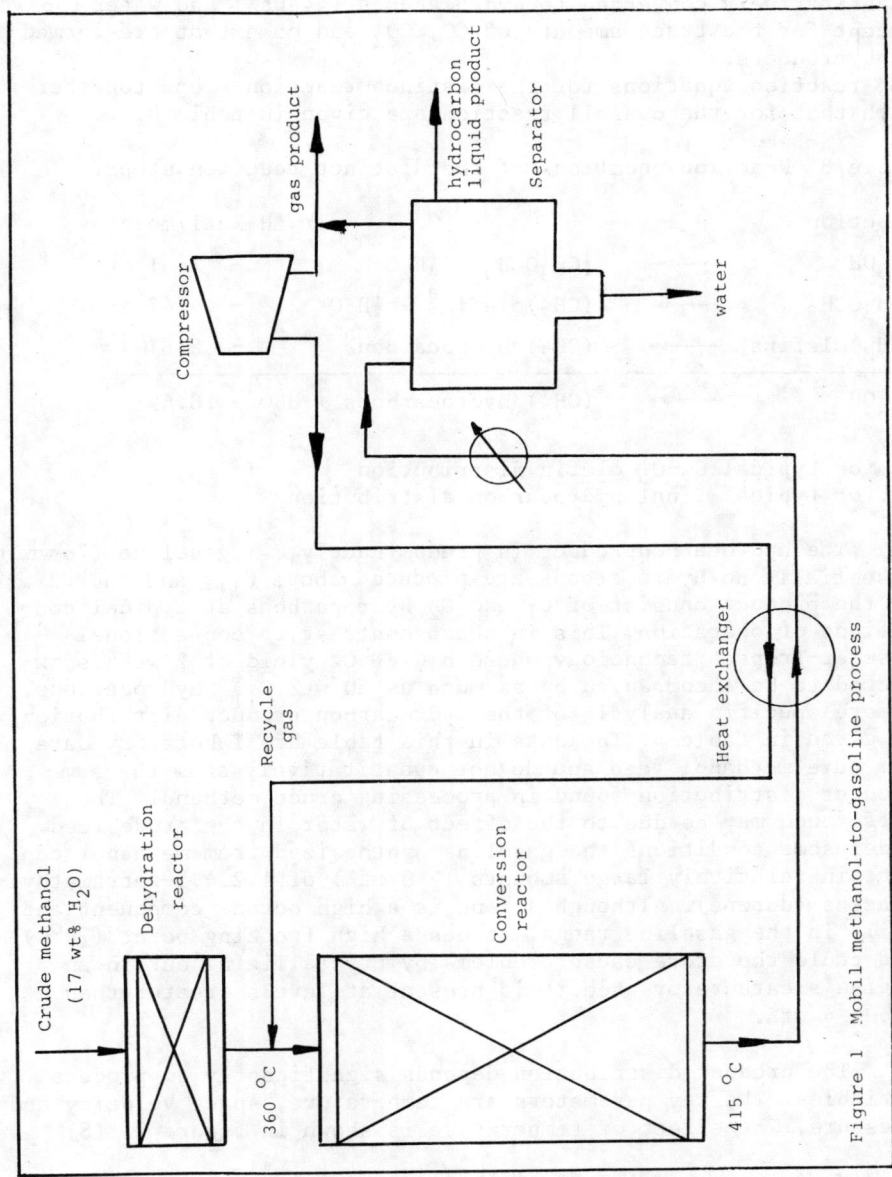

Figure 1 Mobil methanol-to-gasoline process

At these conditions, the original methanol in the crude feet is quantitatively converted to hydrocarbons (44 wt%) and water (56 wt%) except for the trace amounts of CO, CO_2 and coke that are formed as byproducts.

The reaction equations for the distinct reaction steps together with that for the overall reaction are given in table 3.

Table 3. Reaction equations of the distinct reaction steps.

Reaction			ΔH kcal/mole
CH_3OH	$\longrightarrow$	$\tfrac{1}{2}CH_3OCH_3 + \tfrac{1}{2}H_2O$	$-$ 2.41
$\tfrac{1}{2}CH_3OCH_3$	$\longrightarrow$	(CH_2)olefins^a + $\tfrac{1}{2}H_2O$	$-$ 4.47
(CH_2)olefins	$\longrightarrow$	(CH_2) hydrocarbons^b	$-$ 3.81
CH_3OH	$\longrightarrow$	(CH_2) hydrocarbons + H_2O	$-$ 10.69

a. for typical C_2-C_5 olefin distribution
b. for typical final hydrocarbon distribution

The hydrocarbonproduct is predominantly C_5^+ gasoline (76 wt%). Essentially no hydrocarbons are produced above C_{11}, and only 2 wt% of the product consist of C_1 and C_2 hydrocarbons at typical conditions of operation. This in sharp contrast to conventional Fischer-Tropsch technology where a C_1 + C_2 yield of 2 wt% is expected to be accompanied by as much as 50 wt% C_{11}^+ hydrocarbons. A more specific analysis of the hydrocarbon product distribution is given in table 4. The data in this table are laboratory data for pure methanol feed and do not quantitatively show the same product distribution found in processing crude methanol. The difference may be due to the effect of water in the crude feed. Under some conditions the gasoline synthesized from methanol can contain relatively large amounts (> 3 wt%) of 1.2.4.5-tetramethylbenzene (durene). Although durene is a high octane component and boils in the gasoline range, it has a high freezing point (79°C) and could therefore cause problems by crystallizing out in an engine's carburetor when it is present at levels greater than about 4 wt%.

The product distribution depends significantly on process variables. The key parameters are temperature, space velocity and pressure. The effect of temperature is shown in figure 2. (8)

Maximum conversion to C_5^+ hydrocarbons is obtained at a temperature which is just sufficient to give complete conversion of methanol and dimethylether to hydrocarbons (375°-400°C). As the temperature is increased, small amounts of H_2, CO, CO_2 and large amounts of CH_4 and olefins begin to form.

Table 4 Hydrocarbon product distribution for Mobil methanol-to-gasoline process

Reaction conditions

Temperature	^0C	370
LHSV	hr^{-1}	1,0
Pressure	atm	1,0
Conversion	%	100

Hydrocarbon distribution wt%

Methane	1,0		C_6^+ aliphatics	4,3
Ethane	0,6		Benzene	1,7
Ethene	0,5		Toluene	10,5
Propane	16,2		Ethylbenzene	0,8
Propene	1,0		Xylenes	17,2
i-Butane	18,7		C_9 Aromatics	7,5
n-Butane	5,6		C_{10} Aromatics	3,3
Butenes	1,3		C_{11} Aromatics	0,2
i-Pentane	7,8			
n-Pentane	1,3		C_5^+	55,1
Pentenes	0,5		Total Aromatics	41,2

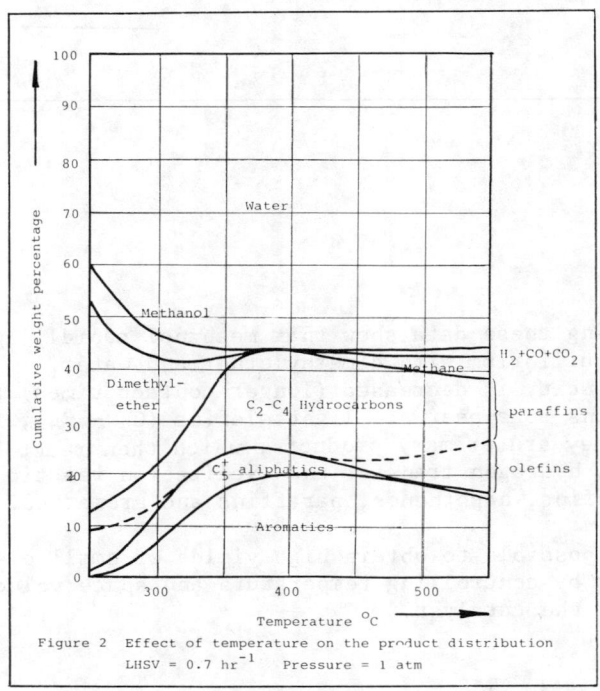

Figure 2 Effect of temperature on the product distribution
LHSV = 0.7 hr^{-1} Pressure = 1 atm

The effect of space velocity is shown in figure 3 (8) where the product distribution for methanol conversion is plotted as a function of contact time (reciprocal of space velocity).

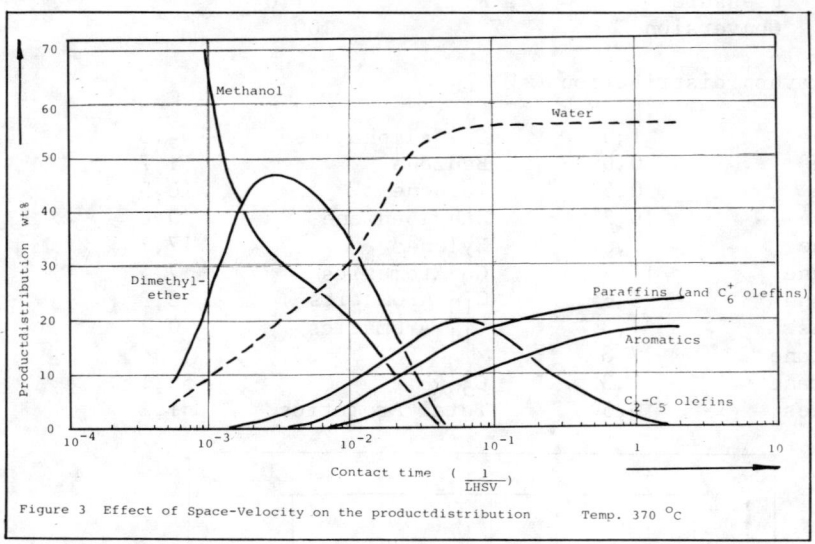

Figure 3 Effect of Space-Velocity on the productdistribution   Temp. 370 °C

Calculations using these data show that methanol rapidly approaches thermodynamic equilibrium with dimethylether and water.
As the space velocity is decreased (longer contact times) the first hydrocarbons to appear are light olefins ($C_2 - C_5$), indicating that they are primary products, which then react by oligomerization, hydrogen transfer and cyclization reactions to longer chain olefins, naphthenes, paraffins and aromatics.

It is also possible to obtain high yields of small olefins and/or aromatics by controlling temperature and space velocity, and by modifying the catalyst.

THE NATURE OF THE CATALYST

The key component in the Mobil process is a shape-selective zeolite catalyst capable of selectively converting methanol or dimethylether to low-molecular-weight hydrocarbons with minimum catalyst deactivation due to thermal degradation and coke deposition.

The class of zeolites which are used in the Mobil technology are intermediate in pore dimensions between the well-known wide-pore faujasites (X and Y) and the narrow-pore zeolites such as A zeolite and erionite. These zeolites exhibit three important characteristics:
- a Si/Al ratio of at least 15
- a crystal density greater than 1.6 g/ml
- and a optimum range for a parameter termed by Mobil workers as the 'constraint index'. (1)
  (1-12 preferably 2-7)

This constraint index is approximated by the ratio of cracking rate constants for n-hexane and 3-methyl pentane over a catalyst at a given set of conditions. The index is a measure of the steric constraints on molecular mobility in the zeolite structure. When there are no diffusivity effects, 3-methyl pentane is expected to crack to a greater rate than n-hexane.
Several zeolite classes meet these critical constraints inclucing ZSM-5 (7), ZSM-11 (9), ZSM-21 (10) and TEA-mordenite.
However it is clearly stated in the Mobil patent literature that the ZSM-5 class of zeolites are the preferred catalysts.
The crystal structure of ZSM-5 zeolite (11) and of the related silica molecular sieve 'silicalite' (12) have been reported recently. The framework of ZSM-5 contains a novel configuration of linked tetrahedra (shown in figure 4), consisting of 8 five-membered rings. These units join through edges to form chains (shown in figure 5).
The chains can be connected to form sheets (shown in figure 6) and the linking of the sheets lead to a threedimensional framework structure with orthorhombic symmetry. This framework contains two intersecting channel systems, one straight and parallel to [010] and the other sinusoidal running parallel to [001] as indicated in figure 7a.

The 10-membered ring openings (clearly visible in figure 6) controlling the channels have an effective diameter of about 7A°. The dimensions of these channels are in good agreement with sorption data (13).

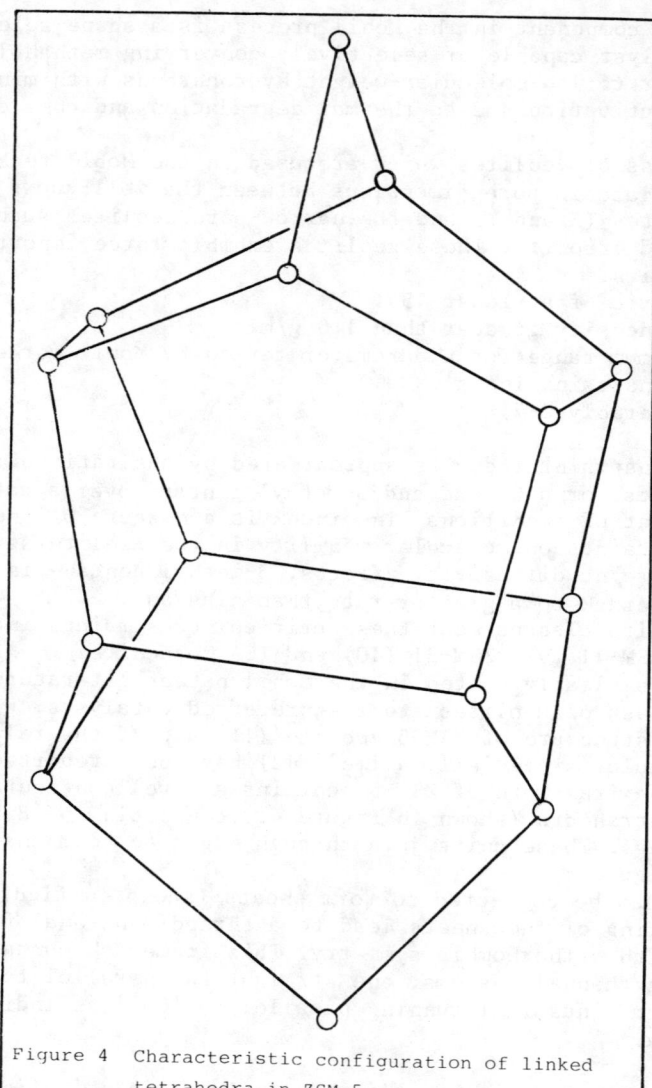

Figure 4 Characteristic configuration of linked tetrahedra in ZSM-5

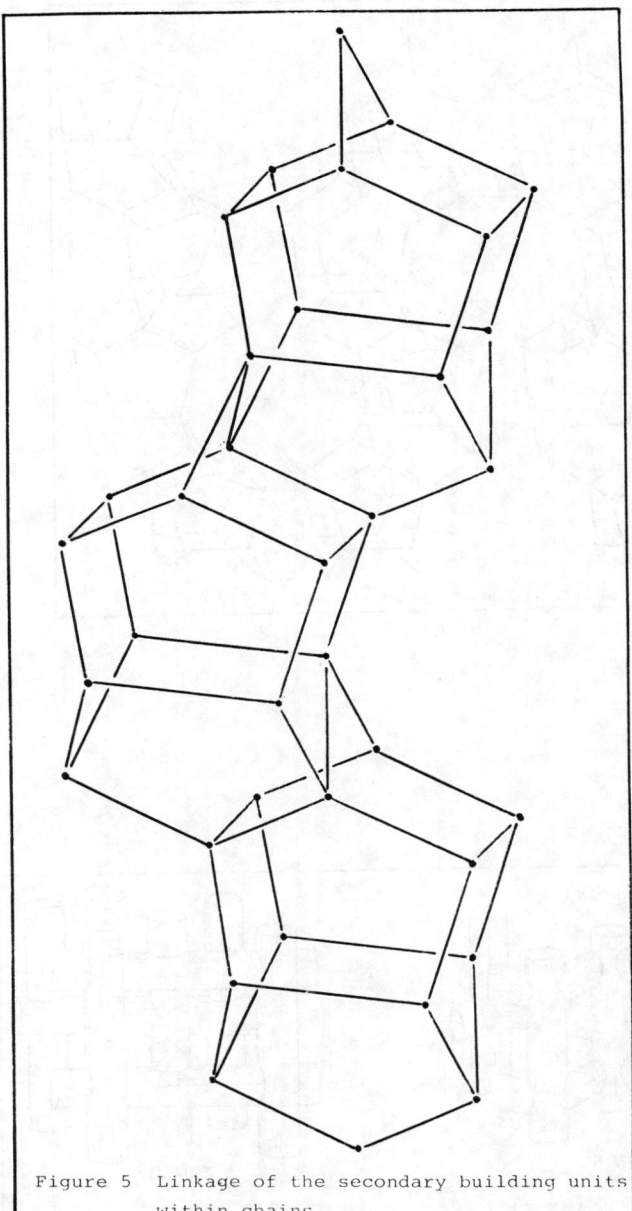

Figure 5 Linkage of the secondary building units within chains

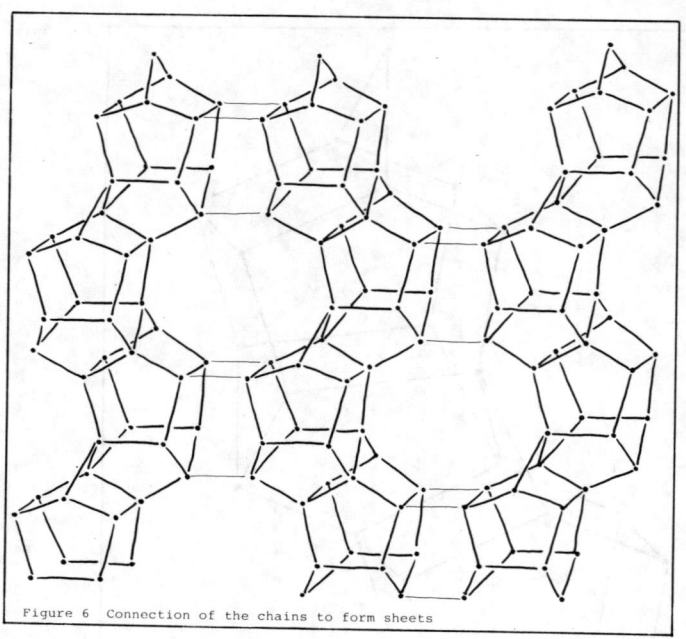
Figure 6 Connection of the chains to form sheets

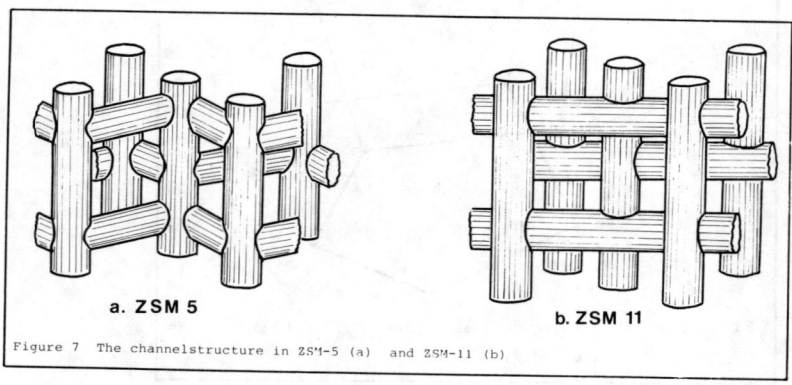

a. ZSM 5

b. ZSM 11

Figure 7 The channelstructure in ZSM-5 (a) and ZSM-11 (b)

Methanol to hydrocarbons by ZSM-5

The crystal structure of ZSM-11 (14) is strongly related to that of ZSM-5. The only difference is the connection of the sheets as shown in figure 6 to a threedimensional network with tetragonal symmetry. As in ZSM-5 the ZSM-11 framework contains two intersecting channel systems with 10-membered ring openings but unlike those in ZSM-5 they are both straight (see figure 7b).
The ZSM-5 zeolite is prepared in the presence of a tetrapropylammonium compound (such as $(CH_3CH_2CH_2)_4NOH$) whereas ZSM-11 zeolite is prepared in the presence of a tetrabutylammonium compound. Apparently the size of the tetra-alkylammonium cation affects the geometry and the resultant pore size of the final zeolite by interactions during the genesis of the crystalline material.

A typical composition of the crystallization mixture for ZSM-5 zeolite is (7):

Compound	mole ratios		
	Na-aluminate solution	SiO_2-gel solution	Total
$(C_3H_7)_4NOH$	–	18	18
NaOH	2.4	–	2.4
Al_2O_3	1.0	–	1.0
SiO_2	–	30	30
H_2O	10	~ 480	~ 490

Crystallization is carried out at $150°C$ during about 5 days.

The resulting zeolite has approximately the following composition

$$Na_2[(C_3H_7)_4N]_4 \ [(AlO_2)_6(SiO_2)_{90}] \cdot nH_2O$$

Upon calcination at $540°C$ in air decomposition of the tetrapropylammonium cation occurs resulting in the very stable zeolite ZSM-5 with formula:

$$Na_2H_4[(AlO_2)_6(SiO_2)_{90}]$$

The sodium content can be lowered by a treatment with a 0.5 molar HCl solution at $80°C$ resulting in the socalled H-ZSM-5.
(During this treatment also some aluminum is extracted causing an increase of the Si/Al ratio).
A typical formula for H-ZSM-5 is:

$$Na_{0.1}H_{4.9}[(AlO_2)_5(SiO_2)_{91}]$$

The strength of the acid sites of H-ZSM-5 can be estimated from the reactivity of paraffins over this material, as is done when determining the 'constraint-index'. The results of this test are given in table 5.

Table 5 The determination of the Constraint-Index of H-ZSM-5

Feed: 50:50 mixture of n-hexane and 3-methylpentane at LHSV:1

mole% C_6 converted in	50% H-ZSM-5/SiO_2 300°C	50% LaY/SiO_2 300°C	500°C
C_2	0.5	0.3	2.9
C_3	12.3	1.0	29.0
C_4	19	2.8	9.4
C_5	7.2	1.9	3.2
2m C_5	1.5	15.4	2.6
3m C_5	40.1	30.6	20.7
n C_6	18.5	48.1	32.2
CI	4.51	–	0.50

$$CI = \frac{\text{rate constant for n-hexane cracking}}{\text{rate constant for 3-methylpentane cracking}}$$

$$= \frac{\log(\text{fraction unreacted n-hexane})}{\log(\text{fraction unreacted 3-methylpentane})}$$

For H-ZSM-5: fraction unreacted n-hexane = $\frac{18.5}{50}$ = 0.37

fraction unreacted 3-methylpentane = $\frac{40.1}{50}$ = 0,80

$$CI = \frac{\log 0,37}{\log 0,80} = 4.51$$

As can be seen in this table already at 300°C about 63% of the n-hexane is converted with H-ZSM-5 whereas with LaY even at 500°C the n-hexane conversion is limited to about 36%. Therefore it seems likely that H-ZSM-5 is much more strongly acidic than LaY. This is also confirmed by IR, microcalorimetric and ESR measurements (14) (15).

MECHANISTIC CONSIDERATIONS

From the effect of temperature and space velocity on the product distribution Chang and Silvestri (8) concluded that the overall reaction sequence is likely

$$2CH_3OH \xrightarrow{-H_2O} CH_3\text{-}O\text{-}CH_3 \xrightarrow{-H_2O} C_2\text{-}C_5 \text{ olefins} \rightarrow \begin{array}{l} C_6^+ \text{ olefins} \\ \text{Paraffins} \\ \text{Naphthenes} \\ \text{Aromatics} \end{array}$$

Methanol to hydrocarbons by ZSM-5

This reaction scheme could be confirmed by ^{13}C-NMR measurements (16) as shown in figure 8.

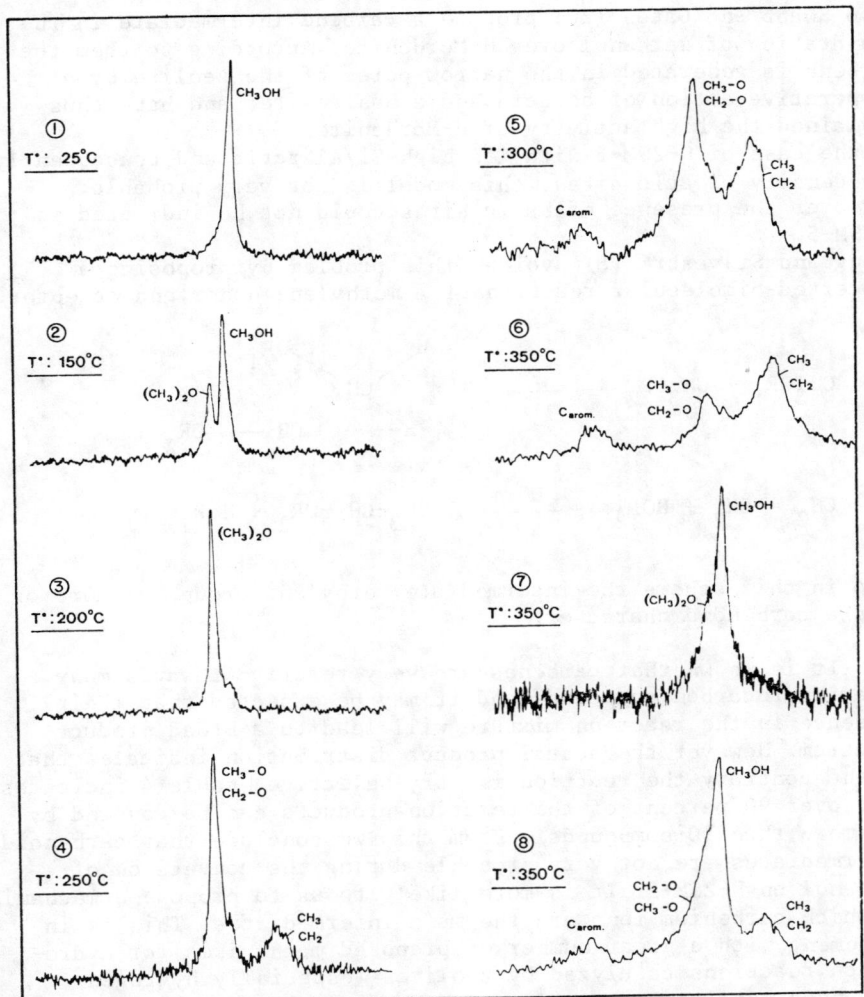

Figure 8 Typical ^{13}C-NMR spectra as observed during the static conversion of methanol

However untill now not much is published about the mechanisms of the various reaction steps.
Several authors described the dehydration of alcohols over alumina (17)(18) and crystalline aluminosilicates (19) (20) (21) (22).
In their study of the formation of stilbenes by the reaction of benzyl-mercaptans over 13X and 4A zeolites, Venuto and Landis (19)

propose a carbene intermediate generated by an α-elimination process and stabilized by the electrostatic field of the zeolite lattice.

Also Swabb and Gates (20) propose a carbene intermediate in the dehydration of methanol over H-Mordenite. According to them the carbene is generated in the narrow pores of the zeolite by a cooperative action of an acid and a basic site; and have thus explained the high activity of H-Mordenite.

In the case of H-ZSM-5 with its high Si/Al ratio and consequently low density of acid sites, this model is not very probable. Moreover the presence of basic sites could not be indicated in H-ZSM-5.

Chang and Silvestri (8) avoided this problem by proposing a concerted bimolecular reaction of a methylene donor and acceptor.

$$CH_3OR_1 + CH_3OR_2 \longrightarrow \begin{array}{c} H \diagdown \quad \diagup OR_1 \\ CH_2 \\ H \dashrightarrow CH_2 \text{---} OR_2 \end{array}$$

$$CH_2 = CH_2 + HOR_2 \xrightarrow{H^+} CH_3\text{-}CH_2\text{-}OR_2 + HOR_1$$

Also in this scheme the intermediate methylene group has more or less a carbenoïd character.

It is known that carbenes are very reactive towards many other hydrocarbon compounds and it may be expected that their presence in the reaction mixture will lead to a broad product spectrum. However the actual product distribution indicates that on the contrary the reaction is very selective (table 4 indicates that over 90 percent of the reaction products can be covered by not more than 10 compounds). From this we conclude that carbenoïd intermediates are not very probable during the conversion of methanol on H-ZSM-5. It is more likely to us to propose a mechanism in which carbenium ions are the main intermediates. This is in agreement with a great number of proposed mechanisms for hydrocarbon reactions catalyzed by zeolites (especially hydrogen exchanged zeolites) e.g. alcohol dehydration, olefin isomerisation, oligomerisation and cyclisation (21 (22).

In the following discussion we will divide the reaction scheme in five separate reaction types
1. the formation of dimethylether
2. the formation of ethene
3. the formation of higher olefins
4. the formation of paraffins and cyclic hydrocarbons
5. the formation of aromatics.

For each of these reaction types a mechanism will be proposed that is based upon the action of the acid sites in H-ZSM-5.

1. The formation of dimethylether

To explain the conversion of methanol to dimethylether we propose a bimolecular Rideal type mechanism (figure 9 eq 1) as already proposed by Jacobs (21).
A slightly different mechanism is proposed by Salvador and Kladnig (23) and is shown in figure 9 eq 2.

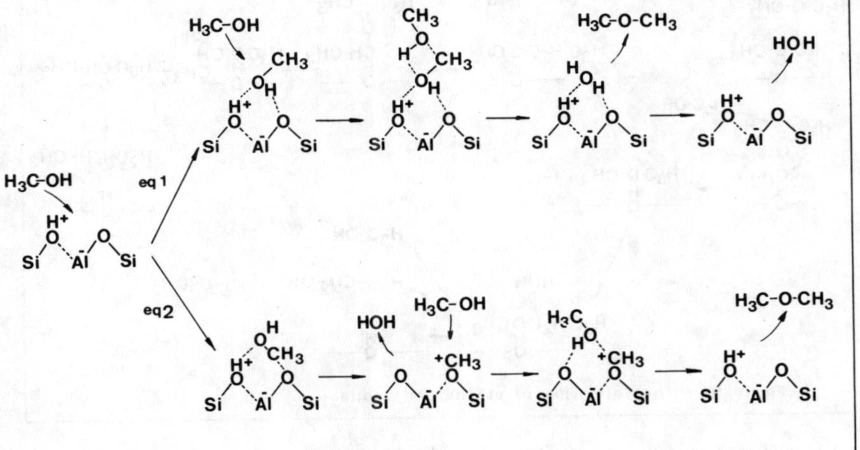

Figure 9 The formation of dimethylether

2. The formation of ethene

The mechanism for this reaction is still in discussion. Possibly the reaction proceeds via an intermolecular reaction as indicated in figure 10.

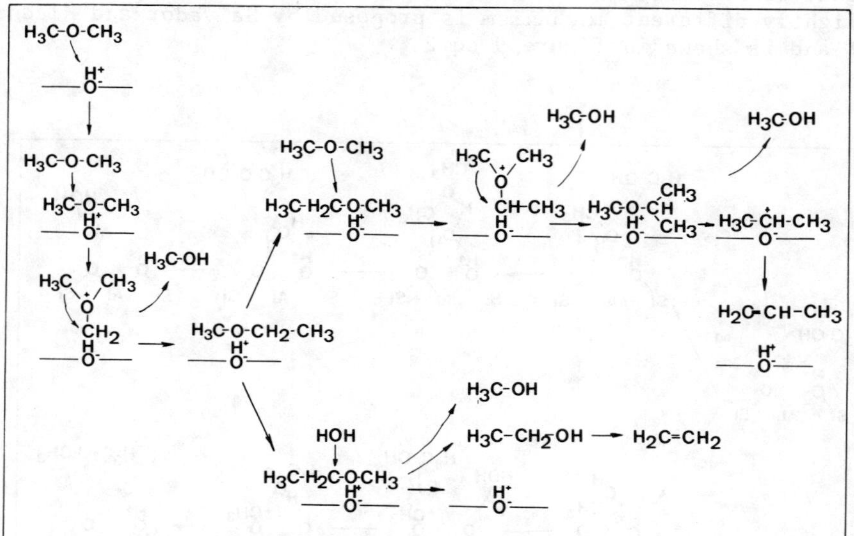

Figure 10 The formation of ethene and propene

3. The formation of higher olefins

When ethene is present an ethylcarbenium ion can be formed by a reaction with the Brønsted-acid sites of the zeolite. This ethylcarbenium ion can react either with a second ethene molecule to form butene (figure 11 eq 3) or with a dimethylether molecule to form propene and methanol (figure 11 eq 4).
The so obtained alkenes can now react further in the same way i.e. propene can form a propylcarbenium ion that can react either with an ethene molecule to form isopentene a with a dimethylether molecule to form isobutene and methanol. In this way a reaction network for the formation of the higher olefins can be built up.

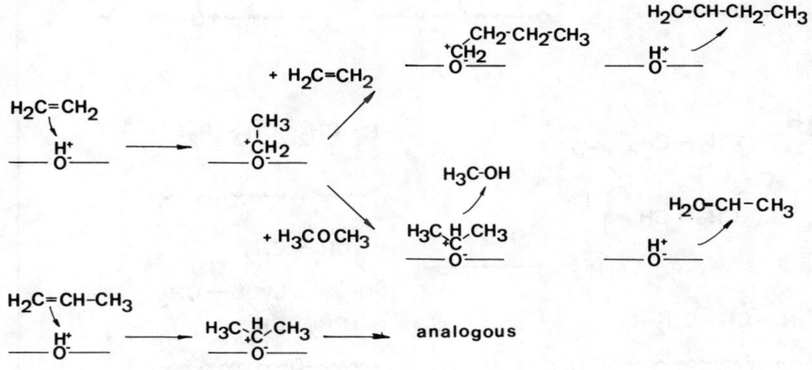

Figure 11 The formation of higher olefins

4. The formation of paraffins and cyclic hydrocarbons

If the process has proceeded so far, that a mixture of small and large olefins is present hydride transfer reactions can occur. This reaction starts with the interaction of an olefin with a Brønsted-acid site of the zeolite to form a carbenium ion. After that a hydride ion is transfered from a second olefin to the carbenium ion to form a parafin and a new carbenium ion (figure 12 eq 5). This reaction is thermodynamically favoured if the hydrogen is transfered from a large to a small olefin because of the relatively high stability of the so formed small paraffin. It is also in accordance with the presence of large amounts (on mole base) of small alkanes (mainly, propane, iso-butane and iso-pentane)

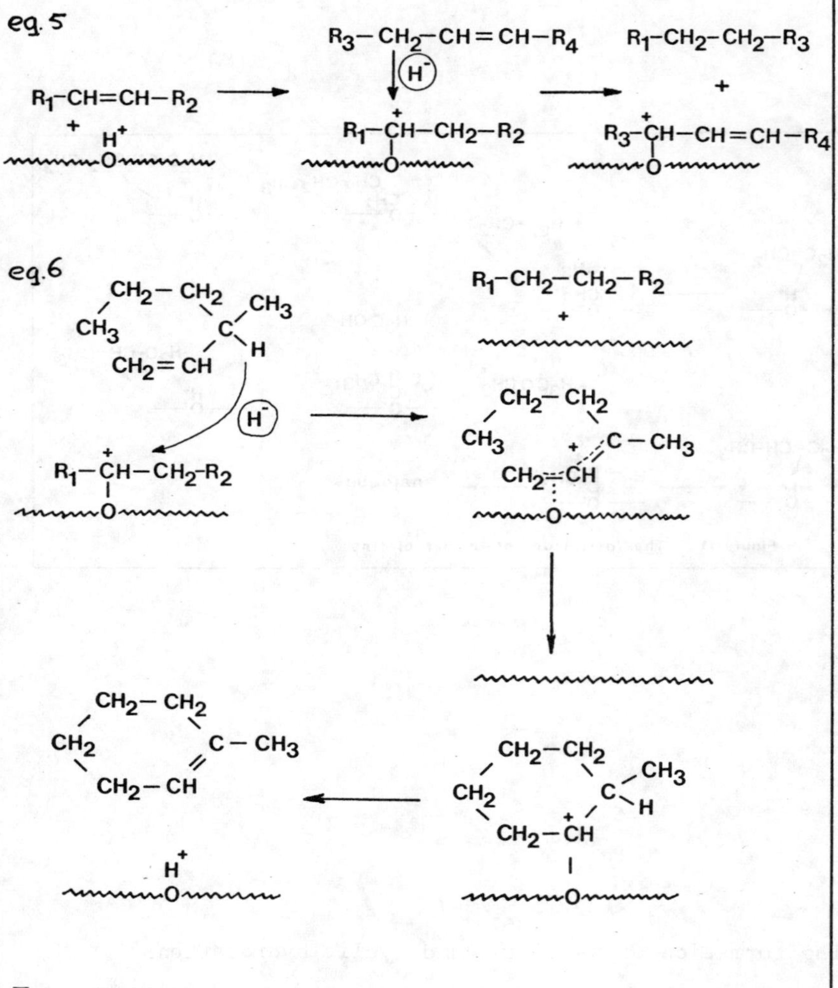

Figure 12
THE FORMATION OF PARAFINS AND CYCLIC HYDROCARBONS

in the reaction product (see table 4). If the olefin from which the hydride ion has been abstracted has 6 or more linearly bounded C-atoms an allylic carbenium ion can be formed that on its turn can be transformed to a cyclic hydrocarbon with one double bond (figure 12 eq 6).

In this cyclization reaction the geometry of the pores may play an important role. The dimensions of the pores (about 7 A°) could favour the formation of six-membered rings.

5. The formation of aromatics

Once these cyclo hexenes have been formed a further 'dehydrogenation' can occur to form the thermodynamically more stable corresponding aromatics. The 'dehydrogenation' in this case is a two-step process.
First a hydride ion is transfered to a (small) carbenium ion, followed by an abstraction of a proton to form a cyclo-hexadien.

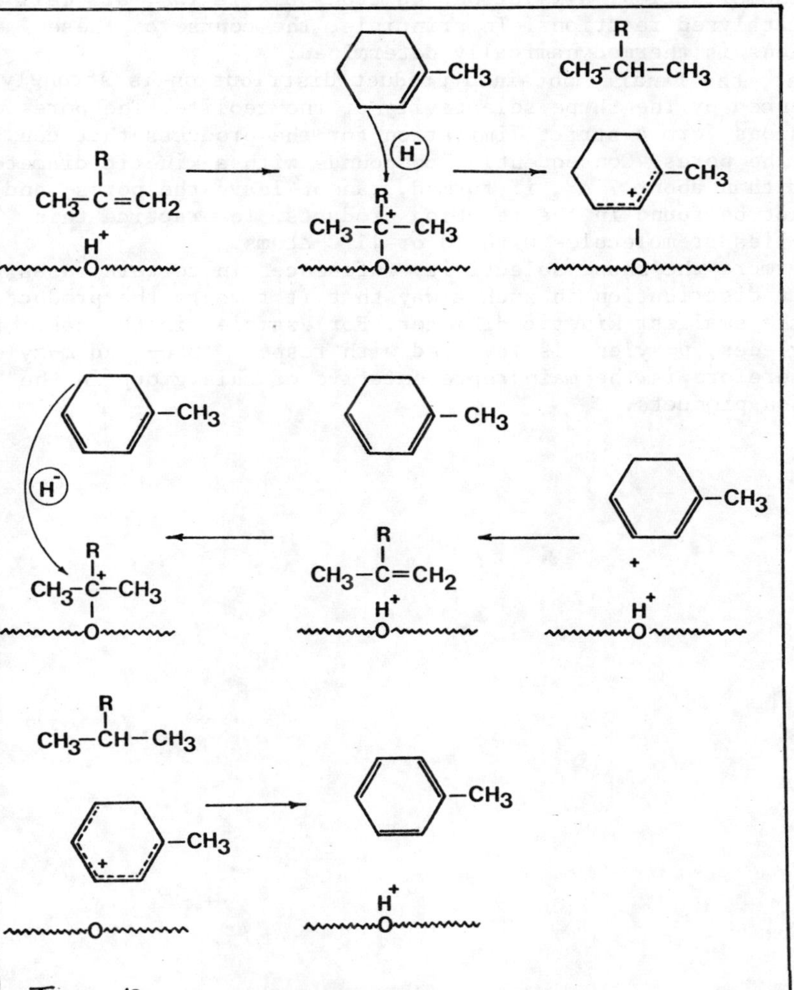

Figure 13 THE FORMATION OF AROMATICS

5. The formation of aromatics

Once these cyclo-hexenes have been formed a further 'dehydrogenation' can occur to form the thermodynamically more stable corresponding aromatics. The 'dehydrogenation' in this case is a two-step process.
First a hydride ion is transfered to a (small carbenium ion, followed by an abstraction of a proton to form a cyclo-hexadiene. In the second step this process is repeated resulting in an aromatic compound (figure 13).

Besides the reactions mentioned above, several other reactions will occur inside the pores. The most important of these reactions are: isomerisation, alkylation and trans-alkylation, all well-known acid catalyzed reactions. In principle, the course of these reactions is thermodynamically determined.
However, the finally obtained product distribution is strongly influenced by the shape selectivity of the zeolite. The pore dimensions form a strict limitation for the products that can leave the pores. Consequently, compounds with a kinetic diameter larger than about 7 A^o, if formed, cannot leave the pores, and will not be found in the reaction products. In practice this limit lies at molecules with 10 or 11 C-atoms.
Furthermore the shape selectivity influences in certain cases the product distribution in such a way that it favours the product with the smallest kinetic diameter. For example, in the case of the xylenes, p-xylene is favoured with respect to o- and m-xylene and therefore is the main representative of this group in the reaction products.

REFERENCES

1. Mobil Oil Corp., Dutch Patent Appl. 7410583, Aug. 6, 1974.
2. S.L. Meisel, J.P. McCullough, C.H. Lechthaler, P.B. Weisz, Chem. Technol. 6, 86 (1976).
3. C.D. Frohning, B. Cornils, Hydrocarbon Processing 53, 11, 143 (1974)
4. A.J. Silvestri, Lecture July 1976, Geleen, Netherlands.
5. J.C.W. Kuo, U.S. Patent 3,931,349, Jan. 6, 1976.
6. J.J. Wise, A.J. Silvestri, Oil and Gas J., Nov. 22, 1976, 140.
7. R.J. Argauer, G.R. Landolt, U.S. Patent 3,702,866, Nov. 14, 1972.
8. C.D. Chang, A.J. Silvestri, J. Catalysis 47, 249 (1977).
9. P. Chu, U.S. Patent 3,709,979, Jan. 9, 1973.
10. C.J. Plank, E.J. Rosinski, M.K. Rubin, U.S. Patent 4,046,856, Sept. 6, 1977.
11. G.T. Kokotailo, S.L. Lawton, D.H. Olson, W.M. Meier, Nature 272, 437 (1978).
12. E.M. Flanigen, J.M. Bennett, R.W. Grose, J.P. Cohen, R.L. Patton, R.M. Kirchner, J.V. Smith, Nature, 271, 512 (1978).
13. S.L. Meisel, J.P. McCullough, J.P. Lechthaler, P.B. Weisz, ACS Meeting, Chicago (1977).
14. A. Auroux, V. Bolis, P. Wierzchowski, P.C. Gravelle, J.C. Vedrine, accepted for publication in J.C.S. Faraday Trans I.
15. J.C. Vedrine, A. Auroux, V. Bolis, P. Dejaifve, C. Naccache, P. Wierzchowski, E.G. Derouane, J.B. Nagy, J.P. Gilson, J.H.C. van Hooff, J.P. van den Berg, J.P. Wolthuizen, accepted for publication in J. Catalysis.
16. E.G. Derouane, J.B. Nagy, P. Dejaifve, J.H.C. van Hooff, B.P. Spekman, J.C. Vedrine, C. Naccache, J. Catalysis 53, 40 (1978).
17. E.I. Heiba, P.S. Landis, J. Catalysis 3, 471 (1964).
18. H. Knözinger, Angew. Chem. 80, 778 (1968).
19. P.B. Venuto, P.S. Landis, J. Catalysis 21, 333 (1971).
20. E.A. Swabb, B.C. Gates, Ind. Eng. Chem. Fuud. 11 (4) 540 (1972).
21. P.A. Jacobs, 'Carboniogenic Activity of Zeolites', Elsevier, Amsterdam (1977).
22. P.B. Venuto, P.S. Landis, Adv. Catalysis 18, 259 (1968).
23. P. Salvador, W. Kladnig, J.C.S. Faraday Trans I 1153, (1977).

LIQUEFIED COAL BY HYDROGENATION*

B. C. Gates

Center for Catalytic Science and Technology
Department of Chemical Engineering
University of Delaware
Newark, Delaware   19711   U.S.A.

1. INTRODUCTION

Using coal directly as either a fuel or chemical feedstock is fraught with difficulties and environmental problems. Coal liquefaction and gasification lead to cleaner products which are potential petroleum replacements [1]. The Bergius process for liquefaction/hydrogenation spawned several descendants with good prospects of commercial application, including the Exxon Donor Solvent (EDS), Gulf Solvent Refined Coal (SRC), and H-Coal^R processes (Table 1) [2]. In all these, coal particles are first slurried in recycled liquid product and are broken up. They then accept hydrogen and give heavy, highly aromatic liquids that contain a lot of sulfur and nitrogen. In the following paragraphs I'd like to discribe what we know of the chemistry of these processes, which is becoming sufficiently well defined to allow us to identify reaction types, illuminate the role of catalysts, and recognize opportunities for controlling product distributions.

2. THE FIRST STEP

Even though there are frequent references to "catalytic liquefaction" of coal, it is clear that the initial break-up of the organic coal matrix cannot be accelerated by an added solid, since

---

*Reproduced with permission from CHEMTECH, February, 1979. Copyright by the American Chemical Society.

Table 1. Processes for liquefaction/hydrogenation of coal

Process	Reactor type(s)	Typical catalyst	Typical liquefaction conditions		Distinctive Characteristics
			Temp., C	Pressure, atm	
Bergius--single stage for liquefaction and hydrogenation	Stirred-batch reactor	Iron oxide	465	200	Severe conditions, throwaway catalyst
SRC--single stage for liquefaction and hydrogenation; recycle of product liquid	Tubular flow reactor	Coal mineral matter	450	140	No added catalyst; lack of specificity in hydrogenation
H-Coal^R--one or more stages; liquefaction of coal and catalytic hydroprocessing of product liquid occur in the same reactor	Continuous flow slurry-bed reactor	Hydroprocessing catalyst like sulfided CoO-MoO$_3$/Al$_2$O$_3$	450	200	Liquefaction and catalytic conversion of products in the same reactor; rapid catalyst aging
EDS--first stage for liquefaction in the presence of hydrogen-rich donor solvent (recycle liquid) and second stage for catalytic hydroprocessing of liquid product.	First stage, continuous tubular flow reactor; second stage, trickle-bed flow reactor	In liquefaction stage, coal mineral matter; in trickle-bed reactor, hydroprocessing catalyst like CoO-MoO$_3$/Al$_2$O$_3$	450	120	Liquefaction in the presence of good hydrogen-donor species for high yields of liquid products; catalytic hydroprocessing of liquids in a separate reactor for selective hydroprocessing with relatively slow catalyst deactivation.

Liquefied coal by hydrogenation

catalysis requires intimate contacting and that cannot be achieved until the break-up has begun. There is strong evidence that there is first a pyrolitic break-up [3-7]. Neavel [6] has measured the kinetics of the conversion of coal particles into gases plus benzene-soluble and pyridine-soluble liquids at 400°C (carefully avoiding the confusion of a long heat-up period). He has collected precise data for reaction times as short as seconds. The results (Figure 1) show that about 10% of a bituminous coal was converted to gases plus benzene-soluble liquids in about 20 seconds, and 90% was converted to gases plus pyridine-soluble components in about 2 minutes. Microscopic examination of the solids at early stages of disintegration [6,7] demonstrates the physical nature of the initial break-up: gaseous pyrolysis products form within seconds, expanding the softened coal particles; organic material also dissolves. The initial rate of coal conversion is independent of whether a good hydrogen donor like tetralin (1,2,3,4-tetrahydronaphthalene) is present (Figure 1). Thus the initial pyrolitic break-up is accelerated neither by an added catalyst nor by a hydrogen-donor solvent. After longer times, however, the donor solvent gives much higher yields than a non-donor solvent like naphthalene, and in the presence of naphthalene, the yield even passes through a maximum (Figure 1). Behavior like this has been observed often and is attributed to formation of products like coke, which are highly resistant to further conversion.

All the results mentioned above--and results obtained with model compounds under conditions similar to those of coal liquefaction [8]--support the idea that the intermediates in the disintegrating organic coal matrix are free radicals. Breaking of C-C and C-O bonds accounts for the following:

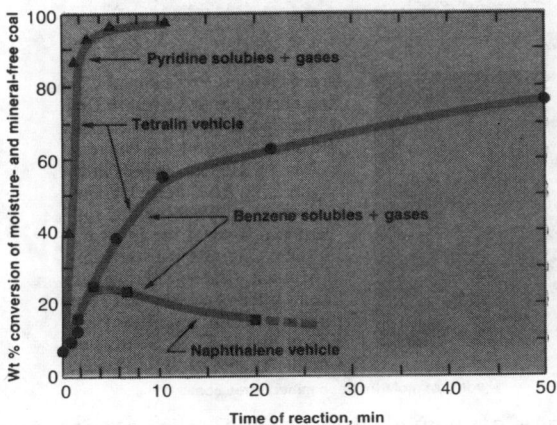

Figure 1. Conversion of a bituminous coal (rank HVCB) at 400°C; comparison of hydrogen-donor solvent (tetralin) with non-donor naphthalene solvent [6].

• evolution of some light products directly split out of the coal matrix,

• production of highly reactive intermediates (free radicals) capable of abstracting hydrogen from molecules like tetralin or from neighboring, hydrogen-rich groups in the coal itself, and

• further reaction of the highly reactive intermediates with one another to give higher-molecular-weight materials, including coke.

The several percent of organic material which finally remain unconverted (Figure 1) are mostly fusinite--macerals roughly resembling charcoal. [The term <u>maceral</u> refers to organic components in the way that the term <u>mineral</u> refers to inorganic components in coal. Macerals are microscopically identifiable products of the coalification of what were originally organs and tissues of plants.] The other macerals, including the predominant one, vitrinite--derived in the coalification process from bark or woody tissue--are almost completely converted [9,10]. Patterns of reactivity in liquefaction depend both on the maceral content and the rank of the coal [9,10] (Figure 2). Lignite is the lowest-rank coal; it has not undergone as much coalification as those of higher ranks, bituminous and anthracite coals. Work has already been reported [9,10a] suggesting how maceral content and rank can be of value in predicting the reactivities of new coals. One should not expect much accuracy in these predictions, however, since the correlations are rough and since coals are far from uniform even within a given seam.

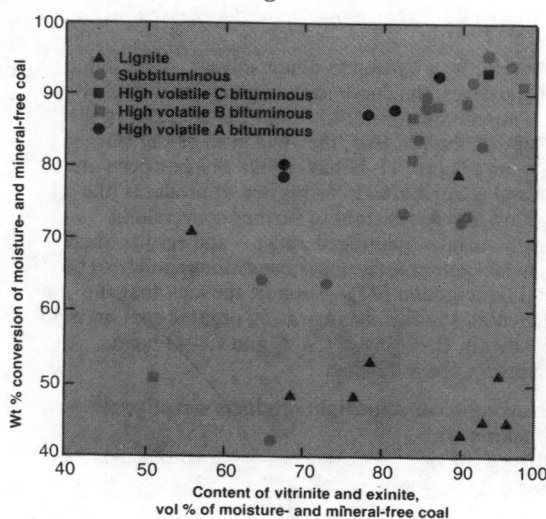

Figure 2. Dependence of conversion in coal liquefaction/hydrogenation on coal rank and maceral content: conversion at 385°C and 238 atm was determined in the presence of hydrogen, partially hydrogenated anthracene oil, and an unspecified catalyst [11].

Only recently have the products of coal liquefaction begun to yield to detailed chemical characterization [12]. Farcasiu et al. [12a] fractionated products by liquid chromatography and identified 10 component classes which contain predominantly one of the following: saturates, aromatics, polar aromatics plus nonbasic sulfur-, nitrogen-, and oxygen-containing compounds, polyphenols, and so forth. The subgroup of component classes soluble in hexane was called oils, the remainder soluble in benzene was called asphaltenes (according to the familiar notion), and the remainder soluble in pyridine was called asphaltols, since it had a high concentration of -OH groups; this last fraction is often referred to as preasphaltenes, since it appears to be an intermediate in asphaltene formation [12a,13]. There are few data to allow comparison of different coals or of the several liquefaction processes in terms of the yields of various product fractions, but it seems to be fruitful to proceed assuming that there might be general validity in representing these fractions as characteristic liquefaction products. Further analytical work can be expected to provide better separation methods, more detailed characterization of the products, and improved definitions of the component classes. High-resolution mass spectrometry and ^{13}C nmr spectroscopy should be especially valuable.

3. CATALYSIS

With an understanding of the pyrolytic coal break-up, we can begin to understand the role of catalysts in liquefaction/hydrogenation processes. Many catalysts have been tested [14] and those most commonly used are hydroprocessing catalysts, typically prepared from a support of porous, high-surface-area (~ 200 m^2/g) γ-Al$_2$O$_3$ with added oxides of molybdenum (or tungsten) and cobalt (or nickel). Catalysts of this class were first applied in coal liquefaction/hydrogenation in prewar Germany, and they have found wide application for trickle-bed hydroprocessing of petroleum, especially hydrodesulfurization. In operation, the catalyst surfaces are converted from oxides into sulfides; these catalysts are the most active known hydrogenation catalysts in use with sulfur-containing compounds. Metals like platinum are intrinsically much more active hydrogenation catalysts than the metal sulfides, but they are strongly poisoned by sulfur and cannot be used in processing of coal or heavier petroleum fractions.

The sulfide catalysts are active for hydrogenation of the unsaturated products liberated in the break-up of the coal matrix. Compounds like naphthalene are transported rapidly to the catalyst surface, where they are converted into compounds like tetralin. The catalytic hydrogenation taking place in a liquefaction reactor replenishes the donor solvent; tetralin and similar hydrocarbons play the role of hydrogen carriers [5,6,8]--they give up hydrogen to the disintegrating coal matrix, return to the catalyst surface, where they react with dissociatively adsorbed hydrogen, and then shuttle back to the coal. Figure 3 is a rough scheme of this cycle.

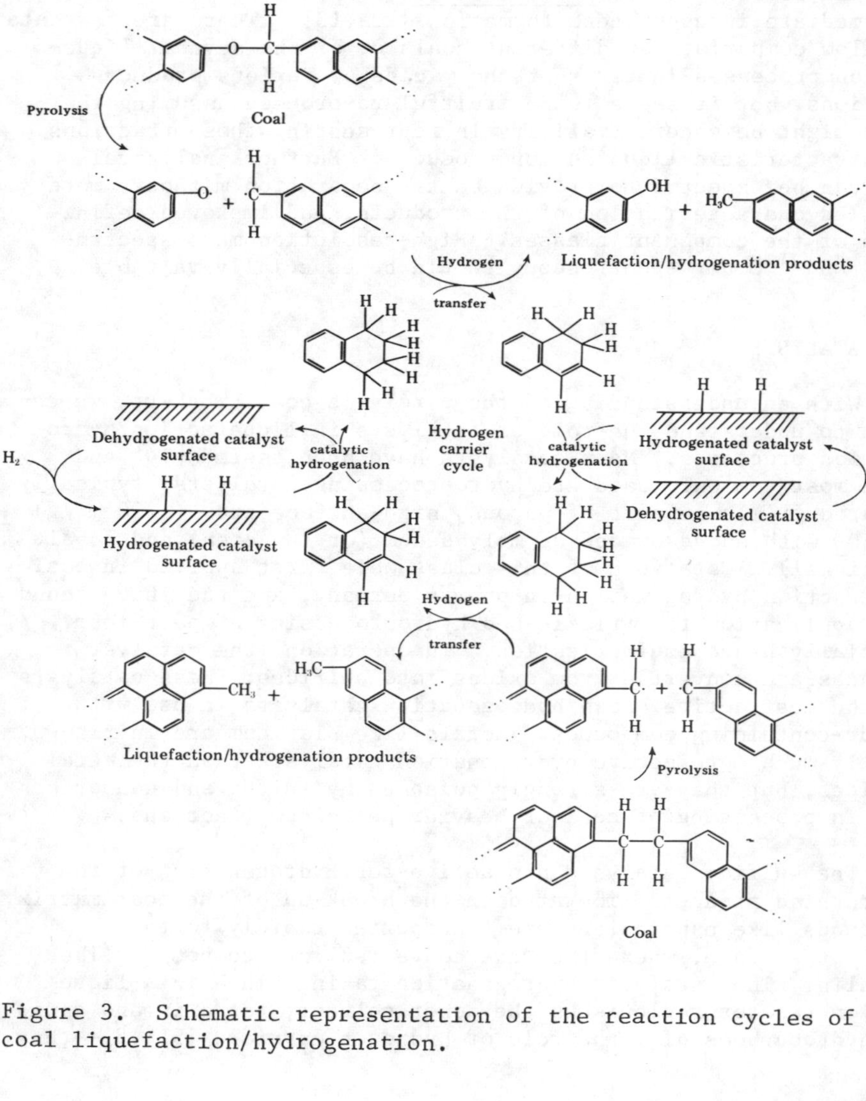

Figure 3. Schematic representation of the reaction cycles of coal liquefaction/hydrogenation.

We can now identify one of the critical questions in process design: how and where should the catalytic hydrogenation be carried out? One possibility, exemplified in the H-Coal^R process, is to include the catalyst in the reactor where the liquefaction takes place. To some extent, this kind of operation is unavoidable, since the mineral matter of coal (including pyrites, clays, and many minor constituents) has some activity as a hydrogenation catalyst [15,16]. Another possibility, exemplified in the EDS process, is to confine the hydroprocessing catalyst within its own reactor and separately optimize the liquefaction and catalytic hydroprocessing operations.

4. DIFFERENCES

The operation of the EDS process is sensitive [17,18] to an undefined "solvent quality index," surmised to be a measure of the reactivity (and perhaps also the capacity) of the recycle liquid which acts as a hydrogen donor in the presence of the liquefying coal. The solvent quality index can be increased by increasing the amount of hydrogen added to the recycle solvent in a trickle-bed catalytic reactor [17]. There is flexibility in the operation with regard to how much hydrogen is transferred to the coal in the liquefaction reactor and how much is transferred to the liquid product in a downstream reactor. The amount of hydrogen transfer required in the liquefaction reactor increases sharply with increased conversion (Figure 4), and for the optimum to be realized, some of the organic part of the coal must remain unconverted. The unconverted solid is separated from the liquid in the EDS process by vacuum distillation; it can be steam gasified to provide some of the hydrogen needed for hydroprocessing.

An essential difference between the H-Coal^R process and the EDS processes is therefore the lack of opportunity in the former for separately optimizing the coal liquefaction and recycle solvent hydrogenation. But it may well be that the H-Coal^R process offers some undetermined advantage in having the hydrogenation catalyst immediately available to the fragments split out from the coal; perhaps a rapid catalytic conversion of some of these fragments leads to desirable products, whereas when the fragments are allowed to react with each other in the absence of catalyst, the result may be undesired products like coke.

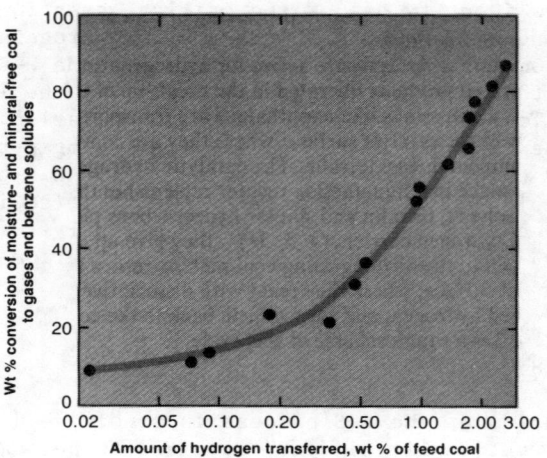

Figure 4. Dependence of coal conversion to benzene solubles and gases on the amount of hydrogen transferred from tetralin donor; bituminous coal (rank HVCB) at 400°C [6].

5. COMPETITION FOR HYDROGEN

The SRC process lacks any added catalyst. Some attempts to increase the processing flexibility have been made in modifying the original SRC I process. In the modified process, called SRC II [19] liquid product is recycled and the contact between the catalytically active mineral matter and the liquid is regulated, providing some control over the hydrogen transfer and product distribution; the typical SRC I product is a friable solid at room temperature, but SRC II operation allows production of liquid, with the consumption of about twice as much hydrogen as SRC I [26]. Reduction of contact time in the SRC reactor minimizes the hydrogen consumption at that point and produces a feed for downstream catalytic hydroprocessing, which may be carried out in a slurry bed like that used in the H-Coal^R process.

The objectives in all these processes include production of clean-burning low-sulfur fuels, and even the SRC process, which uses no added catalyst, removes most of the organic sulfur (as well as almost all of the oxygen and the inorganic sulfur, but not a large fraction of the nitrogen). The EDS and H-Coal^R processes can be even more efficient in producing low-sulfur fuels, since their catalysts may be selective for hydrodesulfurization. For example, during the conversion of dibenzothiophene the sulfided $CoO-MoO_3/Al_2O_3$ catalyst directs the hydrodesulfurization reaction, the splitting out of H_2S (with the formation of biphenyl), almost to the exclusion of the accompanying hydrogenation reactions (Figure 5). There are many other reactants, such as hydrocarbons, however, which react with the expensive hydrogen. The optimum control of the hydrogen consumption between hydrodesulfurization and solvent hydrogenation could probably best be obtained by use of a catalyst like sulfided $CoO-MoO_3/Al_2O_3$ in a separate trickle-bed reactor, as in petroleum refining.

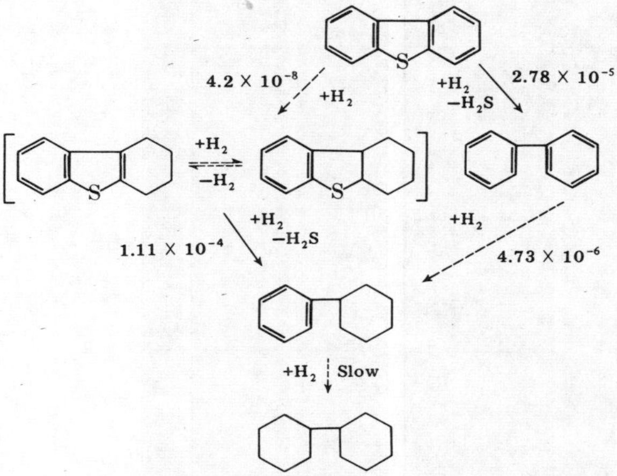

Figure 5. Reaction network for hydrodesulfurization of dibenzothiophene in n-hexadecane carrier oil catalyzed by sulfided $CoO-MoO_3/\gamma-Al_2O_3$ at 300°C and 103 atm.
The numbers next to the arrows are pseudo-first-order rate constants in units of m^3/kg of catalyst·s. The hydrogenation reactions, denoted with broken arrows, are almost negligibly slow in comparison with the hydrodesulfurization reactions [20].

Using the catalyst in a separate reactor instead of including it in the liquefaction reactor offers the further advantage of preventing loss of catalyst activity from accumulation of coal mineral matter (but not of carbonaceous deposits). The mineral deposits in used H-Coal[R] [21] and other [22,23] catalysts have been examined by electron microscopy, electron microprobe analysis, energy dispersive x-ray analysis (EDAX)[21], and other characterization techniques [23]. The deposits form within the catalyst pore structure (Figure 6) and include mineral matter (primarily iron sulfide and clays) and the ever-present carbonaceous material (coke). Titanium deposits also form, probably from conversion of soluble organotitanium components, which are liberated from the coal matrix during liquefaction. The deposits can cause rapid catalyst deactivation by covering the catalyst surface and perhaps blocking the pore mouths. Deposits within the particles can also enlarge cracks and break up the particles. They are a serious disadvantage of the slurry-bed process.

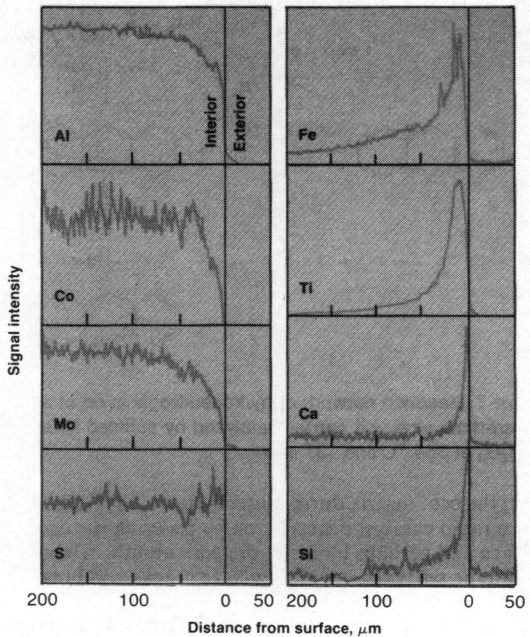

Figure 6. Qualitative composition profiles in a used H-Coal^R catalyst, determined by electron microprobe analysis. Profiles indicate the deposition of mineral matter from coal preferentially near the pore mouths [21].

6. PRODUCT COMPOSITION

The liquids produced from coal are potentially valuable as replacements for petroleum to be used as feedstocks for further refining. Appropriate catalytic refining operations like cracking and hydrocracking, however, require feeds which are almost free of nitrogen-containing compounds, since many of these are basic and poison the catalysts by combining with acidic surface sites. Petroleum feeds for hydrocracking usually undergo catalytic hydrodenitrogenation prior to contacting the hydrocracking catalyst. A similar pretreatment may be applied to coal-liquefaction products [24], but conditions are much more severe than those required for petroleum; the severity results from the relatively high nitrogen contents (as much as a few weight percent) and the lack of reactivity of multi-ring nitrogen-containing compounds, such as acridine. Compare the catalytic hydrodenitrogenation of acridine (Figure 7) with hydrodesulfurization (Figure 5): hydrodenitrogenation takes place with much lower selectivity--the reactant undergoes extensive hydrogenation (which is needed to weaken the C-N bonds) prior to denitrogenation. There are no known catalysts that are able to coordinate nitrogen and selectively break the C-N bonds as they break C-S bonds; development of such catalysts would have a major impact on the processing of coal-derived liquids.

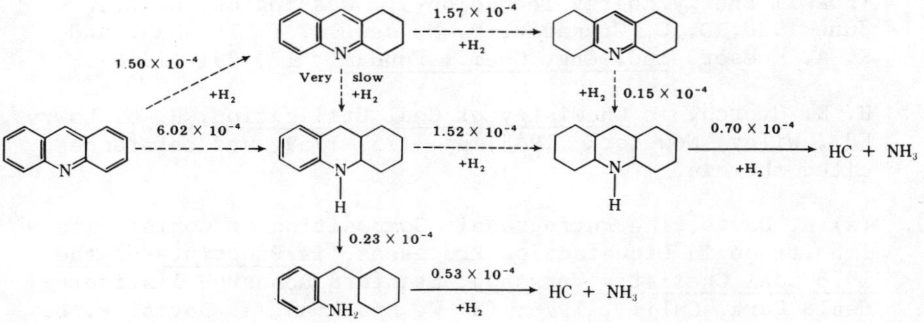

Figure 7. Reaction network of hydrodenitrogenation of acridine in paraffinic white-oil carrier catalyzed by sulfided $NiO-MoO_3/\gamma-Al_2O_3$ at 364°C and 137 atm.
The numbers next to the arrows are pseudo-first-order rate constants in units of m^3/kg of catalyst·s. The hydrogenation reactions, denoted with broken arrows, precede breaking of C-N bonds [25].

REFERENCES

1. A. Verma, CHEMTECH, June 1978, p. 372; October, 1978, p. 626.

2. National Academy of Sciences, Assessment of Technology for Liquefaction of Coal, Washington, D. C., 1977.

3. G. P. Curran, R. T. Struck, and E. Gorin, Ind. Eng. Chem., Process Des. Dev., 6, 166 (1967).

4. W. H. Wiser, L. L. Anderson, S. A. Qader, and G. R. Hill, J. Appl. Chem. Biotechnol., 21, 82 (1971).

5. R. C. Neavel, Coal Plasticity Mechanism: Inferences from Liquefaction Studies, in *Proceedings of the Symposium on Agglomeration and Conversion of Coal*, Morgantown, West Virginia, 1975.

6. R. C. Neavel, *Fuel*, **55**, 237 (1976).

7. J. Guin, A. Tarrer, L. Taylor, J. Prather, S. Green, Jr., *Ind. Eng. Chem. Process Des. Dev.*, **15**, 490 (1976).

8. D. Cronauer, paper presented at symposium Mechanisms of Coal Liquefaction [Proceedings of DOE Project Review Meeting (Fossil Energy/Energy Technology)], Washington, D. C., June 1978; D. C. Cronauer, D. M. Jewell, Y. T. Shah, and K. A. Kueser, *Ind. Eng. Chem., Fundam.*, **17**, 291 (1978).

9. H. H. Storch, in *Chemistry of Coal Utilization*, H. H. Lowry, Ed., Wiley, New York, 1945, pp. 1757-1759, and references cited therein.

10. (a) A. Davis, The Petrographic Composition of Coals: Its Importance in Liquefaction Processes, in *Preprints of the 1976 Coal Chemistry Workshop*, Stanford Research Institute, Menlo Park, Calif., 1976; (b) W. Spackman, A. Davis, P. L. Walker, H. L. Lovell, R. H. Essenhigh, F. J. Vastola, and P. H. Given, The Characteristics of American Coals in Relation to their Conversion into Clean Energy Fuels, U. S. Department of Energy FE-2030-8, 1977, and other reports in this series; (c) P. H. Given, D. C. Cronauer, W. Spackman, H. L. Lovell, A. Davis, and B. Biswas, *Fuel*, 54, 34, 40 (1975).

11. A. Davis, W. Spackman, and P. H. Given, cited in ref. 10(a).

12. (a) M. Farcasiu, T. O. Mitchell, and D. D. Whitehurst, CHEMTECH, Nov. 1977, p. 680; (b) as a leading reference, also see R. G. Ruberto, D. M. Jewell, and D. C. Cronauer, *Fuel*, 57, 575 (1978).

13. F. K. Schweighardt, H. L. Retcofsky, and R. Raymond, *Prepr. ACS Div. Fuel Chem.*, **21** (6), 27 (1976).

14. R. J. Bertolacini, L. C. Gutberlet, D. K. Kim, and K. K. Robinson, *Prepr., ACS Div. Fuel Chem.*, **23** (1), 1 (1978).

15. B. Granoff, M. G. Thomas, P. M. Baca, and G. T. Notes, *Prepr., ACS Div. Fuel Chem.*, **23** (1), 23 (1978).

16. S. Morooka, C. E. Hamrin, Jr., *Chem. Eng. Sci.*, **32**, 125 (1977).

17. L. E. Furlong, E. Effron, L. W. Vernon, and E. L. Wilson, Chem. Eng. Prog., 72 (8), 69 (1976).

18. B. T. Fant, EDS Coal Liquefaction Process Development Phase IIIA, ERDA FE 2353-1, 1977, and succeeding reports.

19. R. P. Anderson, Prepr., ACS Div. Fuel Chem., 22 (6), 132 (1977).

20. M. Houalla, N. K. Nag, A. V. Sapre, D. M. Broderick, and B. C. Gates, AIChE J., 24, 1015 (1978).

21. B. C. Gates, J. R. Katzer, J. H. Olson, H. Kwart, and A. B. Stiles, Kinetics and Mechanism of Desulfurization and Denitrogenation of Coal-Derived Liquids, ERDA FE-2028-7, 1977; M. J. Chiou, and J. H. Olson, to be published.

22. J. J. Stanulonis, B. C. Gates, and J. H. Olson, AIChE J., 22, 576 (1976).

23. P. H. Holloway, G. C. Nelson, Prepr., ACS Div. Petrol. Chem., 22, 1352 (1977); P. H. Holloway, Chemical Studies of the Synthoil Process: Catalyst Deactivation, SAND 78-0056, 1978 (available from National Technical Information Service).

24. R. H. Heck, and T. R. Stein, Prepr., ACS Div. Petrol. Chem., 22, 948 (1977).

25. R. Zawadski, S. S. Shih, J. R. Katzer, and H. Kwart, J. Catal., submitted for publication.

26. Proceedings of the DOE-sponsored symposium, Mineral Matter Effects and Use of Disposable Catalysts in Coal Liquefaction, Albuquerque, N. Mex., June 1978.

27. T. O. Mitchell, R. H. Heck, and D. D. Whitehurst, New Liquefaction Technology by Short Contact Time Processes, presented at AIChE meeting, Miami Beach, November 1978.

This work was supported by the Department of Energy. The helpful comments of Robert Baldwin are gratefully acknowledged.

PROCESS AND CATALYST NEEDS FOR HYDRODENITROGENATION

J. R. KATZER and R. SIVASUBRAMANIAN

Center for Catalytic Science and Technology
Department of Chemical Engineering
University of Delaware
Newark, Delaware 19711

INTRODUCTION

Synthetic fuels represent one of the major alternatives available to supplement declining petroleum reserves. The emergence of a synthetic fuels industry is becoming more and more realistic for several reasons. Some principal reasons among them are

- the desire by the industrialized nations of the world to replace reserves of oil which will soon begin to decline.

- estimated reserves of coal, oil shale, tar sands (the feedstocks for the synthetic fuels industry) could meet energy requirements for hundreds of years.

These synthetic fuels will differ considerably from petroleum feedstocks in their physical and chemical properties. Table 1 presents a comparison of the elemental analyses of several petroleum crudes and liquids derived from coal and oil shale. The hydrogen content of liquids derived from coal is particularly low in comparison with that for petroleum liquids. Oil shale has a hydrogen content similar to that of petroleum crude.

Because of the high nitrogen content of coal and oil shale, liquids derived from them also have a very high nitrogen concentration. Petroleum-based crudes are also becoming "dirtier" from a heteroatom-content point of view, and the increased nitrogen concentrations are becoming an increasing concern. These high

TABLE I

Elemental Analyses of Synthetic Liquids and Petroleum Crudes

Element	SRCI[a]	H-Coal[a]	Synthoil[a]	Colorado[b] shale oil	El Palito[a] No. 6 Fuel Oil	1000 F+[a] West Texas residuum
C, wt%	87.93	89.00	87.62	83.92	86.40	83.88
H, wt%	5.72	7.94	7.97	11.36	11.20	9.97
O, wt%	3.50	2.12	2.08	0.67	0.30	0.48
N, wt%	1.71	0.77	0.97	2.14	0.41	0.40
S, wt%	0.57	0.42	0.43	0.70	1.96	4.19

[a]Ref. 1.
[b]Ref. 2.

concentrations of nitrogen will have to be reduced for the following reasons

- Hydrodenitrogenation is more difficult than hydrodesulfurization.

- NO_x emission restrictions could prohibit using these liquids directly.

- The activity of cracking, reforming, hydrocracking, hydrodesulfurization, hydrodenitrogenation and isomerization catalysts will be severely reduced by the nitrogen-containing compounds.

- High nitrogen concentrations are detrimental to product quality and properties, including stability.

- These synthetic fuels can be quite toxic (carcinogenic) and will require hydrodenitrogenation to reduce the nitrogen content and their toxicity.

Hydrodesulfurization (HDS) of petroleum feedstocks has been extensively studied whereas hydrodenitrogenation (HDN) has become increasingly important only in recent years. Since nitrogen-containing compounds are considerably different in their chemistry than sulfur-containing compounds, catalysts that are optimized for hydrodesulfurization are not optimum for hydrodenitrogenation. While typical hydrodesulfurization catalysts are very selective for sulfur removal, current HDN catalysts are not selective and require high hydrogen consumptions.

Hydrodenitrogenation

Reduction in hydrogen consumption is highly desirable from an economic point of view.

Hydrodenitrogenation is becoming increasingly important as feedstocks, including the emergence of synthetic feedstocks, with higher nitrogen content becoming available. The catalytic chemistry of HDN is different and not well understood and the existing technology has its limitations. Hence a better understanding of the chemistry is needed as are better catalysts and process concepts to optimize hydrodenitrogenation of synthetic feedstocks and heavy petroleum liquids. This article reviews what is known about the catalytic chemistry of hydrodenitrogenation and the catalysts and processes used.

CHEMISTRY OF HYDRODENITROGENATION

Hydrodenitrogenation involves hydrogenolysis of the carbon-nitrogen bonds in nitrogen-containing compounds to give ammonia and the corresponding hydrocarbon. The overall reaction can be represented by nitrogen-containing compounds + $H_2 \xrightarrow{catalyst}$ denitrogenated organic compound + NH_3. To understand the chemistry of hydrodenitrogenation and to develop improved catalysts, it is necessary to have knowledge of the types of nitrogen-containing compounds present in heavy petroleum crudes and in synthetic crudes.

Most of the nitrogen present in these liquids is found in heterocyclic compounds which are resistant to hydrodenitrogenation (Table II). Non-heterocyclic nitrogen-containing compounds are present in liquid fuels in small concentrations and include aliphatic amines, and nitriles. These non-heterocyclic nitrogen-containing compounds are relatively more reactive for hydrodenitrogenation than are heterocyclic nitrogen-containing compounds. The low concentrations and relatively high reactivity of non-heterocyclic nitrogen-containing compounds means that they are of little engineering significance in hydrodenitrogenation. The heterocyclic nitrogen-containing compounds in petroleum and synthetic liquids derived from coal and oil shale can be subdivided into basic or non-basic fractions (Table II).

Several studies of the nature of the nitrogen-containing compounds in petroleum have been made (3-7). Most of these studies found that pyridines, quinolines, pyroles and indoles (one and two-ring compounds) were abundant in the light fractions, whereas large multi-ring nitrogen-containing compounds were concentrated in the heavier fractions. By comparison, less information is available on the nitrogen-containing compounds in synthetic liquids; nitrogen concentra-

TABLE II

Representative Nitrogen-Containing Compounds Found in Petroleum Crude, Shale Oil, and Coal-Derived Liquids[a]

Compound	Formula
Nonheterocyclic compounds:	
Aniline	$C_6H_5NH_2$
Pentylamine	$C_5H_{11}NH_2$
Nonbasic heterocyclic compounds:	
Pyrrole	C_4H_5N
Indole	C_8H_7N
Carbazole	$C_{12}H_9N$
Basic heterocyclic compounds:	
Pyridine	C_5H_5N
Quinoline	C_9H_7N
Indoline	C_8H_9N
Acridine	$C_{13}H_9N$
Benz(a)acridine	$C_{17}H_{11}N$
Benz(c)acridine	$C_{17}H_{11}N$
Dibenz(c,h)acridine	$C_2H_{13}N$

[a] Coal-derived liquids also contain methyl and alkyl substitution on most of the aromatic compounds.

tions are typically four to ten times those in petroleum. Characterization of coal-derived liquids carried out in several laboratories (8-13) indicates that similar types of nitrogen-containing compounds are present in petroleum liquids, coal-derived liquids and shale oil.

Hydrodenitrogenation

Although the nitrogen-containing compounds are generally divided into basic and non-basic compounds, non-basic compounds are converted into basic compounds upon hydrogenation. High molecular weight compounds frequently contain sulfur and oxygen in addition to nitrogen. Since most of the nitrogen-containing compounds are common to both petroleum and synthetic feedstocks, generally the principles of hydrodenitrogenation of petroleum liquids will apply for hydrodenitrogenation of coal-derived liquids and shale oil. However, the synthetic liquids because of their higher nitrogen content, their high aromaticity and low hydrogen content are difficult to hydroprocess and require special consideration.

Hydrodenitrogenation of aromatic nitrogen-containing compounds occurs via a complex reaction network involving hydrogenation of the aromatic rings followed by carbon-nitrogen bond scission, in contrast to hydrodesulfurization which involves mainly direct scission of carbon-sulfur bonds. Although it is generally accepted that nitrogen removal from aromatic-ring compounds takes place through an initial saturation of the ring (hydrogenation) followed by breaking of the carbon-nitrogen bonds (hydrogenolysis), the kinetics of the reaction network have not been established until very recently, particularly for complex nitrogen-containing compounds under industrial hydrotreating conditions. To understand how the catalyst affects the relative rates of hydrogenation and of carbon-nitrogen bond scission in the complex nitrogen removal network, it is necessary to quantitatively define the reaction network and the kinetics of the individual reactions within the network. Knowledge of the reaction networks and reaction kinetics involved in the hydrodenitrogenation of nitrogen-containing compounds present in synthetic liquids and heavy petroleum fractions is essential to the rational design of improved hydrodenitrogenation catalysts.

The reaction network for pyridine-hydrodenitrogenation as demonstrated by McIlvried (14) on sulfided Co-Ni-Mo/Al$_2$O$_3$ and by Sonnemans et al. (15) on unsulfided supported Mo/Al$_2$O$_3$ is

$$\text{pyridine} \underset{cat + 3H_2}{\overset{fast}{\rightleftharpoons}} \text{piperidine} \underset{cat + H_2}{\overset{slow}{\longrightarrow}} H_2N\text{-}C_5H_{11} \underset{cat + H_2}{\overset{fast}{\longrightarrow}} NH_3 + C_5H_{12}$$

Sonnemans et al. (15) found that piperidine can undergo disproportionation reactions that can be quite important in the overall reaction scheme.

Reaction networks become increasingly complex as the nitrogen-containing compound becomes more complex. Shih et al. (16) and Zawadzki et al. (17) defined the reaction networks and associated reaction kinetics for quinoline and acridine hydrodenitrogenation, respectively. Figures 1 and 2 present reaction networks for quinoline and acridine hydrodenitrogenation; the pseudo first-order rate constants for each individual reaction in the network is given on the arrows for the reaction conditions indicated. These results show that hydrodenitrogenation involves two distinct reaction types, hydrogenation and carbon-nitrogen bond scission. Carbon-nitrogen bond scission occurs only in saturated rings whereas carbon-sulfur bonds can be selectively broken without hydrogenating the aromatic ring (18). For hydrodenitrogenation, both hydrogenation and carbon-nitrogen bond scission are kinetically important, and hydrodenitrogenation is not characterized by a rate limiting step.

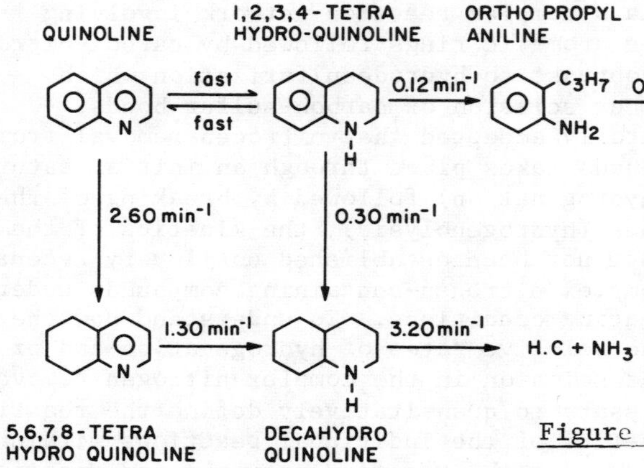

Figure 1

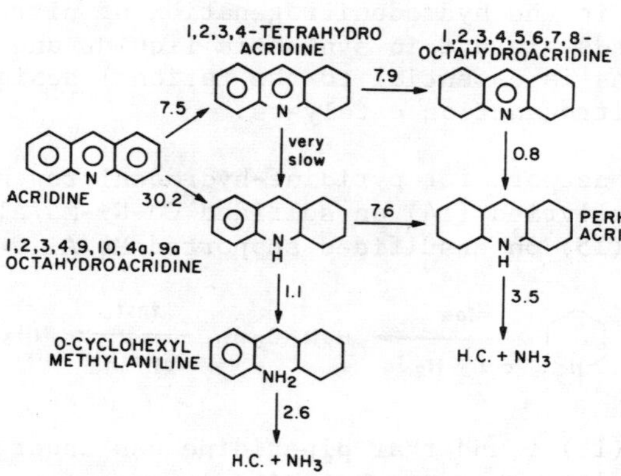

Figure 2

Reaction conditions for Figures 1 and 2: 324°C, 136 atm, presulfided Ni-Mo/Al_2O_3 catalyst, 0.5 wt% catalyst in carrier oil, 0.5 wt% CS_2 in carrier oil; the numbers on the arrows are pseudo-first-order rate constants in grams of oil per gram of catalyst per minute.

Hydrodenitrogenation

These data also clearly show the reason for high hydrogen consumption in hydrodenitrogenation; in addition to having to saturate the aromatic rings associated with the nitrogen-containing compounds other aromatic species in the oil are simultaneously being hydrogenated. Thus the highly aromatic nature of coal-derived liquids results in excessive hydrogen consumption and in high processing costs. Because of the complexity of the hydrodenitrogenation reaction networks, involving hydrogenation and carbon-nitrogen bond scission as distinct reactions, the evaluation and further development of hydrodenitrogenation catalysts can be greatly aided by knowledge of the kinetics of the various intermediate reactions in the network. Further, it is extremely important that the nature of the active sites involved in hydrogenation and in carbon-nitrogen bond scission and the mechanism of these reaction be adequately defined and understood.

Hydrodenitrogenation is typically more difficult than hydrodesulfurization as shown in Table III. The operating conditions and the catalyst used were the same for all three coal-derived liquids. The differences in difficulty of hydrodesulfurization and hydrodenitrogenation and the need to hydrogenate aromatic ring systems in the case of hydrodenitrogenation before carbon-nitrogen bond scission can occur are partially explained by the relative carbon-carbon, carbon-nitrogen, and carbon-sulfur bond strengths in aromatic and saturated systems (Table IV). Carbon-nitrogen bonds are always stronger than carbon-sulfur bonds. Carbon-nitrogen bonds are slightly stronger than carbon-carbon bonds in aromatic systems; the relative bond strengths (carbon-nitrogen vs. carbon-carbon) are reversed upon hydrogenation.

TABLE III

Comparison of Sulfur and Nitrogen Removal from Different Petroleum and Coal-Derived Liquids (19)

Feedstock	Coal-Derived Liquids % removal	
	S	N
Raw anthracene oil	96	63
Hydrocarbonization liquid	70	38
Synthoil liquid	33	15

TABLE IV

Bond Energies between Carbon and Heteroatoms in Polyatomic Molecules (20)

Bond	Energy, kcal/mole	Bond	Energy, kcal/mole
C—H	99	C—N	73
C—C	83	C=N	147
C=C	146	C≡N	213
C≡C	200	C—S	65
N—H	93	C=S	128
		S—H	83

The difficulty of hydrodenitrogenation typically increases with increasing boiling point of the feedstocks. Flinn et al. (21) postulated that the difficulty of nitrogen removal increases with increasing boiling point of the fraction because the increasing size of the non-nitrogen-containing part of the molecule reduces the accessibility to the catalytic surface. However, Sarbak et al. (22) have shown that the pseudo first-order rate constants and relative reactivities of important polynuclear aromatic nitrogen-containing compounds found in coal-derived liquids decrease slightly to four-ring nitrogen-containing compounds and then increase for larger molecules (Table V). These data question the conclusion that one of the reasons hydrodenitrogenation becomes more difficult with increasing boiling range is increasing size of the nitrogen-containing molecule. Since these data were obtained with single compounds dissolved in white oil, other factors may cause decreasing reactivity for hydrodenitrogenation with increasing boiling point for real feedstocks. These factors include:

- formation of nitrogen-containing secondary products during hydrodenitrogenation which are more resistant to nitrogen removal than the original ones.

- competitive adsorption by the higher molecular weight highly-aromatic molecules in the feedstock

- self-inhibition by the high concentration of nitrogen-containing compounds

- more severe catalyst deactivation caused by the higher-molecular weight highly-aromatic compounds in the feedstock.

TABLE V

Relative Reactivities of Nitrogen-Containing Compounds (22)

Compound	Pseudo-first Order Rate Constant, k, (min)$^{-1}$
Quinoline	2.52
Carbazole	2.43
Acridine	1.62
Benz(a)acridine	1.08
Benz(c)acridine	1.54
Dibenzacridine	3.79

Standard Conditions: 367°C, 136 atm, presulfided Ni-Mo/Al$_2$O$_3$, carrier oil white oil, H$_2$S present during reaction in batch autoclave.

Recent studies in our laboratory show that aromatic compounds compete only very weakly with nitrogen-containing compounds for adsorption on the catalyst surface and thus do not strongly inhibit the rate of hydrodenitrogenation (23). However, nitrogen-containing compounds strongly inhibit the rates of other hydroprocessing reactions and also inhibit their own rates of reaction by competitive adsorption. Since higher boiling feedstocks typically contain higher concentrations of nitrogen, this self-inhibiting effect results in the requirement of more severe operating conditions to achieve the same extent of nitrogen removal. Considering these results we infer that this is one of the main reasons that hydrodenitrogenation becomes progressively more difficult with increasing boiling point. The increased severity of catalyst deactivation which results from increasing boiling point and decreasing hydrogen content of the feedstock requires severe operating conditions and is another major reason for the difficulty of hydrodenitrogenation of higher boiling feedstocks.

CATALYSTS AND PROCESSES

Supported metal sulfide catalysts are widely used as hydroprocessing catalysts. The most commonly used supports are gamma alumina and gamma alumina stabilized with minor amounts of silica. Typical metallic combinations used are Co-Mo, Ni-Mo, Ni-Co-Mo and Ni-W. Hydrodesulfurization catalysta are made up of Co-Mo on Al_2O_3 while Ni-Mo on Al_2O_3 is commonly used for hydrodenitrogenation. These catalysts are presulfided in the reactor before their use. Ahuya et al. (24) studied the activity and selectivity of presulfided hydrotreating catalysts and evaluated many combinations of metals and supports. They found that nickel-promoted catalysts are better for hydrodenitrogenation than cobalt promoted catalysts since nickel promotes hydrogenation.

The concentration of the active metals is usually determined by cost and competitive factors. The catalyst activity increases with active metals concentration to a maximum and then decreases. Hence the optimum concentration of active metals involves a balance between obtained activity and cost. Typical hydroprocessing catalysts contain 2-4% Co as CoO or 2-4% Ni as NiO and 8-15% Mo as MoO_3. Beuther et al. (25) reported that commercial preparations consist of cobalt and molybdenum in atomic ratios of 0.1:1.0 to 1.0:1.0; maximum activity was observed for ratios around 0.3:1.0 for Co-Mo catalysts and around 0.6:1.0 for Ni-Mo catalysts.

In addition to the chemical composition of the catalyst, the physical properties of the catalysts such as surface area, average pore diameter and pore volume also have a significant effect on a catalyst's performance. The literature contains conflicting data on the effect of pore size and surface area on the activity of hydrotreating catalysts (26-33). Decreasing pore size could increase the pore diffusion resistance and also restrict larger molecules from reaching the active catalyst surface. For hydrodesulfurization, diffusion resistance may cause the observed catalyst activity to decrease with decreasing pore size. Hydrodenitrogenation, being slower than hydrodesulfurization may not be affected by diffusional limitations, and thus within a reasonable range variations in pore size will not affect the observed catalyst activity. Thus the catalytic activity will be proportional to the available surface area. However, with higher molecular weight nitrogen-containing compounds, decreasing pore size may result in the appearance of diffusional limitations. For liquids containing large amounts of mineral matter, large pores are frequently desirable because they can tolerate more mineral matter deposition before being plugged and thus promote increased catalyst life.

A simplified flow diagram of a hydrogenation process is shown in Figure 3. The liquid feedstock is mixed with hydrogen and passed over a catalyst, a higher temperature and pressure. Hydrogen separated from the product stream is scrubbed to

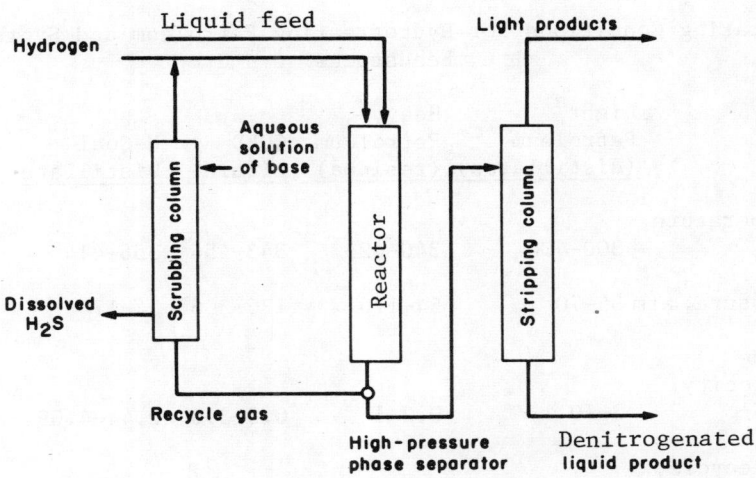

Figure 3. Simplified diagram of a hydrodenitrogenation process.

remove H_2S, NH_3, and light hydrocarbons, and is recycled to the process with added make-up hydrogen. The individual components of the processes include: preheat furnace, reactor, hydrogen recycle facilities and product separation train. The extent of nitrogen removal depends on the nature of the feedstock, the severity of reaction conditions, and the type of catalyst. Table VI presents typical operating conditions used in hydrotreating petroleum and synthetic liquids. It is clear from the table that synthetic liquids and heavy petroleum fractions require more severe operating conditions than those required for lighter petroleum liquids and consume more hydrogen.

Reactors are typically of the trickle-bed type with oil "trickling" down over the catalyst and high pressure hydrogen flowing co-current with the oil. The catalyst bed is fixed, the flow pattern of liquid and gas is essentially plug flow and the liquid holdup is low. The heat effect due to reaction for light petroleum feeds in a single-stage reactor is typically not a major problem, and the trickle-bed reactor is preferred because of its simplicity and because a plug-flow reactor is smaller than a stirred-tank (ebullated-bed) reactor for a given conversion.

For the more severe operating conditions which are required to achieve high degrees of nitrogen removal from coal-derived liquids, shale oils and heavy petroleum residua, multiple-bed

TABLE VI

Operating Conditions for Hydrotreating Petroleum and Synthetic Feedstocks

	Light[a] Petroleum (distillates)	Heavy[a] Petroleum (residua)	COED Liquid[b]	H-Coal[c] Distillate	Shale Oil[d]
Temperature, C	300–400	340–425	343–454	356–414	406–418
Pressure, atm	35–70	54–170	136–306	54–170	160
Space velocity, hr^{-1}	2–10	0.2–1	0.3–3.0	0.44–4.09	0.60
H_2 Recycle, scf/bbl	300–2000	2000–10,000			4906–10308

[a]Ref. 34.
[b]Ref. 35.
[c]Ref. 36.
[d]Ref. 37.

reactors may be required to limit the temperature rise in a given reactor. Cold hydrogen is injected between the beds to reduce the temperature of the reactant stream. For "dirty" feeds a guard bed containing a cheap, disposable material to take out metals, particulate matter and rapid-coking components may be desirable to extend the life of the catalyst in the reactor. Two guard beds may be included so that one can be used alternately while the catalyst in the other is replaced. When the feed is particularly heavy and high in nitrogen and sulfur contents, two reactors and two separate hydrogen recycle stages, one for each reactor, may be required for more efficient process operation. Scrubbing the light gases out of the hydrogen between stages increases the partial pressure of hydrogen in the second reactor for a given total operating pressure and removes NH_3 and H_2S which inhibit the rates of heteroatom (S and N) removal.

In operation, the temperature of trickle-bed reactor is increased with time to compensate for catalyst deactivation and to maintain constant conversion. However, there is an upper limit on the temperature since the rate of hydrocracking and of

Hydrodenitrogenation

coking increases strongly (possibly exponentially) with increasing temperature leading to high heat release rates, to hot spots, to rapid catalyst deactivation, and ultimately to plugging of the reactor. It is most desirable to start the reactor with a freshly regenerated catalyst or with fresh catalyst at the lowest possible temperature because this should result in the lowest initial rate of catalyst deactivation and the longest run time between catalyst regeneration or replacement.

An alternative reactor type is the ebullated-bed reactor; the H-oil process utilizes an ebullated-bed reactor to hydrotreat petroleum residua. However, this type of reactor has received limited application to date. For heavier liquids such as coal-derived liquids and heavy petroleum residua the ebullated-bed reactor may be the reactor of choice because of rapid catalyst deactivation caused by these liquids. The ebullated-bed reactor allows catalyst to be removed continuously and regenerated, if possible, or replaced with fresh catalyst. An ebullated-bed reactor, being essentially a continuous-flow stirred-tank reactor with respect to the liquid, requires a larger reactor volume for a given conversion than a plug-flow trickle-bed reactor; for high conversion this size difference is quite large. Thus, for high conversions ebullated-bed reactors may be staged. Further, the larger reactor volume required for the ebullated-bed reactor can be partially offset by increased operating temperature, but this leads to increased hydrocracking and increased rates of catalyst deactivation. Also, smaller particles (<0.8 mm) of catalyst can be used in an ebullated-bed reactor than are feasible in a trickle-bed reactor, reducing intraparticle mass-transfer limitations, which frequently exist with heavier feeds. The design of both reactor types is straightforward and based on much experience in the petroleum industry. The required data are rather complete "lumped" analysis of the feed, kinetic rate constants for the feed or pilot plant data over the desired conversion range, and catalyst deactivation rates for the feedstock.

PROCESSING PROBLEMS

a) Hydrogen requirements

Hydrodenitrogenation consumes hydrogen in large excess of stoichiometric requirements, since hydrodenitrogenation requires full ring hydrogenation to destroy the aromaticity of the heterocyclic nitrogen-nitrogen compounds prior to nitrogen atom removal. High hydrogen consumption is one of the major costs of hydroprocessing heavy petroleum feeds and synthetic liquids. In a balanced petroleum refinery sufficient hydrogen is available from catalytic reforming to meet the needs of hydrotreatment, and therefore the cost of hydrogen is not as important in considering hydrotreating economics. However, when it becomes necessary to generate hydrogen separately as in the case of coal liquefaction or hydrotreating synthetic feedstocks, hydrogen consumption can become a large fraction of the cost of processing, particularly for low quality liquids, such as coal-derived liquids.

Nelson (38) estimated the stoichiometric amount of hydrogen consumed in removing the sulfur and nitrogen-containing compounds and compared this with the average hydrogen consumption. A one percent reduction in sulfur content requires an average hydrogen consumption of about 73 scf/bbl, whereas a one percent reduction in nitrogen would require a hydrogen consumption of about 300 scf/bbl. Hydrogen consumption increases with increase in pressure and with decrease in space velocity. Table VII summarizes the hydrogen consumptions for hydroprocessing several coal-derived liquids and shale oil. Hydrogen consumption ranges from 65 to 3740 scf/bbl. The low value is for a high quality light liquid which is not characteristic of typical coal-derived liquids and shale oils.

Heck and Stein (36) reported that 700 to 800 scf/bbl of hydrogen was consumed without significant oxygen or nitrogen removal from an SRC recycle solvent. However, for H-Coal distillate which is much more saturated than SRC recycle solvent, substantial heteroatom removal was accomplished with lower hydrogen consumption. Figure 4 shows hydrogen consumption as a function of the amount of nitrogen remaining in a processed shale oil (37). Hydrogen consumption increases as the concentration of nitrogen in the product shale oil is decreased. Large amounts of hydrogen are consumed even for low extents of nitrogen removal, but incremental hydrogen consumption decreases with increase in nitrogen removal.

The behavior observed for coal liquids is just opposite to that observed for shale oil and for hydrodesulfurization in which the differential hydrogen consumption increases with the degree of heteroatom removal; the difference is due to the fact that in hydrodenitrogenation, the heterocyclic ring structures require hydrogenation before carbon-nitrogen bond scission can occur. Thus high hydrogen consumption may occur at low degrees of nitrogen removal; at high degrees of nitrogen removal differential hydrogen consumption may be low because most of the aromatic structures have already been saturated. Simultaneous saturation of aromatics not containing heteroatoms also contributes to high hydrogen consumption and is highly undesirable. This trend is probably enhanced with highly aromatic feeds—such as coal-derived liquids.

TABLE VII
Hydrogen Consumption in Hydrotreating Synthetic Liquids

Feedstock	Hydrogen consumption, scf/bbl	Temperature, °C	Pressure, atm	Feedrate or LHSV, hr^{-1}	S	N	O	Refs.
Oil from Western Kentucky coal	2597-2894	454-469	115-125	150-174 lb/hr	91-96	67-87	84-94	25
SRC recycle solvent (various boiling point cuts)	973-3410	400-479	80-170	0.45-2.11 LHSV	88-98	37-98	36-98	26
H-Coal distillate (various boiling point cuts)	65-1698	400-471	55-170	0.46-4.09 LHSV	64-98	26-98	25-92	26
Oil from Pittsburgh Seam	2096-3735	440-457	120-130	127-153 lb/hr	95-96	62-85	76-85	25
SRC (60/40 blend)	2530-3230	486	190	0.5 LHSV	95	63-79	64-85	37
Synthoil	1840	Not available		0.98 LHSV	96	58	80	28
Shale oil	1616-2175	443-460	147	0.60 LHSV	99	99	-	27

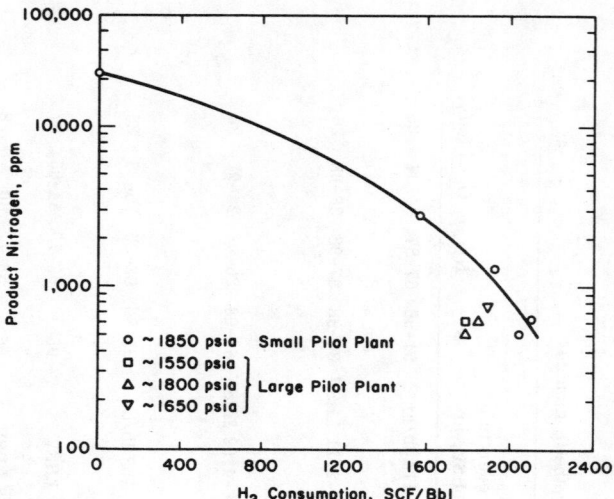

Figure 4. Hydrogen consumption in hydrotreating whole shale oil as a function of product nitrogen concentration (37).

Hydrogen consumption typically increases with increased hydrogen pressure and with increased conversion. Since nitrogen removal from petroleum residua and synthetic feedstocks requires severe operating conditions, hydrogen consumption will be higher than for lighter liquids. The hydrogen consumption problem is further complicated by the fact that projected future fuel needs will require an increasing fraction of high hydrogen content fuels; this is just opposite to the trend in fuel availability. To reduce hydrogen consumption, catalysts are required that can selectively remove nitrogen from aromatic nitrogen-containing compounds without first requiring hydrogenation of the entire ring system. Process requirements combined with catalyst requirements must meet the need for highly selective hydrogen consumption involving only the hydrogen requirement for heteroatom removal and for fuel quality upgrading to achieve the needed product requirements. Small reductions in hydrogen consumption are worth millions of dollars per year at the scale of our fuel consumption.

b) Catalyst Aging

Since catalyst deactivation rates often determine the commercial viability of a process, successful commercial hydrotreating catalysts should have long life. Most of the research on hydrotreating catalysts has dealt with their activity for petroleum feedstocks. Catalyst deactivation rates in petroleum hydroprocessing vary from low for light feeds to high for heavier feeds. Petroleum residue hydrotreating has

Hydrodenitrogenation

advanced in the last fifteen to twenty years with respect to catalyst activity and stability and is practiced commercially. However, synthetic liquids, particularly coal-derived liquids, are much harder to hydroprocess, cause more rapid catalyst deactivation, and represent another step upward in terms of difficulty of hydrotreating.

The rate of catalyst deactivation can be significantly reduced by increasing the hydrogen pressure, and therefore heavier feedstocks require higher hydrogen pressures. Lighter feeds cause deactivation predominantly by coke laydown. These catalysts are easily and almost completely regenerated by burning the coke off at intervals of 6-12 months and may last up to 10 years. Heavier petroleum fractions may contain high concentrations of Ni and V, e.g., in excess of 300 ppm in many Venezuelan crudes. These metals are deposited on the catalyst causing deactivation by covering the active surface and plugging the pores, frequently at the pore mouths thereby reducing mass transfer into the catalyst. For deactivation caused by metals deposition on the catalyst, regeneration techniques have not been developed, and thus for residue processing, catalyst life may be as short as six months. Hence, deactivation caused by metals deposition is more serious than deactivation due to coke deposition if coke deposition is not excessively rapid.

Synthetic liquids contain predominantly different metals than petroleum liquids. Table VIII gives trace metals analysis for three coal-derived liquids, and for El Palito No. 6 fuel oil. The table shows that the three coal-derived liquids have high iron and titanium contents, whereas the petroleum liquid contains significantly higher levels of vanadium and nickel. The No. 6 fuel oil has a higher than typical Ti concentration. In petroleum liquids, the metals are mostly present as organometallic compounds; however, for coal-derived liquids the nature of the metal-containing species has not been clearly defined. Table IX shows the fate of metal components from hydrotreating runs involving Synthoil and an SRC liquid. The table suggests that iron and calcium entered the reactor as particulates, and the majority did not deposit on the catalyst. Titanium, on the other hand, is probably organically complexed, and the majority was reactively deposited on the catalyst.

Estimates of catalyst requirements based on measured deactivation indicate that the H-Coal process would require 1 pound of fresh catalyst per ton of coal processed, whereas the Synthoil process would require 4 pounds of catalyst per ton of coal processed (41). The above estimates were for coal liquefaction. Dabkowski et al. (42) recently estimated the

TABLE VIII

Trace Metals Analyses of Coal-Derived Liquids and Petroleum [36]

Element, ppm	SRC recycle solvent	SRC	H-Coal	Synthoil	El Palito No. 6 fuel oil
Si	0.0	30.0	2.0	1348.0	3.0
Mg	0.2	4.0	1.0	33.0	4.0
Pb	0.3	1.0	0.0	5.0	2.0
Fe	4.4	140.0	20.0	375.0	6.4
Al	1.5	31.0	11.0	886.0	5.0
Ni	0.3	2.1	1.0	1.0	59.0
Ti	11.0	130.0	80.0	150.0	78.0
Ca	0.4	49.0	8.0	27.0	14.0
Na	1.6	100.0	0.8	79.0	19.0
K	0.4	8.0	0.4	116.0	a
V	0.0	6.8	2.6	1.8	275.0

[a]Analysis not determined.

TABLE IX

Fate of Inorganic Components in Coal-Derived Liquids During Hydroprocessing (40)

	Ti	Ca	Fe
Feedstock: Synthoil liquid			
Charged, g of metal	1.41	1.58	1.49
Deposited on catalyst, g	1.46	0.16	0.59
% deposited	104	10	40
Feedstock: SRC product			
Charged, g of metal	3.87	3.41	6.85
Deposited on catalyst, g	2.81	0.54	2.21
% deposited	73	16	32

Hydrodenitrogenation

average catalyst requirements for hydrotreating $400°F^+$ fuel oil from the H-Coal process is 0.21 pounds per barrel of feed and that for hydrotreating solvent refined coal is 0.8 pounds per barrel of feed. Therefore the total catalyst requirement for producing an environmentally acceptable fuel oil by the H-Coal process is about 0.5 lb fresh catalyst per bbl. At a commercial level of operation of 2 million barrels per day (~10% of oil consumption) the projected catalyst requirements range from 1.0 (H-Coal) to 1.6 (SRC) million pounds per day. These large projected catalyst requirements represent rates of production far in excess of current levels and would require metals (Co and Mo) production several times the current total world production. These numbers underscore the serious need for process configuration that do not entail such severe catalyst deactivation and for catalysts that are less susceptible to deactivation and that are regenerable.

Catalyst deactivation depends to a large extent on the environment of the working catalyst and is particularly severe for hydrotreating synthetic liquids. The deactivation of hydrotreating catalysts used in petroleum processing has been well characterized and many advances in improved catalyst durability have been made in the last two decades. Much of this improvement has come from modification and control of surface area and pore size distribution to maximize catalyst life. Sufficient quantitative data on catalyst deactivation during hydroprocessing of synthetic liquids are not available. The metallic constituents of synthetic liquids differ significantly from those of petroleum and the effect of these metals on catalyst deactivation is not well quantified. Generally, if deactivation is due to metals deposition, regeneration of the catalyst is not possible at present, whereas deactivated catalysts due to coke laydown can be regenerated by controlled combustion.

Currently hydrotreating catalysts are deactivated so rapidly in an environment with dissolving coal that it is not possible to see how such a process configuration can be considered acceptable as a commercial coal-liquefaction process. The quality of the liquids produced by most coal-liquefaction technologies are quite low, and these liquids cause rapid deactivation of hydrotreating catalysts. Coal liquids of higher quality can be hydroprocessed including hydrodenitrogenation, but even these liquids require high hydrogen pressures to prevent catalyst deactivation, thereby resulting in high consumption of hydrogen. Since environmentally acceptable synthetic fuels (coal-derived liquids and shale oil) cannot be produced in a single-step process, future developments need to focus on the interplay between the production of these fuels (liquefaction) and their

upgrading. Processes need to be developed that can produce
liquids of better quality so that they may be upgraded with
acceptable rates of catalyst deactivation and acceptable
amounts of hydrogen consumption. The interplay between
catalyst activity, selectivity, and resistance to deactivation and process configuration and operation must remain a
dynamic one as improved catalysts are developed to assure
that the ultimate process-catalyst combination produces the
products desired at a minimum of hydrogen consumption and
cost.

c) Toxicity

Coal-derived liquids, shale oil and heavy petroleum
fractions contain many carcinogenic nitrogen-containing
compounds and in the presence of polynuclear aromatics their
carcinogenicity is frequently increased many fold.
Benzo(a)pyrene, benzo(a)phenanthrene and benzo(a)anthracene
are established carcinogens in coal-derived liquids (43-46),
aromatic petroleum fractions (47,48), and shale oils (49).
Coal-derived liquids are particularly toxic, because of the
high concentration of both nitrogen-containing compounds and
polynuclear aromatics. The extent of toxicity was fully
illustrated during the seven-year operating program of a
300-ton-per-day coal hydrogenation plant at Institute, West
Virginia, during the 1950's. There was a 11% incidence of
skin carcinnoma among the workers in this plant (43-46).
More recent experience demonstrates, however, that coal can
be liquefied safely, and that the coal-derived liquids can
be handled safely with proper hygiene measures and periodic
medical examinations just as carcinogenic streams in current
petroleum refinery operations, such as "cat recycle" oil,
can be handled safely. The carcinogenic nature of synthetic
liquids, particularly coal-derived liquids, will severely
restrict their applications to major combustion facilities
where strict hygienic measures can be maintained and enforced
unless they are effectively detoxified by hydrotreatment,
particularly by hydrodenitrogenation. Thus the broad distribution and utilization of coal-derived liquids will require
extensive hydrodenitrogenation to remove the carcinogenic
nitrogen-containing compounds and reduce somewhat the concentration of polynuclear aromatic hydrocarbons. Again these
needs require catalysts and processes that can selectively
remove nitrogen heteroatoms with minimum hydrogen consumption
and perform no more than the desired amount of hydrogenation
to achieve required product properties. If these highly
toxic synthetic liquids cannot be selectively detoxified,
then extensive hydrocracking and hydrogenation will be required
producing substantially light fuels at a high cost.

CONCLUSIONS AND RECOMMENDATIONS

Current hydrodenitrogenation catalysts achieve nitrogen removal by first hydrogenating the aromatic rings in the nitrogen-containing molecule, particularly the ring containing the nitrogen atom, followed by carbon-nitrogen bond scission. Because hydrodenitrogenation involves both hydrogenation and carbon-nitrogen bond scission (hydrogenolysis) on current catalysts, these catalysts must be bifunctional, having both hydrogenation sites and hydrogenolysis sites. Therefore improved catalysts can be developed by carefully optimizing the hydrogenation activity and the cracking activity of the catalyst. This optimization is typically carried out empirically by determining the total rate of nitrogen removal as a function of catalyst preparation and composition. However, quantitative knowledge of the reaction networks and reaction kinetics for each step in the reaction network provides a firm basis for the rational design and optimization of improved hydrodenitrogenation catalysts. In this way it is possible to determine how catalyst modifications affect the rate of hydrogenation and that of hydrogenolysis. Process optimization can also utilize this type of information because processing conditions affect the two reaction types differently.

It is also critical to determine the nature of the catalyst sites which catalyze these two reaction types (hydrogenation and carbon-nitrogen bond scission) to rationally develop new and improved catalysts. The metal sulfide provides the hydrogenation function in typical commercial hydrotreating catalysts although a detailed understanding of the nature of the active site and of the reaction mechanism is still lacking. The origin of the carbon-nitrogen bond scission (hydrogenolysis) function is not clear. The role of acidity in hydrodenitrogenation catalysts needs to be investigated. We speculate that the hydrogenolysis site may involve an OH^+ group adjacent to a transition metal ion in the surface of the Al_2O_3, that both the proton and the transition metal ion are involved, and that the reaction occurs by a Hoffman E-2, β-elimination reaction. A better understanding of the nature of the site that catalyzes the carbon-nitrogen bond scission reaction would help considerably in developing catalysts which had a more-active carbon-nitrogen bond cracking function; improved catalysts will require a more active carbon-nitrogen bond cracking function because carbon-nitrogen bond scission is probably always kinetically important.

A very pressing need is the development of hydrodenitrogenation catalysts that can selectively coordinate the nitrogen atom and promote its removal without first requiring ring hydrogenation and thus be able to limit expensive hydrogen consumption. Catalysts must be designed that can selectively coordinate and catalyze the carbon-nitrogen bond scission reaction without first requiring full hydrogenation of the aromatic ring, and the catalyst must be stable under industrial operating conditions. The selectivity of hydrodesulfurization catalysts may be related to the fact that they are metal sulfides; thus sulfur atom vacancies on their surfaces are natural locations for the sulfur atom of the sulfur-containing molecule to interact, promoting its removal from the hydrocarbon molecule. Nitrogen having a very considerably different chemistry from that of sulfur can be expected to require a different type of catalyst site. The literature provides little specific information on the directions in catalyst properties and composition that could be most fruitful; major new developments and chemical routes remain to be developed in this area. For instance a Ni-W/zeolite catalyst has been developed which catalyzes hydrodenitrogenation without catalyzing hydrodesulfurization (50). We can only speculate on directions but one guiding principle should be to evaluate materials involving elements that coordinate nitrogen (or nitrogen and carbon) or which can form vacancies into which the nitrogen atom can be coordinated as with sulfur. Because the sulfur level may be lower than the nitrogen level in some coal-derived liquids a number of materials not normally stable in higher concentration H_2S environments may now be possible candidates. Materials of particular interest are transition metal nitrides possibly promoted by other metals. Fe, Ru, Mo and W are particularly good candidates both as the major components and/or as promoters. Mixed transition metal oxides and oxysulfides in addition to materials such as borides and carbides or mixed systems such as Mo_2BC, and $M_2Mo_3O_8$ where M can be Mg, Zn, Co, Mn and other transition metals may show potential (51,52). Boudart, Cusumano, and Levy (53) have reviewed some of the properties of new materials that could be applicable to new catalytic chemistry. Important criteria are thermal stability, chemical stability and particularly stability in the presence of H_2S and coke-forming species. However, thermodynamic information on bulk compounds frequently is of little value when considering highly dispersed materials on a support. Thus a combination of empiricism and enlightened chemical intuition will serve best the development of new catalytic materials. This work should always be coupled with careful network and kinetic analysis to define uniquely the chemical behavior of these new materials in catalytic hydrodenitrogenation.

Catalyst deactivation problems can be extremely severe with coal-derived liquids because of their high aromatic content and coke-forming tendencies. The rate of deactivation can be reduced by increasing the hydrogen partial pressure which, however, may result in excessive hydrogen consumption.

The metallic constituents of synthetic liquids differ significantly from those of petroleum and the effect of these metals on catalyst deactivation is not well quantified. Permanent loss of activity results from metal deposition on hydrotreating catalysts. The development of regeneration procedures for hydrotreating catalysts poisoned by alkali metals, iron, vanadium, titanium, and other materials is needed, and the tolerance of the catalyst materials to such metals should be increased. Catalysts with built-in activity maintenance must be developed. Current methods of increasing the tolerance of hydrotreating catalysts to metal deposition involve modification of the pore-size distribution of the catalyst. A more quantitative understanding of the role of the pore structure in activity maintenance is needed, prediction of the optimum pore structure and pore size distribution is required, and development of techniques to produce desired pore structure and pore size distributions in commercial catalysts is needed.

The effect of catalyst support properties such as surface area, pore volume, and pore size distribution on hydrodenitrogenation activity needs to be better understood. Detailed characterization of the synthetic liquids from coal and oil shale is desirable to better quantify the type and concentration of nitrogen-containing compounds present in these liquids. These studies should also provide the information needed for lumping and process modeling under industrial operating conditions.

Industrial hydrodenitrogenation processes almost invariably use trickle-bed reactors. The fluid dynamics and model description of trickle-bed reactors are complex and have not been adequately characterized. Uncertainties exist as to degree of vaporization of the feed, the number of phases present, the distribution of components between phases at reaction conditions, and the degree of catalyst wetting. Quantitative information on these uncertainties would markedly help in modeling trickle-bed reactors and hydrotreating processes.

The products obtained from most current coal liquefaction processes are very low in quality and cause rapid deactivation of hydroprocessing catalysts. For these feeds, it may be desirable to use an ebullated-bed reactor since the ebullated-bed reactor allows catalysts to be removed continuously and regenerated, if possible, or replaced with fresh catalyst. However, an ebullated-bed reactor requires a larger volume than a trickle-bed reactor to achieve a given conversion. The larger volume required for an ebullated-bed reactor can be partially offset by increased operating temperature which leads to increased rates of catalyst deactivation and increased hydrogen consumption. The ebullated-bed reactor typically achieves the desired conversion at a lower selectivity. Thus the optimum processes for hydrotreating low-quality liquids will probably involve an ebullated-bed reactor to give partial upgrading followed by a higher-selectivity trickle-bed reactor designed to give the desired heteroatom removal and product quality. Therefore, it is essential that the interface between the synthetic liquid formation process and the hydrotreating process remain a dynamic one to achieve the best balance between the two with respect to hydrogen consumption and product quality.

References

1. R. B. Callen, J. G. Bendovaitis, C. A. Simpson, and S. E. Voltz, Ind. Eng. Chem., Prod. Res. Dev. 15, 222 (1976).
2. U. S. Patent 3,523,073.
3. L. R. Snyder, Am. Chem. Soc. Div. Petrol. Chem., Prep. 4 (2), C43 (1970).
4. A. C. Bratton and J. R. Bailey, J. Am. Chem. Soc. 59, 175 (1937).
5. A. C. Nixon and R. E. Thorpe, J. Chem. Eng. Data 7, 429 (1962).
6. D. M. Jewell and G. K. Hartung, J. Chem. Eng. Data 2, 95 (1957).
7. C. F. Brandenburg and D. R. Latham, J. Chem. Eng. Data 13, 391 (1968).
8. S. A. Quader and G. R. Hill, Ind. Eng. Chem. Process Des. Develop., 8, 450 (1969).
9. H. C. Anderson and W. R. K. Wu, Bureau of Mines Bulletin 606, U. S. Dept. of the Interior, 1963.
10. D. McNeil, in Bituminous Materials: Asphalts, Tars and Pitches, ed. A. J. Hoiberg, Interscience Publishing Co., New York, N.Y., 1966, Vol. 3, pp. 139-216.
11. S. E. Scheppele, G. J. Greenwood, and P. A. Benson, Anal. Chem. 49, 1847 (1977).
12. T. Aczel and H. E. Lumpkin, Preprints, Div. Petrol. Chem., Am. Chem. Soc., 22 (3), 911 (1977).
13. J. L. Schultz, R. A. Friedel and A. G. Sharkey, Jr., "Mass Spectrometric Analysis of Coal Tar Distillates and Residues," Bureau of Mines Report of Investigation No. 7000, U.S. Dept. of Interior (1967).
14. H. G. McIlvried, Ind. Eng. Chem. Process Des. Develop. 10, 125 (1971).

15. J. Sonnemans, W. J. Neyens, and P. Mars. J. Catalysis, 34, 230 (1974).
16. S. S. Shih, J. R. Katzer, H. Kwart, and A. B. Stiles, Preprints, Petrol. Div., Am. Chem. Soc., 22 (3) 919 (1977).
17. R. Zawadzki, S. S. Shih, J. R. Katzer, and H. Kwart, "Kinetics of Acridine Hydrodenitrogenation," to be published.
18. M. Houalla, N. K. Nag, A. V. Sapre, D. H. Broderick, and B. C. Gates, AIChE Journal, 24, 1015 (1978).
19. B. L. Crynes, "Catalysts for upgrading coal-derived liquids," Quarterly Report for the Period September 9-December 8, 1977, FE2011-10, prepared for the United States Energy Research and Development Administration, 1977.
20. J. D. Roberts, R. Stewart, and M. C. Caserio, "Organic Chemistry," Benjamin Inc., Menlo Park, California (1971).
21. F. A. Flinn, O. A. Larson, and H. Beuther, Hydrocarbon Processing and Petroleum Refiner, 42 (9), 129 (1963).
22. Z. Sarbak, J. R. Katzer, S. S. Shih, M. Bhinde, unpublished results.
23. M. Bhinde, Ph.D. Thesis, University of Delaware, Newark, Delaware (1979).
24. S. P. Ahuja, M. L. Derrien, and J. F. LePage, Ind. Eng. Chem. Prod. Res. Development, 9, 272 (1970).
25. H. Beuther, R. A. Flinn, and J. B. McKinley, Ind. Eng. Chem., 51, 1349 (1959).
26. J. Y. Livingston, "Hydrotreating Catalyst Properties Do Affect Performance," presented at the 74th National Meeting, AIChE, New Orleans, Louisiana, March 11-15, 1973.
27. D. Van Zoonen and C. Th. Douwes, J. Inst. Petroleum, 49, 385 (1963).
28. D. Satchell, Ph.D. Thesis, Oklahoma State University, Stillwater, Oklahoma (1974).
29. S. M. Kovach, L. J. Castle, J. V. Bennett, and J. T. Schrodt, Ind. Eng. Chem. Prod. Res. Dev., 17, 62 (1978).
30. M. C. Sooter and B. L. Crynes, Ind. Eng. Chem. Prod. Res. Dev., 14, 199 (1975).
31. M. J. Chiou and J. H. Olson, Preprints, Div. Pet. Chem. Am. Chem. Soc., 23, (4), 1421 (1978).
32. C. C. Kang and J. Gendler, Preprints, Div. Pet. Chem. Am. Chem. Soc., 23 (4), 1412 (1978).
33. R. Sivasubramanian and B. L. Crynes, Ind. Eng. Chem. Prod. Res. Develop. 18, 179 (1979). support properties", to be published.
34. B. C. Gates, J. R. Katzer, and G. C. A. Schuit, "Chemistry of Catalytic Processes," McGraw-Hill Book Company, New York, 1979.
35. J. F. Jones, N. J. Brunsvold, H. D. Terzian, L. J. Scotti, F. H. Schoemann, R. C. Merrill, J. D. Alcantara, D. J.

domina, S. J. Romelczyk, and L. Ford, "Char Oil Energy Development," Volume I, Final Report for the period August 18, 1971 through June 30, 1975, FMC Corporation, Princeton, New Jersey (1975).

36. R. H. Heck and T. R. Stein, Preprints, Div. Petrol. Chem. Am. Chem. Soc. 23, 948 (1977).
37. R. F. Sullivan and B. E. Stangeland, Preprints, Div. Petrol. Chem. Am. Chem. Soc. 23 (1) 322 (1977).
38. W. L. Nelson, Oil and Gas Journal, Feb. 28, 1977, p. 126.
39. R. J. Bertolacini, L. C. Gutberlet, D. K. Kim, and K. K. Robinson, Catalyst Development for Coal Liquefaction, Report prepared for EPRI, Palo Alto, California, 1977.
40. A. J. deRosset, G. Tan, J. G. Gatsis, J. P. Shoffner, and R. F. Swensen, "Characterization of Coal Liquids," Final Report, No. FE-2010-09, Submitted to U.S. Dept. of Energy, March 1977.
41. National Academy of Sciences Report, "Assessment of Technology for the Liquefaction of Coal" (1977).
42. M. J. Dabkowski, R. H. Heck, A. V. Perrella, M. Schreiner, Jr., T. R. Stein, Economic Screening Evaluation of Upgrading Coal Liquids to Turbine Fuels, EPRI AF710, Final Report, March 1978.
43. R. J. Sexton, "I - Introductory Statement. The Hazards to Health in Hydrogenation of Coal," Archives of Environmental Health 1 (September 1960).
44. C. S. Weil and N. I. Contra, "II - Carcinogenic Effect of Materials in Skin of Mice," Archives of Environmental Health 1 (September 1960).
45. N. H. Ketcham and R. W. Norton, "III - Industrial Hygiene Studies," Archives of Environmental Health 1 (September (1960).
46. R. J. Sexton, "IV - The Control Program in Clinical Effects," Archives of Environmental Health 1 (September 1960).
47. E. Bingham, A. W. Horton, and R. Tye, "The Carcinogenic Potency of Certain Oils," Archives of Environmental Health 10 (March 1965): 449.
48. Z. Bell and A. W. Norton, "Carcinogen in a Cracked Petroleum Residuum," Archives of Environmental Health 13 (August 1966).
49. W. C. Heuper, Experimental Studies on Concerigenesis of Synthetic Liquid Fuels and Petroleum Substitutes, American Medical Association, Chicago, Illinois (1956).
50. U. S. Patent 3,778,365 (1973).
51. L. E. Toth, "Transition Metal Carbides and Nitrides," Academic Press, New York (1971).
52. S. J. Tauster, J. Catal. 26, 487 (1972).
53. M. Boudart, J. A. Cusumano, and R. B. Levy, "New Catalytic Materials for Coal Liquefaction," Electric Power Research Institute, Report No. RP-415-I, October 30, 1975.